Embryo Implantation and Placental Development

Embryo Implantation and Placental Development

Editor

Luisa Campagnolo

MDPI • Basel • Beijing • Wuhan • Barcelona • Belgrade • Manchester • Tokyo • Cluj • Tianjin

Editor
Luisa Campagnolo
Biomedicine and Prevention
Tor Vergata
Rome
Italy

Editorial Office
MDPI
St. Alban-Anlage 66
4052 Basel, Switzerland

This is a reprint of articles from the Special Issue published online in the open access journal *International Journal of Molecular Sciences* (ISSN 1422-0067) (available at: www.mdpi.com/journal/ijms/special_issues/Embryo_Implantation).

For citation purposes, cite each article independently as indicated on the article page online and as indicated below:

LastName, A.A.; LastName, B.B.; LastName, C.C. Article Title. *Journal Name* **Year**, *Volume Number*, Page Range.

ISBN 978-3-0365-5854-7 (Hbk)
ISBN 978-3-0365-5853-0 (PDF)

Contents

About the Editor

Luisa Campagnolo

Luisa Campagnolo, PhD in Medical Embryology, is an Associate Professor of Histology and Embryology at the University of Rome "Tor Vergata" and Visiting Professor at the Weill Medical College of Cornell University (New York, USA). She has a broad background in Reproductive and Developmental Biology, and research in her laboratory is focused on the identification of molecular pathways regulating embryo implantation and placental development in physiological and pathological conditions. Her work has been supported by several national and international research grants.

Preface to "Embryo Implantation and Placental Development"

Human reproduction has been defined by many as a highly inefficient biological process. This is a consequence of the numerous coordinated and finely regulated molecular pathways which drive the different stages of embryonic development, from fertilization to implantation and placentation. Synchronization between the acquisition of blastocyst competence and uterine receptivity appears to be a limiting step in implantation, and the timely release of growth factors and cytokines is a key basis for a successful pregnancy. Over the last few years, many of the players involved in the regulation of such an intricated process have been identified, and their altered expression in pathological conditions has been recognized in some cases. The role of novel factors is emerging, possibly opening new paths for the interpretation of what is still defined as idiopathic infertility.

Luisa Campagnolo
Editor

International Journal of *Molecular Sciences*

MDPI

Review

Molecular Signaling Regulating Endometrium–Blastocyst Crosstalk

Micol Massimiani [1,2], Valentina Lacconi [1], Fabio La Civita [1], Carlo Ticconi [3], Rocco Rago [4] and Luisa Campagnolo [1,*]

1 Department of Biomedicine and Prevention, University of Rome Tor Vergata, Via Montpellier 1, 00133 Rome, Italy; micol.massimiani@unicamillus.org (M.M.); valelcc@gmail.com (V.L.); fabio.lacivita25@gmail.com (F.L.C.)

2 Saint Camillus International University of Health Sciences, Via di Sant'Alessandro, 8, 00131 Rome, Italy

3 Department of Surgical Sciences, Section of Gynecology and Obstetrics, University Tor Vergata, Via Montpellier, 1, 00133 Rome, Italy; ticconi@uniroma2.it

4 Physiopathology of Reproduction and Andrology Unit, Sandro Pertini Hospital, Via dei Monti Tiburtini 385/389, 00157 Rome, Italy; rocco.rago@aslroma2.it

* Correspondence: campagnolo@med.uniroma2.it; Tel.: +39-06-7259-6154

Received: 27 October 2019; Accepted: 16 December 2019; Published: 18 December 2019

Abstract: Implantation of the embryo into the uterine endometrium is one of the most finely-regulated processes that leads to the establishment of a successful pregnancy. A plethora of factors are released in a time-specific fashion to synchronize the differentiation program of both the embryo and the endometrium. Indeed, blastocyst implantation in the uterus occurs in a limited time frame called the "window of implantation" (WOI), during which the maternal endometrium undergoes dramatic changes, collectively called "decidualization". Decidualization is guided not just by maternal factors (e.g., estrogen, progesterone, thyroid hormone), but also by molecules secreted by the embryo, such as chorionic gonadotropin (CG) and interleukin-1β (IL-1 β), just to cite few. Once reached the uterine cavity, the embryo orients correctly toward the uterine epithelium, interacts with specialized structures, called pinopodes, and begins the process of adhesion and invasion. All these events are guided by factors secreted by both the endometrium and the embryo, such as leukemia inhibitory factor (LIF), integrins and their ligands, adhesion molecules, Notch family members, and metalloproteinases and their inhibitors. The aim of this review is to give an overview of the factors and mechanisms regulating implantation, with a focus on those involved in the complex crosstalk between the blastocyst and the endometrium.

Keywords: implantation; endometrium; blastocyst; embryo; chorionic gonadotropin; progesterone; Notch; cytokines

1. Introduction

Implantation requires a complex crosstalk between the endometrium and the blastocyst and is highly regulated by a variety of factors, such as soluble growth factors, hormones, prostaglandins, adhesion molecules, and the extracellular matrix (ECM) [1–5]. These factors, produced by the receptive endometrium in response to the presence of the blastocyst and vice versa, are able to synchronize the development of the embryo to the blastocyst stage and the differentiation of the uterus to the receptive state [6,7]. This complex network of signaling accounts for implantation being one of the major limiting steps in mammalian reproduction. Indeed, the implantation rate in humans is about 30% per cycle [8,9]. Alterations of these signaling pathways may result in pathological conditions leading to infertility.

The WHO has designated infertility as "a disease of the reproductive system defined by the failure to achieve a clinical pregnancy after 12 months or more of regular unprotected sexual intercourse" [10,11].

Infertility is one of the main health issues in all societies worldwide, with a prevalence of 3.5–16.7% in developed countries and 6.9–9.3% in developing countries [12,13] and may be a consequence of low embryo quality, male problems, or female dysfunctions. Female fertility problems account for 20–35% of infertility cases and may derive from a wide variety of causes such as age, anatomical, endocrine and immunological problems, and several pathological conditions affecting the endometrium [14–19]. These conditions may lead to defects in blastocyst implantation in the maternal uterus, resulting in implantation failure, a common cause of impaired fertility [20]. The term "implantation failure" actually implies a series of conditions in which the embryo does not implant in the maternal endometrium after both spontaneous and in vitro fertilization (IVF) [21]. A condition in which implantation failure occurs after the transfer of three or more good quality embryos is defined recurrent implantation failure (RIF) and it is only applicable to assisted reproductive technology (ART) [21,22]. According to ASRM and ESHRE definitions, RIF is considered a distinct pathological condition from recurrent pregnancy loss [21,23,24].

The present review describes and discusses the molecular mechanisms underlying the implantation process, focusing on factors implicated in the complex blastocyst–endometrium crosstalk, which are crucial for successful implantation. Further research for new factors involved in the dialogue between the blastocyst and the endometrium would allow to reduce the current rates of implantation failure, allowing many couples with infertility problems to reach a successful pregnancy.

2. Preparation of the Endometrium to Implantation

Interaction between the uterus and the blastocyst can only occur during a limited defined period, known as the "window of implantation" (WOI) [25–27]. In humans, this defined period corresponds to the mid-secretory phase, occurring between the 20th and the 24th day of the menstrual cycle, or 6-10 days after the luteinizing hormone (LH) peak [25,28–30]. In this timeframe, the molecular program regulating growth and differentiation of the embryo synchronizes with the molecular program regulating endometrial receptivity. Failure in such synchronization results in failure of the blastocyst to implant. Given the relevance of this stage for the establishment of a successful pregnancy, the WOI is regulated by a wide variety of cytokines, growth factors, prostaglandins, enzymes, and adhesion molecules [31–33].

2.1. Gland Development and Function

During the WOI, the uterine endometrium is affected by morphological changes which favor blastocyst implantation [34]. The epithelial cells present vacuoles to a supranuclear position and glands become more irregular with a papillary appearance. Uterine glands are necessary for embryo implantation. Their major development in several mammalian species, including humans, occurs mainly during postnatal life and starts from invagination of the luminal epithelium [35–37]. At birth, in humans, glands are sparse, and little deepened into the stroma. At puberty, they extend toward the myometrium and form a coiled network of tubules [36]. Animal studies have demonstrated that progesterone treatment during neonatal life impairs gland development and this severely affects fertility, supporting a central role of endometrial glands for embryo implantation [36]. Experimental data suggest that progesterone treatment may affect the expression of genes central to endometrial adenogenesis, including members of the Wnt family [38], whose expression and adenogenic role has been demonstrated in both glands and stroma [39–45]. A central role of glands in implantation is also suggested by loss-of-function studies of genes involved in epithelial morphogenesis and proliferation in mice, for example, ablation of the cell–cell adhesion molecule Cdh1 results in epithelial disorganization and absence of glands in the neonatal uterus, with consequent infertility [46]; moreover, conditional knock-out of Sox17 in the uterus is associated with impaired endometrial adenogenesis and infertility [47].

Endometrial glands produce and secrete a cocktail of molecules, the histotroph, including amino acids, glucose and growth factors, which appear to be involved in embryo survival, trophectoderm activation, endometrial invasion and nourishment of the implanted embryo [48–56]. Leukemia inhibitory factor (LIF) and vascular endothelial growth factor (VEGF) are produced by uterine glands [57]. Interestingly, several studies have reported differences in composition of the histotroph between fertile and infertile women, strengthening the relevance of gland products in supporting embryo implantation and survival [52,58–61]. The role of endometrial glands in pregnancy is not limited to implantation. The connection between glands and the intervillous space of the primitive placenta suggests that carbohydrates and lipids produced by the glands may contribute to nurturing the implanted embryo at least until syncytiotrophoblast cells contact the maternal vessels [56]. In addition, growth factors and hormones secreted by the glands during early pregnancy [50] may be involved in placental morphogenesis, considering that receptors for some of these factors have been identified on trophoblast cells [62–66]. Altogether, these data indicate that endometrial glands are central in the establishment of a successful pregnancy and a deeper understanding of their precise role in implantation is of importance to reveal potential causes of infertility, as well as other reproductive disorders.

2.2. Decidualization

In addition to the changes occurring in the luminal and glandular epithelium, major changes take place also in the endometrial stroma. The endometrial stromal cells undergo a decidual reaction, in which they proliferate and differentiate from fibroblast-like to epithelial-like cells, which will form the maternal decidua. Decidual cells progressively increase in size and number throughout pregnancy, starting from 9.8% of stromal cells in early pregnancy and arrive to 57.8% at term [67]. The acquisition of the epithelial-like phenotype by stromal cells consists in an increase in size, rounding of the nucleus with increased number of nucleoli, accumulation of glycogen, lipid droplets and secretory granules in the cytoplasm, and expansion of rough endoplasmic reticulum and Golgi apparatus [68]. The term "decidua" derives from Latin "de cadere" and means to fall down, so it refers to the fact that the decidualized uterine tissue is lost after parturition. Decidua is mainly formed by decidualized endometrial stromal cells, but also contains hematopoietic cells, macrophages, uterine natural killer (uNK) and monocytes [69,70]. Decidualization starts in the luteal phase, with stromal cells surrounding the spiral arteries in the upper two-thirds of the endometrium, regardless of whether or not the blastocyst is present [71]. Differently from most mammals, decidualization in humans occurs before the embryo reaches the uterine cavity and is driven by the postovulatory rise in progesterone levels and local increase of cyclic adenosine monophosphate (cAMP) production, occurring long before the embryo is ready to implant. In the absence of pregnancy, progesterone levels decrease, and menstrual shedding and cyclic regeneration of the endometrium occur. Decidualization is responsible for embryo quality control, promoting implantation and development, or facilitating early rejection in case, for example, of chromosomally abnormal human embryos [72,73].

2.3. Hormone Signaling

Uterine function is primarily regulated by estrogen and progesterone, which modulate gene expression of luminal and glandular epithelium and stromal cells. These ovarian hormones guide the structural and functional remodeling occurring during decidualization. Estrogen receptor (ER) exists in two forms, ERα and ERβ with distinct biological function, both expressed in the endometrium. ERα is essential for implantation since ERα knockout mice display endometrial hypoplasia and are infertile [74]; mice knockout for ERβ present normal endometrium and appear fertile, suggesting that ERβ may be involved in other aspects of endometrial function [75–77]. During the proliferative phase, high levels of estrogen induce proliferation of the epithelial, stromal, and vascular endothelial cells [78,79]. Indeed, activated ERα induces proliferation of human epithelial cells and decidualization in stromal cells through rapid non-genomic activation of the extracellular signal-regulated kinase/mitogen-activated protein kinase (ERK/MAPK) pathway [80,81]. In addition to this rapid activity, estrogen increases epithelial proliferation by inducing insulin-like growth factor 1 (IGF1). IGF1 is expressed and secreted by the stroma, and by binding its receptor IGF1R in the epithelium, induces the phosphoinositide 3-kinase (PI3K)/ serine/threonine protein kinase (AKT) pathway leading to proliferation [82–84]. Other known targets of estrogen in the endometrium are fibroblast growth factor (FGF)-9, CCAAT enhancer binding protein beta (C/EBPβ) and Mucin 1 (MUC1). FGF-9 is expressed at high levels in the stromal compartment of the endometrium during the late proliferative phase; in vitro FGF-9 stimulates stromal cell proliferation, and expression of FGF-9 in such cells is induced upon 17β-estradiol stimulation [85], suggesting that in vivo estrogen may induce proliferation of stromal cells through the up-regulation of FGF-9. Estrogen-induced proliferation of endometrial cells is also mediated by C/EBPβ, whose pro-proliferative action is exerted on both the endometrial epithelium and stroma through regulation of cyclin-dependent kinases involved in G2 to M transition of mitosis [86]. Estrogen also regulates the expression of the glycoprotein MUC1, which is expressed on the surface of the luminal epithelium to create a protective barrier that has to be removed at implantation to allow embryo attachment [87]. Beyond its activity on cell proliferation, estrogen also induces endometrial expression of leukemia inhibitory factor (LIF), an interleukin-6 family cytokine whose central role for successful implantation and decidualization has been widely reported [88,89] and discussed later in this review. During the proliferative phase, estrogen induces progesterone receptor (PR) in endometrial cells through ERα to determine progesterone responsiveness during the luteal phase, and in turn PR inhibits ERα expression in a negative feedback crucial for endometrial function [90]. Progesterone increases during the secretory phase of the menstrual cycle, inducing decidualization and thus opening the WOI and remains elevated if pregnancy occurs [91]. The effects of progesterone in endometrial cells are mediated by PR, which exists in two isoforms, PR-A and PR-B, transcribed from two promoters of the same gene. Deletion of either PR-A or PR-B demonstrates specific roles of each PR isoform in mediating progesterone actions on the murine uterus. In PR-B knockout mice, progesterone effects mediated by PR-A are sufficient for a normal uterine function, since implantation, pregnancy, and parturition are normal in these mice [92]. On the contrary, in PR-A-knockout mice progesterone actions mediated by PR-B lead to increased hyperplasia of the endometrial epithelium and inflammation, and no decidualization in the endometrial stroma [93]. Taken together these data indicate that PR-A is critical for implantation, and that PR-B is involved in endometrial differentiation. Female mice knockout for both PR-A and PR-B are infertile, showing severely reduced or absent ovulation, uterine hyperplasia, absence of decidualization, severely limited mammary gland development, and an impaired sexual behavior [94]. After progesterone binding, PR activates a series of signal transductions, involving both genomic and non-genomic pathways. The non-genomic response is rapid and based on the interaction with c-Src kinase to induce the pro-proliferative ERK/MAPK and AKT pathways, important for peri-implantation stromal proliferation [90,95,96]. The genomic action of PR involves its translocation into the nucleus and modulation of gene expression [90]. In the uterine epithelium, progesterone promotes the expression of Indian hedgehog (IHH), which in turn induces up-regulation of stromal chicken ovalbumin upstream promoter-transcription factor II (COUPTFII) that regulates stromal bone morphogenetic protein 2 (BMP2) and consequently the decidualization response of the stromal compartment [97–100]. Progesterone-mediated induction of IHH is also responsible for the down-regulation of MUC1 [98]. As for the effect of progesterone on decidualization, it has been demonstrated that progesterone stimulation induces heart and neural crest derivatives expressed 2 (HAND2), a transcription factor, whose down-regulation in mouse and human fibroblast cells is associated to reduction of decidualization markers [101].

The genomic response to progesterone action also regulates the expression of transcription factors of the homeobox family [102]. Homeobox protein-A10 and -A11 (HOXA10, HOXA11), are expressed in stromal and glandular compartments of the endometrium throughout the menstrual cycle and are both essential for pregnancy, since their deletion in mice results in implantation defects [103–105]. Both HOXA10 and HOXA11 have a role in decidualization [104,106]. HOXA10 positively regulates the expression of the decidual marker Insulin Growth Factor Binding Protein 1 (IGFBP1) [107], while HOXA11 normally function as a repressor of the decidual marker prolactin (PRL) gene, however in cooperation with FOXO1A it induces its 3-fold increase [104]. In vitro models have greatly contributed in understanding the role of progesterone in decidualization [108,109]. Treatment of endometrial organ cultures or endometrial derived stromal cells with progesterone induces expression of PRL [109,110], but with higher efficiency if steroid hormones are used in combination with cAMP [111,112]. cAMP alone can induce decidualization of human endometrial stromal cells (HESCs) but for few days only [113–115], since for the stabilization of the process it is necessary the presence of both cAMP and progesterone [112].

In addition to the steroid hormones produced by the ovary, other hormones are involved in the establishment of pregnancy, among which one of the most studied is the chorionic gonadotropin (CG). CG is produced by the trophectoderm of the blastocyst and is one of the main players involved in endometrium–embryo crosstalk at the time of implantation. The ovaries respond to CG, which acts as an agonist of LH, by maintaining the corpus luteum, thus producing the progesterone necessary for the establishment and progression of pregnancy [116]. The responses of the endometrium are multiple, including the inhibition of apoptosis, which usually occurs at the end of the menstrual cycle, by activating anti-apoptotic genes as B-cell lymphoma 2 (BCL-2) [117,118], and the induction of the decidualization process [118–120]. Both epithelial and stromal cells possess the LH/CG receptor (LHCGR), a seven transmembrane G protein-coupled receptor, which shows the highest expression during the secretory phase of the menstrual cycle [119,121,122]. Endometrial epithelial cells respond to CG by expressing cyclooxygenase-2 (COX2) and prostaglandin E synthase (PGES), through the activation of extracellular signal-regulated protein kinases 1/2 (Erk1/2) signaling pathway. The increased production of prostaglandin E2 (PGE2) [122–124] induces cAMP in endometrial stromal cells and promotes their decidualization [124,125]. COX-derived PGE2 plays an important role in the increase of endometrial vascular permeability, which characterizes the inflammatory reaction typical of implantation [126,127]. In endometrial stromal cells CG activates Erk1/2 signaling pathway, thus increasing the expression of the PR and regulating the expression of genes controlling endometrial receptivity [121]. Moreover, in primates, endometrial stromal cells respond to CG and progesterone by activating Notch receptor 1 (NOTCH1) pathway, as discussed later. NOTCH1 induces the expression of α-smooth muscle actin (α-SMA), which positively regulates remodeling of cytoskeleton and the initial changes typical of the decidualization process [128]. Subsequently, a decrease in CG and NOTCH1 levels is necessary for the completion of decidualization, which is accompanied by an increase in the expression of insulin-like growth factor binding protein-1 (IGFBP1) and prolactin (PRL), markers of decidualization [129–131], and a downregulation of LHCGR [120,132–134].

2.4. Role of Pinopodes

One of the major structural changes of the endometrium during the luteal phase is the formation of apical protrusions on the epithelial cells called pinopodes (also known as uterodomes). These dome-like structures are formed in response to progesterone, but regress upon estrogen stimulation [135–138]. The function of pinopodes is not clear. Some authors suggest that pinopodes are responsible of pinocytosis and endocytosis of uterine fluid and macromolecules, which facilitates adhesion of the blastocyst to the endometrium [139,140]; others have suggested that they might be directly involved in blastocyst–endometrial interaction through the expression of adhesion molecules, such as integrins [141–144], or of LIF [145], although co-localization of these molecules and pinopodes has been questioned [146,147].

Pinopodes formation has been initially demonstrated to coincide with the WOI [137], hence their role as potential markers of endometrial receptivity was proposed [148,149]. However, this role is still currently a topic of great debate. Several studies demonstrated that pinopode are present beyond the WOI [150], questioning their utility to identify endometrial receptivity. Moreover, no major differences in the coverage and morphology of pinopodes was observed in endometrial samples from fertile women compared to those of women with recurrent pregnancy loss, suggesting no direct correlation between pinopode density/morphology and pregnancy success [151]. However, recent studies re-evaluated pinopode utility to identify endometrial receptivity, by demonstrating a strong correlation between pinopode quality and pregnancy rate [152–154]. These contrasting results may be explained, at least in part, by sampling variability, and lack of standardization for morphological identification and staging of the pinopodes. As recently reported, computer-assisted analysis of endometrial tissue images could be used to overcome operator subjectivity [140]. It should also be considered that absence or presence of pinopodes might not be the solely parameter to consider for endometrial receptivity, as quality and molecular content of pinopodes could also be of relevance.

2.5. Growth Factor of the EGF Family

Uterine receptivity is also regulated by members of the epidermal growth factor (EGF) family, whose expression pattern in the peri-implantation uterus has been widely investigated in murine models [155–162]. Among the EGF family members, amphiregulin (AREG) has been identified in the luminal epithelium exclusively at the site of blastocyst apposition and its expression appears to correlate initially with the increase of progesterone levels and then with the attachment reaction [157]. Similarly, the expression of heparin binding-EGF (HB-EGF), which is under the control of both estrogen and progesterone [160], requires the presence of a competent blastocyst and it occurs in the luminal epithelium when pinopodes are fully formed at the site of blastocyst apposition [155,161], while epiregulin (EREG) is expressed in both the luminal epithelium and stroma during blastocyst attachment [158]. This unique expression pattern suggests a role for AREG, HB-EGF, and EREG in uterine receptivity and subsequent embryo adhesion. The role of HB-EGF in blastocyst adhesion to the uterus has been further demonstrated in vitro in a co-culture of a mouse cell line synthesizing transmembrane human HB-EGF (TM HB-EGF) and mouse blastocysts. Cells synthesizing TM HB-EGF adhered to mouse blastocysts more than parental cells or cells synthesizing a constitutively secreted form of HB-EGF [163]. These results are confirmed by a more recent study using HB-EGF mutant mice, which demonstrates that maternal deficiency of HB-EGF limits pregnancy success [162].

2.6. NOTCH Signaling Pathway

NOTCH signaling pathway is involved in the regulation of various cellular processes such as cell proliferation, invasion, adhesion, survival, apoptosis and differentiation [164–167]. All four NOTCH receptors, the ligands Jagged1 (JAG1) and Delta-like (DLL) 4 and the target genes hairy enhancer of split (HES) and Hes-related 1 (HEY1) are known to be expressed by the endometrium [168–171]. Several ligands and receptors of the NOTCH signaling pathway are expressed in both the inner cell mass (ICM) and trophectoderm of the human blastocyst [172–174]. NOTCH1 plays an important role in the process of decidualization, by inducing pro-survival signals in the endometrium, thus avoiding apoptosis normally occurring at the end of the menstrual cycle. Hess et al. showed that blastocyst-conditioned medium induces an increase in the expression of NOTCH family members in decidual cells, suggesting a role for this pathway in decidualization [175]. Moreover, it has recently been shown that NOTCH signaling pathway is dysregulated in the endometrium of women with unexplained recurrent pregnancy loss [176]. Activation of NOTCH1 pathway in the endometrium is stimulated by CG and progesterone and leads to increased expression of α-SMA and Forkhead box protein O1 (FOXO1) [1,128,177]. FOXO1, in turns, induces expression of PRL and IGFBP1 and it is essential for the decidualization process [178–182]. NOTCH1 is involved in the inhibition of cAMP/protein kinase A (PKA) signaling pathway [183], so that NOTCH1 has to be downregulated to allow cAMP response of stromal cells. Similar to what described for α-SMA and LHCGR expression, a downregulation of NOTCH1 is necessary for the induction of IGFBP1 and the completion of decidualization [111,120,128].

2.7. Interleukin-1b in Blastocyst–Endometrium Dialogue

Interleukin (IL)-1β is another important factor supporting blastocyst–endometrium dialogue, playing a fundamental role in decidualization of stromal cells and in successful blastocyst implantation. IL-1β is secreted by cytotrophoblast cells isolated from first trimester placenta, while its expression is lower in cultures from second and third trimester placenta [184]. In endometrial stromal cells IL-1β induces the expression of COX2 and PGE2, known to increase the levels of cAMP, which are necessary for decidualization, as above described [185,186]. Moreover, in vivo infusion of IL-1β and CG promotes the expression of IGFBP1 in apical surface stromal cells [133]. It has been demonstrated that inhibition of COX2 in human and baboon endometrial stromal cells is able to block the decidualization induced by IL-1β in the presence of steroid hormones, suggesting that IL-1β acts upstream of COX2 [185]. On the contrary, inhibition of COX2 does not affect decidualization induced by cAMP and steroid hormones, suggesting that cAMP acts downstream of COX2 and PGE2 [185]. Interestingly, cAMP is able to block decidualization induced by IL-1β indicating a negative feedback between IL-1β and cAMP [185,187]. In baboon, IL-1β positively regulates the expression of matrix metalloproteinase (MMP) 3 in endometrial stroma, thus inducing degradation of the ECM. Considering that disruption of the ECM might reflect in cellular cytoskeleton remodeling, IL-1β may play an important role in the decidualization also by promoting cytoskeleton changes typical of this process [188,189]. All these data clearly indicate that IL-1β plays a relevant role in blastocyst–endometrium crosstalk.

2.8. Thyroid Hormone in Endometrial Receptivity

Endometrial receptivity is regulated also by thyroid hormone (TH). Both thyroid hormone and thyroid-stimulating hormone receptors (TR and TSHR, respectively) are expressed in the endometrium with variations during the menstrual cycle [190]. Two of the isoforms of TR, TRα1, and TRβ1, are expressed during the mid-luteal phase in glandular and luminal epithelium, showing an increase during the secretory phase, followed by a drastic decrease. Interestingly, the expression of TRα1 and TRβ1, and also of TRα2 and TSHR, in endometrial cells is concomitant to the appearance of the pinopodes and the establishment of endometrial receptivity. The expression of TRα1, TRβ1, TRα2 and also of type 2 deiodinase (DIO2) is regulated by progesterone. In fact, the administration of mifepristrone, an anti-progestinic drug that makes the endometrium unreceptive and induces menstrual bleeding, reduces the expression of TRα1 and TRα2, while it up-regulates TRβ1 and DIO2 expression, suggesting a role for progesterone in regulating molecules involved in TH synthesis and metabolism [191]. The role of TH pathway in endometrial function is also demonstrated by the observation that hypothyroidism is able to reduce uterine endometrial thickness, and also interferes with estrogenic response of the endometrium [192]. TH regulates endometrial receptivity also by acting on LIF pathway, since TSH induces increased expression of LIF and LIF receptor (LIFR) in endometrial stromal cells obtained from human endometrial samples, suggesting a major role for TSH in the implantation process [190].

2.9. Immune Cells in Implantation

A role for the immune system in embryo implantation has been widely investigated for obvious reasons. The decidua plays a fundamental role in ensuring immune tolerance toward the semi-allogenic conceptus, protecting it from the mother's immune system. Regulatory T cells (Tregs) are CD4+ CD25+ T cells, having the role to suppress the immune response [193]. During early pregnancy, in the decidua there is an increase in Tregs, which produce immunosuppressive cytokines, such as IL-10, for inducing immune tolerance [194–197]. Other cells involved in maternal immune tolerance are the uNK, a particular type of NK cells, which lose their cytotoxic functions during pregnancy andplay a supportive role by enhancing angiogenesis. uNK cells induce immune tolerance by reducing inflammation through interferon-γ (IFN-γ) [198] and by inhibiting the function of T cells through the expression of immunomodulatory molecules such as galectin-1 and glycodelin A [199].

2.10. Endometrial Receptivity Array

Recently, a customized endometrial receptivity array (ERA), containing 238 genes related to endometrial receptivity, was created [200]. These genes, differentially expressed in the receptive phase, encode for factors involved in several biological processes, such as processes related to the immune system, circulation, response to external stimulus, behavior, cell cycle, cell adhesion, anatomical structure development, cell–cell signaling, and mitotic cell cycle [200]. Beside the many above mentioned genes suggested to regulate endometrial receptivity, additional genes have been identified by ERA, highlighting the great complexity of factors regulating implantation. ERA has been suggested as a more accurate and reproducible approach to assess endometrial receptivity compared to histological analysis [201] and its use has been proposed for RIF patients [202]. Considering how critical the molecular signature of the endometrium is for embryo implantation, a test which unequivocally assess if the embryo and the uterus are in synchrony may be of great value to avoid ineffective embryo transfers. However, the utility of ERA in the clinical practice is still debated [203]. More recently, a smaller set of genes has been proposed to assess the receptivity status of the endometrium in biopsies obtained in the secretory phase [204]. It is reasonable to foresee new additional advances in this area, that is of potential great clinical utility in the management of infertile women undergoing IVF, as well as in women with RIF and unexplained RPL.

3. Implantation of the Competent Blastocyst

In order for a healthy pregnancy to proceed, the embryo needs to synchronize its developmental program with endometrial receptivity and to acquire the ability to implant, defined as competence. A competent blastocyst is characterized by distinctive morphological and molecular features, which are discussed in this section.

3.1. Transport, orientation and hatching

3.1.1. Blastocyst Transport and Orientation

After fertilization, the embryo encased in a non-anchored glycocalix, the so-called zona pellucida, which prevents ectopic implantation, descends the Fallopian tube and reaches the uterine cavity, while undergoing profound morphological changes ending in the formation of the blastocyst [205,206]. For successful implantation into the maternal tissues, a correct orientation of the blastocyst towards the uterine wall is needed. In most eutherian mammals, at the time of first contact of the blastocyst with the endometrial epithelium, the ICM of the various embryos has an almost constantly specific orientation toward the uterus. In humans, the ICM faces the uterine wall. This positioning of the ICM usually correlates with the site of trophectoderm attachment to the endometrium, as well as with subsequent development of the fetal membranes and placental structures [207,208]. In rodents, implantation occurs in anti-mesometrial position with the ICM facing the mesometrium [209]. Why, within most species, the ICM of the blastocyst, or the placenta, should be positioned consistently in the same way with respect to the uterine wall is not completely understood. Moreover, how the blastocyst becomes correctly oriented [210,211] or what directs the process has not been well clarified, for even the most commonly-studied mammals. However, it has been postulated that orientation depends on signals from the endometrium rather than from the embryo, since embryo-mimicking structures (beads, bubbles or cells) end up in the position that the embryo would occupy [212–215]. For example, in mice endometrial expression of the transcriptional regulator Rbpj is required to instruct embryo orientation, and its conditional deletion determines loss of ventral-dorsal orientation [216]. A role for endometrial glands in embryo orientation has been also proposed. Indeed, recent data indicate that in mice endometrial gland development is confined to the anti-mesometrial side of the uterus [217]. In consideration of the above discussed essential role of uterine glands in implantation, it can be speculated that glands may drive the anti-mesometrial orientation of the implanting mouse embryo, possibly through the expression of specific factors. In this respect, it has been shown that Wnt signaling activity in the mouse uterus is limited to the anti-mesometrial region and a role for Wnt proteins in anti-mesometrial localization of the implanting embryo has been proposed [217].

3.1.2. Blastocyst Hatching

Embedding of the blastocyst into the maternal endometrium requires hatching from the zona pellucida, which otherwise would prevent adhesion of the embryo to the uterine wall. Blastocyst hatching exposes the trophectoderm and allows the blastocyst to implant in the maternal uterus. The crucial event for blastocyst hatching is the formation of a nick into the zona pellucida, and proteases, such as serine-, cysteine-, and metallo-proteases have been proposed to play a major role in this event depending on the species [218–223]. Cathepsins, belonging to the ubiquitous cysteine proteases family [224], have been demonstrated to be involved in blastocyst hatching and zona lysis in mice; the expression of cathepsin L and P (mRNA and protein) and their natural inhibitor, Cystatin C, has been demonstrated in mouse peri-hatching blastocysts [225]; treatment of golden hamster embryos with Cystatin C is able to block blastocyst hatching [221]. The process of murine blastocyst hatching from the zona pellucida is also regulated by two mouse-specific proteinases, Strypsin (ISP1) and Lysin (ISP2). ISP1 and ISP2 are two related S1-family serine proteinases, which are tandemly localized in a cluster of tryptase genes [226,227]. The ISPs are co-expressed in the mouse preimplantation embryos and in the mouse uterine endometrium during the WOI, indicating that they could play a role in the process of blastocyst implantation [226,228]. Expression of ISP genes is positively regulated by progesterone and TH [219,223,226] and ISPs are secreted by the blastocyst and the endometrial glands into uterine fluid just prior to implantation [229]. The use of antibodies against ISP1/ISP2 abrogate murine embryo hatching and outgrowth, ascribing a crucial role for ISPs in this process [228]. This is further supported by our recent observations using mouse blastocysts cultured in the presence of TH, with or without endometrial cells used as the feeder layer. In the presence of endometrial feeder cells, TH is able to anticipate blastocyst hatching (Figure 1) by upregulating the expression of blastocyst produced ISPs, and to enhance blastocyst outgrowth by upregulating endometrial ISPs and MMPs. On the contrary, in the absence of the endometrial feeder layer, TH does not affect blastocyst hatching, suggesting that TH is one of the players involved in the bidirectional crosstalk between the blastocyst and the endometrium during the WOI [223]. Human homologs of ISPs have not been so far identified, and it is possible that other proteases might be involved in blastocyst hatching in humans.

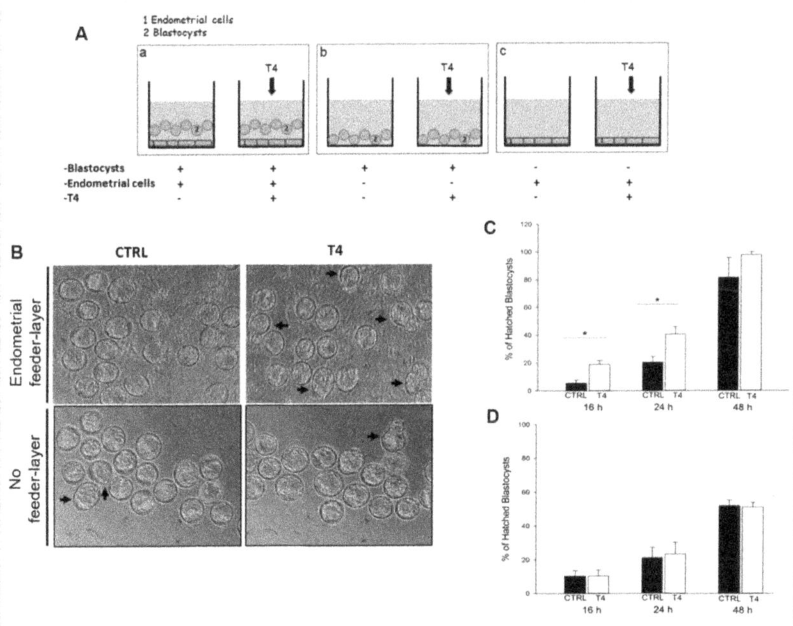

Figure 1. Thyroid hormone (TH) supplementation stimulates mouse blastocyst hatching in vitro. (**A**) Schematic representation of the in vitro model developed to assess TH role in implantation. (**a**) Co-culture of murine blastocysts and endometrial primary cells as the feeder layer; (**b**) blastocysts

cultured on plastic; and (**c**) endometrial cells cultured without blastocysts. (**B**) Representative images of the cultures. Scale bar 50 μm. (**C**,**D**) Graphs summarizing the results shown in B: percent of hatched blastocysts after co-culture on endometrial cells (**C**) or on plastic (**D**). Reproduced with permission from Piccirilli et al. [223].

3.2. Apposition

Histological analysis of uteri of pregnant women allows to recognize three different levels of blastocyst adhesion to the uterine wall, which correspond to the three stages of blastocyst implantation (Figure 2) [230,231].

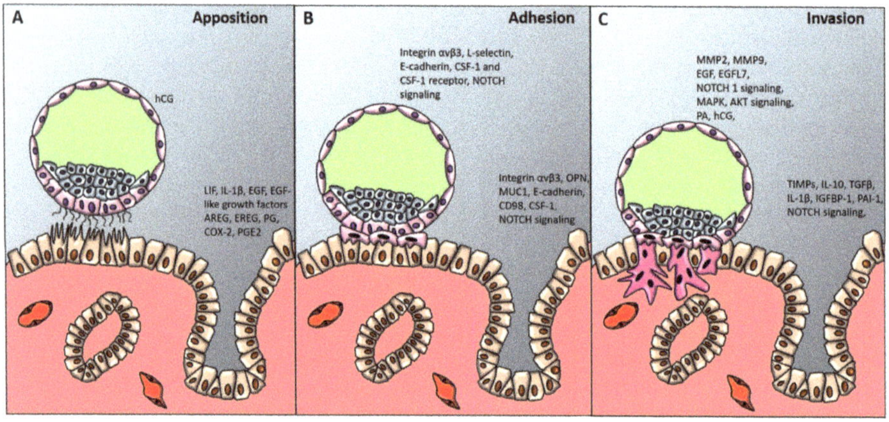

Figure 2. Blastocyst apposition, adhesion and invasion. The diagram shows a preimplantation-stage (**A**, **B**) and invading (**C**) blastocyst (about 9 to 10 days after conception) and the processes and factors required for uterine receptivity and blastocyst apposition (**A**), adhesion (**B**) and invasion (**C**). hCG denotes human chorionic gonadotropin, LIF leukemia inhibiting factor, IL-1β interleukin-1 beta, EGF-like growth factors epidermal growth factor-like growth factors, AREG amphiregulin, EREG epiregulin, PG progesterone, COX-2 cyclooxygenase-2, PGE2 prostaglandin E2, CSF-1 colony stimulating factor-1, NOTCH1 Notch receptor 1, OPN osteopontin, MUC-1 mucin-1, MMPs metalloproteinases, EGFL7 epidermal growth factor-like domain 7, MAPK mitogen activated protein kinase, AKT protein kinase B, PA plasminogen activator, TGFβ transforming growth factor beta, TIMPs tissue inhibitor of metalloproteinases, PAI-1 plasminogen activator inhibitor-1.

3.2.1. LIF Signaling

Blastocyst apposition is the initial stage representing the first physical contact between the blastocyst and the endometrium, in which the blastocyst finds a site for implantation, guided by the maternal endometrium [232,233]. The site of implantation in the human uterus is usually in the upper and posterior part in the midsagittal plane. During blastocyst apposition, the microvilli placed on the apical surface of trophectoderm interdigitate with the pinopodes localized on the apical surface of the uterine epithelium (Figure 2A). These specialized structures support a stable binding between trophoblast and uterine epithelial cells, so that the plasma membranes of these cells are parallel and separated by a distance of 20 nm [234]. The pinopodes secrete LIF [145]. LIF is a cytokine of the IL-6 family, which in the uterus activates the Janus kinases (JAK)-signal transducer and activator of transcription protein (STAT) pathway, and therefore phosphorylates STAT3, whose activation is required for implantation [235,236]. LIF is indispensable for blastocyst implantation. Mice knockout for LIF are infertile. Although able to develop blastocysts, these mice show implantation failure; however successful implantation occurs in surrogate mothers [90]. In Lif-null mice the expression of EGF-like growth factors, such as HB-EGF, AREG, and EREG, which, as previously mentioned, are normally expressed by the luminal epithelium adjacent to the blastocyst and are essential for successful pregnancy, is abolished [237]. Since the defects in decidualization caused by the absence of LIF can be rescued by intrauterine administration of EGF ligand [238], it has been hypothesized that LIF favors blastocyst invasion by reducing the expression of cell–cell junction molecules and proliferation of the stromal cells through activation of EGF signaling pathway [239]. In fertile women, LIF expression increases in the endometrium around the time of implantation, while infertile women express low levels of this factor [240,241].

3.2.2. Chorionic Gonadotropin

Once a competent blastocyst makes contact with the maternal endometrium, a dialogue made of signals and responses between them occurs. One of the most important factors secreted by trophoblast cells is CG. CG is expressed very early by the embryo, since its mRNA can be detected starting from the 6–8 cell stage. The protein is secreted by both zona enclosed or hatched blastocysts, and is independent of blastocyst interaction with the endometrium [242]. During pregnancy, CG is firstly detectable in maternal blood during implantation and then rapidly increases [243]. As discussed before, CG plays a fundamental role in inducing the production of progesterone and mediating the decidualization process, thus allowing implantation of the blastocyst.

3.3. *Adhesion*

3.3.1. Adhesion Molecules Mediating Blastocyst Adhesion

Following apposition, stable adhesion of the blastocyst to the endometrium occurs, mediated by the interaction between several receptors and ligands (Figure 2B). Over the last decades, several of these ligands and receptors have been identified. It has been observed that both the pinopodes of the endometrial epithelium and the trophectoderm of the blastocyst express the integrin $\alpha v\beta 3$, together with the endometrial expression of its ligand, the glycoprotein osteopontin (OPN). Their expression during the WOI suggests a role in implantation [160,244,245], and the binding between integrin $\alpha v\beta 3$ and its ligand OPN might mediate the stable adhesion between the trophoblast and the endometrium [246]. Using an in vitro model of implantation, Genbacev et al. suggested that trophoblast adhesion to the uterine wall is also mediated by L-selectin expressed on the surface of the trophoblast cells, and uterine epithelial oligosaccharide ligands, such as HECA-452 and MECA-79 [247,248]. More recently it has been also demonstrated that the transmembrane glycoprotein MUC1, abundantly expressed at the apical surface of uterine epithelium under the control of progesterone, acts as a scaffold mediating the binding between L-selectin and their ligands [249].

The adhesion of the blastocyst to the endometrium is also promoted by the expression of adhesion molecules, such as cadherins. The presence of endothelial cadherin (E-cadherin) in both the trophoblasts and endometrial epithelium, regulated by progesterone, indicates that it may play an important role in blastocyst adhesion to the endometrium [250]. As trophoblast cells proliferate, differentiate and invade the stroma, they downregulate E-cadherin and increase osteoblast cadherin (OB-cadherin) [251,252]. This temporal expression of OB-cadherin in the endometrial epithelium suggests that this adhesion molecule later mediates trophoblast–endometrium interactions. Blastocyst adhesion is also favored by the expression of the glycoproteic receptor CD98 on the surface of endometrial cells, which is normally involved not only in amino acids transport but also in cell fusion [253,254]. Using two human endometrial cell lines characterized by low and high receptivity, Dominguez et al. demonstrated that CD98 receptor is significantly associated with the receptive phenotype. In human endometrial samples, CD98 expression was spatially restricted to the apical surface of endometrial cells and temporally restricted to the WOI. Treatment of primary endometrial epithelial cells with hCG, 17-β-estradiol, LIF, or EGF increases expression of CD98, greatly enhancing murine blastocyst adhesion, while its siRNA-mediated depletion reduced blastocyst adhesion rate [255].

3.3.2. NOTCH Signaling in Blastocyst Adhesion to the Endometrium

The expression of NOTCH receptors and ligands in the trophectoderm of the blastocyst and that of NOTCH1, DLL4, and JAG1 in the apical surface of the endometrial epithelium during the mid-secretory phase [170,256] would suggest a role for NOTCH signaling in the adhesion of the blastocyst to the epithelium. Indeed, it has been demonstrated that blastocyst-conditioned medium regulates NOTCH1 and JAG1 expression in the endometrial epithelium [256], suggesting that the blastocyst is able to activate NOTCH signaling in the endometrium, thus possibly regulating its receptivity. This is reinforced by the fact that women with primary infertility show a reduced or absent immunostaining for JAG1 in the luminal epithelium during the mid-secretory phase [256].

3.3.3. Colony-Stimulating Factor-1 in Implantation

A role for colony-stimulating factor-1 (CSF-1) in implantation has been proposed. Indeed, supplementation of CSF-1 in cultures of human trophoblast cells promotes their differentiation in syncytiotrophoblast cells and leads to the production of placental lactogen [257]. In addition, supplementation of CSF-1 to cultures of murine blastocyst induces trophoblast outgrowth [258]. However, using osteopetrotic mutant mice, which lack CSF-1, it has been shown that a maternal source of CSF-1 is not necessary for pregnancy, and possibly the fetus can provide a source of CSF-1 which compensate for the absence of maternally produced CSF-1 [259].

3.4. *Invasion*

Finally, in the third stage, invasion occurs, starting with the penetration of highly invasive trophoblast cells in the uterine epithelium (Figure 2C), followed by infiltration in the basement membrane and in the stromal compartment, a process known as "interstitial invasion" [233,260,261]. Besides invading the endometrial stroma, trophoblast cells also migrate down the lumen of maternal spiral arteries, replace the vascular endothelial lining and become embedded in the arterial walls. This process of "endovascular invasion" allows to replace small-caliber, high-resistance vessels with large-caliber, low-resistance vessels, ensuring an adequate blood supply to the fetoplacental unit [262,263]. Defects in trophoblast endovascular invasion of maternal spiral arteries can seriously impair placental function, leading to significant complications of advanced gestation, such as intrauterine growth restriction (IUGR) and preeclampsia [264].

3.4.1. Matrix Metalloproteinases in Blastocyst Invasion

The huge invasive ability of the fetal trophoblast is due to a high production of activated gelatinases, in particular MMPs 2 and 9 [265–267]. Trophoblastic MMPs are regulated in response to IL-1β, tumor necrosis factor alpha (TNFα) IL-1α, macrophage colony-stimulating factor (MCSF), transforming growth factor β (TGFβ), IGFBP1, leptin, hCG, and EGF [124,268–272], which are secreted from different cell types at the feto-maternal interface, such as trophoblasts themselves and endometrial cells, promoting trophoblast invasion. As already mentioned above, the expression of MMPs involved in endometrial invasion by trophoblast cells is also under the control of TH, as TH positively regulates MMP expression by endometrial cells [223].

3.4.2. Epidermal Growth Factor-Like Domain 7

Recently, we demonstrated that migration and invasion of trophoblast cells is regulated by the secreted factor Epidermal growth factor-like domain 7 (EGFL7), a novel NOTCH interactor. EGFL7 activates NOTCH1, MAPK and AKT signaling pathways in both trophoblast cell lines and primary cells [273]. Activation of the NOTCH pathway is important in both interstitial and endovascular invasion by trophoblast cells. In vitro functional assays show that invasion of Matrigel by trophoblast cells overexpressing EGFL7 is impaired in the presence of a γ-secretase inhibitor, normally used to inhibit NOTCH activation [264,273]. NOTCH appears to be also involved in trophoblast endovascular invasion, since uNK, involved in the disruption of endometrial spiral arteries integrity, express NOTCH1 and 2 and maternal cells surrounding spiral arteries express DLL1 [264], and NOTCH activation may lead to arterial wall disruption. These results are further confirmed by the fact that NOTCH pathway is dysregulated in placenta of women affected by preeclampsia [264,274–280], a common pregnancy disorder characterized by an insufficient trophoblast invasion and an inadequate vascular remodeling. In women affected by preeclampsia, the alteration of NOTCH pathway is accompanied by a concomitant altered expression of EGFL7, in both placenta and maternal circulation [274,281].

3.4.3. Endometrial Control of Blastocyst Invasion

In all the placental species the extent of endometrial decidualization is proportional to the invasiveness of the embryo. The human placenta is the most invasive one known so far, and it has been suggested that the unique invasiveness of the human trophoblast could be due to its high production of hyperglycosylated CG isoform, which is maximal in the first weeks of pregnancy [282,283].

In order to limit the extent of trophoblast invasion, both trophoblast and endometrium balance the expression of growth factors, cytokines, and enzymes. As an example, maternal endometrium increases the production of tissue inhibitors of MMPs (TIMPs), due to a spatial and temporal regulation of cytokines and growth factors, such as IL-10 [284], TGFβ and IL-1α [268]. While IL-1α significantly increases the activity of MMP-9 and MCSF increases MMP-9 immunoreactivity, TGFβ inhibits total gelatinolytic activity, MMP-9 activity and immunoreactivity [268]. TIMP-3, which is up-regulated by progesterone, plays a major role in restricting trophoblast invasion by limiting ECM degradation, and its expression has been detected in the fetal extravillous trophoblast, as well as in the maternal endometrial cells [285,286]. On the contrary, by in situ hybridization in implanting mouse embryos no expression was observed for TIMP-1 or TIMP-2 in the embryo proper, trophoblasts, or in the decidua. Weak signals were demonstrated for TIMP-1 only in the circular layer of myometrial smooth muscle and in some uterine stroma cells distant from the site of embryo implantation. Moreover, the expression of TIMP-1 and TIMP-2 is not dependent on the stage of the menstrual cycle [286]. Trophoblast invasion is promoted by the action of the plasminogen activator (PA) system since it is able to promote trophoblast invasion, by converting plasminogen into the active serine protease plasmin, which in turn, degrades ECM [287]. In endometrial cells, TGFβ regulates trophoblast invasion up-regulating the expression of plasminogen activator inhibitor-1 (PAI-1), which is the main inhibitor of urokinase-type plasminogen activator (uPA) [288–290], and decorin, a decidua-derived TGFβ binding proteoglycan, which inhibits proliferation, migration and invasion of trophoblast cells [291]. The blastocyst is completely embedded in the uterine stroma 8 days after fertilization and the site of entry is covered by fibrin, over which the uterine epithelial cells grow [233,292,293].

3.4.4. Blastocyst Competence Profiling in ARTs

Although many of the molecular players involved in the complex process of implantation have been characterized, the selection of competent embryos remains one of the major challenges in ART. A parallel and complementary morphological and molecular profiling analysis of the embryo may represent a successful approach for embryo selection, thus improving IVF outcome. Although morphological characteristics have been significantly associated with euploidy and competence of the embryo [294,295], their evaluation for good quality embryo selection has some limitations, such as operator subjectivity, variability linked to the timing of laboratory observation, culture medium and other culture conditions, hence combined different approaches might be useful [296]. In this respect, metabolomic and proteomic analyses of embryo spent media have been proposed as complementary, non-invasive tools to select embryos with higher implantation ability [297–299]. Limitations derive from the variability of commercial culture media, high metabolic plasticity of the embryos which can adapt to different culture conditions and from the fact that embryo development and metabolism vary under different culture conditions [300]. Recently, novel strategies based on gene expression profiling of trophectoderm biopsies have been developed and have linked gene expression patterns with developmental competence [301–303]. Although complementary approaches may be used to select the best embryos to be transferred, it should be considered that it has been recently proposed that it is the endometrium itself that selectively discriminates between high-and low-quality embryos in order to guarantee a successful implantation. Based on this, it could be envisioned a test in which the endometrium might be used as a "bio-sensor" to avoid transfer of low-quality embryos, which if implanted would be possibly later rejected resulting in a miscarriage [304,305].

4. Conclusions

Over the last decades the research aimed to reveal the biomolecular processes and pathways underlying animal and human implantation has greatly progressed for two major reasons. On one hand, the exciting advances in available technologies allowed to define in depth the factors and the pathways involved in proper implantation. On the other hand, the introduction of ARTs and their spectacular development in response to the increasing clinical demands from infertile patients allowed to better understand the determinants of a successful implantation and of several conditions of reproductive failure. Additionally, the large diffusion of ARTs provided new perspectives for studies on implantation, making available biological samples previously unavailable for research; follicular fluid, granulosa cells, oocytes, embryos, culture medium of embryos, and blastocysts are examples of this.

As a general concept, it has become clear that reproduction in humans can be considered a rather inefficient process and in several ways is different from reproduction in other species. An emerging concept is that the proper molecular crosstalk between endometrium and blastocyst is of paramount relevance to ensure a proper implantation. In this context, the studies on animal models, apart for the above differences, may greatly help to increase current knowledge. The specific roles of blastocyst and endometrium are being discovered, although much progress still has to be done in this field. The final objective of this field of research effort is twofold: (1) to improve the understanding of how reproduction and implantation evolved and differentiated among the species; (2) to offer more and more effective treatment options to patients with infertility, RIF and RPL.

Author Contributions: Conceptualization, L.C. and M.M.; Writing—Original draft preparation, M.M.; Writing—Review and editing, L.C., C.T. and R.R.; visualization, V.L. and F.L.C.; supervision, L.C.; project administration, L.C. and M.M. All authors have read and agreed to the published version of the manuscript.

Funding: This work has been supported by the Grant for Fertility Innovation 2017, funded by Merck KGaA.

Conflicts of Interest: The authors declare no conflict of interest. The funders had no role in the design of the study; in the collection, analyses, or interpretation of data; in the writing of the manuscript, or in the decision to publish the results.

Abbreviations

WOI	Window of implantation
CG	Chorionic gonadotropin
IL	Interleukin
LIF	Leukemia inhibitory factor
IVF	In vitro fertilization
RIF	Recurrent implantation failure
ART	Assisted reproduction technology
ECM	Extracellular matrix
LH	Luteinizing hormone
uNK	Uterine natural killer
cAMP	Cyclic adenosine monophosphate
BCL-2	B-cell lymphoma 2
LHCGR	LH/CG receptor
COX2	Cyclooxygenase-2

PGES	Prostaglandin E synthase
Erk1/2	Extracellular signal-regulated protein kinases 1/2
PGE2	Prostaglandin E2
NOTCH1	Notch receptor 1
PR	Progesterone receptor
α-SMA	α-smooth muscle actin
IGFBP1	Insulin-like growth factor binding protein-1
ER	Oestrogen receptor
PRL	Prolactin
EGF	Epidermal growth factor
AREG	Amphiregulin
HB-EGF	Heparin binding epidermal growth factor
EREG	Epiregulin
JAG1	Jagged1
DLL	Delta-like
HES	Hairy enhancer of split
HEY1	Hes-related 1
ICM	Inner cell mass
FOXO1	Forkhead box protein O1
PKA	Protein kinase A
MMP	Matrix metalloproteinase
TH	Thyroid hormone
TR	Thyroid hormone receptor
TSHR	Thyroid-stimulating hormone receptor
DIO2	Type 2 deiodinase
LIFR	LIF receptor
ERA	Endometrial receptivity array
HESCs	Human endometrial stromal cells
Tregs	Regulatory T cells
IFN-γ	Interferon-γ
ISP1	Strypsin
ISP2	Lysin
JAK	Janus kinases
STAT	Signal transducer and activator of transcription protein
OPN	Osteopontin
MUC1	Mucin 1
E-cadherin	Endothelial cadherin
OB-cadherin	Osteoblast cadherin
CSF-1	Colony-stimulating factor-1
IUGR	Intrauterine growth restriction
TNFα	Tumor necrosis factor α
MCSF	Macrophage colony-stimulating factor
TGFβ	Transforming growth factor β
EGFL7	Epidermal growth factor-like domain 7
TIMPs	Tissue inhibitors of MMPs
PA	Plasminogen activator
PAI-1	Plasminogen activator inhibitor-1
uPA	Urokinase-type plasminogen activator

References

1. Su, R.W.; Fazleabas, A.T. Implantation and Establishment of Pregnancy in Human and Nonhuman Primates. *Adv. Anat. Embryol. Cell Biol.* **2015**, *216*, 189–213. [PubMed]
2. Fazleabas, A.T.; Strakova, Z. Endometrial function: Cell specific changes in the uterine environment. *Mol. Cell. Endocrinol.* **2002**, *186*, 143–147. [CrossRef]

3. Tabibzadeh, S.; Babaknia, A. The signals and molecular pathways involved in implantation, a symbiotic interaction between blastocyst and endometrium involving adhesion and tissue invasion. *Hum. Reprod.* **1995**, *10*, 1579–1602. [CrossRef] [PubMed]
4. Norwitz, E.R.; Schust, D.J.; Fisher, S.J. Implantation and the survival of early pregnancy. *N. Engl. J. Med.* **2001**, *345*, 1400–1408. [CrossRef] [PubMed]
5. Sharkey, A.M.; Smith, S.K. The endometrium as a cause of implantation failure. *Best Pract. Res. Clin. Obstet. Gynaecol.* **2003**, *17*, 289–307. [CrossRef]
6. Paria, B.C.; Huet-Hudson, Y.M.; Dey, S.K. Blastocyst's state of activity determines the "window" of implantation in the receptive mouse uterus. *Proc. Natl. Acad. Sci. USA* **1993**, *90*, 10159–10162. [CrossRef]
7. Psychoyos, A. Endocrine control of egg implantation. In *Handbook of Physiology*; Greep, R.O., Astwood, E.G., Geiger, S.R., Eds.; American Physiological Society: Washington, DC, USA, 1973; pp. 187–215.
8. Kim, S.M.; Kim, J.S. A Review of Mechanisms of Implantation. *Dev. Reprod.* **2017**, *21*, 351–359. [CrossRef]
9. Wilcox, A.J.; Weinberg, C.R.; O'Connor, J.F.; Baird, D.D.; Schlatterer, J.P.; Canfield, R.E.; Armstrong, E.G.; Nisula, B.C. Incidence of early loss of pregnancy. *N. Engl. J. Med.* **1988**, *319*, 189–194. [CrossRef]
10. World Health Organization. Multiple Definitions of Infertility. Available online: https://www.who.int/reproductivehealth/topics/infertility/multiple-definitions/en/ (accessed on 1 October 2019).
11. Gurunath, S.; Pandian, Z.; Richard, A.R.; Bhattacharya, S. Defining infertility a systematic review of prevalence studies. *Hum. Reprod. Update* **2011**, *17*, 575–588. [CrossRef]
12. Boivin, J.; Bunting, L.; Collins, J.; Nygren, K. International estimates of infertility prevalence and treatmentseeking: Potential need and demand for infertility medical care. *Hum. Reprod.* **2007**, *22*, 1506–1512. [CrossRef]
13. Mascarenhas, M.N.; Flaxman, S.R.; Boerma, T.; Vanderpoel, S.; Stevens, G.A. National, regional, and global trends in infertility prevalence since 1990: A systematic analysis of 277 health surveys. *PLoS Med.* **2012**, *9*, e1001356. [CrossRef]
14. European Society of Human Reproduction and Embryology. Available online: https://www.eshre.eu/Press-Room/Resources (accessed on 4 October 2019).
15. National Health Service. Causes of Infertility. Available online: https://www.nhs.uk/conditions/infertility/ (accessed on 4 October 2019).
16. Abrao, M.S.; Muzii, L.; Marana, R. Anatomical causes of female infertility and their management. *Int. J. Gynaecol. Obstet.* **2013**, *123*, S18–S24. [CrossRef] [PubMed]
17. Unuane, D.; Tournaye, H.; Velkeniers, B.; Poppe, K. Endocrine disorders & female infertility. *Best Pract. Res. Clin. Endocrinol. Metab.* **2011**, *25*, 861–873. [PubMed]
18. Mekinian, A.; Cohen, J.; Alijotas-Reig, J.; Carbillon, L.; Nicaise-Roland, P.; Kayem, G.; Daraï, E.; Fain, O.; Bornes, M. Unexplained Recurrent Miscarriage and Recurrent Implantation Failure: Is There a Place for Immunomodulation? *Am. J. Reprod. Immunol.* **2016**, *76*, 8–28. [CrossRef] [PubMed]
19. Ticconi, C.; Pietropolli, A.; Di Simone, N.; Piccione, E.; Fazleabas, A. Endometrial Immune Dysfunction in Recurrent Pregnancy Loss. *Int. J. Mol. Sci.* **2019**, *20*, 5332. [CrossRef] [PubMed]
20. Larsen, E.C.; Christiansen, O.B.; Kolte, A.M.; Macklon, N. New insights into mechanisms behind miscarriage. *BMC Med.* **2013**, *11*, 154. [CrossRef]
21. Bashiri, A.; Halper, K.I.; Orvieto, R. Recurrent Implantation Failure-update overview on etiology, diagnosis, treatment and future directions. *Reprod. Biol. Endocrinol.* **2018**, *16*, 121. [CrossRef]
22. Coughlan, C.; Ledger, W.; Wang, Q.; Liu, F.; Demirol, A.; Gurgan, T.; Cutting, R.; Ong, K.; Sallam, H.; Li, T.C. Recurrent implantation failure: Definition and management. *Reprod. Biomed. Online* **2014**, *28*, 14–38. [CrossRef]
23. Practice Committee of the American Society for Reproductive Medicine. Evaluation and treatment of recurrent pregnancy loss: A committee opinion. *Fertil. Steril.* **2012**, *98*, 1103–1111. [CrossRef]
24. ESHRE Early Pregnancy Guideline Development Group. *Recurrent Pregnancy Loss*; European Society of Human Reproduction and Embryology: Beigem, Belgium, 2017.
25. Psychoyos, A. Uterine receptivity for nidation. *Ann. N.Y. Acad. Sci.* **1986**, *476*, 36–42. [CrossRef]
26. Psychoyos, A. The 'implantation window': Can it be enlarged or displaced? In *Human Reproduction. Current Status/Future Prospect*; Lizuka, R., Semm, K., Eds.; Excerpta Medica: Amsterdam, the Netherlands; New York, NY, USA; Oxford, UK, 1988; pp. 231–232.
27. Ma, W.G.; Song, H.; Das, S.K.; Paria, B.C.; Dey, S.K. Estrogen is a critical determinant that specifies the duration of the window of uterine receptivity for implantation. *Proc. Natl. Acad. Sci. USA* **2003**, *100*, 2963–2968. [CrossRef]

28. Blesa, D.; Ruiz-Alonso, M.; Simon, C. Clinical management of endometrial receptivity. *Semin. Reprod. Med.* **2014**, *32*, 410–413. [CrossRef]
29. Donaghay, M.; Lessey, B.A. Uterine receptivity: Alterations associated with benign gynecological disease. *Semin. Reprod. Med.* **2007**, *25*, 461–475. [CrossRef]
30. Navot, D.; Scott, R.T.; Droesch, K.; Veeck, L.L.; Liu, H.C.; Rosenwaks, Z. The window of embryo transfer and the efficiency of human conception in vitro. *Fertil. Steril.* **1991**, *55*, 114–118. [CrossRef]
31. Franchi, A.; Zaret, J.; Zhang, X.; Bocca, S.; Oehinger, S. Expression of immunomodulatory genes, their protein products and specific ligands/receptors during the window of implantation in the human endometrium. *Mol. Hum. Reprod.* **2008**, *14*, 413–421. [CrossRef]
32. Altmäe, S.; Reimand, J.; Hovatta, O.; Zhang, P.; Kere, J.; Laisk, T.; Saare, M.; Peters, M.; Vilo, J.; Stavreus-Evers, A.; et al. Research resource: Interactome of human embryo implantation: Identification of gene expression pathways, regulation, and integrated regulatory networks. *Mol. Endocrinol.* **2012**, *26*, 203–217. [CrossRef]
33. Koot, Y.E.; Macklon, N.S. Embryo implantation: Biology, evaluation, and enhancement. *Curr. Opin. Obstet. Gynecol.* **2013**, *25*, 274–279. [CrossRef]
34. Lessey, B.A. Assessment of endometrial receptivity. *Fertil. Steril.* **2011**, *96*, 522–529. [CrossRef]
35. Bartol, F.F.; Wiley, A.A.; Floyd, J.G.; Ott, T.L.; Bazer, F.W.; Gray, C.A.; Spencer, T.E. Uterine differentiation as a foundation for subsequent fertility. *J. Reprod. Fertil. Suppl.* **1999**, *54*, 287–302. [CrossRef]
36. Gray, C.A.; Bartol, F.F.; Tarleton, B.J.; Wiley, A.A.; Johnson, G.A.; Bazer, F.W.; Spencer, T.E. Developmental biology of uterine glands. *Biol. Reprod.* **2001**, *65*, 1311–1323. [CrossRef]
37. Spencer, T.E.; Hayashi, K.; Hu, J.; Carpenter, K.D. Comparative developmental biology of the mammalian uterus. *Curr. Top. Dev. Biol.* **2005**, *68*, 85–122.
38. Cooke, P.S.; Ekman, G.C.; Kaur, J.; Davila, J.; Bagchi, I.C.; Clark, S.G.; Dziuk, P.J.; Hayashi, K.; Bartol, F.F. Brief exposure to progesterone during a critical neonatal window prevents uterine gland formation in mice. *Biol. Reprod.* **2012**, *86*, 1–10. [CrossRef]
39. Hayashi, K.; Yoshioka, S.; Reardon, S.N.; Rucker, E.B., III; Spencer, T.E.; Demayo, F.J.; Lydon, J.P.; Maclean, J.A., II. WNTs in the neonatal mouse uterus: Potential regulation of endometrial gland development. *Biol. Reprod.* **2011**, *84*, 308–319. [CrossRef]
40. Miller, C.; Sassoon, D.A. Wnt-7a maintains appropriate uterine patterning during the development of the mouse female reproductive tract. *Development* **1998**, *125*, 3201–3211.
41. Mericskay, M.; Kitajewski, J.; Sassoon, D. Wnt5a is required for proper epithelial-mesenchymal interactions in the uterus. *Development* **2004**, *131*, 2061–2072. [CrossRef]
42. Dunlap, K.A.; Filant, J.; Hayashi, K.; Rucker, E.B., III; Song, G.; Deng, J.M.; Behringer, R.R.; DeMayo, F.J.; Lydon, J.; Jeong, J.W.; et al. Postnatal deletion of Wnt7a inhibits uterine gland morphogenesis and compromises adult fertility in mice. *Biol. Reprod.* **2011**, *85*, 386–896. [CrossRef]
43. Franco, H.L.; Dai, D.; Lee, K.Y.; Rubel, C.A.; Roop, D.; Boerboom, D.; Jeong, J.W.; Lydon, J.P.; Bagchi, I.C.; Bagchi, M.K.; et al. WNT4 is a key regulator of normal postnatal uterine development and progesterone signaling during embryo implantation and decidualization in the mouse. *FASEB J.* **2011**, *4*, 1176. [CrossRef]
44. Jeong, J.W.; Lee, H.S.; Franco, H.L.; Broaddus, R.R.; Taketo, M.M.; Tsay, S.Y.; Lydon, J.P.; DeMayo, F.J. Beta-catenin mediates glandular formation and dysregulation of beta-catenin induces hyperplasia formation in the murine uterus. *Oncogene* **2009**, *28*, 31–40. [CrossRef]
45. Farah, O.; Biechele, S.; Rossant, J.; Dufort, D. Regulation of porcupine-dependent Wnt signaling is essential for uterine development and function. *Reproduction* **2018**, *155*, 93–102. [CrossRef]
46. Reardon, S.N.; King, M.L.; MacLean, J.A.; Mann, J.L.; DeMayo, F.J.; Lydon, J.P.; Hayashi, K. CDH1 is essential for endometrial differentiation, gland development, and adult function in the mouse uterus. *Biol. Reprod.* **2012**, *86*, 1–10. [CrossRef]
47. Guimarães-Young, A.; Neff, T.; Dupuy, A.J.; Goodheart, M.J. Conditional deletion of Sox17 reveals complex effects on uterine adenogenesis and function. *Dev. Biol.* **2016**, *414*, 19–27. [CrossRef]
48. Cheong, Y.; Boomsa, C.; Heijnen, C.; Macklon, N. Uterine secretomics: A window on the maternal-embryo interface. *Fertil. Steril.* **2013**, *99*, 1093–1099. [CrossRef]
49. Salamonsen, L.A.; Edgell, T.; Rombauts, L.J.; Stephens, A.N.; Robertson, D.M.; Rainczuk, A.; Nie, G.; Hannan, N.J. Proteomics of the human endometrium and uterine fluid: A pathway to biomarker discovery. *Fertil. Steril.* **2013**, *99*, 1086–1092. [CrossRef]

50. Hempstock, J.; Cindrova-Davies, T.; Jauniaux, E.; Burton, G.J. Endometrial glands as a source of nutrients, growth factors and cytokines during the first trimester of human pregnancy: A morphological and immunohistochemical study. *Reprod. Biol. Endocrinol.* **2004**, *2*, 58. [CrossRef]
51. Kane, M.T.; Morgan, P.M.; Coonan, C. Peptide growth factors and preimplantation development. *Hum. Reprod.* **1997**, *3*, 137–157. [CrossRef]
52. Hannan, N.J.; Stephens, A.N.; Rainczuk, A.; Hincks, C.; Rombauts, L.J.; Salamonsen, L.A. 2D-DiGE analysis of the human endometrial secretome reveals differences between receptive and nonreceptive states in fertile and infertile women. *J. Proteome Res.* **2010**, *9*, 6256–6264. [CrossRef]
53. Salamonsen, L.A.; Hannan, N.J.; Dimitriadis, E. Cytokines and chemokines during human embryo implantation: Roles in implantation and early placentation. *Semin. Reprod. Med.* **2007**, *25*, 437–444. [CrossRef]
54. Vilella, F.; Ramirez, L.B.; Simon, C. Lipidomics as an emerging tool to predict endometrial receptivity. *Fertil. Steril.* **2013**, *99*, 1100–1106. [CrossRef]
55. Burton, G.J.; Scioscia, M.; Rademacher, T.W. Endometrial secretions: Creating a stimulatory microenvironment within the human early placenta and implications for the aetiopathogenesis of preeclampsia. *J. Reprod. Immunol.* **2011**, *89*, 118–125. [CrossRef]
56. Burton, G.J.; Jauniaux, E.; Charnock-Jones, D.S. Human early placental development: Potential roles of the endometrial glands. *Placenta* **2007**, *28*, S64–S69. [CrossRef]
57. Guzeloglu-Kayisli, O.; Kayisli, U.A.; Taylor, H.S. The role of growth factors and cytokines during implantation: Endocrine and paracrine interactions. *Semin. Reprod. Med.* **2009**, *27*, 62–79. [CrossRef] [PubMed]
58. Boomsma, C.M.; Kavelaars, A.; Eijkemans, M.J.; Lentjes, E.G.; Fauser, B.C.; Heijnen, C.J.; Macklon, N.S. Endometrial secretion analysis identifies a cytokine profile predictive of pregnancy in IVF. *Hum. Reprod.* **2009**, *24*, 1427–1435. [CrossRef] [PubMed]
59. Hannan, N.J.; Paiva, P.; Meehan, K.L.; Rombauts, L.J.; Gardner, D.K.; Salamonsen, L.A. Analysis of fertility-related soluble mediators in human uterine fluid identifies VEGF as a key regulator of embryo implantation. *Endocrinology* **2011**, *152*, 4948–4956. [CrossRef]
60. Heng, S.; Hannan, N.J.; Rombauts, L.J.; Salamonsen, L.A.; Nie, G. PC6 levels in uterine lavage are closely associated with uterine receptivity and significantly lower I a subgroup of women with unexplained infertility. *Hum. Reprod.* **2011**, *26*, 840–846. [CrossRef]
61. Zhang, Y.; Wang, Q.; Wang, H.; Duan, E. Uterine Fluid in Pregnancy: A Biological and Clinical Outlook. *Trends Mol. Med.* **2017**, *23*, 604–614. [CrossRef]
62. Ladines-Llave, C.A.; Maruo, T.; Manalo, A.S.; Mochizuki, M. Cytologic localization of epidermal growth factor and its receptor in developing human placenta varies over the course of pregnancy. *Am. J. Obstet. Gynecol.* **1991**, *165*, 1377–1382. [CrossRef]
63. Mühlhauser, J.; Crescimanno, C.; Kaufmann, P.; Höfler, H.; Zaccheo, D.; Castellucci, M. Differentiation and proliferation patterns in human trophoblast revealed by c-erbB-2 oncogene product and EGF-R. *J. Histochem. Cytochem.* **1993**, *41*, 165–173. [CrossRef]
64. Sharkey, A.M.; King, A.; Clark, D.E.; Burrows, T.D.; Johki, P.P.; Charnock Jones, D.S.; Loke, Y.W.; Smith, S.K. Localization of leukaemia inhibitory factor and its receptor in human placenta thoughout pregnancy. *Biol. Reprod.* **1999**, *60*, 355–364. [CrossRef]
65. Kojima, K.; Kanzaki, H.; Iwai, M.; Hatayama, H.; Fujimoto, M.; Narukawa, S.; Higuchi, T.; Kaneko, Y.; Mori, T.; Fujita, T. Expression of leukaemia inhibitory factor (LIR) receptor in human placenta: A possible role for LIF in the growth and differentiation of trophoblasts. *Hum. Reprod.* **1995**, *10*, 1907–1911. [CrossRef]
66. Cooper, J.C.; Sharkey, A.M.; McLaren, J.; Charnock Jones, D.S.; Smith, S.K. Localization of vascular endothelial growth factor and its receptor, flt, in human placenta and decidua by immunohistochemistry. *J. Reprod. Fertil.* **1995**, *105*, 205–213. [CrossRef]
67. Wewer, U.M.; Faber, M.; Liotta, L.A.; Albrechtsen, R. Immunochemical and ultrastructural assessment of the nature of the pericellular basement membrane of human decidual cells. *Lab. Investig.* **1985**, *53*, 624–633. [PubMed]
68. Wynn, R.M. Ultrastructural development of the human decidua. *Am. J. Obstet. Gynecol.* **1974**, *118*, 652–670. [CrossRef]
69. Dunn, C.L.; Kelly, R.W.; Critchley, H.O. Decidualization of the human endometrial stromal cell: An enigmatic transformation. *Reprod. Biomed. Online* **2003**, *7*, 151–161. [CrossRef]

70. Kim, J.J.; Jaffe, R.C.; Fazleabas, A.T. Blastocyst invasion and the stromal response in primates. *Hum. Reprod.* **1999**, *14*, 45–55. [CrossRef] [PubMed]
71. Ramathal, C.Y.; Bagchi, I.C.; Taylor, R.N.; Bagchi, M.K. Endometrial decidualization: Of mice and men. *Semin. Reprod. Med.* **2010**, *28*, 17–26. [CrossRef] [PubMed]
72. Teklenburg, G.; Salker, M.; Molokhia, M.; Lavery, S.; Trew, G.; Aojanepong, T.; Mardon, H.J.; Lokugamage, A.U.; Rai, R.; Landles, C.; et al. Natural selection of human embryos: Decidualizing endometrial stromal cells serve as sensors of embryo quality upon implantation. *PLoS ONE* **2010**, *5*, e10258. [CrossRef]
73. Macklon, N.S.; Brosens, J.J. The human endometrium as a sensor of embryo quality. *Biol. Reprod.* **2014**, *91*, 98. [CrossRef]
74. Chen, M.; Wolfe, A.; Wang, X.; Chang, C.; Yeh, S.; Radovick, S. Generation and characterization of a complete null estrogen receptor alpha mouse using Cre/LoxP technology. *Mol. Cell. Biochem.* **2009**, *321*, 145–153. [CrossRef]
75. Lee, H.R.; Kim, T.H.; Choi, K.C. Functions and physiological roles of two types of estrogen receptors, ERα and ERβ, identified by estrogen receptor knockout mouse. *Lab. Anim. Res.* **2012**, *28*, 71–76. [CrossRef]
76. Lubahn, D.B.; Moyer, J.S.; Smithies, O.; Golding, T.S.; Couse, J.F.; Korach, K.S. Alteration of reproductive function but not prenatal sexual development after insertional disruption of the mouse estrogen receptor gene. *Proc. Natl. Acad. Sci. USA* **1993**, *90*, 11162–11166. [CrossRef]
77. Hapangama, D.K.; Kamal, A.M.; Bulmer, J.N. Estrogen receptor β: The guardian of the endometrium. *Hum. Reprod. Update* **2015**, *2*, 174–193. [CrossRef] [PubMed]
78. Cha, J.; Sun, X.; Dey, S.K. Mechanisms of implantation: Strategies for successful pregnancy. *Nat. Med.* **2012**, *18*, 1754–1767. [CrossRef] [PubMed]
79. Thomas, K.; De Hertogh, R.; Pizarro, M.; Van Exter, C.; Ferin, J. Plasma LH-HCG, 17-estradiol, estrone and progesterone monitoring around ovulation and subsequent nidation. *Int. J. Fertil.* **1973**, *18*, 65–73. [PubMed]
80. Stefkovich, M.L.; Arao, Y.; Hamilton, K.J.; Korach, K.S. Experimental models for evaluating non-genomic estrogen signaling. *Steroids* **2018**, *133*, 34–37. [CrossRef] [PubMed]
81. Lee, C.H.; Kim, T.H.; Lee, J.H.; Oh, S.J.; Yoo, J.Y.; Kwon, H.S.; Kim, Y.I.; Ferguson, S.D.; Ahn, J.Y.; Ku, B.J.; et al. Extracellular signal-regulated kinase 1/2 signaling pathway is required for endometrial decidualization in mice and human. *PLoS ONE* **2013**, *8*, e75282. [CrossRef] [PubMed]
82. Zhu, L.; Pollard, J.W. Estradiol-17beta regulates mouse uterine epithelial cell proliferation through insulin-like growth factor 1 signaling. *Proc. Natl. Acad. Sci. USA* **2007**, *104*, 15847–15851. [CrossRef]
83. Klotz, D.M.; Hewitt, S.C.; Ciana, P.; Raviscioni, M.; Lindzey, J.K.; Foley, J.; Maggi, A.; DiAugustine, R.P.; Korach, K.S. Requirement of estrogen receptor-alpha in insulin-like growth factor-1 (IGF-1)-induced uterine responses and in vivo evidence for IGF-1/estrogen receptor cross-talk. *J. Biol. Chem.* **2002**, *277*, 8531–8537. [CrossRef]
84. Hewitt, S.C.; Lierz, S.L.; Garcia, M.; Hamilton, K.J.; Gruzdev, A.; Grimm, S.A.; Lydon, J.P.; DeMayo, F.J.; Korach, K.S. A distal super enhancer mediates estrogen-dependent mouse uterine-specific gene transcription of Insulin-like growth factor 1 (Igf1). *J. Biol. Chem.* **2019**, *294*, 9746–9759. [CrossRef]
85. Tsai, S.J.; Wu, M.H.; Chen, H.M.; Chuang, P.C.; Wing, L.Y. Fibroblast growth factor-9 is an endometrial stromal growth factor. *Endocrinology* **2002**, *143*, 2715–2721. [CrossRef]
86. Wang, W.; Li, Q.; Bagchi, I.C.; Bagchi, M.K. The CCAAT/enhancer binding protein beta is a critical regulator of steroid-induced mitotic expansion of uterine stromal cells during decidualization. *Endocrinology* **2010**, *151*, 3929–3940. [CrossRef]
87. Surveyor, G.A.; Gendler, S.J.; Pemberton, L.; Das, S.K.; Chakraborty, I.; Julian, J.; Pimental, R.A.; Wegner, C.C.; Dey, S.K.; Carson, D.D. Expression and steroid hormonal control of Muc-1 in the mouse uterus. *Endocrinology* **1995**, *136*, 3639–3647. [CrossRef] [PubMed]
88. Rosario, G.X.; Stewart, C.L. The Multifaceted Actions of Leukaemia Inhibitory Factor in Mediating Uterine Receptivity and Embryo Implantation. *Am. J. Reprod. Immunol.* **2016**, *75*, 246–255. [CrossRef] [PubMed]
89. Stewart, C.L.; Kaspar, P.; Brunet, L.J.; Bhatt, H.; Gadi, I.; Kontgen, F.; Abbondanzo, S.J. Blastocyst implantation depends on maternal expression of leukaemia inhibitory factor. *Nature* **1992**, *359*, 76–79. [CrossRef] [PubMed]
90. Patel, B.; Elguero, S.; Thakore, S.; Dahoud, W.; Bedaiwy, M.; Mesiano, S. Role of nuclear progesterone receptor isoforms in uterine pathophysiology. *Hum. Reprod. Update* **2015**, *21*, 155–173. [CrossRef]
91. Paulson, R.J. Hormonal induction of endometrial receptivity. *Fertil. Steril.* **2011**, *96*, 530–535. [CrossRef]
92. Mulac-Jericevic, B.; Lydon, J.P.; DeMayo, F.J.; Conneely, O.M. Defective mammary gland morphogenesis in mice lacking the progesterone receptor B isoform. *Proc. Natl. Acad. Sci. USA* **2003**, *100*, 9744–9749. [CrossRef]

93. Mulac-Jericevic, B.; Mullinax, R.A.; DeMayo, F.J.; Lydon, J.P.; Conneely, O.M. Subgroup of reproductive functions of progesterone mediated by progesterone receptor-B isoform. *Science* **2000**, *289*, 1751–1754. [CrossRef]
94. Lydon, J.P.; DeMayo, F.J.; Funk, C.R.; Mani, S.K.; Hughes, A.R.; Montgomery, C.A., Jr.; Shyamala, G.; Conneely, O.M.; O'Malley, B.W. Mice lacking progesterone receptor exhibit pleiotropic reproductive abnormalities. *Genes Dev.* **1995**, *9*, 2266–2278. [CrossRef]
95. Boonyaratanakornkit, V.; Scott, M.P.; Ribon, V.; Sherman, L.; Anderson, S.M.; Maller, J.L.; Miller, W.T.; Edwards, D.P. Progesterone receptor contains a proline-rich motif that directly interacts with SH3 domains and activates c-Src family tyrosine kinases. *Mol. Cell* **2001**, *8*, 269–280. [CrossRef]
96. Vallejo, G.; La Greca, A.D.; Tarifa-Reischle, I.C.; Mestre-Citrinovitz, A.C.; Ballare, C.; Beato, M.; Saragueta, P. CDC2 mediates progestin initiated endometrial stromal cell proliferation: A PR signaling to gene expression independently of its binding to chromatin. *PLoS ONE* **2014**, *9*, e97311. [CrossRef]
97. Lee, D.K.; Kurihara, I.; Jeong, J.W.; Lydon, J.P.; DeMayo, F.J.; Tsai, M.J.; Tsai, S.Y. Suppression of ERalpha activity by COUP-TFII is essential for successful implantation and decidualization. *Mol. Endocrinol.* **2010**, *24*, 930–940. [CrossRef] [PubMed]
98. Takamoto, N.; Zhao, B.; Tsai, S.Y.; DeMayo, F.J. Identification of Indian hedgehog as a progesterone-responsive gene in the murine uterus. *Mol. Endocrinol.* **2002**, *16*, 2338–2348. [CrossRef] [PubMed]
99. Lee, K.; Jeong, J.; Kwak, I.; Yu, C.T.; Lanske, B.; Soegiarto, D.W.; Toftgard, R.; Tsai, M.J.; Tsai, S.; Lydon, J.P.; et al. Indian hedgehog is a major mediator of progesterone signaling in the mouse uterus. *Nat. Genet.* **2006**, *38*, 1204–1209. [CrossRef] [PubMed]
100. Kurihara, I.; Lee, D.K.; Petit, F.G.; Jeong, J.; Lee, K.; Lydon, J.P.; DeMayo, F.J.; Tsai, M.J.; Tsai, S.Y. COUP-TFII mediates progesterone regulation of uterine implantation by controlling ER activity. *PLoS Genet.* **2007**, *3*, e102. [CrossRef]
101. Huyen, D.V.; Bany, B.M. Evidence for a conserved function of heart and neural crest derivatives expressed transcript 2 in mouse and human decidualization. *Reproduction* **2011**, *142*, 353–368. [CrossRef]
102. Du, H.; Taylor, H.S. The Role of Hox Genes in Female Reproductive Tract Development, Adult Function, and Fertility. *Cold Spring Harb. Perspect. Med.* **2015**, *6*, a023002. [CrossRef]
103. Benson, G.V.; Lim, H.; Paria, B.C.; Satokata, I.; Dey, S.K.; Maas, R.L. Mechanisms of reduced fertility in Hoxa-10 mutant mice: Uterine homeosis and loss of maternal Hoxa-10 expression. *Development* **1996**, *122*, 2687–2696.
104. Lim, H.; Ma, L.; Ma, W.G.; Maas, R.L.; Dey, S.K. Hoxa-10 regulates uterine stromal cell responsiveness to progesterone during implantation and decidualization in the mouse. *Mol. Endocrinol.* **1999**, *13*, 1005–1017. [CrossRef]
105. Gendron, R.L.; Paradis, H.; Hsieh-Li, H.M.; Lee, D.W.; Potter, S.S.; Markoff, E. Abnormal uterine stromal and glandular function associated with maternal reproductive defects in Hoxa-11 null mice. *Biol. Reprod.* **1997**, *56*, 1097–1105. [CrossRef]
106. Lynch, V.J.; Brayer, K.; Gellersen, B.; Wagner, G.P. HoxA-11 and FOXO1A cooperate to regulate decidual prolactin expression: Towards inferring the core transcriptional regulators of decidual genes. *PLoS ONE* **2009**, *9*, e6845. [CrossRef]
107. Kim, J.J.; Taylor, H.S.; Akbas, G.E.; Foucher, I.; Trembleau, A.; Foucher, I.; Trembleau, A.; Jaffe, R.C.; Fazleabas, A.T.; Unterman, T.G. Regulation of insulin-like growth factor binding protein-1 promoter activity by FKHR and HOXA10 in primate endometrial cells. *Biol. Reprod.* **2003**, *68*, 24–30. [CrossRef] [PubMed]
108. Maslar, I.A.; Ansbacher, R. Effects of progesterone on decidual prolactin production by organ cultures of human endometrium. *Endocrinology* **1986**, *118*, 2102–2108. [CrossRef] [PubMed]
109. Daly, D.C.; Maslar, I.A.; Riddick, D.H. Prolactin production during in vitro decidualization of proliferative endometrium. *Am. J. Obstet. Gynecol.* **1983**, *145*, 672–678. [CrossRef]
110. Tabanelli, S.; Tang, B.; Gurpide, E. In vitro decidualization of human endometrial stromal cells. *J. Steroid Biochem. Mol. Biol.* **1992**, *42*, 337–344. [CrossRef]
111. Kim, J.J.; Jaffe, R.C.; Fazleabas, A.T. Comparative studies on the in vitro decidualization process in the baboon (Papio anubis) and human. *Biol. Reprod.* **1998**, *59*, 160–168. [CrossRef]
112. Brosens, J.J.; Hayashi, N.; White, J.O. Progesterone receptor regulates decidual prolactin expression in differentiating human endometrial stromal cells. *Endocrinology* **1999**, *140*, 4809–4820. [CrossRef]

113. Telgmann, R.; Maronde, E.; Taskén, K.; Gellersen, B. Activated protein kinase A is required for differentiation-dependent transcription of the decidual prolactin gene in human endometrial stromal cells. *Endocrinology* **1997**, *138*, 929–937. [CrossRef]
114. Samalecos, A.; Reimann, K.; Wittmann, S.; Schulte, H.M.; Brosens, J.J.; Bamberger, A.M.; Gellersen, B. Characterization of a novel telomerase-immortalized human endometrial stromal cell line, St-T1b. *Reprod. Biol. Endocrinol.* **2009**, *7*, 76. [CrossRef]
115. Popovici, R.M.; Kao, L.C.; Giudice, L.C. Discovery of new inducible genes in in vitro decidualized human endometrial stromal cells using microarraytechnology. *Endocrinology* **2000**, *141*, 3510–3513. [CrossRef]
116. Hirose, T. Exogenous stimulation of corpus luteum formation in the rabbit: Influence of extracts of human placenta, decidua, fetus, hydatid mole, and corpus luteum on the rabbit gonad. *J. Jpn. Gynecol. Soc.* **1920**, *16*, 1055.
117. Lovely, L.P.; Fazleabas, A.T.; Fritz, M.A.; McAdams, D.G.; Lessey, B.A. Prevention of endometrial apoptosis: Randomized prospective comparison of human chorionic gonadotropin versus progesterone treatment in the luteal phase. *J. Clin. Endocrinol. Metab.* **2005**, *90*, 2351–2356. [CrossRef] [PubMed]
118. Jasinska, A.; Strakova, Z.; Szmidt, M.; Fazleabas, A.T. Human chorionic gonadotropin and decidualization in vitro inhibits cytochalasin-D-induced apoptosis in cultured endometrial stromal fibroblasts. *Endocrinology* **2006**, *147*, 4112–4121. [CrossRef] [PubMed]
119. Reshef, E.; Lei, Z.M.; Rao, C.V.; Pridham, D.D.; Chegini, N.; Luborsky, J.L. The presence of gonadotropin receptors in nonpregnant human uterus, human placenta, fetal membranes, and decidua. *J. Clin. Endocrinol. Metab.* **1990**, *70*, 421–430. [CrossRef] [PubMed]
120. Cameo, P.; Szmidt, M.; Strakova, Z.; Mavrogianis, P.; Sharpe-Timms, K.L.; Fazleabas, A.T. Decidualization regulates the expression of the endometrial chorionic gonadotropin receptor in the primate. *Biol. Reprod.* **2006**, *75*, 681–689. [CrossRef] [PubMed]
121. Tapia-Pizarro, A.; Archiles, S.; Argandoña, F.; Valencia, C.; Zavaleta, K.; Cecilia Johnson, M.; González-Ramos, R.; Devoto, L. hCG activates Epac-Erk1/2 signaling regulating Progesterone Receptor expression and function in human endometrial stromal cells. *Mol. Hum. Reprod.* **2017**, *23*, 393–405. [CrossRef] [PubMed]
122. Banerjee, P.; Sapru, K.; Strakova, Z.; Fazleabas, A.T. Chorionic gonadotropin regulates prostaglandin E synthase via a phosphatidylinositol 3-kinase-extracellular regulatory kinase pathway in a human endometrial epithelial cell line: Implications for endometrial responses for embryo implantation. *Endocrinology* **2009**, *150*, 4326–4337. [CrossRef]
123. Zhou, X.L.; Lei, Z.M.; Rao, C.V. Treatment of human endometrial gland epithelial cells with chorionic gonadotropin/luteinizing hormone increases the expression of the cyclooxygenase-2 gene. *J. Clin. Endocrinol. Metab.* **1999**, *84*, 3364–3377. [CrossRef]
124. Srisuparp, S.; Strakova, Z.; Brudney, A.; Mukherjee, S.; Reierstad, S.; Hunzicker-Dunn, M.; Fazleabas, A.T. Signal transduction pathways activated by chorionic gonadotropin in the primate endometrial epithelial cells. *Biol. Reprod.* **2003**, *68*, 457–464. [CrossRef]
125. Tanaka, N.; Miyazaki, K.; Tashiro, H.; Mizutani, H.; Okamura, H. Changes in adenylyl cyclase activity in human endometrium during the menstrual cycle and in human decidua during pregnancy. *J. Reprod. Fertil.* **1993**, *98*, 33–39. [CrossRef]
126. Van der Weiden, R.M.; Helmerhorst, F.M.; Keirse, M.J. Influence of prostaglandins and platelet activating factor on implantation. *Hum. Reprod.* **1991**, *6*, 436–442. [CrossRef]
127. Lim, H.; Paria, B.C.; Das, S.K.; Dinchuk, J.E.; Langenbach, R.; Trzaskos, J.M.; Dey, S.K. Multiple female reproductive failures in cyclooxygenase 2-deficient mice. *Cell* **1997**, *91*, 197–208. [CrossRef]
128. Afshar, Y.; Miele, L.; Fazleabas, A.T. Notch1 is regulated by chorionic gonadotropin and progesterone in endometrial stromal cells and modulates decidualization in primates. *Endocrinology* **2012**, *153*, 2884–2896. [CrossRef] [PubMed]
129. Christian, M.; Pohnke, Y.; Kempf, R.; Gellersen, B.; Brosens, J.J. Functional association of PR and CCAAT/enhancer-binding protein beta isoforms: Promoter-dependent cooperation between PR-B and liverenriched inhibitory protein, or liver-enriched activatory protein and PR-A in human endometrial stromal cells. *Mol. Endocrinol.* **2002**, *16*, 141–154. [PubMed]
130. Gao, J.; Mazella, J.; Tang, M.; Tseng, L. Ligand-activated progesterone receptor isoform hPR-A is a stronger transactivator than hPR-B for the expression of IGFBP-1 (insulin-like growth factor binding protein-1) in human endometrial stromal cells. *Mol. Endocrinol.* **2000**, *14*, 1954–1961. [CrossRef] [PubMed]

131. Gellersen, B.; Brosens, I.A.; Brosens, J.J. Decidualization of the human endometrium: Mechanisms, functions, and clinical perspectives. *Semin. Reprod. Med.* **2007**, *25*, 445–453. [CrossRef]
132. Christensen, S.; Verhage, H.G.; Nowak, G.; de Lanerolle, P.; Fleming, S.; Bell, S.C.; Fazleabas, A.T.; Hild-Petito, S. Smooth muscle myosin II and alpha smooth muscle actin expression in the baboon (Papio anubis) uterus is associated with glandular secretory activity and stromal cell transformation. *Biol. Reprod.* **1995**, *53*, 598–608. [CrossRef]
133. Strakova, Z.; Mavrogianis, P.; Meng, X.; Hastings, J.M.; Jackson, K.S.; Cameo, P.; Brudney, A.; Knight, O.; Fazleabas, A.T. In vivo infusion of interleukin-1beta and chorionic gonadotropin induces endometrial changes that mimic early pregnancy events in the baboon. *Endocrinology* **2005**, *146*, 4097–4104. [CrossRef]
134. Tarantino, S.; Verhage, H.G.; Fazleabas, A.T. Regulation of insulin-like growth factor-binding proteins in the baboon (Papio anubis) uterus during early pregnancy. *Endocrinology* **1992**, *130*, 2354–2362.
135. Develioglu, O.H.; Hsiu, J.G.; Nikas, G.; Toner, J.P.; Oehninger, S.; Jones, H.W., Jr. Endometrial estrogen and progesterone receptor and pinopode expression in stimulated cycles of oocyte donors. *Fertil. Steril.* **1999**, *71*, 1040–1047. [CrossRef]
136. Stavreus-Evers, A.; Nikas, G.; Sahlin, L.; Eriksson, H.; Landgren, B.M. Formation of pinopodes in human endometrium is associated with the concentrations of progesterone and progesterone receptors. *Fertil. Steril.* **2001**, *76*, 782–791. [CrossRef]
137. Martel, D.; Monier, M.N.; Roche, D.; Psychoyos, A. Hormonal dependence of pinopode formation at the uterine luminal surface. *Hum. Reprod.* **1991**, *6*, 597–603. [CrossRef] [PubMed]
138. Singh, M.M.; Trivedi, R.N.; Chauhan, S.C.; Srivastava, V.M.; Makker, A.; Chowdhury, S.R.; Kamboj, V.P. Uterine estradiol and progesterone receptor concentration, activities of certain antioxidant enzymes and dehydrogenases and histoarchitecture in relation to time of secretion of nidatory estrogen and high endometrial sensitivity in rat. *J. Steroid Biochem. Mol. Biol.* **1996**, *59*, 215–224. [CrossRef]
139. Parr, M.B.; Parr, E.L. Uterine luminal epithelium: Protrusions mediate endocytosis, not apocrine secretion, in the rat. *Biol. Reprod.* **1974**, *11*, 220–233. [CrossRef] [PubMed]
140. Matson, B.C.; Pierce, S.L.; Espenschied, S.T.; Holle, E.; Sweatt, I.H.; Davis, E.S.; Tarran, R.; Young, S.L.; Kohout, T.A.; van Duin, M.; et al. Adrenomedullin improves fertility and promotes pinopodes and cell junctions in the peri-implantatio endometrium. *Biol. Reprod.* **2017**, *97*, 466–477. [CrossRef] [PubMed]
141. Lessey, B.A. Two pathways of progesterone action in the human endometrium: Implications for implantation and contraception. *Steroids* **2003**, *68*, 809–815. [CrossRef]
142. Peyghambari, F.; Salehnia, M.; Forouzandeh Moghadam, M.; Rezazadeh Valujerdi, M.; Hajizadeh, E. The correlation between the endometrial integrins and osteopontin expression with pinopodes development in ovariectomized mice in response to exogenous steroids hormones. *Iran. Biomed. J.* **2010**, *14*, 109–119.
143. Liu, S.; Hua, T.; Xin, X.; Shi, R.; Chi, S.; Wang, H. Altered expression of hormone receptor, integrin β3 and pinopode in the endometrium of luteal phase defect women. *Gynecol. Endocrinol.* **2017**, *33*, 315–319. [CrossRef]
144. Qian, Z.-D.; Weng, Y.; Wang, C.F.; Huang, L.L.; Zhu, X.M. Research on the expression of integrin β3 and leukaemia inhibitory factor in the decidua of women with cesarean scar pregnancy. *BMC Pregnancy Childbirth* **2017**, *17*, 84. [CrossRef]
145. Kabir-Salmani, M.; Nikzad, H.; Shiokawa, S.; Akimoto, Y.; Iwashita, M. Secretory role for human pinopodes (pinopods): Secretion of LIF. *Mol. Hum. Reprod.* **2005**, *11*, 553–559. [CrossRef]
146. Mikołajczyk, M.; Skrzypczak, J.; Wirstlein, P. No correlation between pinopod formation and LIF and MMP2 expression in endometrium during implantation window. *Folia Histochem. Cytobiol.* **2011**, *49*, 615–621. [CrossRef]
147. Creus, M.; Ordi, J.; Fábregues, F.; Casamitjana, R.; Ferrer, B.; Coll, E.; Vanrell, J.A.; Balasch, J. alphavbeta3 integrin expression and pinopod formation in normal and out-of-phase endometria of fertile and infertile women. *Hum. Reprod.* **2002**, *17*, 2279–2286. [CrossRef] [PubMed]
148. Nikas, G.; Aghajanova, L. Endometrial pinopodes: Some more understanding on human implantation? *Reprod. Biomed.* **2002**, *4*, 18–23. [CrossRef]
149. Lopata, A.; Bentin-Ley, U.; Enders, A. "Pinopodes" and implantation. *Rev. Endocr. Metab. Disord.* **2002**, *3*, 77–86. [CrossRef] [PubMed]

150. Quinn, C.; Ryan, E.; Claessens, E.A.; Greenblatt, E.; Hawrylyshyn, P.; Cruickshank, B.; Hannam, T.; Dunk, C.; Casper, R.F. The presence of pinopodes in the human endometrium does not delineate the implantation window. *Fertil. Steril.* **2007**, *87*, 1015–1021. [CrossRef]
151. Xu, B.; Sun, X.; Li, L.; Wu, L.; Zhang, A.; Feng, Y. Pinopodes, leukemia inhibitory factor, integrin-beta3, and mucin-1 expression in the peri-implantation endometrium of women with unexplained recurrent pregnancy loss. *Fertil. Steril.* **2012**, *98*, 389–395. [CrossRef]
152. Qiong, Z.; Jie, H.; Yonggang, W.; Bin, X.; Jing, Z.; Yanping, L. Clinical validation of pinopode as a marker of endometrial receptivity: A randomized controlled trial. *Fertil. Steril.* **2017**, *108*, 513–517. [CrossRef]
153. Jin, X.Y.; Zhao, L.J.; Luo, D.H.; Liu, L.; Dai, Y.D.; Hu, X.X.; Wang, Y.Y.; Lin, X.; Hong, F.; Li, T.C.; et al. Pinopode score around the time of implantation is predictive of successful implantation following frozen embryo transfer in hormone replacement cycles. *Hum. Reprod.* **2017**, *32*, 2394–2403. [CrossRef]
154. Aunapuu, M.; Kibur, P.; Jarveots, T.; Arend, A. Changes in Morphology and Presence of Pinopodes in Endometrial Cells during the Luteal Phase in Women with Infertility Problems: A Pilot Study. *Medicina* **2018**, *54*, 69. [CrossRef]
155. Das, S.K.; Wang, X.N.; Paria, B.C.; Damm, D.; Abraham, J.A.; Klagsbrun, M.; Andrews, G.K.; Dey, S.K. Heparin-binding EGF-like growth factor gene is induced in the mouse uterus temporally by the blastocyst solely at the site of its apposition: A possible ligand for interaction with blastocyst EGF-receptor in implantation. *Development* **1994**, *120*, 1071–1083.
156. Lim, H.; Dey, S.K.; Das, S.K. Differential expression of the erbB2 gene in the periimplantation mouse uterus: Potential mediator of signaling by epidermal growth factor-like growth factors. *Endocrinology* **1997**, *138*, 1328–1337. [CrossRef]
157. Das, S.K.; Chakraborty, I.; Paria, B.C.; Wang, X.N.; Plowman, G.; Dey, S.K. Amphiregulin is an implantation-specific and progesterone-regulated gene in the mouse uterus. *Mol. Endocrinol.* **1995**, *9*, 691–705. [PubMed]
158. Das, S.K.; Das, N.; Wang, J.; Lim, H.; Schryver, B.; Plowman, G.D.; Dey, S.K. Expression of beta cellulin and epiregulin genes in the mouse uterus temporally by the blastocyst solely at the site of its apposition is coincident with the "window" of implantation. *Dev. Biol.* **1997**, *190*, 178–190. [CrossRef] [PubMed]
159. Lim, H.; Das, S.K.; Dey, S.K. erbB genes in the mouse uterus: Cell-specific signaling by epidermal growth factor (EGF) family of growth factors during implantation. *Dev. Biol.* **1998**, *204*, 97–110. [CrossRef] [PubMed]
160. Lessey, B.A. Adhesion molecules and implantation. *J. Reprod. Immunol.* **2002**, *55*, 101–112. [CrossRef]
161. Stavreus-Evers, A.; Aghajanova, L.; Brismar, H.; Eriksson, H.; Landgren, B.M.; Hovatta, O. Co-existence of heparin-binding epidermal growth factor-like growth factor and pinopodes in human endometrium at the time of implantation. *Mol. Hum. Reprod.* **2002**, *8*, 765–769. [CrossRef]
162. Xie, H.; Wang, H.; Tranguch, S.; Iwamoto, R.; Mekada, E.; Demayo, F.J.; Lydon, J.P.; Das, S.K.; Dey, S.K. Maternal heparin-binding-EGF deficiency limits pregnancy success in mice. *Proc. Natl. Acad. Sci. USA* **2007**, *104*, 18315–18320. [CrossRef]
163. Raab, G.; Kover, K.; Paria, B.C.; Dey, S.K.; Ezzell, R.M.; Klagsbrun, M. Mouse preimplantation blastocysts adhere to cells expressing the transmembrane form of heparin-binding EGF-like growth factor. *Development* **1996**, *122*, 637–645.
164. Artavanis-Tsakonas, S.; Rand, M.D.; Lake, R.J. Notch signaling: Cell fate control and signal integration in development. *Science* **1999**, *284*, 770–776. [CrossRef]
165. Bray, S.J. Notch signaling: A simple pathway becomes complex. *Mol. Cell Biol.* **2006**, *7*, 678–689.
166. Leong, K.G.; Karsan, A. Recent insights into the role of Notch signaling in tumorigenesis. *Blood* **2006**, *107*, 2223–2233. [CrossRef]
167. Rizzo, P.; Miao, H.; D'Souza, G.; Osipo, C.; Song, L.L.; Yun, J.; Zhao, H.; Mascarenhas, J.; Wyatt, D.; Antico, G.; et al. Cross-talk between notch and the estrogen receptor in breast cancer suggests novel therapeutic approaches. *Cancer. Res.* **2008**, *68*, 5226–5235. [CrossRef] [PubMed]
168. Cobellis, L.; Caprio, F.; Trabucco, E.; Mastrogiacomo, A.; Coppola, G.; Manente, L.; Colacurci, N.; De Falco, M.; De Luca, A. The pattern of expression of Notch protein members in normal and pathological endometrium. *J. Anat.* **2008**, *213*, 464–472. [CrossRef] [PubMed]
169. Mitsuhashi, Y.; Horiuchi, A.; Miyamoto, T.; Kashima, H.; Suzuki, A.; Shiozawa, T. Prognostic significance of Notch signaling molecules and their involvement in the invasiveness of endometrial carcinoma cells. *Histopathology* **2012**, *6*, 826–837. [CrossRef] [PubMed]

170. Mazella, J.; Liang, S.; Tseng, L. Expression of Delta-like protein 4 in the human endometrium. *Endocrinology* **2008**, *149*, 15–19. [CrossRef]
171. Mikhailik, A.; Mazella, J.; Liang, S.; Tseng, L. Notch ligand-dependent gene expression in human endometrial stromal cells. *Biochem. Biophys. Res. Commun.* **2009**, *388*, 479–482. [CrossRef]
172. Adjaye, J.; Huntriss, J.; Herwig, R.; BenKahla, A.; Brink, T.C.; Wierling, C.; Hultschig, C.; Groth, D.; Yaspo, M.L.; Picton, H.M.; et al. Primary differentiation in the human blastocyst: Comparative molecular portraits of inner cell mass and trophectoderm cells. *Stem Cells* **2005**, *23*, 1514–1525. [CrossRef]
173. Aghajanova, L.; Shen, S.; Rojas, A.M.; Fisher, S.J.; Irwin, J.C.; Giudice, L.C. Comparative transcriptome analysis of human trophectoderm and embryonic stem cell-derived trophoblasts reveal key participants in early implantation. *Biol. Reprod.* **2012**, *86*, 1–21. [CrossRef]
174. Wang, Q.T.; Piotrowska, K.; Ciemerych, M.A.; Milenkovic, L.; Scott, M.P.; Davis, R.W.; Zernicka-Goetz, M. A genome-wide study of gene activity reveals developmental signaling pathways in the preimplantation mouse embryo. *Dev. Cell* **2004**, *6*, 133–144. [CrossRef]
175. Hess, A.P.; Hamilton, A.E.; Talbi, S.; Dosiou, C.; Nyegaard, M.; Nayak, N.; Genbecev-Krtolica, O.; Mavrogianis, P.; Ferrer, K.; Kruessel, J.; et al. Decidual stromal cell response to paracrine signals from the trophoblast: Amplification of immune and angiogenic modulators. *Biol. Reprod.* **2007**, *76*, 102–117. [CrossRef]
176. Strug, M.R.; Su, R.W.; Kim, T.H.; Mauriello, A.; Ticconi, C.; Lessey, B.A.; Young, S.L.; Lim, J.M.; Jeong, J.W.; Fazleabas, A.T. RBPJ mediates uterine repair in the mouse and is reduced in women with recurrent pregnancy loss. *FASEB J.* **2018**, *32*, 2452–2466. [CrossRef]
177. Strug, M.R.; Su, R.; Young, J.E.; Dodds, W.G.; Shavell, V.I.; Díaz-Gimeno, P.; Ruíz-Alonso, M.; Simón, C.; Lessey, B.A.; Leach, R.E.; et al. Intrauterine human chorionic gonadotropin infusion in oocyte donors promotes endometrial synchrony and induction of early decidual markers for stromal survival: A randomized clinical trial. *Hum. Reprod.* **2016**, *31*, 1552–1561. [CrossRef] [PubMed]
178. Brar, A.K.; Handwerger, S.; Kessler, C.A.; Aronow, B.J. Gene induction and categorical reprogramming during in vitro human endometrial fibroblast decidualization. *Physiol. Genom.* **2001**, *7*, 135–148. [CrossRef] [PubMed]
179. Christian, M.; Zhang, X.; Schneider-Merck, T.; Unterman, T.G.; Gellersen, B.; White, J.O.; Brosens, J.J. Cyclic AMP induced forkhead transcription factor, FKHR, cooperates with CCAAT/enhancer-binding protein beta in differentiating human endometrial stromal cells. *J. Biol. Chem.* **2002**, *277*, 20825–20832. [CrossRef] [PubMed]
180. Buzzio, O.L.; Lu, Z.; Miller, C.D.; Unterman, T.G.; Kim, J.J. FOXO1A differentially regulates genes of decidualization. *Endocrinology* **2006**, *147*, 3870–3876. [CrossRef] [PubMed]
181. Grinius, L.; Kessler, C.; Schroeder, J.; Handwerger, S. Forkhead transcription factor FOXO1A is critical for induction of human decidualization. *J. Endocrinol.* **2006**, *189*, 179–187. [CrossRef] [PubMed]
182. Labied, S.; Kajihara, T.; Madureira, P.A.; Fusi, L.; Jones, M.C.; Higham, J.M.; Varshochi, R.; Francis, J.M.; Zoumpoulidou, G.; Essafi, A.; et al. Progestins regulate the expression and activity of the forkhead transcription factor FOXO1 in differentiating human endometrium. *Mol. Endocrinol.* **2006**, *20*, 35–44. [CrossRef] [PubMed]
183. Hallaq, R.; Volpicelli, F.; Cuchillo-Ibanez, I.; Hooper, C.; Mizuno, K.; Uwanogho, D.; Causevic, M.; Asuni, A.; To, A.; Soriano, S.; et al. The Notch intracellular domain represses CRE-dependent transcription. *Cell. Signal.* **2015**, *27*, 621–629. [CrossRef]
184. Librach, C.; Feigenbaum, S.; Bass, K.; Cui, T.; Verastas, N.; Sadovsky, Y.; Quigley, J.; French, D.; Fisher, S. Interleukin-1 beta regulates human cytotrophoblast metalloproteinase activity and invasion in vitro. *J. Biol. Chem.* **1994**, *269*, 17125–17131.
185. Strakova, Z.; Srisuparp, S.; Fazleabas, A.T. Interleukin-1beta induces the expression of insulin-like growth factor binding protein-1 during decidualization in the primate. *Endocrinology* **2000**, *141*, 4664–4670. [CrossRef]
186. Fazleabas, A.T.; Kim, J.J.; Strakova, Z. Implantation: Embryonic signals and the modulation of the uterine environment–a review. *Placenta* **2004**, *25*, 26–31. [CrossRef]
187. Strakova, Z.; Srisuparp, S.; Fazleabas, A.T. IL-1beta during in vitro decidualization in primate. *J. Reprod. Immunol.* **2002**, *55*, 35–47. [CrossRef]
188. Strakova, Z.; Szmidt, M.; Srisuparp, S.; Fazleabas, A.T. Inhibition of matrix metalloproteinases prevents the synthesis of insulin-like growth factor binding protein-1 during decidualization in the baboon. *Endocrinology* **2003**, *144*, 5339–5346. [CrossRef] [PubMed]

189. Fazleabas, A.T.; Bell, S.C.; Fleming, S.; Sun, J.; Lessey, B.A. Distribution of integrins and the extracellular matrix proteins in the baboon endometrium during the menstrual cycle and early pregnancy. *Biol. Reprod.* **1997**, *56*, 348–356. [CrossRef] [PubMed]
190. Aghajanova, L.; Stavreus-Evers, A.; Lindeberg, M.; Landgren, B.M.; Skjoldebrand Sparre, L.; Hovatta, O. Thyroid-stimulating hormone receptor and thyroid hormone receptors are involved in human endometrial physiology. *Fertil. Steril.* **2011**, *95*, 230–237. [CrossRef]
191. Catalano, R.D.; Critchley, H.O.; Heikinheimo, O.; Baird, D.T.; Hapangama, D.; Sherwin, J.R.A.; Charnock-Jones, D.S.; Smith, S.K.; Sharkey, A.M. Mifepristone induced progesterone withdrawal reveals novel regulatory pathways in human endometrium. *Mol. Hum. Reprod.* **2007**, *13*, 641–654. [CrossRef]
192. Wakim, A.N.; Polizotto, S.L.; Buffo, M.J.; Marrero, M.A.; Burholt, D.R. Thyroid hormones in human follicular fluid and thyroid hormone receptors in human granulosa cells. *Fertil. Steril.* **1993**, *59*, 1187–1190. [CrossRef]
193. Campbell, D.J.; Koch, M.A. Phenotypical and functional specialization of FOXP3+ regulatory T cells. *Nat. Rev. Immunol.* **2011**, *11*, 119–130. [CrossRef]
194. Tilburgs, T.; Roelen, D.L.; van der Mast, B.J.; de Groot-Swings, G.M.; Kleijburg, C.; Scherjon, S.A.; Claas, F.H. Evidence for a selective migration of fetus-specific CD4 + CD25bright regulatory T cells from the peripheral blood to the decidua in human pregnancy. *J. Immunol.* **2008**, *180*, 5737–5745. [CrossRef]
195. Xiong, H.; Zhou, C.; Qi, G. Proportional changes of CD4 + CD25 + Foxp3+ regulatory T cells in maternal peripheral blood during pregnancy and labor at term and preterm. *Clin. Investig. Med.* **2010**, *33*, 422. [CrossRef]
196. Hara, M.; Kingsley, C.I.; Niimi, M.; Read, S.; Turvey, S.E.; Bushell, A.R.; Morris, P.J.; Powrie, F.; Wood, K.J. IL-10 is required for regulatory T cells to mediate tolerance to alloantigens in vivo. *J. Immunol.* **2001**, *166*, 3789–3796. [CrossRef]
197. Robertson, S.A.; Care, A.S.; Moldenhauer, L.M. Regulatory T cells in embryo implantation and the immune response to pregnancy. *J. Clin. Investig.* **2018**, *128*, 4224–4235. [CrossRef] [PubMed]
198. Fu, B.; Li, X.; Sun, R.; Tong, X.; Ling, B.; Tian, Z.; Wei, H. Natural killer cells promote immune tolerance by regulating inflammatory TH17 cells at the human maternal-fetal interface. *Proc. Natl. Acad. Sci. USA* **2013**, *110*, 231–240. [CrossRef] [PubMed]
199. Koopman, L.A.; Kopcow, H.D.; Rybalov, B.; Boyson, J.E.; Orange, J.S.; Schatz, F.; Masch, R.; Lockwood, C.J.; Schachter, A.D.; Park, P.J.; et al. Human decidual natural killer cells are a unique NK cell subset with immunomodulatory potential. *J. Exp. Med.* **2003**, *198*, 1201–1212. [CrossRef] [PubMed]
200. Díaz-Gimeno, P.; Horcajadas, J.A.; Martínez-Conejero, J.A.; Esteban, F.J.; Alamá, P.; Pellicer, A.; Simón, C. A genomic diagnostic tool for human endometrial receptivity based on the transcriptomic signature. *Fertil. Steril.* **2011**, *95*, 50–60. [CrossRef] [PubMed]
201. Díaz-Gimeno, P.; Ruiz-Alonso, M.; Blesa, D.; Bosch, N.; Martínez-Conejero, J.A.; Alamá, P.; Garrido, N.; Pellicer, A.; Simón, C. The accuracy and reproducibility of the endometrial receptivity array is superior to histology as a diagnostic method for endometrial receptivity. *Fertil. Steril.* **2013**, *99*, 508–517. [CrossRef]
202. Ruiz-Alonso, M.; Blesa, D.; Díaz-Gimeno, P.; Gómez, E.; Fernández-Sánchez, M.; Carranza, F.; Carrera, J.; Vilella, F.; Pellicer, A.; Simón, C. The endometrial receptivity array for diagnosis and personalized embryo transfer as a treatment for patients with repeated implantation failure. *Fertil. Steril.* **2013**, *100*, 818–824. [CrossRef]
203. Bassil, R.; Casper, R.; Samara, N.; Hsieh, T.B.; Barzilay, E.; Orvieto, R.; Haas, J. Does the endometrial receptivity array really provide personalized embryo transfer? *J. Assist. Reprod. Genet.* **2018**, *35*, 1301–1305. [CrossRef]
204. Enciso, M.; Carrascosa, J.P.; Sarasa, J.; Martínez-Ortiz, P.A.; Munné, S.; Horcajadas, J.A.; Aizpurua, J. Development of a new comprehensive and reliable endometrial receptivity map (ER Map/ER Grade) based on RT-qPCR gene expression analysis. *Hum. Reprod.* **2018**, *33*, 220–228. [CrossRef]
205. Croxatto, H.B.; Ortiz, M.E.; Diaz, S.; Hess, R.; Balmaceda, J.; Croxato, H.D. Studies on the duration of egg transport by the human oviduct. II. Ovum location at various intervals following luteninizing hormone peak. *Am. J. Obstet. Gynecol.* **1978**, *132*, 629–634. [CrossRef]
206. Buster, J.E.; Bustillo, M.; Rodi, I.A.; Cohen, S.W.; Hamilton, M.; Simon, J.A.; Thorneycroft, I.H.; Marshall, J.R. Biologic and morphologic development of donated human ova recovered by nonsurgical uterine lavage. *Am. J. Obstet. Gynecol.* **1985**, *153*, 211–217. [CrossRef]

207. Mossman, H.W. Orientation and site of attachment of the blastocyst: A comparative study. In *Biology of the Blastocyst*; Blandau, R.J., Ed.; University of Chicago Press: Chicago, IL, USA, 1971; pp. 49–57.
208. Rasweiler, J.J., 4th; Badwaik, N.K. Relationships between orientation of the blastocyst during implantation, position of the chorioallantoic placenta, and vascularization of the uterus in the noctilionoid bats Carollia perspicillata and Noctilio sp. *Placenta* **1999**, *20*, 241–255. [CrossRef] [PubMed]
209. Kirby, D.R.; Potts, D.M.; Wilson, I.B. On the orientation of the implanting blastocyst. *J. Embryol. Exp. Morphol.* **1967**, *17*, 527–532. [PubMed]
210. Gardner, R.L. Location and orientation of implantation. In *Establishing a Successful Human Pregnancy*; Edwards, R.G., Ed.; Raven Press: New York, NY, USA, 1990; pp. 225–238.
211. Rasweiler, J.J., IV; Badwaik, N.K. Unusual aspects of inner cell mass formation, endoderm differentiation, Reichert's membrane development, and amniogenesis in the lesser bulldog bat, Noctilio albiventris. *Anat. Rec.* **1996**, *246*, 293–304. [CrossRef]
212. Paria, B.C.; Ma, W.; Tan, J.; Raja, S.; Das, S.K.; Dey, S.K.; Hogan, B.L. Cellular and molecular responses of the uterus to embryo implantation can be elicited by locally applied growth factors. *Proc. Natl. Acad. Sci. USA* **2001**, *98*, 1047–1052. [CrossRef]
213. Beer, A.E.; Billingham, R.E. Implantation, transplantation, and epithelial-mesenchymal relationships in the rat uterus. *J. Exp. Med.* **1970**, *132*, 721–736. [CrossRef]
214. McLaren, A. Stimulus and response during early pregnancy in the mouse. *Nature* **1969**, *221*, 739–741. [CrossRef]
215. Hetherington, C.M. Induction of deciduomata in the mouse by carbon dioxide. *Nature* **1968**, *219*, 863–864. [CrossRef]
216. Zhang, S.; Kong, S.; Wang, B.; Cheng, X.; Chen, Y.; Wu, W.; Wang, Q.; Shi, J.; Zhang, Y.; Wang, S.; et al. Uterine Rbpj is required for embryonic-uterine orientation and decidual remodeling via Notch pathway-independent and -dependent mechanisms. *Cell Res.* **2014**, *24*, 925–942. [CrossRef]
217. Goad, J.; Ko, Y.A.; Kumar, M.; Syed, S.M.; Tanwar, P.S. Differential Wnt signalling activity limits epithelial gland development to the anti-mesometrial side of the mouse uterus. *Dev. Biol.* **2017**, *423*, 138–151. [CrossRef]
218. Kimie, Y.; Rika, S.; Eriko, H.; Shunzo, K.; Yoshihiro, K.; Ken, K.; Motonori, H.; Hitoshi, S. Trypsin-like hatching enzyme of mouse blastocysts: Evidence for its participation in hatching process before zona shedding of embryos. *Dev. Growth Differ.* **1994**, *36*, 149–154.
219. O'Sullivan, C.M.; Liu, S.Y.; Karpinka, J.B.; Rancourt, D.E. Embryonic hatching enzyme strypsin/ISP1 is expressed with ISP2 in endometrial glands during implantation. *Mol. Reprod. Dev.* **2002**, *62*, 328–334. [CrossRef] [PubMed]
220. Perona, R.M.; Wassarman, P.M. Mouse blastocysts hatch in vitro by using a trypsin like proteinase associated with cells of mural trophectoderm. *Dev. Biol.* **1986**, *114*, 42–52. [CrossRef]
221. Sireesha, G.V.; Mason, R.W.; Hassanein, M.; Tonack, S.; Navarrete Santos, A.; Fischer, B.; Seshagiri, P.B. Role of cathepsins in blastocyst hatching in the golden hamster. *Mol. Hum. Reprod.* **2008**, *14*, 337–346. [CrossRef] [PubMed]
222. Mishra, A.; Seshagiri, P.B. Evidence for the involvement of species-specific embryonic protease in zona dissolution of hamster blastocysts. *Mol. Hum. Reprod.* **2000**, *6*, 1005–1012. [CrossRef] [PubMed]
223. Piccirilli, D.; Baldini, E.; Massimiani, M.; Camaioni, A.; Salustri, A.; Bernardini, R.; Centanni, M.; Ulisse, S.; Moretti, C.; Campagnolo, L. Thyroid hormone regulates protease expression and activation of Notch signaling in implantation and embryo development. *J. Endocrinol.* **2018**, *236*, 1–12. [CrossRef] [PubMed]
224. Dickinson, D.P. Cysteine peptidases of mammals: Their biological roles and potential effects in the oral cavity and other tissues in health and disease. *Crit. Rev. Oral. Biol. Med.* **2002**, *13*, 238–275. [CrossRef] [PubMed]
225. Afonso, S.; Romagnano, L.; Babiarz, B. The expression and function of cystatin C and cathepsin B and cathepsin L during mouse embryo implantation and placentation. *Development* **1997**, *124*, 3415–3425.
226. O'Sullivan, C.M.; Liu, S.Y.; Rancourt, S.L.; Rancourt, D.E. Regulation of the strypsinrelated proteinase ISP2 by progesterone in endometrial gland epithelium during implantation in mice. *Reproduction* **2001**, *122*, 235–244. [CrossRef]
227. Sharma, N.; Kumar, R.; Renaux, B.; Saifeddine, M.; Nishikawa, S.; Mihara, K.; Ramachandran, R.; Hollenberg, M.D.; Rancourt, D.E. Implantation serine proteinase 1 exhibits mixed substrate specificity that silences signaling via proteinase activated receptors. *PLoS ONE* **2011**, *6*, e27888. [CrossRef]

228. Sharma, N.; Liu, S.; Tang, L.; Irwin, J.; Meng, G.; Rancourt, D.E. Implantation Serine Proteinases heterodimerize and are critical in hatching and implantation. *BMC Dev. Biol.* **2006**, *11*, 6–61.
229. O'Sullivan, C.M.; Tang, L.; Xu, H.; Liu, S.; Rancourt, D.E. Origin of the murine implantation serine proteinase subfamily. *Mol. Reprod. Dev.* **2004**, *69*, 126–136. [CrossRef] [PubMed]
230. Lindenberg, S. Experimental studies on the initial trophoblast endometrial interaction. *Dan. Med. Bull.* **1991**, *38*, 371–380. [PubMed]
231. Hertig, A.T.; Rock, J.; Adams, E.C. A description of 34 human ova within the first 17 days of development. *Am. J. Anat.* **1956**, *98*, 435–493. [CrossRef] [PubMed]
232. Sharma, A.; Kumar, P. Understanding implantation window, a crucial phenomenon. *J. Hum. Reprod. Sci.* **2012**, *5*, 2–6.
233. Bischof, P.; Campana, A. A model for implantation of the human blastocyst and early placentation. *Hum. Reprod. Update* **1996**, *2*, 262–270. [CrossRef]
234. Denker, H.W. Implantation: A cell biological paradox. *J. Exp. Zool.* **1993**, *266*, 541–558. [CrossRef]
235. Cheng, J.G.; Chen, J.R.; Hernandez, L.; Alvord, W.G.; Stewart, C.L. Dual control of LIF expression and LIF receptor function regulate Stat3 activation at the onset of uterine receptivity and embryo implantation. *Proc. Natl. Acad. Sci. USA* **2001**, *98*, 8680–8685. [CrossRef]
236. Catalano, R.; Johnson, M.H.; Campbell, E.A.; Charnock-Jones, D.S.; Smith, S.K.; Sharkey, A.M. Inhibition of Stat3 activation in the endometrium prevents implantation: A nonsteroidal approach to contraception. *Proc. Natl. Acad. Sci. USA* **2005**, *102*, 8585–8590. [CrossRef]
237. Song, H.; Lim, H.; Das, S.K.; Paria, B.C.; Dey, S.K. Dysregulation of EGF family of growth factors and COX-2 in the uterus during the preattachment and attachment reactions of the blastocyst with the luminal epithelium correlates with implantation failure in LIF-deficient mice. *Mol. Endocrinol.* **2000**, *14*, 1147–1161. [CrossRef]
238. Pawar, S.; Starosvetsky, E.; Orvis, G.D.; Behringer, R.R.; Bagchi, I.C.; Bagchi, M.K. STAT3 regulates uterine epithelial remodeling and epithelial-stromal crosstalk during implantation. *Mol. Endocrinol.* **2013**, *27*, 1996–2012. [CrossRef]
239. Hantak, A.M.; Bagchi, I.C.; Bagchi, M.K. Role of uterine stromal-epithelial crosstalk in embryo implantation. *Int. J. Dev. Biol.* **2014**, *58*, 139–146. [CrossRef] [PubMed]
240. Laird, S.M.; Tuckerman, E.M.; Dalton, C.F.; Dunphy, B.C.; Li, T.C.; Zhang, X. The production of leukaemia inhibitory factor by human endometrium: Presence in uterine flushings and production by cells in culture. *Hum. Reprod.* **1997**, *12*, 569–574. [CrossRef] [PubMed]
241. Hambartsoumian, E. Endometrial leukemia inhibitory factor (LIF) as a possible cause of unexplained infertility and multiple failures of implantation. *Am. J. Reprod. Immunol.* **1998**, *39*, 137–143. [CrossRef] [PubMed]
242. Srisuparp, S.; Strakova, Z.; Fazleabas, A.T. The role of chorionic gonadotropin (CG) in blastocyst implantation. *Arch. Med. Res.* **2001**, *32*, 627–634. [CrossRef]
243. Alfthan, H.; Stenman, U.H. Pathophysiological importance of various molecular forms of human choriogonadotropin. *Mol. Cell. Endocrinol.* **1996**, *125*, 107–120. [CrossRef]
244. Aplin, J.D.; Spanswick, C.; Behzad, F.; Kimber, S.J.; Vicovac, L. Integrins beta 5, beta 3 and alpha v are apically distributed in endometrial epithelium. *Mol. Hum. Reprod.* **1996**, *2*, 527–534. [CrossRef]
245. Apparao, K.B.; Murray, M.J.; Fritz, M.A.; Meyer, W.R.; Chambers, A.F.; Truong, P.R.; Lessey, B.A. Osteopontin and its receptor alphavbeta(3) integrin are coexpressed in the human endometrium during the menstrual cycle but regulated differentially. *J. Clin. Endocrinol. Metab.* **2001**, *86*, 4991–5000.
246. Reddy, K.V.; Mangale, S.S. Integrin receptors: The dynamic modulators of endometrial function. *Tissue Cell* **2003**, *35*, 260–273. [CrossRef]
247. Genbacev, O.D.; Prakobphol, A.; Foulk, R.A.; Krtolica, A.R.; Ilic, D.; Singer, M.S.; Yang, Z.Q.; Kiessling, L.L.; Rosen, S.D.; Fisher, S.J. Trophoblast L-selectin-mediated adhesion at the maternal-fetal interface. *Science* **2003**, *299*, 405–408. [CrossRef]
248. Foulk, R.A.; Zdravkovic, T.; Genbacev, O.; Prakobphol, A. Expression of L-selectin ligand MECA-79 as a predictive marker of human uterine receptivity. *J. Assist. Reprod. Genet.* **2007**, *24*, 316–321. [CrossRef]
249. Carson, D.D.; Julian, J.; Lessey, B.A.; Prakobphol, A.; Fisher, S.J. MUC1 is a scaffold for selectin ligands in the human uterus. *Front. Biosci.* **2006**, *11*, 2903–2908. [CrossRef] [PubMed]

250. Rowlands, T.M.; Symonds, J.M.; Farookhi, R.; Blaschuk, O.W. Cadherins: Crucial regulators of structure and function in reproductive tissues. *Rev. Reprod.* **2000**, *5*, 53–61. [CrossRef] [PubMed]
251. Shih, I.M.; Hsu, M.Y.; Oldt, R.J., III; Herlyn, M.; Gearhart, J.D.; Kurman, R.J. The Role of E-cadherin in the Motility and Invasion of Implantation Site Intermediate Trophoblast. *Placenta* **2002**, *23*, 706–715. [CrossRef] [PubMed]
252. MacCalman, C.D.; Furth, E.E.; Omigbodun, A.; Bronner, M.; Coutifaris, C.; Strauss, J.F., III. Regulated expression of cadherin-11 in human epithelial cells: A role for cadherin-11 in trophoblast-endometrium interactions? *Dev. Dyn.* **1996**, *206*, 201–211. [CrossRef]
253. Chillaron, J.; Roca, R.; Valencia, A.; Zorzano, A.; Palacin, M. Heteromeric amino acid transporters: Biochemistry, genetics, and physiology. *Am. J. Physiol. Renal Physiol.* **2001**, *281*, F995–F1018. [CrossRef] [PubMed]
254. Tsurudome, M.; Ito, Y. Function of fusion regulatory proteins (FRPs) in immune cells and virus-infected cells. *Crit. Rev. Immunol.* **2000**, *20*, 167–196. [CrossRef]
255. Dominguez, F.; Simon, C.; Quinonero, A.; Ramirez, M.A.; Gonzalez-Munoz, E.; Burghardt, H.; Cervero, A.; Martinez, S.; Pellicer, A.; Palacin, M.; et al. Human endometrial CD98 is essential for blastocyst adhesion. *PLoS ONE* **2010**, *5*, 13380. [CrossRef]
256. Cuman, C.; Menkhorst, E.M.; Rombauts, L.J.; Holden, S.; Webster, D.; Bilandzic, M.; Osianlis, T.; Dimitriadis, E. Preimplantation human blastocysts release factors that differentially alter human endometrial epithelial cell adhesion and gene expression relative to IVF success. *Hum. Reprod.* **2013**, *28*, 1161–1171. [CrossRef]
257. Garcia-Lloret, M.; Morrish, D.W.; Guilbert, L.J. Functional expression of CSF-1 receptors on normal human trophoblast. In Proceedings of the Third European Placental Group Meeting, Dourdan, France, 27–30 September 1989.
258. Haimovici, F.; Anderson, D.J. Cytokines and growth factors in implantation. *Microsc. Res. Tech.* **1993**, *25*, 201–207. [CrossRef]
259. Pollard, J.W.; Hunt, J.S.; Wiktor-Jedrzejczak, W.; Stanley, E.R. A pregnancy defect in the osteopetrotic (opop) mouse demonstrates the requirement for CSF-1 in female fertility. *Dev. Biol.* **1991**, *148*, 273–283. [CrossRef]
260. Pijnenborg, R.; Bland, J.M.; Robertson, W.B.; Dixon, G.; Brosens, I. The pattern of interstitial trophoblastic invasion of the myometrium in early human pregnancy. *Placenta* **1981**, *2*, 303–316. [CrossRef]
261. Giudice, L.C. Potential biochemical markers of uterine receptivity. *Hum. Reprod.* **1999**, *14*, 3–16. [CrossRef] [PubMed]
262. Burrows, T.D.; King, A.; Loke, Y. Trophoblast migration during human placental implantation. *Hum. Reprod. Update* **1996**, *2*, 307–321. [CrossRef] [PubMed]
263. Pijnenborg, R.; Robertson, W.B.; Brosens, I.; Dixon, G. Trophoblast invasion and establishment of haemochorial placentation in man and laboratory animals. *Placenta* **1981**, *2*, 71–91. [CrossRef]
264. Hunkapiller, N.M.; Gasperowicz, M.; Kapidzic, M.; Plaks, V.; Maltepe, E.; Kitajewski, J.; Cross, J.C.; Fisher, S. A role for Notch signaling in trophoblast endovascular invasion and in the pathogenesis of pre-eclampsia. *Development* **2011**, *138*, 2987–2998. [CrossRef]
265. Shimonovitz, S.; Hurwitz, A.; Dushnik, M.; Anteby, E.; GevaEldar, T.; Yagel, S. Developmental regulation of the expression of 72 and 92 kd type IV collagenases in human trophoblasts: A possible mechanism for control of trophoblast invasion. *Am. J. Obstet. Gynecol.* **1994**, *171*, 832–838. [CrossRef]
266. Cañete-Soler, R.; Gui, Y.H.; Linask, K.K.; Muschel, R.J. Developmental expression of MMP-9 (gelatinase B) mRNA in mouse embryos. *Dev. Dyn.* **1995**, *204*, 30–40. [CrossRef]
267. Huppertz, B.; Kertschanska, S.; Demir, A.Y.; Frank, H.G.; Kaufmann, P. Immunohistochemistry of matrix metalloproteinases (MMP), their substrates, and their inhibitors (TIMP) during trophoblast invasion in the human placenta. *Cell Tissue Res.* **1998**, *291*, 133–148. [CrossRef]
268. Meisser, A.; Chardonnens, D.; Campana, A.; Bischof, P. Effects of tumour necrosis factor-alpha, interleukin-1 alpha, macrophage colony stimulating factor and transforming growth factor beta on trophoblastic matrix metalloproteinases. *Mol. Hum. Reprod.* **1999**, *5*, 252–260. [CrossRef]
269. Bischof, P.; Meisser, A.; Campana, A.; Tseng, L. Effects of deciduaconditioned medium and insulin-like growth factor binding protein-1 on trophoblastic matrix metalloproteinases and their inhibitors. *Placenta* **1998**, *19*, 457–464. [CrossRef]
270. Castellucci, M.; De Matteis, R.; Meisser, A.; Cancello, R.; Monsurro, V.; Islami, D.; Sarzani, R.; Marzioni, D.; Cinti, S.; Bischof, P. Leptin modulates extracellular matrix molecules and metalloproteinases: Possible implications for trophoblast invasion. *Mol. Hum. Reprod.* **2000**, *6*, 951–958. [CrossRef] [PubMed]

271. Licht, P.; Russu, V.; Wildt, L. On the role of human chorionic gonadotropin (hCG) in the embryo-endometrial microenvironment: Implications for differentiation and implantation. *Semin. Reprod. Med.* **2001**, *19*, 37–47. [CrossRef] [PubMed]
272. Qiu, Q.; Yang, M.; Tsang, B.K.; Gruslin, A. EGF-induced trophoblast secretion of MMP-9 and TIMP-1 involves activation of both PI3K and MAPK signaling pathways. *Reproduction* **2004**, *128*, 355–363. [CrossRef] [PubMed]
273. Massimiani, M.; Vecchione, L.; Piccirilli, D.; Spitalieri, P.; Amati, F.; Salvi, S.; Ferrazzani, S.; Stuhlmann, H.; Campagnolo, L. Epidermal growth factor-like domain 7 (EGFL7) promotes migration and invasion of human trophoblast cells through activation of MAPK, PI3K and NOTCH signaling pathways. *Mol. Hum. Reprod.* **2015**, *21*, 435–451. [CrossRef]
274. Lacko, L.A.; Massimiani, M.; Sones, J.L.; Hurtado, R.; Salvi, S.; Ferrazzani, S.; Davisson, R.L.; Campagnolo, L.; Stuhlmann, H. Novel expression of EGFL7 in placental trophoblast and endothelial cells and its implication in preeclampsia. *Mech. Dev.* **2014**, *133*, 163–176. [CrossRef]
275. Taki, A.; Abe, M.; Komaki, M.; Oku, K.; Iseki, S.; Mizutani, S.; Morita, I. Expression of angiogenesis-related factors and inflammatory cytokines in placenta and umbilical vessels in pregnancies with preeclampsia and chorioamnionitis/funisitis. *Congenit. Anom.* **2012**, *52*, 97–103. [CrossRef]
276. Meng, T.; Chen, H.; Sun, M.; Wang, H.; Zhao, G.; Wang, X. Identification of differential gene expression profiles in placentas from preeclamptic pregnancies versus normal pregnancies by DNA microarrays. *OMICS* **2012**, *16*, 301–311. [CrossRef]
277. Sahin, Z.; Acar, N.; Ozbey, O.; Ustunel, I.; Demir, R. Distribution of Notch family proteins in intrauterine growth restriction and hypertension complicated human term placentas. *Acta Histochem.* **2011**, *113*, 270–276. [CrossRef]
278. Løset, M.; Mundal, S.B.; Johnson, M.P.; Fenstad, M.H.; Freed, K.A.; Lian, I.A.; Eide, I.P.; Bjørge, L.; Blangero, J.; Moses, E.K.; et al. A transcriptional profile of the decidua in preeclampsia. *Am. J. Obstet. Gynecol.* **2011**, *204*, 84.e1–84.e27. [CrossRef]
279. Sitras, V.; Paulssen, R.H.; Grønaas, H.; Leirvik, J.; Hanssen, T.A.; Vårtun, A.; Acharya, G. Differential placental gene expression in severe preeclampsia. *Placenta* **2009**, *30*, 424–433. [CrossRef]
280. Cobellis, L.; Mastrogiacomo, A.; Federico, E.; Schettino, M.T.; De Falco, M.; Manente, L.; Coppola, G.; Torella, M.; Colacurci, N.; De Luca, A. Distribution of Notch protein members in normal and preeclampsia complicated placentas. *Cell Tissue Res.* **2007**, *330*, 527–534. [CrossRef] [PubMed]
281. Massimiani, M.; Lacko, L.A.; Burke Swanson, C.S.; Salvi, S.; Argueta, L.B.; Moresi, S.; Ferrazzani, S.; Gelber, S.E.; Baergen, R.N.; Toschi, N.; et al. Increased circulating levels of Epidermal Growth Factor-like Domain 7 in pregnant women affected by preeclampsia. *Transl. Res.* **2019**, *207*, 19–29. [CrossRef] [PubMed]
282. Fournier, T. Human chorionic gonadotropin: Different glycoforms and biological activity depending on its source of production. *Ann. Endocrinol.* **2016**, *77*, 75–81. [CrossRef] [PubMed]
283. Cole, L.A. hCG, the wonder of today's science. *Reprod. Biol. Endocrinol.* **2012**, *10*, 24. [CrossRef] [PubMed]
284. Roth, I.; Fisher, S. IL-10 is an autocrine inhibitor of human placental cytotrophoblast MMP-9 production and invasion. *Dev. Biol.* **1999**, *205*, 194–204. [CrossRef]
285. Higuchi, T.; Kanzaki, H.; Nakayama, H.; Fujimoto, M.; Hatayama, H.; Kojima, K.; Iwai, M.; Mori, T.; Fujita, J. Induction of tissue inhibitor of metalloproteinase 3 gene expression during in vitro decidualization of human endometrial stromal cells. *Endocrinology* **1995**, *136*, 4973–4981. [CrossRef]
286. Reponen, P.; Leivo, I.; Sahlberg, C.; Apte, S.S.; Olsen, B.R.; Thesleff, I.; Tryggvason, K. 92-kDa type IV collagenase and TIMP-3, but not 72-kDa type IV collagenase or TIMP-1 or TIMP-2, are highly expressed during mouse embryo implantation. *Dev. Dyn.* **1995**, *202*, 388–396. [CrossRef]
287. Aflalo, E.D.; Sod-Moriah, U.A.; Potashnik, G.; Har-Vardi, I. Differences in the implantation rates of rat embryos developed in vivo and in vitro: Possible role for plasminogen activators. *Fertil. Steril.* **2004**, *81*, 780–785. [CrossRef]
288. Schatz, F.; Aigner, S.; Papp, C.; Toth-Pal, E.; Hausknecht, V.; Lockwood, C.J. Plasminogen activator activity during decidualization of human endometrial stromal cells is regulated by plasminogen activator inhibitor 1. *J. Clin. Endocrinol. Metab.* **1995**, *80*, 2504–2510.
289. Simón, C.; Gimeno, M.J.; Mercader, A.; Francés, A.; Garcia Velasco, J.; Remohí, J.; Polan, M.L.; Pellicer, A. Cytokines-adhesion molecules-invasive proteinases. The missing paracrine/autocrine link in embryonic implantation? *Mol. Hum. Reprod.* **1996**, *2*, 405–424. [CrossRef]

290. Karmakar, S.; Das, C. Regulation of trophoblast invasion by IL-1β and TGF-β1. *Am. J. Reprod. Immunol.* **2002**, *48*, 210–219. [CrossRef] [PubMed]
291. Iacob, D.; Cai, J.; Tsonis, M.; Babwah, A.; Chakraborty, C.; Bhattacharjee, R.N.; Lala, P.K. Decorin-mediated inhibition of proliferation and migration of the human trophoblast via different tyrosine kinase receptors. *Endocrinology* **2008**, *149*, 6187–6197. [CrossRef] [PubMed]
292. Aplin, J.D.; Haigh, T.; Lacey, H.; Chen, C.P.; Jones, C.J. Tissue interactions in the control of trophoblast invasion. *J. Reprod. Fertil. Suppl.* **2000**, *55*, 57–64. [PubMed]
293. Su, R.W.; Strug, M.R.; Joshi, N.R.; Jeong, J.W.; Miele, L.; Lessey, B.A.; Young, S.L.; Fazleabas, A.T. Decreased Notch pathway signaling in the endometrium of women with endometriosis impairs decidualization. *J. Clin. Endocrinol. Metab.* **2015**, *100*, 433–442. [CrossRef] [PubMed]
294. Majumdar, G.; Majumdar, A.; Verma, I.C.; Upadhyaya, K.C. Relationship between morphology, euploidy and implantation potential of cleavage and blastocyst stage embryos. *J. Hum. Reprod. Sci.* **2017**, *10*, 49. [PubMed]
295. Gardner, D.K.; Balaban, B. Assessment of human embryo development using morphological criteria in an era of time-lapse, algorithms and 'OMICS': Is looking good still important? *Mol. Hum. Reprod.* **2016**, *22*, 704–718. [CrossRef]
296. ALPHA Scientists In Reproductive Medicine; ESHRE Special Interest Group Embryology. Istanbul consensus workshop on embryo assessment: Proceedings of an expert meeting. *Reprod. Biomed. Online* **2011**, *22*, 632–646. [CrossRef]
297. Seli, E.; Sakkas, D.; Scott, R.; Kwok, S.C.; Rosendahl, S.M.; Burns, D.H. Noninvasive metabolomic profiling of embryo culture media using Raman and near-infrared spectroscopy correlates with reproductive potential of embryos in women undergoing in vitro fertilization. *Fertil. Steril.* **2007**, *88*, 1350–1357. [CrossRef]
298. Scott, R.; Seli, E.; Miller, K.; Sakkas, D.; Scott, K.; Burns, D.H. Noninvasive metabolomic profiling of human embryo culture media using Raman spectroscopy predicts embryonic reproductive potential: A prospective blinded pilot study. *Fertil. Steril.* **2008**, *90*, 77–83. [CrossRef]
299. Vergouw, C.G.; Botros, L.L.; Roos, P.; Lens, J.W.; Schats, R.; Hompes, P.G.A.; Burns, D.H.; Lambalk, C.B. Metabolomic profiling by near-infrared spectroscopy as a tool to assess embryo viability: A novel, non-invasive method for embryo selection. *Hum. Reprod.* **2008**, *23*, 1499–1504. [CrossRef]
300. Krisher, R.L.; Schoolcraft, W.B.; Katz-Jaffe, M.G. Omics as a window to view embryo viability. *Fertil. Steril.* **2015**, *103*, 333–341. [CrossRef] [PubMed]
301. Ntostis, P.; Kokkali, G.; Iles, D.; Huntriss, J.; Tzetis, M.; Picton, H.; Pantos, K.; Miller, D. Can trophectoderm RNA analysis predict human blastocyst competency? *Syst. Biol. Reprod. Med.* **2019**, *65*, 312–325. [CrossRef] [PubMed]
302. Kirkegaard, K.; Villesen, P.; Jensen, J.M.; Hindkjær, J.J.; Kølvraa, S.; Ingerslev, H.J.; Lykke-Hartmann, K. Distinct differences in global gene expression profiles in non-implanted blastocysts and blastocysts resulting in live birth. *Gene* **2015**, *571*, 212–220. [CrossRef] [PubMed]
303. Jones, G.M.; Cram, D.S.; Song, B.; Kokkali, G.; Pantos, K.; Trounson, A.O. Novel strategy with potential to identify developmentally competent IVF blastocysts. *Hum. Reprod.* **2008**, *23*, 1748–1759. [CrossRef] [PubMed]
304. Weimar, C.H.; Kavelaars, A.; Brosens, J.J.; Gellersen, B.; de Vreeden-Elbertse, J.M.; Heijnen, C.J.; Macklon, N.S. Endometrial stromal cells of women with recurrent miscarriage fail to discriminate between high- and low-quality human embryos. *PLoS ONE* **2012**, *7*, e41424. [CrossRef]
305. Brosens, J.J.; Salker, M.S.; Teklenburg, G.; Nautiyal, J.; Salter, S.; Lucas, E.S.; Steel, J.H.; Christian, M.; Chan, Y.W.; Boomsma, C.M.; et al. Uterine selection of human embryos at implantation. *Sci. Rep.* **2014**, *4*, 3894. [CrossRef]

International Journal of
Molecular Sciences

MDPI

Review

Endometrial Decidualization: The Primary Driver of Pregnancy Health

Shu-Wing Ng [1,2,*], Gabriella A. Norwitz [3], Mihaela Pavlicev [4], Tamara Tilburgs [5,6], Carlos Simón [7,8,9] and Errol R. Norwitz [1,2,*]

1 Department of Obstetrics & Gynecology, Tufts University School of Medicine, Boston, MA 02111, USA
2 Mother Infant Research Institute, Tufts Medical Center, Boston, MA 02111, USA
3 School of Public Health, University of Michigan, Ann Arbor, MI 48109, USA; gnorwitz@umich.edu
4 Department of Theoretical Biology, University of Vienna, 1010 Vienna, Austria; mihaela.pavlicev@univie.ac.at
5 Division of Immunobiology and Center for Inflammation and Tolerance, Cincinnati Children's Hospital, Cincinnati, OH 45229, USA; tamara.tilburgs@cchmc.org
6 Department of Pediatrics, University of Cincinnati College of Medicine, Cincinnati, OH 45229, USA
7 Department of Obstetrics & Gynecology, Valencia University and INCLIVA, 46010 Valencia, Spain; carlos.simon@uv.es
8 Department of Obstetrics & Gynecology, Beth Israel Deaconess Medical Center, Harvard University, Boston, MA 02215, USA
9 Igenomix Foundation and INCLIVA, 46010 Valencia, Spain
* Correspondence: sng1@tuftsmedicalcenter.org (S.-W.N.); enorwitz@tuftsmedicalcenter.org (E.R.N.); Tel.: +617-636-8950 (S.-W.N.); +617-636-2382 (E.R.N.)

Received: 24 April 2020; Accepted: 5 June 2020; Published: 8 June 2020

Abstract: Interventions to prevent pregnancy complications have been largely unsuccessful. We suggest this is because the foundation for a healthy pregnancy is laid prior to the establishment of the pregnancy at the time of endometrial decidualization. Humans are one of only a few mammalian viviparous species in which decidualization begins during the latter half of each menstrual cycle and is therefore independent of the conceptus. Failure to adequately prepare (decidualize) the endometrium hormonally, biochemically, and immunologically in anticipation of the approaching blastocyst—including the downregulation of genes involved in the pro- inflammatory response and resisting tissue invasion along with the increased expression of genes that promote angiogenesis, foster immune tolerance, and facilitate tissue invasion—leads to abnormal implantation/placentation and ultimately to adverse pregnancy outcome. We hypothesize, therefore, that the primary driver of pregnancy health is the quality of the soil, not the seed.

Keywords: endometrium; decidualization; adverse pregnancy outcome; preconception

1. Introduction

Many complications that manifest clinically in the first trimester—such as miscarriage—or in the latter half of pregnancy—including preeclampsia, preterm birth (PTB), fetal growth restriction (FGR), and gestational diabetes (GDM)—have their origins early in gestation with abnormalities in implantation and placentation [1–8]. Despite exhaustive research and a vastly improved understanding of the molecular and cellular mechanisms responsible for implantation/placentation, interventions to prevent these complications have been largely unsuccessful. In this monograph, we suggest this is because the foundation for pregnancy health is laid down earlier than previously appreciated during the preconception period at the time of endometrial decidualization. Humans are one of only a few mammalian viviparous species in which decidualization starts during the latter half of each menstrual cycle and is therefore independent of the conceptus [9–11]. This implies that the health of a pregnancy is determined even before the blastocyst arrives. Once a pregnancy is established, its destiny has already been determined and it is too late to intervene effectively. Stated differently, pregnancy complications are not two-stage disorders as conventionally understood with abnormal implantation/placentation leading to clinical disease, but rather three-stage disorders starting with abnormal endometrial decidualization that predates the arrival of the blastocyst leading thereafter to abnormal implantation/placentation and ultimately to clinical disease [5]. We hypothesize therefore that the primary driver of pregnancy health is the quality of the soil, not the seed.

2. Biological Continuum of Adverse Pregnancy Outcome

While we divide pregnancy disorders into distinct categories, much of this classification is arbitrary for the purposes of description and study. Delivery at 19 weeks 6 days of gestation is defined as a miscarriage, whereas delivery one day later is a premature birth. In reality, these conditions occur along a continuum. They have common and interrelated risk factors. For example, a woman with a history of a prior unexplained PTB at 28 weeks is at increased risk of spontaneous PTB in a subsequent pregnancy, but is also at increased risk of preeclampsia and FGR in future pregnancies [4]. These disorders also have overlapping biomarkers [12]. Moreover, deficient spiral artery remodeling has been linked with a spectrum of obstetrical syndromes, including pre- eclampsia, FGR, PTB, premature rupture of membranes (PPROM), abortion, and fetal death [4,8,13]. Taken together, these observations suggest that adverse pregnancy events occur along a biological continuum and likely have a common underlying pathophysiology.

3. Implantation and Placentation

Implantation is critical to survival of a species, but this process in humans has a surprisingly high failure rate. Maximal fecundity (the likelihood of getting pregnant each cycle) peaks at 30% [2,14]. Only 50% of conceptions advance beyond 20 weeks of gestation and, of all unsuccessful pregnancies, 75% represent a failure of implantation [2,14–16]. Even among eutherian (placental) mammals, humans are unique. Among other features, human pregnancy has the most invasive type of placentation (hemochorial), early recognition of the fetal allograft by the maternal immune system, and a long gestational length [9,11].

The factors regulating implantation have been reviewed in detail elsewhere [2,7,17]. Briefly, as in other mammals, human implantation likely involves three steps: (i) apposition (initial adhesion, which is unstable), (ii) attachment (stable adhesion), and (iii) invasion, which occurs in two phases or waves. The 'first wave' of trophoblast invasion occurs between days 7 and 10 post-conception, starting shortly after the blastocyst hatches out of the zona pellucida. During this time, the blastocyst actively invades the tissues of the uterus. By day 10 postconception, the blastocyst is completely buried within the endometrial lining. For the next few weeks, the placenta is not yet hemochorial [18] and the blastocyst is fed by secretions from the endometrial glands (histiotrophic support) under conditions that are both hypoxic and hypoglycemic. Indeed, high levels of oxygen or glucose at this stage will damage the developing embryo. At 8–10 weeks of gestation, the placental extravillous cytotrophoblast cells (EVCTs) change their adhesion molecule expression and stream out of the placental villi to invade the full thickness of the decidualized endometrium (decidua) and the inner third of the myometrium. These cells invade the maternal spiral arteries, attracted in part by the high oxygen tension [19] and by active recruitment by uterine natural killer (uNK) cells and macrophages [8,17,20], and remodel these vessels by destroying the muscle layer and replacing the endothelial lining with a pseudo-endothelium of fetal origin. This process—known as the 'second wave' of trophoblast invasion—is usually complete by 18 weeks of gestation and is critical for the establishment of the definitive uteroplacental circulation. Interestingly, the initiation of vascular remodeling precedes the trophoblast invasion of the spiral arteries and is likely initiated by resident uNK cells [21,22]. As pregnancy progresses, the 120–140 small, tortuous maternal spiral arteries that supply each placenta need to dilate enormously to accommodate the increasing demands of the fetoplacental unit. The placenta is a high-volume, low-resistance organ. At term, almost one-fifth of the maternal cardiac output (approximately 800mL) passes through the placenta every minute. If this remodeling of the maternal spiral arteries from narrow lumen, tortuous vessels with a thick muscle layer to wide, thin-walled, funnel-shaped vessels is not adequate—a pathological hallmark known as shallow endovascular invasion [1,8]—the feto- placental unit will outgrow its blood supply, resulting in placental dysfunction and ultimately in clinical disease.

4. It's the Quality of the Soil, Not the Seed

Successful implantation is the end result of a complex molecular interaction between two separate components: a viable blastocyst and an appropriately primed endometrium [2,6,23]. Both are important, but do they contribute equally to reproductive disorders? Much attention has focused on the blastocyst, and, indeed, many early miscarriages do result from karyotypic abnormalities within the blastocyst [24]. However, there is increasing evidence to suggest that appropriate priming of the soil (endometrium)—a process known as decidualization—may contribute more to reproductive disorders than the quality of the seed (embryo). Observations in support of this argument include:

- A critical period of time exists within each menstrual cycle—known as the 'window of implantation'—in which the endometrium is maximally receptive to the blastocyst. This period is personalized and implantation outside of this 24–36 h window will result in an absolute failure to establish a pregnancy or in suboptimal implantation increasing the risk of a range of downstream adverse pregnancy events [25,26].
- In contrast to the 'window of implantation' in the endometrium, embryos generated by IVF can be transferred into the uterus any time between days 2 and 7 postconception [27].
- Embryos can be frozen and thawed multiple times prior to transfer.
- Oocyte donor embryos (that are entirely allogeneic) will implant successfully [28,29].
- Pregnancy outcomes appear to be better in frozen rather than fresh cycles [30]. A plausible explanation might be that the hormonal manipulations used to prepare the endometrium for existing cryopreserved embryos are more favorable to the endometrium than protocols used in fresh cycles, which are designed primarily to maximize the number of oocytes retrieved.
- Lastly, although the presence of a decidua is not an absolute requirement for implantation since the blastocyst can implant in the fallopian tube, the cervix, or even into the vasculature of the bowel in the case of extrauterine intraabdominal ectopic pregnancies, such pregnancies are rarely healthy and, if they do go past 20 weeks, have a high rate of complications.

Taken together, these data suggest that the endometrial "window of implantation" is independent of the blastocyst and that the embryo is not the rate-limiting factor for implantation, but rather the synchronization between them.

5. The Decidua as an Anatomically Distinct Autocrine/Paracrine Organ

The decidua is the maternal tissue most intimately associated with the fetoplacental unit and serves a critical role as an endocrine and immunological organ. The process of implantation/placentation results in the formation of three decidual regions, which are anatomically and functionally distinct. This review focuses on that region that underlies the placenta, known as the decidua basalis (or decidua placentalis) (Figure 1).

The endometrium/decidua is a complex, dynamic, heterogeneous tissue made up of multiple cell types. Moreover, its cellular composition changes in a predictable fashion during the menstrual cycle and throughout the course of pregnancy in response to changes in systemic and local hormones. These cellular changes have been reviewed in detail elsewhere [17,31–37]. Importantly, the endometrium/decidua is rich in immune cells, particularly uNK cells and macrophages, which originate in the bone marrow and track selectively via the bloodstream to the uterine lining. In the first 20 weeks of pregnancy, uNK cells and macrophages play a critical role in mediating the process of spiral artery transformation by inducing initial structural changes, secreting a number of cytokines and chemokines, and promoting the actions of EVCTs [38–40]. They also protect against placental infection [41,42]. Another distinct and functionally important group of cells comprises the decidual stromal fibroblast cells (DSCs), which make up 10–30% of decidual cells in the first trimester and up to 60–70% of cells in term decidua (discussed below).

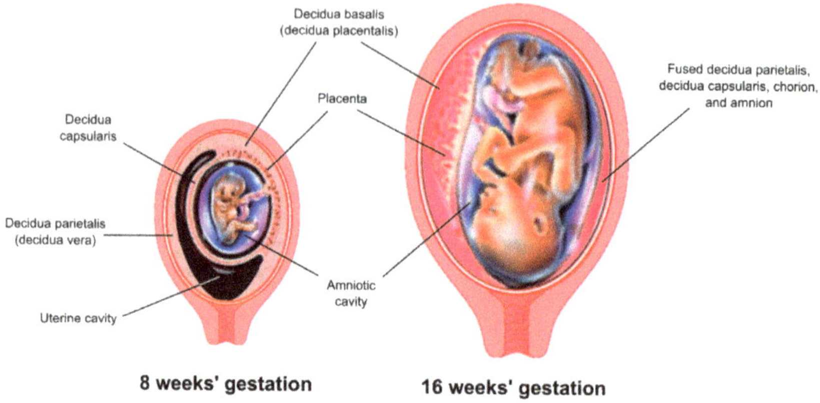

Figure 1. Anatomic arrangement of the decidua. Three anatomically and functionally discrete regions of the decidua are defined based on their relationship to the fetoplacental unit. The decidua basalis (or decidua placentalis) lies directly below the placenta. The decidua capsularis covers (encapsulates) the developing embryo as it grows and expands into the uterine cavity. The decidua parietalis (or decidua vera) lines the uterus remote from the placenta. A virtual space exists between the decidua capsularis and decidua parietalis until about 15–16 weeks of gestation, at which point these two tissues come together and fuse.

6. Endometrial Decidualization

Decidualization refers to the functional and morphological changes that occur within the endometrium to form the decidual lining into which the blastocyst implants. These changes include the recruitment of leukocytes and, importantly, the differentiation of endometrial stromal fibroblast cells (ESCs) into DSCs. It is the ability of ESCs to differentiate into this alternative cell state that appears to be the key element in the decidual transformation. DSCs are not simply modified ESCs; they are a distinct cell type resulting from terminal differentiation and the genetic reprogramming of ESCs. This reprogramming includes the downregulation of genes involved in the pro- inflammatory response and in resisting tissue invasion along with increased expression of genes that promote cellular proliferation, foster tolerance, and facilitate tissue invasion (discussed below). DSCs originated early in the stem lineage of placental mammals [43,44] and their evolution coincided precisely in evolutionary history with the appearance of invasive placentation [11,45].

7. Evolution of the Decidua

Decidualization is widespread among eutherian mammals and is perhaps best understood as a maternal solution to accommodate the invasive trophoblast. However, the presence of trophoblast within the uterine cavity does not always result in invasive placentation. In some placental mammals, most notably in hoofed animals such as the pig, placentation is superficial despite the fact that trophoblast cells retain the ability to invade ectopically [46]. Such species evolved an alternative and yet equally successful strategy to tolerate the presence of the hemi-allogeneic fetal allograft, namely maternal resistance to invasion, resulting in non-invasive placentation and the lack of endometrial decidualization. The fact that different mammalian viviparous species have evolved alternative solutions to the challenge of invading trophoblast highlights the important role that the maternal–fetal interaction plays in determining the resulting pregnancy phenotype, both across and within species. The functional importance of decidualization in human pregnancy is incompletely understood, but it appears to play a critical role in facilitating the active embedding of the conceptus [47], in the negative selection of nonviable embryos [48], in determining the optimal window of implantation [49,50], and in uterine hemostasis [51,52]. Proper decidualization controls conception and the course of pregnancy and is a critical determinant of pregnancy success in humans [2]. In non-menstruating species, the embryo controls this process by delaying implantation [53].

8. Timing of Decidualization

In most mammals that exhibit decidualization, the uterine reaction that transforms the endometrium into decidua is triggered by the arrival of the blastocyst. In contrast, the endometrium in humans, anthropoid primates, and a few non-primate species (including several species of bats, elephant shrews, and the spiny mouse [54,55]) undergoes decidualization extemporaneously in every menstrual cycle and not as a reaction to the presence of a blastocyst. In the absence of a conceptus, decidualization in humans ends with shedding of the upper layer of the decidualized endometrium (i.e., menstruation), a process that is triggered by programmed progesterone withdrawal at the end of the luteal phase. This cyclic decidualization is notably more complex than simple epithelial changes during the estrous cycle, which can be observed in the reproductive tract of most placental mammals. The teleological advantage of cyclic decidualization remains unclear, but it is interesting to note that all species with cyclic decidualization share a particularly invasive type of hemochorial placentation as well as a long gestation [9].

9. Master Regulators of Decidualization

In order to promote optimal implantation/placentation and a healthy pregnancy, the endometrium must be optimally primed hormonally, biochemically, and immunologically during the luteal phase of the menstrual cycle (Figure 2).

- Hormonal factors. During the follicular phase of the menstrual cycle, estrogen production by ovarian granulosa cells causes the endometrium to proliferate and thicken. The major driver of decidualization is progesterone, which is produced by the corpus luteum of the ovary following ovulation. In the absence of a conceptus, the corpus luteum is programmed to regress in 14 days, resulting in systemic progesterone withdrawal and menstruation. In the presence of a pregnancy, production of human chorionic gonadotropin (hCG) by trophoblast cells prevents luteolysis, thereby maintaining progesterone production until the placenta takes over this functionality at 5–7 weeks of gestation [56]. Moreover, the local production of hormones such as relaxin and corticotropin-releasing hormone (CRH) in response to the hCG surge establishes an autocrine/paracrine regulatory loop to enhance intracellular cAMP levels in ESCs, promote decidualization, and support implantation and early pregnancy [57–59].
- Biochemical factors. There is increasing evidence to suggest that biochemical/metabolic factors are important in decidualization. For example, lipid mediators such as lysophosphatidic acid (LPA) are produced by uterine epithelium [60] and regulate heparin-binding epidermal growth factor (HB-EGF) [61] and epidermal growth factor receptor (EGFR) signaling as well as cyclooxygenase 2 (COX2) [62] and thereby prostaglandin E2 (PGE2) production, which together with interferon-γ control the spatial decidualization of ESCs [63,64]. Other autocrine/paracrine factors—including interleukins, such as IL-1β, IL-11, and leukemia inhibitory factor (LIF) [65–69] as well as transforming growth factor-beta (TGF-β superfamily members such as activin, TGF-β1, bone morphogenesis protein 2 (BMP2), and left–right determination factor 2 (LEFTY2) [70–73]—also appear to be important in sustaining the decidualization process, promoting cAMP and extracellular matrix (ECM) signaling, regulating angiogenesis,

and supporting embryo implantation. Glucose also serves as a metabolic signal for decidualization, providing a link between glycemic control and cellular oxidative stress (discussed below).

- Immunological factors. The importance of the immunological priming of the endometrium is becoming increasingly apparent. While this is driven, in part, by intrinsic factors, including a range of endocrine and autocrine/paracrine signals [3,17,35], extrinsic factors are likely also involved. One such factor is exposure to seminal fluid both prior to and around the time of implantation [74–76]. Interestingly, this exposure does not have to be local. Exposure to paternal antigen via nonvaginal routes can also prime the endometrium immunologically [77]. Although the mechanism responsible for this priming effect is not clear, seminal fluid contains soluble and exosome-borne signaling agents that promote leukocyte recruitment and generation of regulatory T cells (Treg cells) which suppress inflammation, promote vascular adaptation, and foster tolerance towards fetal antigens [78]. This mechanism could shed light on a number of well-recognized risk factors for the 'great obstetrical syndromes' that have thus far defied explanation. Why is it that nulliparity, young maternal age, IVF conception, the use of donor sperm, the short length of cohabitation, short inter-pregnancy interval, and the use of barrier contraception are risk factors for conditions such as preeclampsia and PTB? Could the common factor be a lack of exposure to protective seminal fluid? Recent data suggest that intercourse during IVF treatment cycles improves implantation success and pregnancy health [79], which is consistent with the hypothesis that exposure to seminal fluid promotes healthy decidualization and implantation.

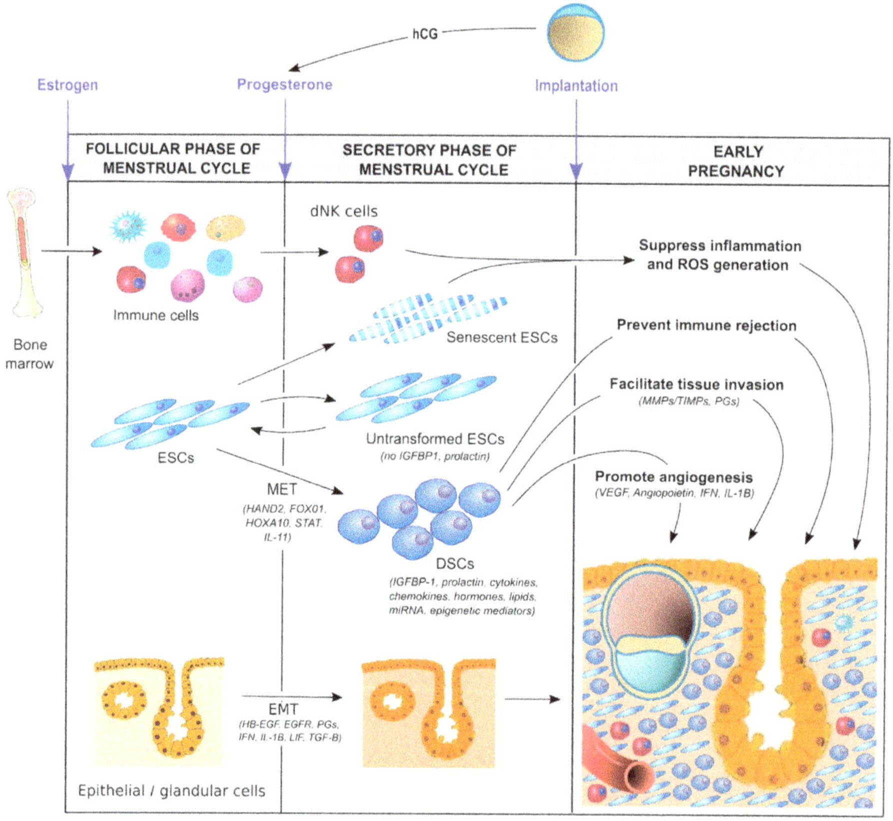

Figure 2. Molecular pathways involved in decidualization.

10. Molecular Regulation of Decidualization

Proper decidualization is a critical determinant of pregnancy success. The endometrium must be optimally primed prior to and shortly after the arrival of the blastocyst (Figure 2). In addition to the aforementioned maternal hormonal, metabolic, and immunological factors, a number of other local factors are involved (discussed below). Some of these local factors are of embryonic origin, such as lactate, relaxin, CRH, and hCG [57–59,80–83], although a detailed discussion of these embryo-derived factors in perpetuating the process of decidualization is beyond the scope of this review.

During decidualization, differentiating ESCs carry a molecular signature of mesenchymal–epithelial transition (MET) as they are reprogrammed to become DSCs with widespread changes in gene expression, including the induction of such genes as HOXA10, HOXA11, FOXO1, WNT4, IGFBP1, and prolactin (PRL) [84–87]. Many of these are known upstream regulators of genes critical for implantation and placental development [17,18]. The signal transduction pathways involved in the genetic reprogramming and terminal differentiation of ESCs into DSCs (summarized in Figure 3) can be classified into several categories:

- Genomic progesterone signaling pathways mediated by the nuclear progesterone receptor (nPGR). nPGR is the dominant member of the 3-ketosteroid nuclear receptor family that responds to progesterone and cyclic AMP/protein kinase A (cAMP/PKA) signaling during decidualization [88,89]. A recent study that employed both RNA-sequencing and PGR chromatin-immunoprecipitation (ChIP)-sequencing of endometrium during the window of implantation showed that the PGR signaling network is made up of multiple different classical signaling pathways and involves numerous downstream regulators [90], including Indian hedgehog (IHH) [91], heart and neural crest derivatives-expressed (HAND2) [92], transcription factors Forkhead Box O1 (FOXO1) [93], SPR-related HMG-box gene 17 (SOX17) [94] and signal transducers and activators of transcription (STAT) transcription factor members (STAT1, STAT3, STAT5) [95], Notch signaling [96], insulin receptor substrate 2 (IRS2) [97], BMP2 and WNT signaling [72], HOXA10 [98], CCAAT/enhancer-binding protein β (CEBPB) [99], EGFR [100], mammalian target of rapamycin complex 1 (MTORC1) [101], and the tumor necrosis factor alpha-nuclear factor kappa-light-chain-enhancer of activated B cells' (TNFα/NFκβ) pathway [102]. These pathways play an important role in the embryo–uterine, epithelial–stromal, and stromal–immune cell crosstalk that occurs in the peri-implantation period and is responsible for such functions as EMT, insulin resistance, focal adhesion, trophoblast invasion, regulation of the complement and coagulation cascade, cytokine-cytokine receptor interactions, xenobiotics metabolism, inflammatory response, ECM receptor interaction, angiogenesis and vasculature development, apoptosis, cytoskeleton remodeling, and the secretion of glycogen and other decidualization markers, such as PRL and insulin-like binding factor (IGFBP1). In a proteome and secretome screening study of in vitro decidualized ESCs, Garrido-Gomez et al. [103] reported that, in addition to PRL and IGFBP1, a number of other secreted decidualization markers might be involved in the attendant angiogenesis, including platelet/endothelial cell adhesion molecule-1 (PECAM-1) and myeloid progenitor inhibitory factor-1 (MPIF-1). In another study of 23 secreted factors derived from primary ESCs prior to ART, coordinated and synchronized changes in the secretome were associated with successful implantation, whereas cultures from the failed implantation group typically demonstrated a disordered secretome profile [104].

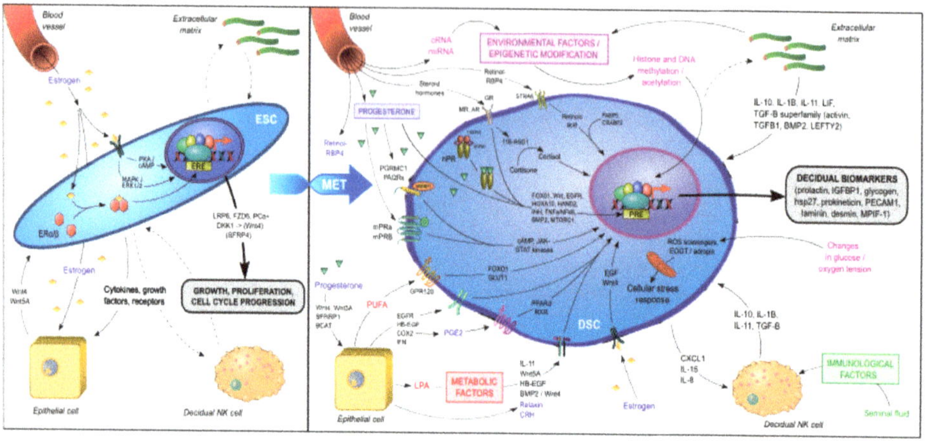

Figure 3. Signal transduction pathways involved in the genetic reprogramming and terminal differentiation of endometrial stromal fibroblast cells (ESCs) to decidual stromal fibroblast cells (DSCs).

- Non-genomic progesterone functions not mediated by nPGR. Recent studies have revealed the presence of membrane-associated putative progesterone-binding proteins, such as PGR membrane component 1 and 2 (PGRMC1, PGRMC2) [105] and progestin and adiponectin receptors (PAQRs) [106], in cycling endometrium and pregnancy tissues that can rapidly activate downstream signal transduction pathways to mediate non-genomic functions of progesterone, including interacting with PGR [107] and other steroid receptors [108], regulating endometrial receptivity [109], and triggering and promoting parturition [110,111]. The functional importance of these membrane-associated proteins in decidualization remains unknown. However, in addition to nPGR, other members of the 3-keto-steroid nuclear receptor family—such as glucocorticoid receptor (GR), mineralocorticoid receptor (MR), and androgen receptor (AR)—have also been found to play an important role in decidualization. In an in vitro decidualization model in which ESCs were induced with 8-bromo-cAMP (8-Br-cAMP) and medroxyprogesterone acetate (MPA), Cloke et al. demonstrated that AR regulated the expression of a distinct decidual gene network with a preponderance of upregulated genes being involved in cytoskeletal organization and cell motility and repressed genes being involved in cell cycle regulation [112]. Moreover, Kuroda et al. reported that progesterone/cAMP induction of ESCs increased expression of the 11β-hydroxysteroid dehydrogenase type 1 (11β-HSD1) enzyme, which converts inert cortisone to active cortisol and thus contributed to the metabolic regulation in decidualizing ESCs [113]. Taken together, these data suggest that the decidualization of ESCs involves the integration of multiple nonredundant signaling networks in response to progesterone stimulation.
- Metabolic regulators. The increased 11β-HSD1 expression and activity associated with ESC decidualization leads to a decrease in GR and reciprocal increase in MR expression [113]. The upregulation of MR-dependent genes, in turn, affects lipid droplet biogenesis and retinoid metabolism. For example, 11β-HSD1 upregulates dehydrogenase/reductase 3 (DHRS3) expression, which promotes retinol storage in lipid droplets [113]. Retinoic acid (RA) is essential in the maintenance of pregnancy and its metabolism is tightly controlled at the maternal–fetal interface [114]. The decidualization of ESCs increases the expression of retinol-binding protein 4 (RBP4) and cytochrome P450 26A1 (CYP26A1) involved in RA metabolism and downregulates the expression of the pro-apoptotic RA nuclear receptor (RAR) [115]. The lipid mediator LPA also regulates EGFR signaling, COX2 expression, and prostaglandin signaling for the spatial decidualization of ESCs [63,64]. COX2 in turn activates uterine peroxisome

proliferator-activated receptor-delta (PPAR-δ) and retinoid X receptor (RXR), which are critical regulators of decidualization and implantation [116]. Omega-3 polyunsaturated fatty acids have been shown in numerous animal and clinical studies to be beneficial for pregnancy outcome [117]. The receptor GPR120, a member of the rhodopsin family of G protein-coupled receptors, mediates potent anti-inflammatory and insulin- sensitizing effects [118]. Huang et al. showed that GPR120 could promote decidualization by upregulating FOXO1 and glucose transporter-1 (GLUT1) expression, glucose uptake, and pentose-phosphate pathway activation in ESCs [119].

Decidualization results in vascular remodeling with fluctuations in oxygen tension and the generation of reactive oxygen species (ROS). DSCs are programmed to resist a range of cellular stress signals to maintain the integrity of the feto–maternal interface and survival of the conceptus. Several molecular mechanisms have been implicated, including the inhibition of stress pathways such as c-Jun N-terminal kinase [120], attenuated inositol trisphosphate signaling [121], resistance to microRNA-mediated gene silencing [122], and the upregulation of free radical scavengers [123]. Another implicated pathway involves O-GlcNAcylation, a post-translational modification that links glucose sensing to cellular stress resistance. Muter et al. reported that the upregulation of the glycosyltransferase enzyme, EGF domain-specific O-linked N-acetylglucosamine transferase (EOGT), in decidualizing ESCs is responsible for the N-acetyl-glucosamine modification of a number of secreted and membrane-associated proteins involved in glucose and fatty acid metabolism [124]. Finally, in a uterine-specific *p53*-ablation PTB mouse model, decreased mitochondrial β-oxidation and ATP-production led to changes in lipid signaling and premature senescence of the decidua with subsequent PTB and/or stillbirth [125,126]. The inhibition of mTORC1 activity using rapamycin in this $p53^{-/-}$ murine model attenuated the premature decidual senescence and rescued the PTB phenotype [127].

- MicroRNA (miRNA) and epigenetic regulation. Using the miRNA profiling of ESC primary cultures before and after in vitro decidualization, Estella et al. reported an upregulation of 26 miRNAs and the downregulation of miR-96, miR-135b, miR-181 and miR-183 [128]. The addition of miR-96 and miR-135b in decidualizing ESCs decreased the expression of FOXO1 and HOXA10 as well as IGFBP-1 secretion [128]. In another study, Jimenez et al. reported that the upregulation of the miR-200 family during in vitro decidualization of ESCs correlated with the downregulation of IHH signaling and expression of the EMT regulator, ZEB1 [129]. Similar studies have demonstrated the functional importance also of miR-181a [130], miR-542-3p [131], and miR-194-3p [132] in decidualization. While individual miRNAs can regulate a range of target genes, there is growing evidence that endometrial cells undergo genome-wide chromatin remodeling for the access of transcription factors or epigenetic modifiers during decidualization [133,134]. In particular, the expression of the histone methyltransferase Enhancer of Zeste Homolog 2 (EZH2) appears to be reduced in endometrium beginning in the mid-secretory phase of the menstrual cycle and specifically in decidualizing ESCs [135]. The knockdown of *Ezh2* in decidualizing human ESCs resulted in reduced levels of trimethylated lysine 27 of histone 3 (H3K27me3), a repressive histone mark for silenced genes, in the proximal promoter regions of the *PRL* and *IGFBP1* genes, with a reciprocal enhancement of histone acetylation and concomitant higher expression of these two gene products [136]. A recent combined H3K27me3 ChIP-Seq and RNA-Seq analysis of mouse decidual cells harvested at different gestation stages confirmed the H3K27me3-induced transcriptional silencing of target genes that specifically suppress inflammation and contractile function in early gestation. In late gestation, genome-wide H3K27me3 demethylation was observed, thereby allowing de-repression and target gene upregulation to lead to the onset of labor [136]. Moreover, the pharmacological inhibition of H23K27 demethylation was able to inhibit labor and delivery while maintaining pup viability in a PTB murine model [136], thereby demonstrating the functional importance of this molecular mechanism. These data are consistent with the hypothesis that parturition in humans is nothing more than a delayed menstruation [11]. Although intriguing, it should be noted that the function of EZH2 and genome-wide chromatin remodeling in the process of human decidualization and implantation remains unclear. Additional studies are needed to further investigate these epigenetic regulatory mechanisms within the various uterine compartments and their association with pregnancy outcome.

11. Decidualization Resistance and Pregnancy Complications

"Decidualization resistance" refers to the inability of the maternal compartment to undergo these decidualization changes leading to aberrations in implantation/placentation and adverse pregnancy outcome. The hypothesis that pregnancy health is shaped prior to implantation/ placentation and is determined by the health of the decidua may shed light on the reproductive phenotype of a number of important clinical disorders.

- **Preeclampsia** is a pregnancy-specific disorder characterized by new-onset hypertension and maternal end-organ damage after 20 weeks' gestation that complicates 5–7% of all pregnancies. The pathologic hallmark is shallow trophoblast invasion and a failure of spiral artery remodeling. The current model posits that the primary cause of this suboptimal endovascular invasion is a failure of the decidua to tolerate and/or facilitate trophoblast invasion. In support of this hypothesis, Garrido-Gomez et al. identified a transcriptomic fingerprint characterizing a decidualization defect in the endometrium of women with a history of severe preeclampsia that is linked to impaired cytotrophoblast invasion. Moreover, this defect was detected at the time of delivery and persisted for years thereafter [137]. The decidualization of ESCs is mediated, at least in part, by annexin A2 (ANXA2) and the maternal deficiency of the *ANXA2* gene contributes to shallow decidual invasion by placental cytotrophoblast cells [138]. These findings highlight the maternal contribution to the pathogenesis of severe preeclampsia.
- **Recurrent pregnancy loss**, defined as 3 or more consecutive miscarriages, is a condition experienced by 1–2% of all couples. The cause is poorly understood. It is widely attributed to either repeated chromosomal instability in the conceptus or ill-defined uterine factors. Recent studies suggest that such women have impaired cyclic decidualization that predisposes to pregnancy failure by disrupting the maternal response to hormonal signaling leading to the dysregulation of decidualization markers, including, among others, prolactin, prokineticin, and the genes *DIO2* and *SCARA5* [139,140].
- A similar mechanism may account for the increased risk of adverse pregnancy events in women with poorly controlled **pregestational diabetes**. In such women, strict glycemic control around the time of conception has been shown to reduce rates of miscarriage and birth defects (diabetic embryology) and to improve overall pregnancy outcome [141].

12. Prevention of Pregnancy Complications

To date, interventions to prevent late-onset pregnancy complications have been largely unsuccessful. There is currently no effective intervention to prevent GDM, FGR, or stillbirth. Efforts to prevent PTB have been similarly disappointing. Despite initial enthusiasm regarding the use of 17α-hydroxyprogesterone caproate (17P) supplementation to prevent PTB in patients at high risk due to one or more prior unexplained spontaneous PTBs based on a single clinical trial [142], follow-up studies have failed to show significant benefit [143]. If there exist a subset of women who do benefit from 17P supplementation—and it is not clear that there is—the mode of action is most likely through a direct anti-inflammatory effect of progesterone on the decidua rather than by supporting suboptimal circulating progesterone concentrations [111,144]. A plausible explanation for the extent of this failure is to be found in the current model, which posits that the foundation for adverse pregnancy events is laid down before pregnancy with abnormalities in endometrial decidualization. Once the pregnancy is established, it is too late to intervene effectively.

The one possible exception to our inability to prevent late pregnancy complications is low-dose aspirin prophylaxis to prevent preeclampsia in patients at high risk [145]. Aspirin is a nonselective inhibitor of COX enzyme, which suppresses the production of the primary pro-inflammatory mediators (PGF2α and PGE2) in the maternal decidua, thereby promoting immune tolerance and facilitating trophoblast invasion. However, it needs to be given early in pregnancy around the time of the second wave of trophoblast invasion and the protective effect is limited.

13. Future Direction

The notion that the health of a pregnancy is shaped by the success of endometrial decidualization and predates the establishment of the pregnancy may explain the mechanism by which pre-pregnancy lifestyle interventions—such as weight loss, strict glycemic control, and preconception folic acid supplementation [146]—improve pregnancy outcomes, namely by improving decidualization. Once the molecular pathways responsible for optimal decidualization have been characterized, existing drugs can be tested or orphan drugs repurposed and their effect on decidualization studied both in vitro (by sampling the endometrium using a variety of minimally invasive gynecological procedures or isolating decidualized cells from menstrual blood) and in vivo in nonpregnant menstrual cycles. The role of the reproductive tract microbiome in decidualization/implantation is not well understood and is another potential avenue for intervention. This approach provides an exciting opportunity for future innovation since any such treatment would be initiated and likely even completed before the pregnancy is established.

14. Conclusions

Although both a viable blastocyst and appropriately primed endometrium are necessary for successful implantation/placentation, we posit that the optimal priming (decidualization) of the endometrium is the most critical determinant of pregnancy success. Decidualization involves the recruitment of leukocytes and, critically, the genetic reprogramming of ESCs to DSCs that suppress inflammation, promote vascular adaptation, and foster tolerance to fetal antigens. Since decidualization in humans occurs during every menstrual cycle in the absence of the conceptus, pregnancy health is determined even before the blastocyst arrives. A better understanding of the molecular and cellular mechanisms responsible for endometrial decidualization will improve our understanding of the factors influencing pregnancy success and provide opportunities for novel approaches to diagnosis and treatment.

Author Contributions: Conceptualization, S.-W.N., M.P., T.T., C.S., and E.R.N.; writing—original draft preparation, S.-W.N., G.A.N., and E.R.N.; writing—review and editing, M.P., T.T., and C.S.; visualization, S.-W.N., G.A.N., and E.R.N. All authors have read and agreed to the published version of the manuscript.

Funding: This research received no external funding.

Conflicts of Interest: Carlos Simón is Founder and Head of the Scientific Advisory Board of Igenomix.

Abbreviations

11β-HSD1	11β-hydroxysteroid dehydrogenase type 1
AR	androgen receptor
BCAT	branched chain amino acid transferase
BMP2	bone morphogenesis protein 2
COX2	cyclooxygenase 2
CRABP2	cellular retinoic acid binding protein 2
CRH	corticotropin-releasing hormone
CXCL1	C-X-C motif chemokine ligand 2
DKK1	Dickkopf WNT signaling pathway inhibitor 1
DSC	decidual stromal cell

EGF	epidermal growth factor
EGFR	epidermal growth factor receptor
EOGT	EGF domain specific O-linked N-acetylglucosamine transferase
ERK	extracellular- signal-regulated kinase
ESC	endometrial stromal cell
FABP5	fatty acid binding protein 5
FOXO1	transcription factors Forkhead Box O1
FZD5	Frizzled 5
GR	glucocorticoid receptor
HAND2	heart and neural crest derivatives-expressed 2
HB-EGF	heparin binding EGF-like growth factor
HSP	heat shock protein
IFN	interferon
IGFBP1	insulin-like growth factor binding protein 1
IL-1β	interleukin-1β
IHH	Indian hedgehog
LEFTY2	left-right determination factor 2
LIF	leukemia inhibitory factor
LPA	lysophosphatidic acid
LRP6	LDL receptor related protein 6
MAPK	mitogen activated protein kinase
MET	mesenchymal-epithelial transition
miRNA	microRNA
MPIF-1	myeloid progenitor inhibitory factor-1
mPR	membrane progesterone receptor
MR	mineralocorticoid receptor
MTORC1	mammalian target of rapamycin complex 1
NK cells	natural killer cells
nPR	nuclear progesterone receptor
PAQRs	progestin and adiponectin receptors
PCa+	prostate cancer a protein
PECAM-1	platelet-endothelial cell adhesion molecule-1
PGE2	prostaglandin E2
PGRMC1	PGR membrane component 1
PPAR-δ	peroxisome proliferator-activated receptor-delta
PUFA	polyunsaturated fatty acids
RBP4	retinol-binding protein 4
ROS	reactive oxygen species
RXR	retinoid X receptor
SFRP4	secreted frizzled-related protein 4
TGF-β	transforming growth factor-β
TNFα/NFκβ	tumor necrosis factor alpha-nuclear factor kappa-light-chain-enhancer of activated B cells
DC	dendritic cells
dNK	decidual natural killer cells
DSCs	decidual stromal cells
EGFR	epidermal growth factor receptor
ESCs	endometrial stromal cells
HB-EGF	heparin binding EGF-like growth factor
EMT	epithelial-mesenchymal transition
hCG	human chorionic gonado-tropin
IGFBP1	isulin-like growth factor binding protein 1
IFN	interferon
IL-1β	interleukin-1β
ROS	reactive oxygen species
LIF	leukemia inhibitory factor
MET	mesenchymal-epithelial transition

miRNA	microRNA
MMP	matrix metalloproteinase
PGS	prostaglandins
TGF-β	transforming growth factor-β
TIMP	tissue inhibitor of meatalloproteinase
VEGF	vascular endothelial growth factor

References

1. Cross, J.C.; Werb, Z.; Fisher, S.J. Implantation and the placenta: Key pieces of the development puzzle. *Science* **1994**, *266*, 1508–1518. [CrossRef] [PubMed]
2. Norwitz, E.R.; Schust, D.J.; Fisher, S.J. Implantation and the survival of early pregnancy. *N. Engl. J. Med.* **2001**, *345*, 1400–1408. [CrossRef] [PubMed]
3. Norwitz, E.R. Defective implantation and placentation: Laying the blueprint for pregnancy complications. *Reprod. Biomed. Online* **2006**, *13*, 591–599. [CrossRef]
4. Brosens, I.; Pijnenborg, R.; Vercruysse, L.; Romero, R. The "Great Obstetrical Syndromes" are associated with disorders of deep placentation. *Am. J. Obstet. Gynecol.* **2011**, *204*, 193–201. [CrossRef] [PubMed]
5. Redman, C.W.; Sargent, I.L. Immunology of pre-eclampsia. *Am. J. Reprod. Immunol.* **2010**, *63*, 534–543. [CrossRef]
6. Cha, J.; Sun, X.; Dey, S.K. Mechanisms of implantation: Strategies for successful pregnancy. *Nat. Med.* **2012**, *18*, 1754–1767. [CrossRef]
7. Redman, C.W.; Tannetta, D.S.; Dragovic, R.A.; Gardiner, C.; Southcombe, J.H.; Collett, G.P.; Sargent, I.L. Review: Does size matter? Placental debris and the pathophysiology of pre-eclampsia. *Placenta* **2012**, *33*, S48–S54. [CrossRef]
8. Brosens, I.; Puttemans, P.; Benagiano, G. Placental bed research: I. The placental bed: From spiral arteries remodeling to the great obstetrical syndromes. *Am. J. Obstet. Gynecol.* **2019**, *221*, 437–456. [CrossRef]
9. Emera, D.; Romero, R.; Wagner, G. The evolution of menstruation: A new model for genetic assimilation: Explaining molecular origins of maternal responses to fetal invasiveness. *Bioessays* **2012**, *34*, 26–35. [CrossRef]
10. Jarrell, J. The significance and evolution of menstruation. *Best Pract. Res. Clin. Obstet. Gynaecol.* **2018**, *50*, 18–26. [CrossRef]
11. Pavlicev, M.; Norwitz, E.R. Human parturition: Nothing more than a delayed menstruation. *Reprod. Sci.* **2018**, *25*, 166–173. [CrossRef] [PubMed]
12. Dugoff, L.; Society for Maternal-Fetal Medicine. First- and second-trimester maternal serum markers for aneuploidy and adverse obstetric outcomes. *Obstet. Gynecol.* **2010**, *115*, 1052–1061. [CrossRef] [PubMed]
13. Burton, G.J.; Jauniaux, E. Pathophysiology of placental-derived fetal growth restriction. *Am. J. Obstet. Gynecol.* **2018**, *218*, S745–S761. [CrossRef] [PubMed]
14. Wilcox, A.J.; Weinberg, C.R.; O'Connor, J.F.; Baird, D.D.; Schlatterer, J.P.; Canfield, R.E.; Armstrong, E.G.; Nisula, B.C. Incidence of early loss of pregnancy. *N. Engl. J. Med.* **1988**, *319*, 189–194. [CrossRef] [PubMed]
15. Zinaman, M.J.; Clegg, E.D.; Brown, C.C.; O'Connor, J.; Selevan, S.G. Estimates of human fertility and pregnancy loss. *Fertil. Steril.* **1996**, *65*, 503–509. [CrossRef]
16. Simón, C.; Valbuena, D. Embryonic implantation. *Ann. Endocrinol.* **1999**, *60*, 134–136.
17. Harris, L.K.; Benagiano, M.; D'Elios, M.M.; Brosens, I.; Benagiano, G. Placental bed research: II. Functional and immunological investigations of the placental bed. *Am. J. Obstet. Gynecol.* **2019**, *221*, 457–469. [CrossRef]
18. Jauniaux, E.; Poston, L.; Burton, G.J. Placental-related diseases of pregnancy: Involvement of oxidative stress and implications in human evolution. *Hum. Reprod. Update* **2006**, *12*, 747–755. [CrossRef]
19. Red-Horse, K.; Zhou, Y.; Genbacev, O.; Prakobphol, A.; Foulk, R.; McMaster, M.; Fisher, S.J. Trophoblast differentiation during embryo implantation and formation of the maternal-fetal interface. *J. Clin. Investig.* **2004**, *114*, 744–754. [CrossRef]
20. Gaynor, L.M.; Colucci, F. Uterine natural killer cells: Functional distinctions and influence on pregnancy in humans and mice. *Front. Immunol.* **2017**, *8*, 467. [CrossRef]
21. Pijnenborg, R.; Bland, J.M.; Robertson, W.B.; Brosens, I. Uteroplacental arterial changes related to interstitial trophoblast migration in early human pregnancy. *Placenta* **1983**, *4*, 397–413. [CrossRef]

22. Craven, C.M.; Morgan, T.; Ward, K. Decidual spiral artery remodelling begins before cellular interaction with cytotrophoblasts. *Placenta* **1998**, *19*, 241–252. [CrossRef]
23. Su, R.W.; Fazleabas, A.T. Implantation and establishment of pregnancy in human and nonhuman primates. *Adv. Anat. Embryol. Cell Biol.* **2015**, *216*, 189–213. [CrossRef]
24. Blue, N.R.; Page, J.M.; Silver, R.M. Genetic abnormalities and pregnancy loss. *Semin. Perinatol.* **2019**, *43*, 66–73. [CrossRef] [PubMed]
25. Valdes, C.T.; Schutt, A.; Simon, C. Implantation failure of endometrial origin: It is not pathology, but our failure to synchronize the developing embryo with a receptive endometrium. *Fertil. Steril.* **2017**, *108*, 15–18. [CrossRef] [PubMed]
26. Franasiak, J.M.; Ruiz-Alonso, M.; Scott, R.T.; Simón, C. Both slowly developing embryos and a variable pace of luteal endometrial progression may conspire to prevent normal birth in spite of a capable embryo. *Fertil. Steril.* **2016**, *105*, 861–866. [CrossRef]
27. Glujovsky, D.; Farquhar, C.; Quinteiro Retamar, A.M.; Alvarez Sedo, C.R.; Blake, D. Cleavage stage versus blastocyst stage embryo transfer in assisted reproductive technology. *Cochrane Database Syst. Rev.* **2016**, *6*, CD002118. [CrossRef]
28. Van der Hoorn, M.L.; Lashley, E.E.; Bianchi, D.W.; Claas, F.H.; Schonkeren, C.M.; Scherjon, S.A. Clinical and immunologic aspects of egg donation pregnancies: A systematic review. *Hum. Reprod. Update* **2010**, *16*, 704–712. [CrossRef]
29. Van Bentem, K.; Bos, M.; van der Keur, C.; Brand-Schaaf, S.H.; Haasnoot, G.W.; Roelen, D.L.; Eikmans, M.; Heidt, S.; Claas, F.H.J.; Lashley, E.E.L.O.; et al. The development of preeclampsia in oocyte donation pregnancies is related to the number of fetal-maternal HLA class II mismatches. *J. Reprod. Immunol.* **2020**, *137*, 103074. [CrossRef]
30. Wei, D.; Liu, J.Y.; Sun, Y.; Shi, Y.; Zhang, B.; Liu, J.Q.; Tan, J.; Liang, X.; Cao, Y.; Wang, Z.; et al. Frozen versus fresh single blastocyst transfer in ovulatory women: A multicentre, randomised controlled trial. *Lancet* **2019**, *393*, 1310–1318. [CrossRef]
31. Bulmer, J.N.; Morrison, L.; Longfellow, M.; Ritson, A.; Pace, D. Granulated lymphocytes in human endometrium: Histochemical and immunohistochemical studies. *Hum. Reprod.* **1991**, *6*, 791–798. [CrossRef] [PubMed]
32. Starkey, P.M.; Sargent, I.L.; Redman, C.W. Cell populations in human early pregnancy decidua: Characterization and isolation of large granular lymphocytes by flow cytometry. *Immunology* **1988**, *65*, 129–134.
33. Vince, G.S.; Starkey, P.M.; Jackson, M.C.; Sargent, I.L.; Redman, C.W. Flow cytometric characterisation of cell populations in human pregnancy decidua and isolation of decidual macrophages. *J. Immunol. Methods* **1990**, *132*, 181–189. [CrossRef]
34. Du, H.; Taylor, H.S. Contribution of bone marrow-derived stem cells to endometrium and endometriosis. *Stem. Cells* **2007**, *25*, 2082–2086. [CrossRef] [PubMed]
35. Evans, J.; Salamonsen, L.A.; Winship, A.; Menkhorst, E.; Nie, G.; Gargett, C.E.; Dimitriadis, E. Fertile ground: Human endometrial programming and lessons in health and disease. *Nat. Rev. Endocrinol.* **2016**, *12*, 654–667. [CrossRef]
36. Fox, C.; Morin, S.; Jeong, J.W.; Scott, R.T., Jr.; Lessey, B.A. Local and systemic factors and implantation: What is the evidence? *Fertil. Steril.* **2016**, *105*, 873–884. [CrossRef]
37. Yang, F.; Zheng, Q.; Jin, L. Dynamic function and composition changes of immune cells during normal and pathological pregnancy at the maternal-fetal interface. *Front. Immunol.* **2019**, *10*, 2317. [CrossRef]
38. Houser, B.L.; Tilburgs, T.; Hill, J.; Nicotra, M.L.; Strominger, J.L. Two unique human decidual macrophage populations. *J. Immunol.* **2011**, *186*, 2633–2642. [CrossRef]
39. Lash, G.E.; Pitman, H.; Morgan, H.L.; Innes, B.A.; Agwu, C.N.; Bulmer, J.N. Decidual macrophages: Key regulators of vascular remodeling in human pregnancy. *J. Leukoc. Biol.* **2016**, *100*, 315–325. [CrossRef]
40. Van der Zwan, A.; Bi, K.; Norwitz, E.R.; Crespo, A.C.; Claas, F.H.J.; Strominger, J.L.; Tilburgs, T. Mixed signature of activation and dysfunction allows human decidual CD8(+) T cells to provide both tolerance and immunity. *Proc. Natl. Acad. Sci. USA* **2018**, *115*, 385–390. [CrossRef]
41. Crespo, A.C.; Strominger, J.L.; Tilburgs, T. Expression of KIR2DS1 by decidual natural killer cells increases their ability to control placental HCMV infection. *Proc. Natl. Acad. Sci. USA* **2016**, *113*, 15072–15077. [CrossRef] [PubMed]

42. Erlebacher, A. Immunology of the maternal-fetal interface. *Ann. Rev. Immunol.* **2013**, *31*, 387–411. [CrossRef] [PubMed]
43. Kin, K.; Maziarz, J.; Chavan, A.R.; Kamat, M.; Vasudevan, S.; Birt, A.; Emera, D.; Lynch, V.J.; Ott, T.L.; Pavlicev, M.; et al. The transcriptomic evolution of mammalian pregnancy: Gene expression innovations in endometrial stromal fibroblasts. *Genome Biol. Evol.* **2016**, *8*, 2459–2473. [CrossRef] [PubMed]
44. Chavan, A.R.; Bhullar, B.A.; Wagner, G.P. What was the ancestral function of decidual stromal cells? A model for the evolution of eutherian pregnancy. *Placenta* **2016**, *40*, 40–51. [CrossRef]
45. Carter, A.M. Evolution of placental function in mammals: The molecular basis of gas and nutrient transfer, hormone secretion, and immune responses. *Physiol. Rev.* **2012**, *92*, 1543–1576. [CrossRef]
46. Samuel, C.A. The development of pig trophoblast in ectopic sites. *J. Reprod. Fertil.* **1971**, *27*, 494–495. [CrossRef]
47. Gellersen, B.; Reimann, K.; Samalecos, A.; Aupers, S.; Bamberger, A.M. Invasiveness of human endometrial stromal cells is promoted by decidualization and by trophoblast-derived signals. *Hum. Reprod.* **2010**, *25*, 862–873. [CrossRef]
48. Teklenburg, G.; Salker, M.; Molokhia, M.; Lavery, S.; Trew, G.; Aojanepong, T.; Mardon, H.J.; Lokugamage, A.U.; Rai, R.; Landles, C.; et al. Natural selection of human embryos: Decidualizing endometrial stromal cells serve as sensors of embryo quality upon implantation. *PLoS ONE* **2010**, *5*, e10258. [CrossRef]
49. Hirota, Y.; Daikoku, T.; Tranguch, S.; Xie, H.; Bradshaw, H.B.; Dey, S.K. Uterine-specific p53 deficiency confers premature uterine senescence and promotes preterm birth in mice. *J. Clin. Investig.* **2010**, *120*, 803–815. [CrossRef]
50. Cha, J.; Hirota, Y.; Dey, S.K. Sensing senescence in preterm birth. *Cell Cycle* **2012**, *11*, 205–206. [CrossRef]
51. Lockwood, C.J.; Krikun, G.; Schatz, F. The decidua regulates hemostasis in human endometrium. *Semin. Reprod. Endocrinol.* **1999**, *17*, 45–51. [CrossRef] [PubMed]
52. Schatz, F.; Guzeloglu-Kayisli, O.; Arlier, S.; Kayisli, U.A.; Lockwood, C.J. The role of decidual cells in uterine hemostasis, menstruation, inflammation, adverse pregnancy outcomes and abnormal uterine bleeding. *Hum. Reprod. Update* **2016**, *22*, 497–515. [CrossRef] [PubMed]
53. Das, S.K.; Lim, H.; Paria, B.C.; Dey, S.K. Cyclin D3 in the mouse uterus is associated with the decidualization process during early pregnancy. *J. Mol. Endocrinol.* **1999**, *22*, 91–101. [CrossRef] [PubMed]
54. Rasweiler, J.J. Spontaneous decidual reactions and menstruation in the black mastiff bat, Molossus ater. *Am. J. Anat.* **1991**, *191*, 1–22. [CrossRef] [PubMed]
55. Bellofiore, N.; Ellery, S.J.; Mamrot, J.; Walker, D.W.; Temple-Smith, P.; Dickinson, H. First evidence of a menstruating rodent: The spiny mouse (Acomys cahirinus). *Am. J. Obstet. Gynecol.* **2017**, *216*, 40.e1–40.e11. [CrossRef] [PubMed]
56. Carp, H.J.A. Progestogens in luteal support. *Horm. Mol. Biol. Clin. Investig.* **2020**. [CrossRef]
57. Palejwala, S.; Tseng, L.; Wojtczuk, A.; Weiss, G.; Goldsmith, L.T. Relaxin gene and protein expression and its regulation of procollagenase and vascular endothelial growth factor in human endometrial cells. *Biol. Reprod.* **2002**, *66*, 1743–1748. [CrossRef]
58. Gravanis, A.; Stournaras, C.; Margioris, A.N. Paracrinology of endometrial neuropeptides: Corticotropin-releasing hormone and opioids. *Semin. Reprod. Endocrinol.* **1999**, *17*, 29–38. [CrossRef]
59. Einspanier, A.; Lieder, K.; Husen, B.; Ebert, K.; Lier, S.; Einspanier, R.; Unemori, E.; Kemper, M. Relaxin supports implantation and early pregnancy in the marmoset monkey. *Ann. N. Y. Acad. Sci.* **2009**, *1160*, 140–146. [CrossRef]
60. Aikawa, S.; Kano, K.; Inoue, A.; Wang, J.; Saigusa, D.; Nagamatsu, T.; Hirota, Y.; Fujii, T.; Tsuchiya, S.; Taketomi, Y.; et al. Autotaxin-lysophosphatidic acid-LPA3 signaling at the embryo-epithelial boundary controls decidualization pathways. *EMBO J.* **2017**, *36*, 2146–2160. [CrossRef]
61. Chobotova, K.; Karpovich, N.; Carver, J.; Manek, S.; Gullick, W.J.; Barlow, D.H.; Mardon, H.J. Heparin-binding epidermal growth factor and its receptors mediate decidualization and potentiate survival of human endometrial stromal cells. *J. Clin. Endocrinol. Metab.* **2005**, *90*, 913–919. [CrossRef] [PubMed]
62. Lim, H.; Paria, B.C.; Das, S.K.; Dinchuk, J.E.; Langenbach, R.; Trzaskos, J.M.; Dey, S.K. Multiple female reproductive failures in cyclooxygenase 2-deficient mice. *Cell* **1997**, *91*, 197–208. [CrossRef]

63. Milne, S.A.; Perchick, G.B.; Boddy, S.C.; Jabbour, H.N. Expression, localization, and signaling of PGE(2) and EP2/EP4 receptors in human nonpregnant endometrium across the menstrual cycle. *J. Clin. Endocrinol. Metab.* **2001**, *86*, 4453–4459. [CrossRef] [PubMed]
64. Christian, M.; Marangos, P.; Mak, I.; McVey, J.; Barker, F.; White, J.; Brosens, J.J. Interferon-gamma modulates prolactin and tissue factor expression in differentiating human endometrial stromal cells. *Endocrinology* **2001**, *142*, 3142–3151. [CrossRef]
65. Dimitriadis, E.; Salamonsen, L.A.; Robb, L. Expression of interleukin-11 during the human menstrual cycle: Coincidence with stromal cell decidualization and relationship to leukaemia inhibitory factor and prolactin. *Mol. Hum. Reprod.* **2000**, *6*, 907–914. [CrossRef]
66. Tamura, M.; Sebastian, S.; Yang, S.; Gurates, B.; Fang, Z.; Bulun, S.E. Interleukin-1beta elevates cyclooxygenase-2 protein level and enzyme activity via increasing its mRNA stability in human endometrial stromal cells: An effect mediated by extracellularly regulated kinases 1 and 2. *J. Clin. Endocrinol. Metab.* **2002**, *87*, 3263–3273. [CrossRef]
67. Dey, S.K.; Lim, H.; Das, S.K.; Reese, J.; Paria, B.C.; Daikoku, T.; Wang, H. Molecular cues to implantation. *Endocr. Rev.* **2004**, *25*, 341–373. [CrossRef]
68. Cullinan, E.B.; Abbondanzo, S.J.; Anderson, P.S.; Pollard, J.W.; Lessey, B.A.; Stewart, C.L. Leukemia inhibitory factor (LIF) and LIF receptor expression in human endometrium suggests a potential autocrine/paracrine function in regulating embryo implantation. *Proc. Natl. Acad. Sci. USA* **1996**, *93*, 3115–3120. [CrossRef]
69. Shuya, L.L.; Menkhorst, E.M.; Yap, J.; Li, P.; Lane, N.; Dimitriadis, E. Leukemia inhibitory factor enhances endometrial stromal cell decidualization in humans and mice. *PLoS ONE* **2011**, *6*, e25288. [CrossRef]
70. Jones, R.L.; Findlay, J.K.; Farnworth, P.G.; Robertson, D.M.; Wallace, E.; Salamonsen, L.A. Activin A and inhibin A differentially regulate human uterine matrix metalloproteinases: Potential interactions during decidualization and trophoblast invasion. *Endocrinology* **2006**, *147*, 724–732. [CrossRef] [PubMed]
71. Kim, M.R.; Park, D.W.; Lee, J.H.; Choi, D.S.; Hwang, K.J.; Ryu, H.S.; Min, C.K. Progesterone-dependent release of transforming growth factor-beta1 from epithelial cells enhances the endometrial decidualization by turning on the Smad signalling in stromal cells. *Mol. Hum. Reprod.* **2005**, *11*, 801–808. [CrossRef]
72. Li, Q.; Kannan, A.; Wang, W.; Demayo, F.J.; Taylor, R.N.; Bagchi, M.K.; Bagchi, I.C. Bone morphogenetic protein 2 functions via a conserved signaling pathway involving Wnt4 to regulate uterine decidualization in the mouse and the human. *J. Biol. Chem.* **2007**, *282*, 31725–31732. [CrossRef]
73. Tang, M.; Naidu, D.; Hearing, P.; Handwerger, S.; Tabibzadeh, S. LEFTY, a member of the transforming growth factor-beta superfamily, inhibits uterine stromal cell differentiation: A novel autocrine role. *Endocrinology* **2010**, *151*, 1320–1330. [CrossRef]
74. Robertson, S.A.; Bromfield, J.J.; Tremellen, K.P. Seminal 'priming' for protection from pre-eclampsia-a unifying hypothesis. *J. Reprod. Immunol.* **2003**, *59*, 253–265. [CrossRef]
75. Lane, M.; Robker, R.L.; Robertson, S.A. Parenting from before conception. *Science* **2014**, *345*, 756–760. [CrossRef]
76. Saftlas, A.F.; Rubenstein, L.; Prater, K.; Harland, K.K.; Field, E.; Triche, E.W. Cumulative exposure to paternal seminal fluid prior to conception and subsequent risk of preeclampsia. *J. Reprod. Immunol.* **2014**, *101-102*, 104–110. [CrossRef]
77. Koelman, C.A.; Coumans, A.B.; Nijman, H.W.; Doxiadis, I.I.; Dekker, G.A.; Claas, F.H. Correlation between oral sex and a low incidence of preeclampsia: A role for soluble HLA in seminal fluid? *J. Reprod. Immunol.* **2000**, *46*, 155–166. [CrossRef]
78. Guerin, L.R.; Moldenhauer, L.M.; Prins, J.R.; Bromfield, J.J.; Hayball, J.D.; Robertson, S.A. Seminal fluid regulates accumulation of FOXP3+ regulatory T cells in the preimplantation mouse uterus through expanding the FOXP3+ cell pool and CCL19-mediated recruitment. *Biol. Reprod.* **2011**, *85*, 397–408. [CrossRef]
79. Robertson, S.A.; Sharkey, D.J. Seminal fluid and fertility in women. *Fertil. Steril.* **2016**, *106*, 511–519. [CrossRef]
80. Nimbkar-Joshi, S.; Rosario, G.; Katkam, R.R.; Manjramkar, D.D.; Metkari, S.M.; Puri, C.P.; Sachdeva, G. Embryo-induced alterations in the molecular phenotype of primate endometrium. *J. Reprod. Immunol.* **2009**, *83*, 65–71. [CrossRef]
81. Fouladi-Nashta, A.A.; Jones, C.J.; Nijjar, N.; Mohamet, L.; Smith, A.; Chambers, I.; Kimber, S.J. Characterization of the uterine phenotype during the peri-implantation period for LIF-null, MF1 strain mice. *Dev. Biol.* **2005**, *281*, 1–21. [CrossRef]

82. Gardner, D.K. Lactate production by the mammalian blastocyst: Manipulating the microenvironment for uterine implantation and invasion? *Bioessays* **2015**, *37*, 364–371. [CrossRef]
83. Thouas, G.A.; Dominguez, F.; Green, M.P.; Vilella, F.; Simon, C.; Gardner, D.K. Soluble ligands and their receptors in human embryo development and implantation. *Endocr. Rev.* **2015**, *36*, 92–130. [CrossRef]
84. Zhang, X.H.; Liang, X.; Liang, X.H.; Wang, T.S.; Qi, Q.R.; Deng, W.B.; Sha, A.G.; Yang, Z.M. The mesenchymal-epithelial transformation during in vitro decidualization. *Reprod. Sci.* **2013**, *20*, 354–360. [CrossRef]
85. Owusu-Akyaw, A.; Krishnamoorthy, K.; Goldsmith, L.T.; Morelli, S.S. The role of mesenchymal-epithelial transition in endometrial function. *Hum. Reprod. Update* **2019**, *25*, 114–133. [CrossRef]
86. Liu, J.-L.; Wang, T.-S. Systematic analysis of the molecular mechanism underlying decidualization using a text mining approach. *PLoS ONE* **2015**, *10*, e0134585. [CrossRef]
87. Popovici, R.M.; Betzler, N.K.; Krause, M.S.; Luo, M.; Jauckus, J.; Germeyer, A.; Bloethner, S.; Schlotterer, A.; Kumar, R.; Strowitzki, T.; et al. Gene expression profiling of human endometrial-trophoblast interaction in a coculture model. *Endocrinology* **2006**, *147*, 5662–5675. [CrossRef]
88. Gellersen, B.; Brosens, J.J. Cyclic decidualization of the human endometrium in reproductive health and failure. *Endocr. Rev.* **2014**, *35*, 851–905. [CrossRef]
89. Wu, S.P.; Li, R.; DeMayo, F.J. Progesterone receptor regulation of uterine adaptation for pregnancy. *Trends Endocrinol. Metab.* **2018**, *29*, 481–491. [CrossRef]
90. Chi, R.A.; Wang, T.; Adams, N.; Wu, S.P.; Young, S.L.; Spencer, T.E.; DeMayo, F. Human endometrial transcriptome and progesterone receptor cistrome reveal important pathways and epithelial regulators. *J. Clin. Endocrinol. Metab.* **2020**, *105*. [CrossRef]
91. Matsumoto, H.; Zhao, X.; Das, S.K.; Hogan, B.L.; Dey, S.K. Indian hedgehog as a progesterone-responsive factor mediating epithelial-mesenchymal interactions in the mouse uterus. *Dev. Biol.* **2002**, *245*, 280–290. [CrossRef] [PubMed]
92. Li, Q.; Kannan, A.; DeMayo, F.J.; Lydon, J.P.; Cooke, P.S.; Yamagishi, H.; Srivastava, D.; Bagchi, M.K.; Bagchi, I.C. The antiproliferative action of progesterone in uterine epithelium is mediated by Hand2. *Science* **2011**, *331*, 912–916. [CrossRef] [PubMed]
93. Vasquez, Y.M.; Wang, X.; Wetendorf, M.; Franco, H.L.; Mo, Q.; Wang, T.; Lanz, R.B.; Young, S.L.; Lessey, B.A.; Spencer, T.E.; et al. FOXO1 regulates uterine epithelial integrity and progesterone receptor expression critical for embryo implantation. *PLoS Genet.* **2018**, *14*, e1007787. [CrossRef]
94. Wang, X.; Li, X.; Wang, T.; Wu, S.P.; Jeong, J.W.; Kim, T.H.; Young, S.L.; Lessey, B.A.; Lanz, R.B.; Lydon, J.P.; et al. SOX17 regulates uterine epithelial-stromal cross-talk acting via a distal enhancer upstream of Ihh. *Nat. Commun.* **2018**, *9*, 4421. [CrossRef]
95. Dimitriadis, E.; Stoikos, C.; Tan, Y.L.; Salamonsen, L.A. Interleukin 11 signaling components signal transducer and activator of transcription 3 (STAT3) and suppressor of cytokine signaling 3 (SOCS3) regulate human endometrial stromal cell differentiation. *Endocrinology* **2006**, *147*, 3809–3817. [CrossRef]
96. Afshar, Y.; Miele, L.; Fazleabas, A.T. Notch1 is regulated by chorionic gonadotropin and progesterone in endometrial stromal cells and modulates decidualization in primates. *Endocrinology* **2012**, *153*, 2884–2896. [CrossRef]
97. Vassen, L.; Wegrzyn, W.; Klein-Hitpass, L. Human insulin receptor substrate-2 (IRS-2) is a primary progesterone response gene. *Mol. Endocrinol.* **1999**, *13*, 485–494. [CrossRef]
98. Lim, H.; Ma, L.; Ma, W.G.; Maas, R.L.; Dey, S.K. Hoxa-10 regulates uterine stromal cell responsiveness to progesterone during implantation and decidualization in the mouse. *Mol. Endocrinol.* **1999**, *13*, 1005–1017. [CrossRef]
99. Kannan, A.; Fazleabas, A.T.; Bagchi, I.C.; Bagchi, M.K. The transcription factor C/EBPbeta is a marker of uterine receptivity and expressed at the implantation site in the primate. *Reprod. Sci.* **2010**, *17*, 434–443. [CrossRef]
100. Large, M.J.; Wetendorf, M.; Lanz, R.B.; Hartig, S.M.; Creighton, C.J.; Mancini, M.A.; Kovanci, E.; Lee, K.F.; Threadgill, D.W.; Lydon, J.P.; et al. The epidermal growth factor receptor critically regulates endometrial function during early pregnancy. *PLoS Genet.* **2014**, *10*, e1004451. [CrossRef]
101. Baek, M.O.; Song, H.I.; Han, J.S.; Yoon, M.S. Differential regulation of mTORC1 and mTORC2 is critical for 8-Br-cAMP-induced decidualization. *Exp. Mol. Med.* **2018**, *50*, 1–11. [CrossRef]

102. Sugino, N.; Karube-Harada, A.; Taketani, T.; Sakata, A.; Nakamura, Y. Withdrawal of ovarian steroids stimulates prostaglandin F2alpha production through nuclear factor-kappaB activation via oxygen radicals in human endometrial stromal cells: Potential relevance to menstruation. *J. Reprod. Dev.* **2004**, *50*, 215–225. [CrossRef]
103. Garrido-Gomez, T.; Dominguez, F.; Lopez, J.A.; Camafeita, E.; Quinonero, A.; Martinez-Conejero, J.A.; Pellicer, A.; Conesa, A.; Simon, C. Modeling human endometrial decidualization from the interaction between proteome and secretome. *J. Clin. Endocrinol. Metab.* **2011**, *96*, 706–716. [CrossRef]
104. Peter Durairaj, R.R.; Aberkane, A.; Polanski, L.; Maruyama, Y.; Baumgarten, M.; Lucas, E.S.; Quenby, S.; Chan, J.K.Y.; Raine-Fenning, N.; Brosens, J.J.; et al. Deregulation of the endometrial stromal cell secretome precedes embryo implantation failure. *Mol. Hum. Reprod.* **2017**, *23*, 478–487. [CrossRef] [PubMed]
105. Pru, J.K.; Clark, N.C. PGRMC1 and PGRMC2 in uterine physiology and disease. *Front. Neurosci.* **2013**, *7*, 168. [CrossRef] [PubMed]
106. Fernandes, M.S.; Pierron, V.; Michalovich, D.; Astle, S.; Thornton, S.; Peltoketo, H.; Lam, E.W.; Gellersen, B.; Huhtaniemi, I.; Allen, J.; et al. Regulated expression of putative membrane progestin receptor homologues in human endometrium and gestational tissues. *J. Endocrinol.* **2005**, *187*, 89–101. [CrossRef]
107. Karteris, E.; Zervou, S.; Pang, Y.; Dong, J.; Hillhouse, E.W.; Randeva, H.S.; Thomas, P. Progesterone signaling in human myometrium through two novel membrane G protein-coupled receptors: Potential role in functional progesterone withdrawal at term. *Mol. Endocrinol.* **2006**, *20*, 1519–1534. [CrossRef]
108. Thomas, P.; Pang, Y.; Dong, J. Enhancement of cell surface expression and receptor functions of membrane progestin receptor alpha (mPRalpha) by progesterone receptor membrane component 1 (PGRMC1): Evidence for a role of PGRMC1 as an adaptor protein for steroid receptors. *Endocrinology* **2014**, *155*, 1107–1119. [CrossRef]
109. Garrido-Gomez, T.; Quinonero, A.; Antunez, O.; Diaz-Gimeno, P.; Bellver, J.; Simon, C.; Dominguez, F. Deciphering the proteomic signature of human endometrial receptivity. *Hum. Reprod.* **2014**, *29*, 1957–1967. [CrossRef]
110. Wu, W.; Shi, S.Q.; Huang, H.J.; Balducci, J.; Garfield, R.E. Changes in PGRMC1, a potential progesterone receptor, in human myometrium during pregnancy and labour at term and preterm. *Mol. Hum. Reprod.* **2011**, *17*, 233–242. [CrossRef]
111. Mesiano, S.; Wang, Y.; Norwitz, E.R. Progesterone receptors in the human pregnancy uterus: Do they hold the key to birth timing? *Reprod. Sci.* **2011**, *18*, 6–19. [CrossRef]
112. Cloke, B.; Huhtinen, K.; Fusi, L.; Kajihara, T.; Yliheikkila, M.; Ho, K.K.; Teklenburg, G.; Lavery, S.; Jones, M.C.; Trew, G.; et al. The androgen and progesterone receptors regulate distinct gene networks and cellular functions in decidualizing endometrium. *Endocrinology* **2008**, *149*, 4462–4474. [CrossRef]
113. Kuroda, K.; Venkatakrishnan, R.; Salker, M.S.; Lucas, E.S.; Shaheen, F.; Kuroda, M.; Blanks, A.; Christian, M.; Quenby, S.; Brosens, J.J. Induction of 11beta-HSD 1 and activation of distinct mineralocorticoid receptor- and glucocorticoid receptor-dependent gene networks in decidualizing human endometrial stromal cells. *Mol. Endocrinol.* **2013**, *27*, 192–202. [CrossRef]
114. Han, B.C.; Xia, H.F.; Sun, J.; Yang, Y.; Peng, J.P. Retinoic acid-metabolizing enzyme cytochrome P450 26a1 (cyp26a1) is essential for implantation: Functional study of its role in early pregnancy. *J. Cell. Physiol.* **2010**, *223*, 471–479. [CrossRef]
115. Ozaki, R.; Kuroda, K.; Ikemoto, Y.; Ochiai, A.; Matsumoto, A.; Kumakiri, J.; Kitade, M.; Itakura, A.; Muter, J.; Brosens, J.J.; et al. Reprogramming of the retinoic acid pathway in decidualizing human endometrial stromal cells. *PLoS ONE* **2017**, *12*, e0173035. [CrossRef]
116. Lim, H.; Gupta, R.A.; Ma, W.G.; Paria, B.C.; Moller, D.E.; Morrow, J.D.; DuBois, R.N.; Trzaskos, J.M.; Dey, S.K. Cyclo-oxygenase-2-derived prostacyclin mediates embryo implantation in the mouse via PPARdelta. *Genes Dev.* **1999**, *13*, 1561–1574. [CrossRef]
117. Lager, S.; Ramirez, V.I.; Acosta, O.; Meireles, C.; Miller, E.; Gaccioli, F.; Rosario, F.J.; Gelfond, J.A.L.; Hakala, K.; Weintraub, S.T.; et al. Docosahexaenoic acid supplementation in pregnancy modulates placental cellular signaling and nutrient transport capacity in obese women. *J. Clin. Endocrinol. Metab.* **2017**, *102*, 4557–4567. [CrossRef]
118. Oh, D.Y.; Talukdar, S.; Bae, E.J.; Imamura, T.; Morinaga, H.; Fan, W.; Li, P.; Lu, W.J.; Watkins, S.M.; Olefsky, J.M. GPR120 is an omega-3 fatty acid receptor mediating potent anti-inflammatory and insulin-sensitizing effects. *Cell* **2010**, *142*, 687–698. [CrossRef]

119. Huang, J.; Xue, M.; Zhang, J.; Yu, H.; Gu, Y.; Du, M.; Ye, W.; Wan, B.; Jin, M.; Zhang, Y. Protective role of GPR120 in the maintenance of pregnancy by promoting decidualization via regulation of glucose metabolism. *EBioMedicine* **2019**, *39*, 540–551. [CrossRef]
120. Leitao, B.; Jones, M.C.; Fusi, L.; Higham, J.; Lee, Y.; Takano, M.; Goto, T.; Christian, M.; Lam, E.W.F.; Brosens, J.J. Silencing of the JNK pathway maintains progesterone receptor activity in decidualizing human endometrial stromal cells exposed to oxidative stress signals. *FASEB J.* **2010**, *24*, 1541–1551. [CrossRef]
121. Muter, J.; Brighton, P.J.; Lucas, E.S.; Lacey, L.; Shmygol, A.; Quenby, S.; Blanks, A.M.; Brosens, J.J. Progesterone-dependent induction of phospholipase c-related catalytically inactive protein 1 (prip-1) in decidualizing human endometrial stromal cells. *Endocrinology* **2016**, *157*, 2883–2893. [CrossRef]
122. Shah, K.M.; Webber, J.; Carzaniga, R.; Taylor, D.M.; Fusi, L.; Clayton, A.; Brosens, J.J.; Hartshorne, G.; Christian, M. Induction of microRNA resistance and secretion in differentiating human endometrial stromal cells. *J. Mol. Cell Biol.* **2013**, *5*, 67–70. [CrossRef]
123. Kajihara, T.; Jones, M.; Fusi, L.; Takano, M.; Feroze-Zaidi, F.; Pirianov, G.; Mehmet, H.; Ishihara, O.; Higham, J.M.; Lam, E.W.F.; et al. Differential expression of FOXO1 and FOXO3a confers resistance to oxidative cell death upon endometrial decidualization. *Mol. Endocrinol.* **2006**, *20*, 2444–2455. [CrossRef]
124. Muter, J.; Alam, M.T.; Vrljicak, P.; Barros, F.S.V.; Ruane, P.T.; Ewington, L.J.; Aplin, J.D.; Westwood, M.; Brosens, J.J. The glycosyltransferase EOGT regulates adropin expression in decidualizing human endometrium. *Endocrinology* **2018**, *159*, 994–1004. [CrossRef]
125. Burnum, K.E.; Hirota, Y.; Baker, E.S.; Yoshie, M.; Ibrahim, Y.M.; Monroe, M.E.; Anderson, G.A.; Smith, R.D.; Daikoku, T.; Dey, S.K. Uterine deletion of Trp53 compromises antioxidant responses in the mouse decidua. *Endocrinology* **2012**, *153*, 4568–4579. [CrossRef]
126. Lanekoff, I.; Cha, J.; Kyle, J.E.; Dey, S.K.; Laskin, J.; Burnum-Johnson, K.E. Trp53 deficient mice predisposed to preterm birth display region-specific lipid alterations at the embryo implantation site. *Sci. Rep.* **2016**, *6*, 33023. [CrossRef]
127. Hirota, Y.; Cha, J.; Yoshie, M.; Daikoku, T.; Dey, S.K. Heightened uterine mammalian target of rapamycin complex 1 (mTORC1) signaling provokes preterm birth in mice. *Proc. Natl. Acad. Sci. USA* **2011**, *108*, 18073–18078. [CrossRef]
128. Estella, C.; Herrer, I.; Moreno-Moya, J.M.; Quinonero, A.; Martinez, S.; Pellicer, A.; Simon, C. miRNA signature and Dicer requirement during human endometrial stromal decidualization in vitro. *PLoS ONE* **2012**, *7*, e41080. [CrossRef] [PubMed]
129. Jimenez, P.T.; Mainigi, M.A.; Word, R.A.; Kraus, W.L.; Mendelson, C.R. miR-200 regulates endometrial development during early pregnancy. *Mol. Endocrinol.* **2016**, *30*, 977–987. [CrossRef] [PubMed]
130. Zhang, Q.; Zhang, H.; Jiang, Y.; Xue, B.; Diao, Z.; Ding, L.; Zhen, X.; Sun, H.; Yan, G.; Hu, Y. MicroRNA-181a is involved in the regulation of human endometrial stromal cell decidualization by inhibiting Kruppel-like factor 12. *Reprod. Biol. Endocrinol.* **2015**, *13*, 23. [CrossRef] [PubMed]
131. Tochigi, H.; Kajihara, T.; Mizuno, Y.; Mizuno, Y.; Tamaru, S.; Kamei, Y.; Okazaki, Y.; Brosens, J.J.; Ishihara, O. Loss of miR-542-3p enhances IGFBP-1 expression in decidualizing human endometrial stromal cells. *Sci. Rep.* **2017**, *7*, 40001. [CrossRef]
132. Pei, T.; Liu, C.; Liu, T.; Xiao, L.; Luo, B.; Tan, J.; Li, X.; Zhou, G.; Duan, C.; Huang, W. miR-194-3p represses the progesterone receptor and decidualization in eutopic endometrium from women with endometriosis. *Endocrinology* **2018**, *159*, 2554–2562. [CrossRef]
133. Munro, S.K.; Farquhar, C.M.; Mitchell, M.D.; Ponnampalam, A.P. Epigenetic regulation of endometrium during the menstrual cycle. *Mol. Hum. Reprod.* **2010**, *16*, 297–310. [CrossRef]
134. Grimaldi, G.; Christian, M.; Quenby, S.; Brosens, J.J. Expression of epigenetic effectors in decidualizing human endometrial stromal cells. *Mol. Hum. Reprod.* **2012**, *18*, 451–458. [CrossRef]
135. Grimaldi, G.; Christian, M.; Steel, J.H.; Henriet, P.; Poutanen, M.; Brosens, J.J. Down-regulation of the histone methyltransferase EZH2 contributes to the epigenetic programming of decidualizing human endometrial stromal cells. *Mol. Endocrinol.* **2011**, *25*, 1892–1903. [CrossRef]
136. Nancy, P.; Siewiera, J.; Rizzuto, G.; Tagliani, E.; Osokine, I.; Manandhar, P.; Dolgalev, I.; Clementi, C.; Tsirigos, A.; Erlebacher, A. H3K27me3 dynamics dictate evolving uterine states in pregnancy and parturition. *J. Clin. Investig.* **2018**, *128*, 233–247. [CrossRef]

137. Garrido-Gomez, T.; Dominguez, F.; Quiñonero, A.; Diaz-Gimeno, P.; Kapidzic, M.; Gormley, M.; Ona, K.; Padilla-Iserte, P.; McMaster, M.; Genbacev, O.; et al. Defective decidualization during and after severe preeclampsia reveals a possible maternal contribution to the etiology. *Proc. Natl. Acad. Sci. USA* **2017**, *114*, E8468–E8477. [CrossRef]
138. Garrido-Gomez, T.; Quiñonero, A.; Dominguez, F.; Rubert, L.; Perales, A.; Hajjar, K.A.; Simon, C. Preeclampsia: A defect in decidualization is associated with deficiency of Annexin A2. *Am. J. Obstet. Gynecol.* **2020**, *222*, e1–e376. [CrossRef]
139. Salker, M.; Teklenburg, G.; Molokhia, M.; Lavery, S.; Trew, G.; Aojanepong, T.; Mardon, H.J.; Lokugamage, A.U.; Rai, R.; Landles, C.; et al. Natural selection of human embryos: Impaired decidualization of endometrium disables embryo-maternal interactions and causes recurrent pregnancy loss. *PLoS ONE* **2010**, *5*, e10287. [CrossRef]
140. Lucas, E.S.; Vrljicak, P.; Muter, J.; Diniz-da-Costa, M.M.; Brighton, P.J.; Kong, C.S.; Lipecki, J.; Fishwick, K.J.; Odendaal, J.; Ewington, L.J.; et al. Recurrent pregnancy loss is associated with a pro-senescent decidual response during the peri-implantation window. *Commun. Biol.* **2020**, *3*, 37. [CrossRef]
141. American College of Obstetricians and Gynecologists. Pregestational diabetes mellitus. ACOG Practice Bulletin No. 201. *Obstet. Gynecol.* **2018**, *132*, 1514–1516. [CrossRef] [PubMed]
142. Meis, P.J.; Klebanoff, M.; Thom, E.; Dombrowski, M.P.; Sibai, B.; Moawad, A.H.; Spong, C.Y.; Hauth, J.C.; Miodovnik, M.; Varner, M.W.; et al. Prevention of recurrent preterm delivery by 17 alpha-hydroxyprogesterone caproate. *N. Engl. J. Med.* **2003**, *348*, 2379–2385. [CrossRef] [PubMed]
143. Blackwell, S.C.; Gyamfi-Bannerman, C.; Biggio, J.R., Jr.; Chauhan, S.P.; Hughes, B.L.; Louis, J.M.; Manuck, T.A.; Miller, H.S.; Das, A.F.; Saade, G.R.; et al. 17-OHPC to prevent recurrent preterm birth in singleton gestations (PROLONG Study): A multicenter, international, randomized double-blind trial. *Am. J. Perinatol.* **2020**, *37*, 127–136. [CrossRef]
144. Keelan, J.A. Intrauterine inflammatory activation, functional progesterone withdrawal, and the timing of term and preterm birth. *J. Reprod. Immunol.* **2018**, *125*, 89–99. [CrossRef]
145. American College of Obstetricians and Gynecologists. Low-dose aspirin use during pregnancy. ACOG Committee Opinion No. 743. *Obstet. Gynecol.* **2018**, *132*, e44–e52. [CrossRef]
146. Williams, P.J.; Bulmer, J.N.; Innes, B.A.; Broughton Pipkin, F. Possible roles for folic acid in the regulation of trophoblast invasion and placental development in normal early human pregnancy. *Biol. Reprod.* **2011**, *84*, 1148–1153. [CrossRef]

International Journal of *Molecular Sciences*

MDPI

Article

Endothelial Jagged1 Antagonizes Dll4/Notch Signaling in Decidual Angiogenesis during Early Mouse Pregnancy

Nicole M. Marchetto [1], Salma Begum [1], Tracy Wu [1], Valerie O'Besso [2], Christina C. Yarborough [3], Nuriban Valero-Pacheco [4,5], Aimee M. Beaulieu [4,5], Jan K. Kitajewski [6], Carrie J. Shawber [7] and Nataki C. Douglas [1,5,*]

1 Department of Obstetrics, Gynecology and Reproductive Health, Rutgers New Jersey Medical School, Newark, NJ 07103, USA; nm949@njms.rutgers.edu (N.M.M.); sb1802@njms.rutgers.edu (S.B.); tw472@njms.rutgers.edu (T.W.)
2 Rutgers New Jersey Medical School, Newark, NJ 07103, USA; valerie.obesso@gmail.com
3 School of Graduate Studies, Rutgers Biomedical and Health Sciences, Newark, NJ 07103, USA; ccy21@gsbs.rutgers.edu
4 Department of Microbiology, Biochemistry, and Molecular Genetics, Rutgers Biomedical and Health Sciences, Newark, NJ 07103, USA; pv164@njms.rutgers.edu (N.V.-P.); ab1550@njms.rutgers.edu (A.M.B.)
5 Center for Immunity and Inflammation, Rutgers Biomedical and Health Sciences, Newark, NJ 07103, USA
6 Department of Physiology & Biophysics, University of Illinois Chicago, Chicago, IL 60612, USA; kitaj@uic.edu
7 Department of Obstetrics and Gynecology, Division of Reproductive Sciences, Columbia University College of Physicians and Surgeons, New York, NY 10032, USA; cjs2002@cumc.columbia.edu
* Correspondence: nataki.douglas@rutgers.edu

Received: 30 June 2020; Accepted: 31 August 2020; Published: 5 September 2020

Abstract: Maternal spiral arteries and newly formed decidual capillaries support embryonic development prior to placentation. Previous studies demonstrated that Notch signaling is active in endothelial cells of both decidual capillaries and spiral arteries, however the role of Notch signaling in physiologic decidual angiogenesis and maintenance of the decidual vasculature in early mouse pregnancy has not yet been fully elucidated. We used the *Cdh5-Cre*^{ERT2}; *Jagged1(Jag1)*$^{flox/flox}$ (*Jag1ΔEC*) mouse model to delete Notch ligand, *Jag1*, in maternal endothelial cells during post-implantation, pre-placentation mouse pregnancy. Loss of endothelial *Jag1* leads to increased expression of Notch effectors, *Hey2* and *Nrarp*, and increased endothelial Notch signaling activity in areas of the decidua with remodeling angiogenesis. This correlated with an increase in Dll4 expression in capillary endothelial cells, but not spiral artery endothelial cells. Consistent with increased Dll4/Notch signaling, we observed decreased VEGFR2 expression and endothelial cell proliferation in angiogenic decidual capillaries. Despite aberrant Dll4 expression and Notch activation in *Jag1ΔEC* mutants, pregnancies were maintained and the decidual vasculature was not altered up to embryonic day 7.5. Thus, Jag1 functions in the newly formed decidual capillaries as an antagonist of endothelial Dll4/Notch signaling during angiogenesis, but Jag1 signaling is not necessary for early uterine angiogenesis.

Keywords: Notch; Jag1; Dll4; endothelial cells; decidua; angiogenesis; capillaries; spiral arteries

1. Introduction

During early pregnancy, from embryo implantation to placentation, formation of new capillaries and physiologic remodeling of uterine spiral arteries (SpAs), which are distal branches of the maternal uterine arteries, are essential for normal embryonic growth and development [1–3]. Abnormal uterine vascular development early in pregnancy can set off a "ripple effect", resulting in poor placentation, reduced placental function and adverse downstream effects, such as abnormal growth and development of the embryo and pregnancy failure by mid-gestation [1,4]. Thus, understanding the pathways that regulate pre-placentation vascular development in the uterus will help to better elucidate the pathways that impact maintenance and success of pregnancy at later stages.

In mice, uterine vascular changes are initiated by embryo implantation. Embryo implantation, which occurs by embryonic day (E) 4.5, triggers uterine stromal cell decidualization and concomitant decidual angiogenesis [1,2,5–8]. Blood vessels grow into the uterus on the mesometrial side creating a mesometrial-anti-mesometrial polarity of the implantation site and dividing the implantation site into two regions, mesometrial region and anti-mesometrial region [1,7] (Figure 1A–C). The mesometrial region can be further divided into the mesometrial pole and central region. In the central region of the mesometrial decidua, sprouting angiogenesis is most pronounced at E4.5 and remodeling angiogenesis occurs from E6.5–E8.5 [7]. In the anti-mesometrial region, angiogenesis begins at E4.5 and initiates endothelial cell (EC) proliferation, which peaks by E6.5, and results in a greater than 2-fold increase in vascular density at E8.5, as compared to E4.5 [7]. From E4.5–E7.5, decidual angiogenesis creates a rich, new capillary network that supports the growing pregnancy prior to placentation.

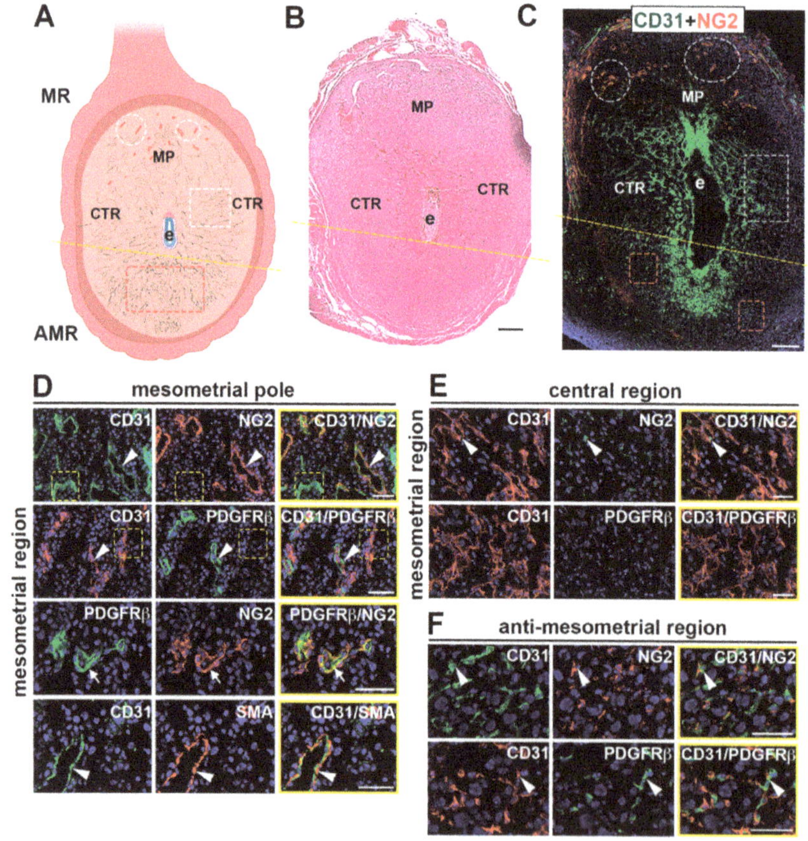

Figure 1. Characterization of the E7.5 uterine vasculature in wildtype mice. (**A**) Schematic representation of an E7.5 implantation site (created with BioRender.com). The implantation site is divided (represented

by the dashed line) into two regions, the anti-mesometrial region (AMR) and the mesometrial region (MR), which is further divided into the mesometrial pole (MP) and central region (CTR). Labeled are spiral arteries (white ellipses) in the MP, capillaries (white rectangles) in the CTR, capillaries (red rectangles) in the AMR and the embryo (e). (**B**) H&E of an implantation site at E7.5 showing the embryo and central parts of the decidua. (**C**) An E7.5 implantation site double stained for the endothelial cell marker, CD31 and pericyte marker, NG2. Ellipses identify NG2$^+$ SpAs at the MP. Rectangles identify CD31$^+$ capillaries in the central region (white rectangle) and the AMR (red rectangle). (**D–F**) High magnification images of the vessels of the MR at the MP and CTR and AMR. Sections were double stained to detect expression of CD31 and mural cell markers, NG2, PDGFRβ or SMA. Merged images are outlined in yellow. CD31$^+$ endothelial cells (ECs) are present throughout the implantation site in all three regions. (**D**) The vasculature in the MP is composed of SpAs seen as CD31$^+$ ECs covered by NG2$^+$, PDGFRβ$^+$ and SMA$^+$ mural cells (white arrowheads) and capillaries comprised of ECs that are not associated with NG2, PDGFRβ or SMA (yellow rectangles). NG2 and PDGFRβ are co-expressed (white arrow). (**E**) In the CTR, expression of NG2 is sparse and expression of PDGFRβ was not detected. Few CD31$^+$ capillaries are associated with NG2$^+$ mural cells (white arrowhead). (**F**) In the AMR, capillaries are comprised of CD31$^+$ ECs closely associated with NG2$^+$/PDGFRβ$^+$ mural cells (white arrowheads). Scale bars = 200 μm in (**B**,**C**) and 50 μm in (**D–F**).

Placenta formation begins at E8.5 and is complete by mid-gestation, or E10.5 [1,2]. Transformation of the maternal SpAs, initiates at E6.5 and is the key vascular change occurring from E8.5–E10.5. Embryo-derived trophoblasts migrate from the implantation site to the SpAs and mediate SpA remodeling [3,9,10]. SpAs, like most arteries, are comprised of two key cell types, ECs, which make up the inner lining of the vessel, and mural cells, which include pericytes and vascular smooth muscle cells, that surround the endothelium and are essential for maintenance of vascular integrity [11–13]. During the process of SpA remodeling, trophoblasts intercalate the vessels, replacing both ECs and vascular smooth muscle cells, and transform constricted artery-like structures into dilated and enlarged vein-like SpAs, which will support adequate blood flow through to the placenta [1,3,14,15]. Prior to mid-gestation, E10.5, two distinct vascular events, decidual angiogenesis and SpA remodeling, are essential for normal placentation and placental function.

The Vascular Endothelial Growth Factor (VEGF) and Notch signaling pathways are interconnected and together have been shown to be critical for vascular development of the retina, embryo and ovary [16–19]. VEGF signaling is essential for vascular proliferation and expansion in decidual angiogenesis and is necessary to support early pregnancy [5,7]. Experimental mouse models with targeted deletions of Notch signaling pathway members possess a variety of vascular defects that cause embryonic lethality by mid-gestation [20–26], but loss of Notch signaling in decidual angiogenesis and vascular remodeling in early pregnancy has not been fully elucidated.

Notch proteins are expressed in ECs and mural cells. Vascular ECs express receptors, Notch1 and Notch4, and ligands delta-like (Dll) 1, Dll4, Jagged1 (Jag1) and Jag2 [23,27–31]. Mural cells have been previously shown to express Notch1, Notch3 and Jag1 [26,27,32–34]. Ligand binding of Notch triggers release of the Notch intracellular domain (NICD), which translocates into the nucleus and promotes transcription of downstream effectors of Notch signaling, including members of the Hairy/Enhancer of Split related with YRPW motif (*Hey*) families and Notch-regulated ankyrin repeat-containing protein (*Nrarp*) [29,35,36]. Notch activation can be elicited by ligands expressed within the same cell (*cis*-activation) as the Notch receptor or on an adjacent cell (*trans*-activation) [37]. Alternatively, blockade of an inhibitory *cis*-interaction between receptor and ligand can drive Notch activation [38].

We have previously shown that Dll4 is expressed in the ECs of SpAs, while Jag1 is expressed in both ECs and pericytes of SpAs [27]. Consistent with ligand expression, Notch signaling is active in ECs in decidual capillaries and SpAs [27,39]. Herein, we investigated the role of endothelial Jag1/Notch signaling in the formation and maintenance of the maternal decidual vasculature in early mouse pregnancy. Mice carrying an EC-specific, tamoxifen inducible Cre recombinase, *Cdh5-Cre*^{ERT2} [40] and a *Jagged1*$^{flox/flox}$ alleles [41] were used to achieve cell-type specific deletion of *Jag1* during decidual angiogenesis. In order to assess the decidual vasculature after embryo implantation and prior to placentation, we evaluated pregnancies at E7.5, a stage of pregnancy when decidual angiogenesis is nearing completion and decidual vascular density is greatest.

2. Results

2.1. Characterization of Vasculature in the Mouse Implantation Site at E7.5

To characterize the vasculature in the peri-implantation uterus, we immunostained wild-type implantation sites with EC marker, CD31, and mural cell markers, platelet-derived growth factor receptor beta (PDGFRβ), neural/glial antigen 2 (NG2), and α-smooth muscle actin (SMA) [12,42–44]. Visualization of the whole implantation site demonstrates that NG2 expression is enriched around the SpAs at the mesometrial pole of the mesometrial region while it is variably expressed in decidual vessels in the central region and anti-mesometrial region (Figure 1C). At the mesometrial pole (Figure 1D), mural cells expressing NG2, PDGFRβ and SMA are adjacent to but not overlapping with $CD31^+$ ECs, consistent with these vessels being the SpAs. In contrast, capillaries at the mesometrial pole of the mesometrial region are comprised of $CD31^+$ ECs that lack SMA^+ mural cell coverage. In the central region of the mesometrial region, the expression of NG2 is sparse (Figure 1E). In the anti-mesometrial region, $NG2^+$ and $PDGFR\beta^+$ cells are closely associated with $CD31^+$ ECs, while co-expression is not noted (Figure 1F). SMA expression was not detected in the central region or anti-mesometrial region. The morphology and pattern of expression of the $NG2^+$ and $PDGFR\beta^+$ cells in the anti-mesometrial region, and lack of SMA expression suggest that these mural cells are pericytes. Together, these data show regional differences in the vasculature at E7.5; mature, non-angiogenic resident SpAs with their associated mural cells are present in the mesometrial decidua, whereas angiogenic capillary networks make up the central and anti-mesometrial regions of the decidua.

2.2. Characterization of Notch Expression in the Decidual Vasculature

To determine expression of Notch proteins in the decidual vasculature, implantation sites were double stained with antibodies to detect Notch1 and Notch4 with respect to ECs and mural cells. In the mesometrial region, $CD31^+$ capillary ECs at the mesometrial pole and in the central region express Notch1 and Notch4 (Figure 2A,B). In SpAs at the mesometrial pole, $PDGFR\beta^+$ mural cells are adjacent to ECs that express both Notch1 and Notch4 (Figure 2A). In the anti-mesometrial region, Notch1 is expressed in $CD31^+$ ECs surrounded by Notch1 negative $PDGFR\beta^+$ mural cells (Figure 2C). Notch4 expression was not detected in the anti-mesometrial region (data not shown).

Double staining for Jag1 or Dll4 and a vascular cell marker revealed that these Notch ligands are dynamically expressed in the decidual vasculature. At the mesometrial pole of the mesometrial region, $CD31^+$ ECs in capillaries and SpAs express both Jag1 and Dll4 (Figure 3A,D). A subset of the $NG2^+$ and $PDGFR\beta^+$ mural cells in the SpAs also express Jag1, but not Dll4 (Figure 3A,D). In the central region of the mesometrial region, expression of Jag1 is not detected (Figure 3B), while Dll4 is expressed in a perinuclear, non-vascular pattern in decidual stromal cells (Figure 3E). In the anti-mesometrial region, Dll4 is expressed in $CD31^+$ ECs and $Dll4^+$ cells are associated with $PDGFR\beta^+$ mural cells. Expression of Jag1 is scant in the anti-mesometrial region but is found in $CD31^+$ ECs and surrounding $PDGFR\beta^+$ mural cells. Together, these data demonstrate that capillary ECs in the mesometrial region and anti-mesometrial region and SpA ECs co-express Notch1, as well as Jag1 and Dll4, suggesting these proteins mediate Notch signaling to regulate decidual angiogenesis and SpA remodeling.

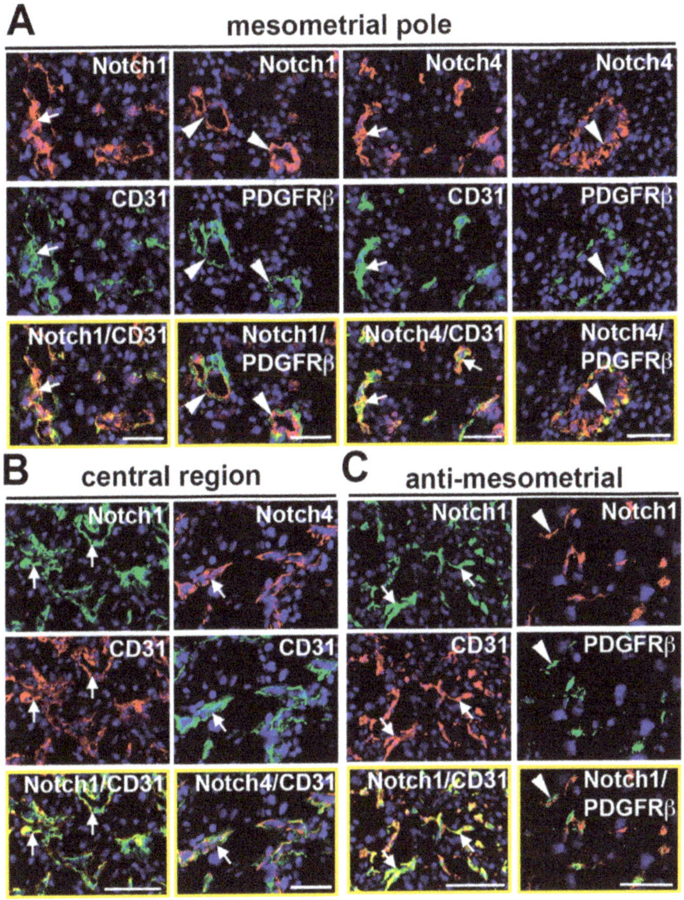

Figure 2. Expression of receptors, Notch1 and Notch4, in the E7.5 decidual vasculature. High magnification images of sections double stained to detect expression of Notch1 or Notch4 and the endothelial cell marker, CD31 or mural cell marker, PDGFRββ. Merged images are outlined in yellow. (**A**) In the MP, Notch1 and Notch4 are expressed in CD31$^+$ SpA ECs (white arrows). PDGFRβ$^+$ mural cells are associated with both Notch1$^+$ and Notch4$^+$ ECs in the SpAs (white arrowheads). (**B**) In the CTR, CD31$^+$ capillary ECs express Notch1 and Notch4 (white arrows). (**C**) In the AMR, CD31$^+$ capillary ECs express Notch1 (white arrows) and PDGFRβ$^+$ pericytes are associated with Notch1$^+$ cells (white arrowhead). AMR = anti-mesometrial; CTR = central region; MP = mesometrial pole. Scale bars = 50 μm.

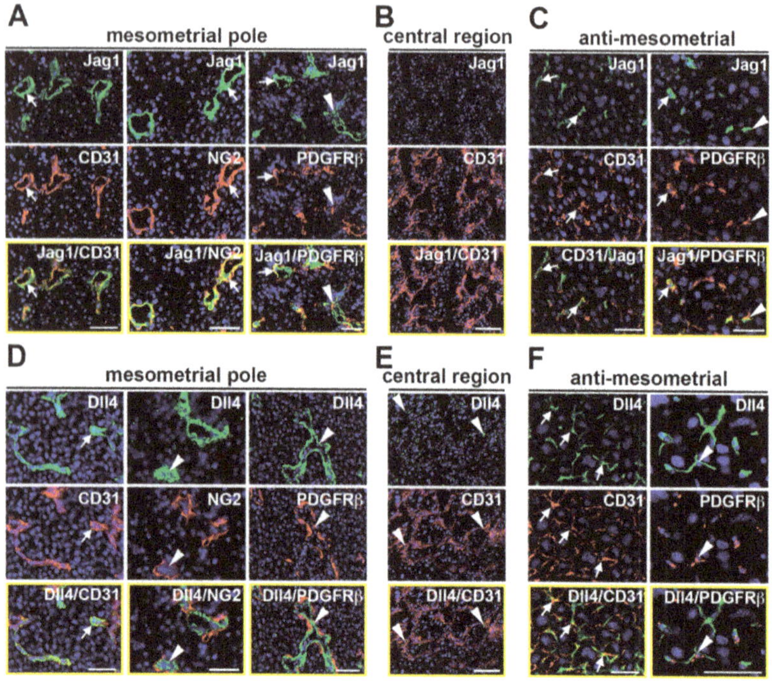

Figure 3. Expression of Notch ligands, Jag1 and Dll4, in the E7.5 decidual vasculature. High magnification images of sections double stained to detect expression of Jag1 or Dll4 and the endothelial cell marker, CD31 or mural cell marker, PDGFRβ. (**A**) In MP SpAs, Jag1 is expressed in $CD31^+$ ECs (white arrow) and in $NG2^+$ and $PDGFR\beta^+$ mural cells (white arrows). Some $Jag1^+$ ECs are closely associated with $PDGFR\beta^+/Jag1^-$ mural cells (white arrowheads). (**B**) Expression of Jag1 is not detected in the central region. (**C**) In the AMR, expression of Jag1 is sparse. Jag1 is expressed in $CD31^+$ ECs and $PDGFR\beta^+$ mural cells (white arrows). $Jag1^+$ cells are closely associated with $PDGFR\beta^+$ mural cells (white arrowheads). (**D**) In MP SpAs, Dll4 is expressed in $CD31^+$ ECs (white arrows). $NG2^+$ and $PDGFR\beta^+$ mural cells are closely associated with $Dll4^+$ cells (white arrowheads). (**E**) In the CTR, Dll4 is expressed in a punctate pattern (white arrowheads). (**F**) In the AMR, Dll4 is expressed in $CD31^+$ ECs (white arrows). $Dll4^+$ cells are associated with $PDGFR\beta^+$ mural cells (white arowheads). AMR = anti-mesometrial; CTR = central region; MP = mesometrial pole. Scale bars = 50 μm.

2.3. EC-Specific Deletion of Jag1 in Early Pregnancy

To determine the role of endothelial Jag1 in the decidual vasculature in early pregnancy, we used a tamoxifen inducible driver, *Cdh5-Cre*^{ERT2}, to express Cre recombinase in the endothelium and delete *Jag1* during decidual angiogenesis. Tamoxifen was administered at embryo implantation, E4.5, and pregnancies and uterine phenotypes were assessed at E7.5, the end of the active period of decidual angiogenesis (Figure 4A). Notch proteins and ligands are expressed in the ovarian endothelium, including in ECs of corpora lutea [45] (Supplementary Figure S1). Given the requirement for progesterone secretion from ovarian corpora lutea to maintain early pregnancy, progesterone supplementation was initiated at E4.5 to overcome potential defects in *Jag1ΔEC* mutant ovaries. To determine the recombination efficiency of the *Cdh5-Cre*^{ERT2} line using this tamoxifen regimen, *Cdh5-Cre*^{ERT2} mice were crossed to the *ROSA26 tdTomato* reporter strain (Figure 4B). Expression of fluorescent tdTomato protein throughout the implantation site is observed in a pattern similar to CD31 in the mesometrial region and anti-mesometrial region (Figure 1C). Implantation sites were stained with endothelial cell marker, CD31, or mural cell marker, NG2, to assess tdTomato expression with respect to ECs and mural cells. Co-expression is observed with CD31$^+$ ECs, but not with NG2$^+$ mural cells (Figure S2). tdTomato is expressed in 94–98% of CD31$^+$ ECs in the mesometrial and anti-mesometrial regions and in 90% of NG2$^+$ SpAs in the mesometrial region (Figure S2). These data demonstrate efficient recombination in decidual ECs of capillaries and SpAs at E7.5.

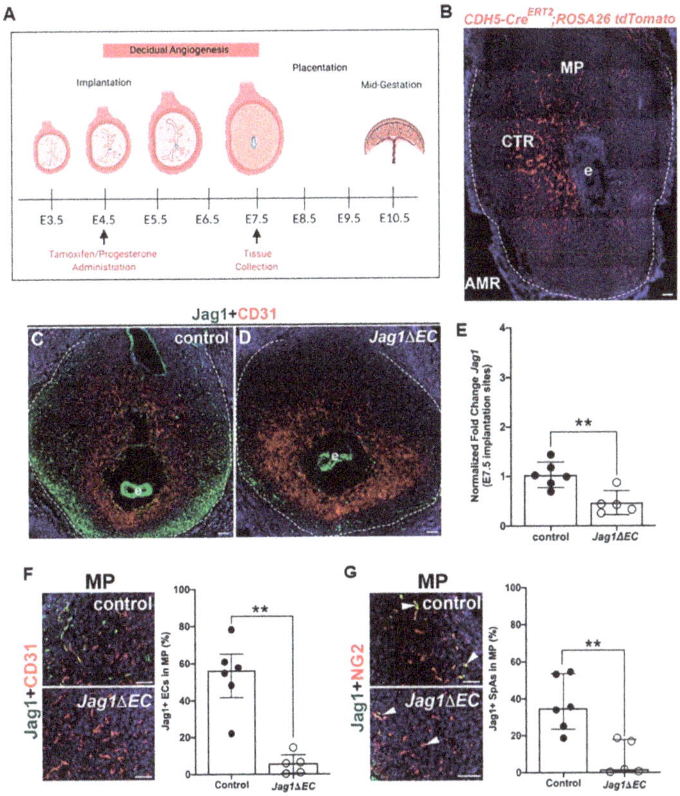

Figure 4. Tamoxifen-induced Cre recombination decreases expression of Jag1 in *Cdh5-Cre*^{ERT2};*Jag1*$^{flox/flox}$ (*Jag1ΔEC*) pregnancy. (**A**) Schematic of the mouse gestation timeline highlighting key processes involved, and experimental timeline (created with BioRender.com). At E4.5, tamoxifen is administered to induce Cre recombination and progesterone was administered to minimize secondary angiogenic defects in the uterus due to vascular defects in the ovaries. Pregnancies and uterine phenotype were assessed at E7.5. (**B**) Representative image of an implantation site at E7.5 from the *Cdh5-Cre*^{ERT2}; *ROSA26 tdTomato* reporter. The decidua is within the dashed white. tdTomato expression, indicating Cre-induced recombination, is detected in a vascular pattern throughout the decidua. Dashed line denotes MR and AMR. (**C**,**D**) Representative images of implantation sites of *Cdh5-Cre*^{ERT2} control and *Jag1ΔEC* pregnancies stained for Jag1 and CD31. Expression of Jag1 is reduced in the decidua (outlined with dashed white lines), but not the embryo of *Jag1ΔEC* mutants. (**E**) qRT-PCR determination of *Jag1* expression in implantation sites from control (n = 6) and *Jag1ΔEC* (n = 5) pregnancies. The relative expression level of *Jag1* was compared to β-actin. *Jag1* was significantly decreased in *Jag1ΔEC* mutants as compared to *Cdh5-Cre*^{ERT2} controls. (**F**) Representative images of the MP of control and *Jag1ΔEC* implantation sites double stained for EC marker CD31 and Jag1. At the MP, expression of Jag1 in all CD31$^+$ ECs is reduced in *Jag1ΔEC* mutants (n = 6) as compared to *Cdh5-Cre*^{ERT2} controls (n = 5). (**G**) Representative images of the MP of control and *Jag1ΔEC* implantation sites double stained for Jag1

and NG2. Expression of Jag1 in the ECs of the NG2$^+$ surrounded SpAs (white arrowheads) is reduced in *Jag1ΔEC* mutants (n = 5) as compared to *Cdh5-Cre^{ERT2}* controls (n = 6). AMR= anti-mesometrial region; CTR = central region; e = embryo; MP = mesometrial pole. Scale bars = 100 μm in (**B–D**,**F**,**G**). Data shown as median + IQR; ** $p < 0.01$.

To evaluate the role of endothelial Jag1 in post-implantation, pre-placentation pregnancy, uteri from pregnant *Jag1ΔEC* and *Cdh5-Cre^{ERT2}* control females were isolated at E7.5. A decrease in Jag1 is observed in the decidua ECs of *Jag1ΔEC* uteri as compared to controls (Figure 4C,D). In contrast, analysis of the uteri from *Jag1ΔEC* pregnancies demonstrates similar expression of Jag1 in embryos of control and *Jag1ΔEC* mutants. qRT-PCR performed on whole implantation sites shows a significant decrease in *Jag1* expression in *Jag1ΔEC* mutants as compared to controls (Figure 4E). To quantify Jag1 expression in ECs and SpAs at the mesometrial pole, implantation sites were stained to detect Jag1 in CD31$^+$ ECs and Jag1 expressed in NG2$^+$ SpAs (Figure 4F,G). Expression of Jag1 is significantly lower in all CD31$^+$ ECs (Figure 4F,G) in *Jag1ΔEC* pregnancies when compared to controls, demonstrating similar efficacy of *Cdh5-Cre^{ERT2}*-mediated deletion of *Jag1* in capillary and SpA ECs, as seen for the *ROSA26 tdTomato* reporter, and confirming that our model systems works.

2.4. Loss of Endothelial Jag1 Does Not Impact Pregnancy at E7.5

To determine if EC-specific loss of *Jag1* impacts pregnancy and embryo development at E7.5, implantation sites were collected and counted, and embryo morphology was assessed in H&E stained sections. Litter size for *Cdh5-Cre^{ERT2}* control and *Jag1ΔEC* pregnancies is similar (Figure 5A). Embryo development involves the formation and disappearance of morphologic structures during gestation [46]. To determine embryo morphology with respect to embryonic age, all tissue sections containing embryos were scored for the presence or absence of the following morphologic structures: primitive streak, allantois, cranial neural fold and somites (Figure 5B,C). These morphologic "landmarks" are common to gastrulating wild-type embryos between E6.5 and E8.5, and the timing of their appearance is used to compare stage of embryo development. Comparison of *Cdh5-Cre^{ERT2}* control and *Jag1ΔEC* pregnancies revealed that embryos from *Cdh5-Cre^{ERT2}* control and *Jag1ΔEC* pregnancies progressed to a similar stage of development (Figure 5D).

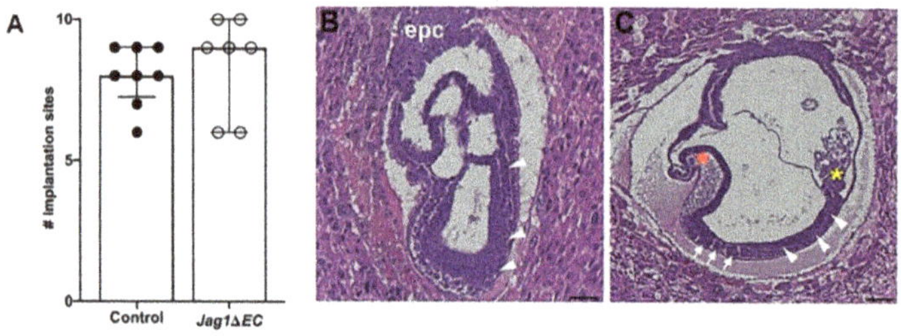

D

Comparison of embryo development at E7.5

Trait	Expected at E7.5	Adherence to expected (# animals with trait/ # animals scored)		Fisher's Exact p-value
		Control	Jag1ΔEC	
Primitive Streak	Yes	100% (8/8)	100% (6/6)	n/a
Allantois	Yes	0% (0/8)	33.3% (2/6)	0.165
Cranial Neural Fold	No	100% (8/8)	66.7% (4/6)	0.165
Somites	No	100% (8/8)	83.3% (5/6)	0.429

Figure 5. EC-specific loss of *Jag1* does not disrupt pregnancy progression or litter size at E7.5. (**A**) Number of implantation sites at E7.5 is similar between controls (n = 8) and *Jag1ΔEC* pregnancies

(n = 7). (**B**,**C**) Representative H&E stained embryos, showing morphologic structures including the primitive streak (arrowheads), cranial neural fold (red asterisk), allantois (yellow asterisk) and somites (white arrows), observed in embryos at E7.5 (**B**) and E8.5 (**C**). (**D**) The presence of the primitive streak, allantois and cranial neural fold structures is similar in control and *Jag1ΔEC* pregnancies. Data were analyzed using Fisher's exact test, comparing morphologic structures in control and *Jag1ΔEC* pregnancies to those expected in pregnancies at E7.5 [46].

2.5. Expression of Notch Ligand, Dll4, and Notch Effectors Is Increased in Jag1ΔEC Pregnancies at E7.5

To assess the impact of loss of endothelial *Jag1* on expression of Dll4, *Jag1ΔEC* and *Cdh5-Cre*^{ERT2} uteri were double stained for CD31 or NG2, and Dll4. The percentage of CD31$^+$ ECs expressing Dll4 was quantified in the mesometrial and anti-mesometrial regions. At the mesometrial pole and in the anti-mesometrial region, Dll4 expression is significantly increased in the CD31$^+$ capillary ECs in *Jag1ΔEC* pregnancies compared to control pregnancies (Figure 6A,B). Dll4 expression is increased but not significantly, in the ECs of the NG2$^+$ cell covered SpAs (Figure 6C). These data show that loss of *Jag1* in ECs leads to increased EC Dll4 expression. To understand the impact of EC-specific loss of *Jag1* and increased EC Dll4 expression on Notch signaling, we assessed the expression of downstream effectors of Notch signaling. Total RNA was isolated from *Jag1ΔEC* mutant and control whole implantation sites, from which myometrium was removed. Expression of direct Notch target genes was determined by qRT-PCR. Expression of Notch effectors, *Hey2* and *Nrarp*, is significantly increased in *Jag1ΔEC* mutants relative to control pregnancies. These data suggest increased Dll4/Notch signaling in the implantation sites of *Jag1ΔEC* mutants (Figure 6D).

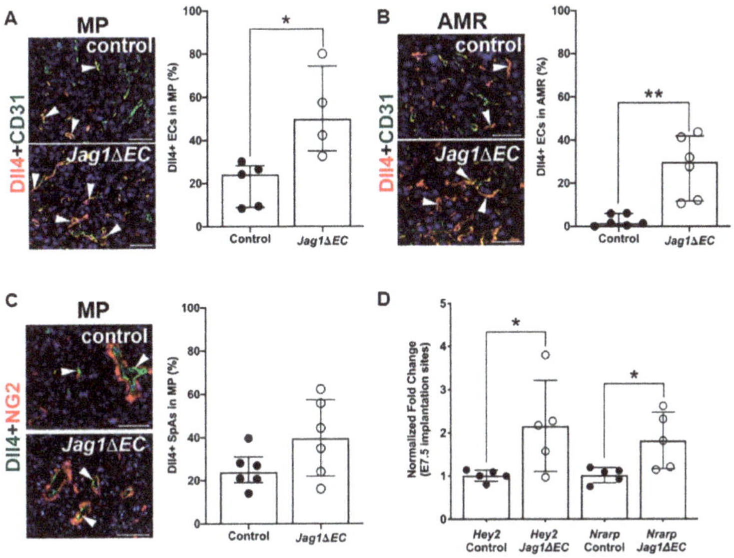

Figure 6. Expression of Notch ligand, Dll4, and Notch effectors are increased in ECs of *Jag1ΔEC* decidua. (**A–C**) Expression of Dll4 in capillaries and SpAs in *Cdh5-Cre*^{ERT2} control and *Jag1ΔEC* pregnancies was determined by double staining IF for Dll4 and CD31 or NG2. Representative images of the MP (**A**) and AMR (**B**) of control and *Jag1ΔEC* implantation sites. (**A**) Dll4 expression is increased in CD31$^+$ ECs at the MP of *Jag1ΔEC* mutants (n = 4) as compared to *Cdh5-Cre*^{ERT2} controls (n = 5). (**B**) Dll4 expression is

increased in CD31$^+$ capillary ECs in the AMR of *Jag1ΔEC* mutants (n = 6) as compared to *Cdh5-Cre^{ERT2}* controls (n = 6). (**C**) Representative images of the MP of control and *Jag1ΔEC* implantation sites double stained for Dll4 and NG2. Expression of Dll4 in the ECs of the NG2$^+$ surrounded SpAs is unchanged in *Jag1ΔEC* mutants (n = 6) as compared to *Cdh5-Cre^{ERT2}* controls (n = 6). (**D**) qRT-PCR determination of Notch effector gene expression in implantation sites from control (n = 5) and *Jag1ΔEC* (n = 5) pregnancies. The relative expression level of each gene was compared to β-actin. *Nrarp* and *Hey2* are significantly increased in *Jag1ΔEC* mutants as compared to *Cdh5-Cre^{ERT2}* controls. AMR = anti-mesometrial region; MP = mesometrial pole. Scale bars = 50 μm. Data shown are median +IQR. * $p < 0.05$, ** $p < 0.01$.

2.6. *Jag1/Notch Signaling Regulates Angiogenic Gene Expression and Endothelial Proliferation in the Anti-Mesometrial Decidua*

Dll4/Notch signaling has been shown in the retina to suppress VEGFR2 expression leading to reduced EC proliferation [17]. To evaluate the impact of EC-specific loss of *Jag1* on endothelial Dll4/Notch signaling, we determined the expression of Notch1 with an antibody against the cytoplasmic domain, the Notch1 ICD (N1ICD). We also determined expression of VEGFR2 and EC proliferation. *Jag1* inactivation increases nuclear expression of N1ICD in anti-mesometrial capillary ECs. This increase in nuclear Notch1 expression is consistent with an increase in expression of Notch1 protein and in EC Notch1 signaling in *Jag1ΔEC* mutants (Figure 7A). We found N1ICD expression in cells adjacent to Dll4$^+$ cells in the capillaries of *Jag1ΔEC* mutants, suggesting *trans*-activation of Notch1 by Dll4 (Figure 7B). The percentage of CD31$^+$ ECs in the anti-mesometrial decidua expressing Ki67 (Figure 7C) and VEGFR2 (Figure 7D) is significantly decreased in *Jag1ΔEC* mutants relative to controls. In contrast, expression of VEGFR2 in the myometrium (Figure 7D, black arrowheads) is similar in *Jag1ΔEC* mutants and controls. Together, the data show that increased Notch signaling is associated with decreased VEGFR2 expression and EC proliferation in the anti-mesometrial decidua.

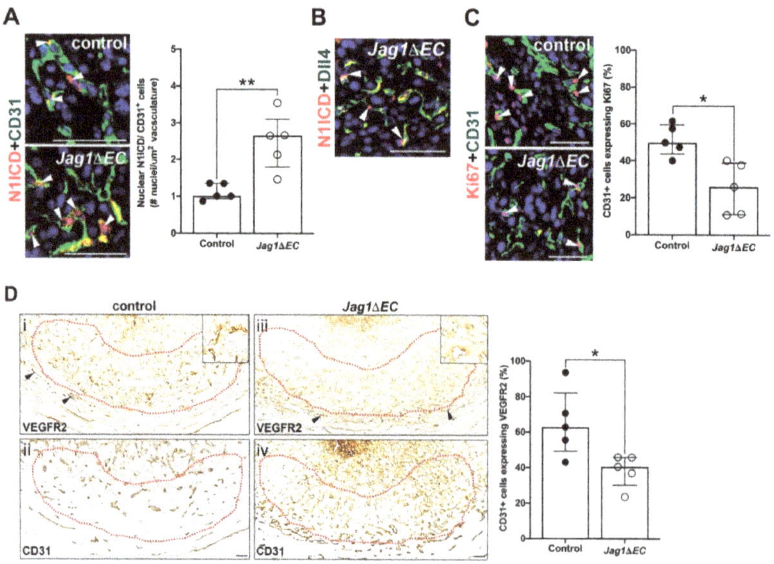

Figure 7. Increased Notch signaling in the anti-mesometrial decidua decreases EC proliferation and expression of VEGFR2. (**A–C**) High magnification images of sections double stained to detect expression of N1ICD and CD31, Dll4 or Ki67. (**A**) Expression of the N1ICD in CD31$^+$ ECs in the AMR of *Cdh5-Cre^{ERT2}* control and *Jag1ΔEC* pregnancies was determined by double staining IF for N1ICD and CD31. Representative images of control and *Jag1ΔEC* implantation sites are shown. Nuclear expression of N1ICD (white arrowheads) is increased in CD31$^+$ ECs in the AMR of *Jag1ΔEC* mutants (n = 5) as

compared to *Cdh5-Cre*^{ERT2} controls (n = 5). (**B**) N1ICD and Dll4 are expressed in adjacent cells in the AMR (white arrowheads) of *Jag1ΔEC* pregnancies. (**C**) EC proliferation in the AMR of *Cdh5-Cre*^{ERT2} control and *Jag1ΔEC* pregnancies was determined by double staining IF for EC proliferation marker, Ki67 and CD31. Representative images of control and *Jag1ΔEC* implantation sites are shown. Ki67 expression (white arrowheads) is decreased in CD31$^+$ ECs in the AMR of *Jag1ΔEC* mutants (n = 5) as compared to *Cdh5-Cre*^{ERT2} controls (n = 5). (**D**) Expression of VEGFR2 with respect to CD31$^+$ ECs in the AMR was determined by comparison of expression of VEGFR2 and CD31 in adjacent sections, in like-regions, of implantation sites from *Cdh5-Cre*^{ERT2} control and *Jag1ΔEC* pregnancies. Representative images of the AMR highlighting areas (decidua within dashed red lines) used to measure signal density from control and *Jag1ΔEC* implantation sites are shown (i–iv). VEGFR2 expression is decreased in CD31$^+$ capillary ECs in the anti-mesometrial decidua of *Jag1ΔEC* mutants (n = 5) as compared to *Cdh5-Cre*^{ERT2} controls (n = 5). Insets highlight VEGFR2 expression in decidual vessels. Black arrowheads highlight VEGFR2 expression in the myometrial vessels, which is similar in *Jag1ΔEC* mutants and controls. AMR = anti-mesometrial region; N1ICD = Notch1 intracellular domain. Scale bars = 50 μm. Data shown as median + IQR; * $p < 0.05$, ** $p < 0.01$.

2.7. EC-Specific Loss of *Jag1* Does Not Impact CD31$^+$ EC Content in Pregnant Uteri at E7.5

To better understand the effect of loss of *Jag1* on decidual vasculature, flow cytometry was done to quantify the proportion of CD31$^+$CD45$^-$ ECs in the implantation site and myometrium of *Jag1ΔEC* and control pregnancies at E7.5 (Figure 8A). Analysis of the uteri from nonpregnant C57BL/6 mice and pregnant C57BL/6 mice showed that an increase CD31$^+$CD45$^-$ EC content was the highest in the decidua of pregnant C57BL/6 mice at E7.5 (Figure 8B). Quantification of the percentages of CD31$^+$CD45$^-$ ECs in the decidua and myometrium of E7.5 *Jag1ΔEC* and *Cdh5-Cre*^{ERT2} control pregnancies revealed no significant differences (Figure 8C). Thus, despite loss of *Jag1* in ECs, the overall percentage of decidual CD31$^+$ ECs is similar in *Jag1ΔEC* and control pregnancies.

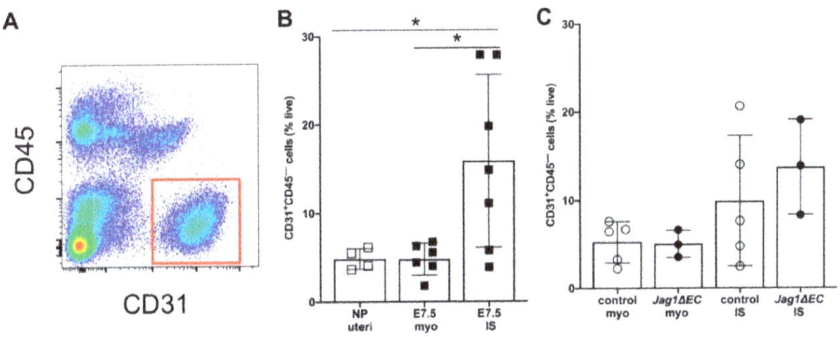

Figure 8. EC-specific loss of *Jag1* does not impact the frequency of CD31$^+$ ECs in implantation sites. (**A**) Representative dot plot of flow cytometry (FCM) data showing gating strategy to identify CD31$^+$ CD45$^-$ EC populations in implantation sites from wild-type C57BL/6 mice at E7.5. (**B**) Results of FCM analysis to assess CD31$^+$CD45$^-$ cell populations in uteri from nonpregnant C57BL/6 mice, and in the myometrium and implantation sites from pregnant C57BL/6 mice at E7.5. The percentage of CD31$^+$CD45$^-$ ECs is significantly higher in implantation sites compared to the myometrium or nonpregnant uteri. (**C**) Results of FCM analysis to determine the percentage of CD31$^+$CD45$^-$ ECs in the myometrium and implantation sites of *Cdh5-Cre*^{ERT2} control and *Jag1ΔEC* pregnancies at E7.5. The percentage of CD31$^+$CD45$^-$ ECs in the myometrium and implantation sites of *Cdh5-Cre*^{ERT2} control and *Jag1ΔEC* pregnancies is similar. IS = implantation site; myo = myometrium; NP = nonpregnant; each group contains n = 3–5 individual mice. Data shown as median + IQR; * $p < 0.05$.

2.8. Vasculature Is Similar in Implantation Sites of Control and *Jag1ΔEC* Pregnancies

The recruitment and differentiation of mural cells, into pericytes and vascular smooth muscle cells, promotes vessel stabilization and is essential for vascular integrity [11,12,47,48]. We have shown that Jag1 is expressed in the ECs of SpAs at the mesometrial pole and in a subset of decidual mural cells (Figure 3A, [27]). To assess whether loss of *Jag1* in ECs impacts vascular density or mural cell content, implantation sites from *Jag1ΔEC* mutant and *Cdh5-Cre*^{ERT2} control pregnancies were stained for expression of CD31 and mural cell markers, PDGFRβ, NG2, and SMA (Figure S3).

Implantation sites stained for CD31 were quantified for percentage CD31$^+$ ECs in the mesometrial pole and central region in the mesometrial region and anti-mesometrial region. In all regions, CD31$^+$ expression is similar in *Jag1ΔEC* mutant and control pregnancies, suggesting that loss of EC-specific *Jag1* does not impact decidual vascular density (Figure 9A–C). To assess whether loss of *Jag1* in ECs impacts mural cell content, implantation sites were analyzed for the expression of mural cell markers. No difference in the expression of NG2, PDGFRβ or SMA at the mesometrial pole of the mesometrial region (Figure 9D–F) was observed, suggesting that loss of endothelial *Jag1* does not impact mural cells in SpAs. Together, these data suggest that EC *Jag1* is not essential for decidual angiogenesis or maintenance of the SpA until E7.5.

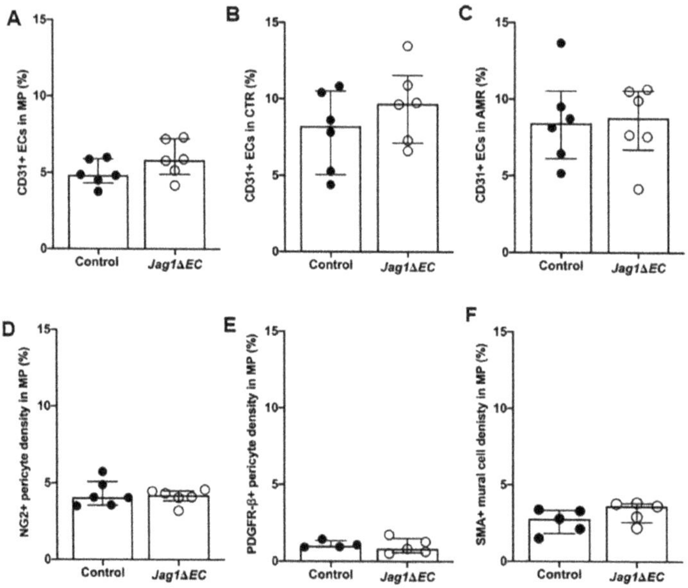

Figure 9. EC-specific loss of *Jag1* does not impact endothelial and mural cell density. (**A–C**) Expression of CD31$^+$ was quantified to determine blood vessel density in each region of the implantation site. (**A**) Density of CD31$^+$ ECs was similar in *Cdh5-Cre*^{ERT2} control and *Jag1ΔEC* pregnancies in the MP (A) and CTR (**B**) of the MR and in the AMR (**C**). (**D–F**) Implantation sites stained for mural markers, NG2, PDGFRβ or SMA are assessed for mural cell content at the MP. Expression of NG2 (**D**), PDGFRβ (**E**) and SMA (**F**) is similar in *Cdh5-Cre*^{ERT2} control and *Jag1ΔEC* pregnancies in the MP. AMR = anti-mesometrial region; CTR = central region; MP = mesometrial pole; MR = mesometrial region. Data shown as median + IQR.

3. Discussion

In this report, we focused on the role the endothelial Jag1 in decidual vasculature in the post-implantation, pre-placentation mouse uterus. When *Jag1* was deleted from the endothelium at E4.5 and pregnancies were assessed at E7.5, the start of placentation, we found that *Jag1ΔEC* mutants maintain pregnancies with no change in embryo development. Loss of *Jag1* in SpA and capillary ECs resulted in a loss of Jag1 expression, but increased expression of Notch effectors, *Hey2* and *Nrarp*, consistent with a loss of Jag1 leading to increased Notch signaling activity. This correlated with an increase in Dll4 expression in capillary ECs, but not SpA ECs, in the mesometrium and in capillary ECs in the anti-mesometrium, suggesting that the increased Notch signaling activity was restricted to capillary ECs throughout the implantation site. In the anti-mesometrial decidua, increased expression of Notch1 and Notch signaling in angiogenic capillary ECs, leads to decreased endothelial VEGFR2 expression and EC proliferation. Despite aberrant Dll4 expression and Notch signaling activity, decidual vasculature development up to E7.5 in *Jag1ΔEC* mutants was not altered in the time window we evaluated.

Angiogenesis occurs in the setting of both physiologic and pathologic growth and is largely controlled by pro-angiogenic signaling pathways, such as the VEGF and Notch pathways [16,19,49–51]. In the post-implantation, pre-placentation pregnant mouse uterus, physiologic angiogenesis is a dynamic process that involves sprouting and remodeling angiogenesis in the decidua that is regulated by VEGF/VEGFR2 [5,7,52] and Dll4/Notch signaling [53]. At E7.5, we found that the central region of the mesometrial decidua contains newer, immature, capillary networks with few, if any, associated mural cells, as there was little to no expression of NG2, PDGFRβ or SMA. In contrast, the ECs associated with $NG2^+$, $PDGFR\beta^+$ pericytes in the anti-mesometrial region are consistent with the remodeling phase of angiogenesis [11,54]. Together, our findings agree with previous literature in suggesting that the central region and anti-mesometrial region are regions of early and active angiogenesis at E7.5.

Our evaluation of Notch proteins and ligands at E7.5 supports and extends our previous findings in the pre-placentation decidua [27,39]. We find Notch1 and Notch4 in the ECs of angiogenic capillaries in the central region, whereas, Notch1, Dll4 and Jag1, but not Notch4, are expressed in the ECs of remodeling capillaries in the anti-mesometrial region. Co-expression of Jag1 and Dll4 in sites of neovascularization and angiogenesis suggests that both these ligands could function in anti-mesometrial region ECs during decidual angiogenesis. In contrast, in the central region, Dll4 expression was detected in stromal, but not endothelial cells, whereas Jag1 expression was not detected in any central region cell-type. Further studies are needed to identify other Notch ligands, such as Jag2 or Dll1, that support angiogenesis and Notch signaling activity in the central region [27,39].

In mouse retinal vascular development [17], in skin wound healing [55] and in coronary plexus formation [56], EC-specific Jag1 has been shown to antagonize Dll4/Notch signaling. Disruption of endothelial *Jag1* leads to disinhibition of Dll4/Notch signaling, followed by increase in Dll4 expression and Notch activation, decreased *VEGFR2* and vascular density, and decreased EC proliferation [17,55]. Further, in wound angiogenesis, it has been proposed that a feedforward signaling mechanism maintains increased Dll4 expression [55]. We observed that loss of EC-specific *Jag1* increases Dll4 and Notch1 expression, as well as increases Notch signaling activity in areas of the decidua with remodeling angiogenesis. Consistent with increased Dll4/Notch signaling, we observed a decrease in VEGFR2 expression and EC proliferation, suggesting that Jag1 antagonizes Dll4/Notch signaling in angiogenic decidual capillaries. Our analysis does not distinguish the type of cellular interaction, *cis-* or *trans-*, that underlies Notch signaling activation in angiogenic decidual capillaries. Further studies are needed to determine if loss of a Jag1/Notch *trans*-interaction promotes Dll4/Notch *cis*-activation or *trans*-activation in anti-mesometrial decidual angiogenesis.

At E7.5, we did not observe a change in vascular density in the anti-mesometrial or central regions. Given that Jag1 is not expressed in central region ECs, the lack of change in Dll4 expression or vascular density in the central region, was not unexpected. In the anti-mesometrial region, where increased expression of EC Dll4 is observed, we predicted an increase in vascular density, as was previously reported with loss of EC-specific *Jag1* [17]. Had we found a change in vascular density, perhaps we would also have seen a change in the progression or survival of the pregnancy. Here, we show that the embryo can tolerate disturbances in the Jag1/Notch signaling by utilizing other members of the signaling pathway. Thus, while loss of Dll4/Notch signaling disrupts decidual angiogenesis [53], loss of endothelial Jag1, which leads to increased Dll4/Notch signaling, does not appear to be essential for early uterine angiogenesis prior to placentation.

SpAs in mice and humans are the distal branches of the uterine artery and are essential for bringing nutrients to the maternal fetal interface before development of the placenta. These arteries undergo important phases of circumferential remodeling, in part triggered by uNK cells, between E6.5 and E10.5 [3,9]. SpA remodeling involves the initial thickening of vascular smooth muscle lined vessel walls between E6.5 and E7.5 followed by their gradual thinning from E8.5 to E9.5. Trophoblast invasion of the SpAs mediates loss of smooth muscle coverage and the breakdown of extracellular matrix, resulting in thin walled vessels with wide lumens that allow for low resistance, high capacity maternal blood flow through the placenta to meet the demands of the growing embryo [3,9,57]. We show that at E7.5, the resident SpAs are still closely associated with mural cells that express NG2, PDGFRβ and SMA, which is consistent with mature vessels in the mesometrial pole that are not undergoing angiogenesis and have not, as yet, been remodeled. SpA ECs express Notch1 and Notch4 and Notch ligands, Jag1 and Dll4. In the pre-implantation uterus, we have previously shown that both Notch3 and Jag1 are expressed in mural cells in vessels in the stroma [27]. Whereas Jag1 is expressed in mural cells in vessels in the decidua, we did not find expression of Notch3 in post-implantation decidual vasculature at E6.5 [27] or E7.5 (data not shown). Thus, endothelial Jag1 and/or Dll4 could activate Notch1 and Notch4 in adjacent ECs to promote Notch signaling in SpA ECs, whereas SpA mural cells which do not express Notch proteins cannot be activated by endothelial Jag1 [39]. We found that loss of EC-specific *Jag1* does not result in increased Dll4 in SpAs in the mesometrial pole, suggesting no compensation or disinhibition of Dll4 with Jag1 loss, as was seen in the anti-mesometrial region capillaries. We also found that loss of EC-specific *Jag1* does not change vascular density or mural cell content in the SpAs. Together, our findings suggest Jag1/Notch signaling in SpA ECs is not essential for maintenance of decidual SpAs from implantation to E7.5. The impact of loss of EC-specific *Jag1* in SpAs may become evident later in gestation, nearing the completion of SpA transformation in the placenta.

Taken together, while we saw no vascular phenotype at E7.5 with loss of EC-specific *Jag1*, our data show that Jag1 and Dll4 mediate Notch signaling activity during decidual angiogenesis. Increased Dll4 in decidual capillary ECs is associated with increased Notch signaling, consistent with loss of inhibitory function of EC Jag1, which has been previously described with EC-specific loss of *Jag1* in retinal angiogenesis and wound healing [17,55]. Thus, our data identify another vascular bed, the newly formed uterine decidual capillaries, in which Jag1 functions as an antagonist of EC Dll4/Notch signaling during angiogenesis.

4. Materials and Methods

4.1. Animals

The Institutional Animal Care and Use Committee at Rutgers-New Jersey Medical School approved the studies (PROTO2018000086; October, 29, 2018). The experiments were designed to determine the role of endothelial Jag1/Notch signaling in the formation and function of uterine decidual blood vessels during early pregnancy development, prior to placentation. To delete Jag1 in the endothelium, *Cdh5*-Cre^{ERT2}, an endothelial specific, tamoxifen inducible *Cre* transgenic strain (gift from Ralf Adams [38]) were crossed with *Jagged1$^{flox/flox}$* mice [39]. To validate our protocol for tamoxifen administration with the *Cdh5-Cre^{ERT2}* driver, tomato reporter mice *ROSA26 tdTomato* (B6.Cg-Gt(ROSA)26Sortm14(CAG-tdTomato)Hze/J, Jackson Laboratories) were used. *Cdh5-Cre^{ERT2}* mice were bred to *Jagged1$^{flox/flox}$* mice or *ROSA26 tdTomato* mice, thereby generating *Cdh5-Cre;Jagged1$^{fl/fl}$* (*Jag1ΔEC*) mutant females and *Cdh5-Cre^{ERT2}*; *ROSA26 tdTomato* reporter females. *Cdh5-Cre^{ERT2}* female mice were used as controls. All strains were maintained on a C57BL/6 background.

Jag1ΔEC or control females, 8–12 weeks of age, were mated to C57BL/6 males for their first pregnancy. Identification of a vaginal plug in the morning was interpreted as successful mating. Noon was designated as embryonic day (E) 0.5. At E4.5, tamoxifen (0.1mg/g body weight (Sigma, Milwakee,

WI, USA) was administered by oral gavage. Since the impact of loss of endothelial Jag1/Notch signaling on ovarian function and progesterone (P4) secretion is not known, P4 replacement was performed. Two extended-release P4 capsules (15 mg P4 per capsule, 21-d release, 4-mm diameter; Innovative Research of America, Sarasota, FL, USA) were placed subcutaneously. Pregnant female mice were euthanized at E7.5 and implantation sites without myometrium were isolated.

4.2. *Quantitative Reverse Transcription-PCR*

The myometrium was removed from implantation sites (IS) and total RNA was extracted from individual control and *Jag1ΔEC* mutant IS using TRIzol (Invitrogen, Carlbad, CA, USA). Total RNA was treated with DNAse I (Invitrogen), reverse transcribed using qScript cDNA Supermix (Quanta Biosciences, VWR, Radnor, PA, USA) and gene expression were determined by quantitative (q) RT-PCR using the QuantiNova SYBR Green PCR Kit (Qiagen, Frederick, MD, USA) and the following primer sequences: *Jagged1* forward 5′-CTGCTTGAATGGGGGTCACT-3′, *Jagged1* reverse 5′-GCAGCTGTCAATCACTTCGC-3′; *Dll4* forward 5′-GTTGCCCTTCAATTTCACCT-3′, *Dll4* reverse 5′-AGCCTTGGATGATGATTTGG-3; *Hey2* forward 5′-AAGCGCCCTTGTGAGGAAAC-3′ *Hey2* reverse 5′-GGTAGTTGTCGGTGAATTGGAC-3′; *Nrarp* forward 5′-GCG TGG TTA TGG GAG AAA GAT-3′, *Nrarp* reverse 5′-GGG AGA GGA AAA GAG GAA TGA-3′. Relative expression levels were quantified using the $2^{-\Delta\Delta Ct}$ method and are expressed as fold change normalized to β-actin forward 5′-CGT GAA AAG ATG ACC CAG ATC-3′ and β-actin reverse 5′-CAC AGC CTG GAT GGC TAC GT-3′.

4.3. *Analysis of Embryo Morphology*

For histologic examination, implantation sites were fixed in Bouin's solution (Sigma), paraffin embedded, section at 10 μm and stained with hematoxylin and eosin (H&E). For each implantation site, all sections having an implantation chamber were assessed by 4 different observers for the presence or absence of morphologic traits, including primitive streak, allantois, cranial neural fold and somites, that reflect the stage of embryonic development [44]. For each pregnant dam, two implantation sites were scored.

4.4. *Immunofluorescence and Immunohistochemistry*

Implantation sites with myometrium attached for immunofluorescence (IF) and immunohistochemistry (IHC) were fixed in 4% paraformaldehyde at 4 °C, infiltrated with 30% sucrose in PBS, embedded in Tissue-Tek®O.C.T™ Compound (Sakura Fine Technical Co. Ltd., VWR, Radnor, PA, USA), and cryosectioned at 7 μm. Interembryonic regions and central parts of the decidua were confirmed by H&E staining of every 5th section. For each implantation site, immunostaining was performed, as previously described [27,45] on at least 3 sections that were equally spaced along central parts of the decidua. The specificity of Notch protein and ligand primary antibodies was previously determined [27,58]. Slides stained with secondary antibody alone served as negative controls for IF and IHC staining. Antibodies are listed in the Supplementary Table S1. For colorimetric IHC, biotinylated anti-goat (Vector BA-5000, 1:400) or biotinylated anti-rat (Vector BA-4001, 1:200), the avidin/biotin blocking kit (Vector SP-2001), the Vectastain ABC kit and DAB substrate kit (Vector SK-4100) were used. For IF, Vectashield containing 40, 6-diamidino-2-phenylindole (DAPI) (Vector) was used for nuclear visualization and mounting. For each experiment, 2–3 implantation sites per pregnant dam were analyzed.

4.5. *Flow Cytometry*

Implantation sites were separated from myometrium and tissue finely minced in 100 uL of ice cold PBS. The finely minced tissue was digested in RPMI 1640 (Gibco) containing 1 mg/mL Collagenase A (Roche) and 0.1 mg/mL DNase I (Roche) for 30 min at 37 °C with horizontal shaking at 250 rpm. Mechanical dissociation by repeated passage through an 18 G needle was performed twice at 15 min intervals during enzymatic digestion. Ice cold RPMI containing 3% FBS (Atlanta Biologicals, R&D

Systems, Minneapolis, MN, USA; RPMI-3)) was added and cells were filtered through a 100 μM nylon cell strainer.

Cells were stained in RPMI-3 for 30 min on ice in the dark with the following fluorochrome-conjugated anti-mouse antibodies: APC/Fire™ 750 anti-CD31 (BioLegend, clone MEC13.3, San Diego, CA, USA) and Super Bright 600 anti-CD45 (eBioscience, clone 30-F11, Carlsbad, CA, USA). Cells were washed twice and then resuspended in RPMI-3 containing 1μg/mL DAPI ((ThermoFisher, Carlsbad, CA, USA) and 2 mM EDTA (Corning, Carlsbad, CA, USA). Flow cytometry was performed using an Attune NxT (ThermoFisher, Carlsbad, CA, USA), LSRII (Becton, Dickinson and Company Sparks, MD, USA), or Fortessa (Becton, Dickinson and Company Sparks, MD, USA) flow cytometer, and analysis was performed using FlowJo™ for Mac v10.6.1 software (Becton, Dickinson and Company Sparks, MD, USA). Live endothelial cells were identified as $DAPI^{-}CD45^{-}CD31^{+}$ cells.

4.6. Imaging

Images of H&E, IF and IHC samples were taken with the Keyence BZ-X710 All-in-One fluorescence microscope (Keyence, Osaka, Japan). Standard filters were used to image DAPI, Alexa Fluor 488 and Alexa Fluor 594. Images were taken using the 10×, 20×, and 40× objectives. Images were captured using Keyence BZ-X Viewer version 01.03 software.

4.7. Morphometric Analyses

Morphometric measurements of protein expression and blood vessel densities were determined on tissue immunostained for endothelial cell marker, CD31, mural cell markers NG2, PDGFRβ and SMA, and Notch ligands Jagged1 or Dll4. The mesometrial region and anti-mesometrial region of each implantation site were identified and further divided into mesometrial pole, defined as the area extending from the internal border of the myometrium to the uterine lumen, and the central region, defined as the areas of the implantation site lateral to and flanking the embryo and implantation chamber (Figure 1A,B). For blood vessel analysis, CD31 signal density was measured in 5 random 0.025 mm^2 areas of the decidua in these three regions using ImageJ Software Version 2.0.0 (mesometrial pole, central region and anti-mesometrial region) [7]. Expression of Jag1 and Dll4 was determined by measurements of the signal densities and divided by $CD31^+$ signal density or NG2+ or $PDGFR\beta^+$ signal density in 5 random 0.025 mm^2 areas of the decidua. To quantify expression of N1ICD with respect to CD31, number of $N1ICD^+$ nuclei was determined for 5 random 0.037 mm^2 areas of anti-mesometrial decidua. The CD31 signal density was determined for each of these areas, and number of nuclei per square micrometer of $CD31^+$ ECs was calculated. For analysis of VEGFR2 in $CD31^+$ ECs, signal densities were evaluated in adjacent sections immunostained with VEGFR2 or CD31. The signal density was measured in 5 random 0.025mm^2 areas of the decidua in like-regions for both VEGFR2 and CD31 and signal densities of VEGFR2 were divided by that of $CD31^+$ cells. Sections from 2–3 implantation sites per each mouse were examined and were used for the analyses.

4.8. Statistics

Statistical analyses were performed with Prism version 8.4.1 (GraphPad, La Jolla, CA, USA). We analyzed these data to identify outliers and removed all outliers prior to performing statistical comparisons. Normality was determined using the Shapiro–Wilk test. Significant differences between medians were determined by unpaired Mann–Whitney U test. For normally distributed data, significant differences between means were determined by unpaired t test and ANOVA. Binary data were compared using Fisher's exact test. Data are presented as median + IQR data or mean ± standard deviation (SD). Statistical significance was defined as $p < 0.05$.

Supplementary Materials: Supplementary Materials can be found at http://www.mdpi.com/1422-0067/21/18/6477/s1.

Author Contributions: Conception, C.J.S., J.K.K., N.C.D., N.M.M.; methodology, A.M.B., C.J.S., N.C.D., N.M.M., S.B.; validation, C.J.S., C.C.Y., N.C.D., N.M.M., T.W., V.O.; formal analysis, C.C.Y., N.C.D., N.M.M.; investigation, C.C.Y., N.C.D., N.M.M., N.V.-P., S.B., T.W., V.O.; resources, A.M.B., J.K.K., N.C.D.; writing—original draft preparation C.J.S., N.C.D., N.M.M., N.V.-P.; writing—review and editing C.J.S., J.K.K., N.C.D., N.M.M.; visualization, N.C.D., N.M.M.; supervision, A.M.B., C.J.S., J.K.K., N.C.D.; funding acquisition, N.C.D. All authors have read and agreed to the published version of the manuscript.

Funding: This work was supported by National Institutes of Health, grant number: R01HL127013.

Acknowledgments: We thank Jason Butler and Ralf Adams for their donations of the experimental mice for generation of our colony. We thank Naiche Adler for her oversight on the development and layout of the project and figures. We thank Sara Morelli and Pranela Rameshwar who provided insightful comments and assistance with design of the project. All schematic representations were created with BioRender.com.

Conflicts of Interest: The authors declare no conflict of interest.

References

1. Cha, J.; Sun, X.; Dey, S.K. Mechanisms of implantation: Strategies for successful pregnancy. *Nat. Med.* **2012**, *18*, 1754–1767. [CrossRef] [PubMed]
2. Wang, H.; Dey, S.K. Roadmap to embryo implantation: Clues from mouse models. *Nat. Rev. Genet.* **2006**, *7*, 185–199. [CrossRef] [PubMed]
3. Charalambous, F.; Elia, A.; Georgiades, P. Decidual spiral artery remodeling during early post-implantation period in mice: Investigation of associations with decidual uNK cells and invasive trophoblast. *Biochem. Biophys. Res. Commun.* **2012**, *417*, 847–852. [CrossRef] [PubMed]
4. Woods, L.; Perez-Garcia, V.; Hemberger, M. Regulation of Placental Development and its Impact on Fetal Growth-New Insights From Mouse Models. *Front. Endocrinol. (Lausanne)* **2018**, *9*, 570. [CrossRef]
5. Douglas, N.C.; Tang, H.; Gomez, R.; Pytowski, B.; Hicklin, D.J.; Sauer, C.M.; Kitajewski, J.; Sauer, M.V.; Zimmermann, R.C. Vascular Endothelial Growth Factor Receptor 2 (VEGFR-2) Functions to Promote Uterine Decidual Angiogenesis During Early Pregnancy in the Mouse. *Endocrinology* **2009**, *150*, 3845–3854. [CrossRef]
6. Ramathal, C.Y.; Bagchi, I.C.; Taylor, R.N.; Bagchi, M.K. Endometrial decidualization: Of mice and men. *Semin. Reprod. Med.* **2010**, *28*, 17–26. [CrossRef]
7. Kim, M.; Park, H.J.; Seol, J.W.; Jang, J.Y.; Cho, Y.S.; Kim, K.R.; Choi, Y.; Lydon, J.P.; Demayo, F.J.; Shibuya, M.; et al. VEGF-A regulated by progesterone governs uterine angiogenesis and vascular remodelling during pregnancy. *EMBO Mol. Med.* **2013**, *5*, 1415–1430. [CrossRef]
8. Plaisier, M. Decidualisation and angiogenesis. *Best Pract. Res. Clin. Obstet. Gynaecol.* **2011**, *25*, 259–271. [CrossRef]
9. Elia, A.; Charalambous, F.; Georgiades, P. New phenotypic aspects of the decidual spiral artery wall during early post-implantation mouse pregnancy. *Biochem. Biophys. Res. Commun.* **2011**, *416*, 211–216. [CrossRef]
10. Silva, J.F.; Serakides, R. Intrauterine trophoblast migration: A comparative view of humans and rodents. *Cell Adhes. Migr.* **2016**, *10*, 88–110. [CrossRef]
11. Bergers, G.; Song, S. The role of pericytes in blood-vessel formation and maintenance. *Neuro. Oncol.* **2005**, *7*, 452–464. [CrossRef] [PubMed]
12. Ozerdem, U.; Grako, K.A.; Dahlin-Huppe, K.; Monosov, E.; Stallcup, W.B. NG2 proteoglycan is expressed exclusively by mural cells during vascular morphogenesis. *Dev. Dyn.* **2001**, *222*, 218–227. [CrossRef] [PubMed]
13. Hellstrom, M.; Gerhardt, H.; Kalen, M.; Li, X.; Eriksson, U.; Wolburg, H.; Betsholtz, C. Lack of pericytes leads to endothelial hyperplasia and abnormal vascular morphogenesis. *J. Cell. Biol.* **2001**, *153*, 543–553. [CrossRef] [PubMed]
14. Cross, J.C.; Hemberger, M.; Lu, Y.; Nozaki, T.; Whiteley, K.; Masutani, M.; Adamson, S.L. Trophoblast functions, angiogenesis and remodeling of the maternal vasculature in the placenta. *Mol. Cell. Endocrinol.* **2002**, *187*, 207–212. [CrossRef]

15. Lima, P.D.; Zhang, J.; Dunk, C.; Lye, S.J.; Croy, B.A. Leukocyte driven-decidual angiogenesis in early pregnancy. *Cell. Mol. Immunol.* **2014**, *11*, 522–537. [CrossRef]
16. Benedito, R.; Hellstrom, M. Notch as a hub for signaling in angiogenesis. *Exp. Cell. Res.* **2013**, *319*, 1281–1288. [CrossRef]
17. Benedito, R.; Roca, C.; Sorensen, I.; Adams, S.; Gossler, A.; Fruttiger, M.; Adams, R.H. The Notch Ligands Dll4 and Jagged1 Have Opposing Effects on Angiogenesis. *Cell* **2009**, *137*, 1124–1135. [CrossRef]
18. Hellstrom, M.; Phng, L.K.; Hofmann, J.J.; Wallgard, E.; Coultas, L.; Lindblom, P.; Alva, J.; Nilsson, A.K.; Karlsson, L.; Gaiano, N.; et al. Dll4 signalling through Notch1 regulates formation of tip cells during angiogenesis. *Nature* **2007**, *445*, 776–780. [CrossRef]
19. Blanco, R.; Gerhardt, H. VEGF and Notch in tip and stalk cell selection. *Cold Spring Harb. Perspect. Med.* **2013**, *3*, a006569. [CrossRef]
20. Krebs, L.T.; Xue, Y.; Norton, C.R.; Shutter, J.R.; Maguire, M.; Sundberg, J.P.; Gallahan, D.; Closson, V.; Kitajewski, J.; Callahan, R.; et al. Notch signaling is essential for vascular morphogenesis in mice. *Genes Dev.* **2000**, *14*, 1343–1352.
21. Krebs, L.T.; Shutter, J.R.; Tanigaki, K.; Honjo, T.; Stark, K.L.; Gridley, T. Haploinsufficient lethality and formation of arteriovenous malformations in Notch pathway mutants. *Genes Dev.* **2004**, *18*, 2469–2473. [CrossRef] [PubMed]
22. Gale, N.W.; Dominguez, M.G.; Noguera, I.; Pan, L.; Hughes, V.; Valenzuela, D.M.; Murphy, A.J.; Adams, N.C.; Lin, H.C.; Holash, J.; et al. Haploinsufficiency of Delta-like 4 ligand results in embryonic lethality due to major defects in arterial and vascular development. *Proc. Natl. Acad. Sci. USA* **2004**, *101*, 15949–15954. [CrossRef] [PubMed]
23. Villa, N.; Walker, L.; Lindsell, C.E.; Gasson, J.; Iruela-Arispe, M.L.; Weinmaster, G. Vascular expression of Notch pathway receptors and ligands is restricted to arterial vessels. *Mech. Dev.* **2001**, *108*, 161–164. [CrossRef]
24. Shutter, J.R.; Scully, S.; Fan, W.; Richards, W.G.; Kitajewski, J.; Deblandre, G.A.; Kintner, C.R.; Stark, K.L. Dll4, a novel Notch ligand expressed in arterial endothelium. *Genes Dev.* **2000**, *14*, 1313–1318. [PubMed]
25. Xue, Y.; Gao, X.; Lindsell, C.E.; Norton, C.R.; Chang, B.; Hicks, C.; Gendron-Maguire, M.; Rand, E.B.; Weinmaster, G.; Gridley, T. Embryonic lethality and vascular defects in mice lacking the Notch ligand Jagged1. *Hum. Mol. Genet.* **1999**, *8*, 723–730. [CrossRef]
26. Liu, H.; Zhang, W.; Kennard, S.; Caldwell, R.B.; Lilly, B. Notch3 is critical for proper angiogenesis and mural cell investment. *Circ. Res.* **2010**, *107*, 860–870. [CrossRef]
27. Shawber, C.J.; Lin, L.; Gnarra, M.; Sauer, M.V.; Papaioannou, V.E.; Kitajewski, J.K.; Douglas, N.C. Vascular Notch proteins and Notch signaling in the peri-implantation mouse uterus. *Vasc. Cell.* **2015**, *7*, 9. [CrossRef]
28. Liu, Z.J.; Shirakawa, T.; Li, Y.; Soma, A.; Oka, M.; Dotto, G.P.; Fairman, R.M.; Velazquez, O.C.; Herlyn, M. Regulation of *Notch1* and *Dll4* by vascular endothelial growth factor in arterial endothelial cells: Implications for modulating arteriogenesis and angiogenesis. *Mol. Cell. Biol.* **2003**, *23*, 14–25. [CrossRef]
29. Phng, L.K.; Gerhardt, H. Angiogenesis: A Team Effort Coordinated by Notch. *Dev. Cell.* **2009**, *16*, 196–208. [CrossRef]
30. Uyttendaele, H.; Marazzi, G.; Wu, G.; Yan, Q.; Sassoon, D.; Kitajewski, J. *Notch4/int-3*, a mammary proto-oncogene, is an endothelial cell-specific mammalian *Notch* gene. *Development* **1996**, *122*, 2251–2259.
31. Leimeister, C.; Schumacher, N.; Gessler, M. Expression of Notch pathway genes in the embryonic mouse metanephros suggests a role in proximal tubule development. *Gene Expr. Patterns* **2003**, *3*, 595–598. [CrossRef]
32. Kofler, N.M.; Cuervo, H.; Uh, M.K.; Murtomaki, A.; Kitajewski, J. Combined deficiency of Notch1 and Notch3 causes pericyte dysfunction, models cadasil, and results in arteriovenous malformations. *Sci. Rep.* **2015**, *5*, 16449. [CrossRef] [PubMed]
33. Loomes, K.M.; Taichman, D.B.; Glover, C.L.; Williams, P.T.; Markowitz, J.E.; Piccoli, D.A.; Baldwin, H.S.; Oakey, R.J. Characterization of Notch receptor expression in the developing mammalian heart and liver. *Am. J. Med. Genet.* **2002**, *112*, 181–189. [CrossRef]
34. Joutel, A.; Andreux, F.; Gaulis, S.; Domenga, V.; Cecillon, M.; Battail, N.; Piga, N.; Chapon, F.; Godfrain, C.; Tournier-Lasserve, E. The ectodomain of the Notch3 receptor accumulates within the cerebrovasculature of CADASIL patients. *J. Clin. Investig.* **2000**, *105*, 597–605. [CrossRef] [PubMed]

35. Krebs, L.T.; Deftos, M.L.; Bevan, M.J.; Gridley, T. The *Nrarp* Gene Encodes an Ankyrin-Repeat Protein That Is Transcriptionally Regulated by the Notch Signaling Pathway. *Dev. Biol.* **2001**, *238*, 110–119. [CrossRef]
36. Schweisguth, F. Regulation of Notch Signaling Activity. *Curr. Biol.* **2004**, *14*, R129–R138. [CrossRef]
37. Nandagopal, N.; Santat, L.A.; Elowitz, M.B. *Cis*-activation in the Notch signaling pathway. *Elife* **2019**, *8*, e37880. [CrossRef]
38. Del Alamo, D.; Rouault, H.; Schweisguth, F. Mechanism and significance of *cis*-inhibition in Notch signaling. *Curr. Biol.* **2011**, *21*, R40–R47. [CrossRef]
39. Levin, H.I.; Sullivan-Pyke, C.S.; Papaioannou, V.E.; Wapner, R.J.; Kitajewski, J.K.; Shawber, C.J.; Douglas, N.C. Dynamic maternal and fetal Notch activity and expression in placentation. *Placenta* **2017**, *55*, 5–12. [CrossRef]
40. Sorensen, I.; Adams, R.H.; Gossler, A. DLL1-mediated Notch activation regulates endothelial identity in mouse fetal arteries. *Blood* **2009**, *113*, 5680–5688. [CrossRef]
41. Brooker, R.; Hozumi, K.; Lewis, J. Notch ligands with contrasting functions: Jagged1 and Delta1 in the mouse inner ear. *Development* **2006**, *133*, 1277–1286. [CrossRef] [PubMed]
42. Lindblom, P.; Gerhardt, H.; Liebner, S.; Abramsson, A.; Enge, M.; Hellstrom, M.; Backstrom, G.; Fredriksson, S.; Landegren, U.; Nystrom, H.C.; et al. Endothelial PDGF-B retention is required for proper investment of pericytes in the microvessel wall. *Genes Dev.* **2003**, *17*, 1835–1840. [CrossRef] [PubMed]
43. Lertkiatmongkol, P.; Liao, D.; Mei, H.; Hu, Y.; Newman, P.J. Endothelial functions of PECAM-1 (CD31). *Curr. Opin. Hematol.* **2016**, *23*, 253–259. [CrossRef] [PubMed]
44. Owens, G.K. Regulation of differentiation of vascular smooth muscle cells. *Physiol. Rev.* **1995**, *75*, 487–517. [CrossRef] [PubMed]
45. Vorontchikhina, M.A.; Zimmermann, R.C.; Shawber, C.J.; Tang, H.; Kitajewski, J. Unique patterns of Notch1, Notch4 and Jagged1 expression in ovarian vessels during folliculogenesis and corpus luteum formation. *Gene Expr. Patterns* **2005**, *5*, 701–709. [CrossRef]
46. Downs, K.M.; Davies, T. Staging of gastrulating mouse embryos by morphological landmarks in the dissecting microscope. *Development* **1993**, *118*, 1255–1266.
47. Fouillade, C.; Monet-Lepretre, M.; Baron-Menguy, C.; Joutel, A. Notch signalling in smooth muscle cells during development and disease. *Cardiovasc. Res.* **2012**, *95*, 138–146. [CrossRef]
48. Stratman, A.N.; Pezoa, S.A.; Farrelly, O.M.; Castranova, D.; Dye, L.E., 3rd; Butler, M.G.; Sidik, H.; Talbot, W.S.; Weinstein, B.M. Interactions between mural cells and endothelial cells stabilize the developing zebrafish dorsal aorta. *Development* **2017**, *144*, 115–127. [CrossRef]
49. Carmeliet, P.; Moons, L.; Luttun, A.; Vincenti, V.; Compernolle, V.; De Mol, M.; Wu, Y.; Bono, F.; Devy, L.; Beck, H.; et al. Synergism between vascular endothelial growth factor and placental growth factor contributes to angiogenesis and plasma extravasation in pathological conditions. *Nat. Med.* **2001**, *7*, 575–583. [CrossRef]
50. Dufraine, J.; Funahashi, Y.; Kitajewski, J. Notch signaling regulates tumor angiogenesis by diverse mechanisms. *Oncogene* **2008**, *27*, 5132–5137. [CrossRef]
51. Shawber, C.J.; Kitajewski, J. Notch function in the vasculature: Insights from zebrafish, mouse and man. *Bioessays* **2004**, *26*, 225–234. [CrossRef] [PubMed]
52. Burri, P.H.; Hlushchuk, R.; Djonov, V. Intussusceptive angiogenesis: Its emergence, its characteristics, and its significance. *Dev. Dyn.* **2004**, *231*, 474–488. [CrossRef] [PubMed]
53. Garcia-Pascual, C.M.; Ferrero, H.; Zimmermann, R.C.; Simon, C.; Pellicer, A.; Gomez, R. Inhibition of Delta-like 4 mediated signaling induces abortion in mice due to deregulation of decidual angiogenesis. *Placenta* **2014**, *35*, 501–508. [CrossRef] [PubMed]
54. Gerhardt, H.; Betsholtz, C. Endothelial-pericyte interactions in angiogenesis. *Cell Tissue Res.* **2003**, *314*, 15–23. [CrossRef] [PubMed]
55. Pedrosa, A.R.; Trindade, A.; Fernandes, A.C.; Carvalho, C.; Gigante, J.; Tavares, A.T.; Dieguez-Hurtado, R.; Yagita, H.; Adams, R.H.; Duarte, A. Endothelial Jagged1 antagonizes Dll4 regulation of endothelial branching and promotes vascular maturation downstream of Dll4/Notch1. *Arterioscler. Thromb. Vasc. Biol.* **2015**, *35*, 1134–1146. [CrossRef]
56. Travisano, S.I.; Oliveira, V.L.; Prados, B.; Grego-Bessa, J.; Pineiro-Sabaris, R.; Bou, V.; Gomez, M.J.; Sanchez-Cabo, F.; MacGrogan, D.; de la Pompa, J.L. Coronary arterial development is regulated by a Dll4-Jag1-EphrinB2 signaling cascade. *Elife* **2019**, *8*, e49977. [CrossRef]

57. Rennie, M.Y.; Whiteley, K.J.; Adamson, S.L.; Sled, J.G. Quantification of gestational changes in the uteroplacental vascular tree reveals vessel specific hemodynamic roles during pregnancy in mice. *Biol. Reprod.* **2016**, *95*, 43. [CrossRef]
58. Murtomaki, A.; Uh, M.K.; Choi, Y.K.; Kitajewski, C.; Borisenko, V.; Kitajewski, J.; Shawber, C.J. Notch1 functions as a negative regulator of lymphatic endothelial cell differentiation in the venous endothelium. *Development* **2013**, *140*, 2365–2376. [CrossRef]

International Journal of *Molecular Sciences*

Article

Effect of Placenta-Derived Mesenchymal Stromal Cells Conditioned Media on an LPS-Induced Mouse Model of Preeclampsia

Anna Maria Nuzzo [1], Laura Moretti [1], Paolo Mele [2], Tullia Todros [1], Carola Eva [2] and Alessandro Rolfo [1,*]

1 Department of Surgical Sciences, University of Turin, 10126 Turin, Italy; a.nuzzo@unito.it (A.M.N.); l.moretti@unito.it (L.M.); tullia.todros@unito.it (T.T.)
2 Department of Neurosciences, Neurosciences Institute Cavalieri Ottolenghi (NICO), San Luigi Hospital, University of Turin, Orbassano, 10043 Turin, Italy; paolo.mele@unito.it (P.M.); carola.eva@unito.it (C.E.)
* Correspondence: alessandro.rolfo@unito.it; Tel.: +39-01-1670-7804

Abstract: We tested the pro-angiogenic and anti-inflammatory effects of human placenta-derived mesenchymal stromal cells (hPDMSCs)-derived conditioned media (CM) on a mouse model of preeclampsia (PE), a severe human pregnancy-related syndrome characterized by maternal hypertension, proteinuria, endothelial damage, inflammation, often associated with fetal growth restriction (FGR). At d11 of pregnancy, PE was induced in pregnant C57BL/6N mice by bacterial lipopolysaccharide (LPS) intravenous injection. At d12, 300 μL of unconditioned media (control group) or 300 μL PDMSCs-CM (CM group) were injected. Maternal systolic blood pressure was measured from 9 to 18 days of pregnancy. Urine protein content were analyzed at days 12, 13, and 17 of pregnancy. At d19, mice were sacrificed. Number of fetuses, FGR, fetal reabsorption, and placental weight were evaluated. Placentae were analyzed for sFlt-1, IL-6, and TNF-α gene and protein expressions. No FGR and/or reabsorbed fetuses were delivered by PDMSCs-CM-treated PE mice, while five FGR fetuses were found in the control group accompanied by a lower placental weight. PDMSCs-CM injection significantly decreased maternal systolic blood pressure, proteinuria, sFlt-1, IL-6, and TNF-α levels in PE mice. Our data indicate that hPDMSCs-CM can reverse PE-like features during pregnancy, suggesting a therapeutic role for hPDMSCs for the treatment of preeclampsia.

Keywords: placenta-derived mesenchymal stromal cells; preeclampsia; mouse model; placenta

Citation: Nuzzo, A.M.; Moretti, L.; Mele, P.; Todros, T.; Eva, C.; Rolfo, A. Effect of Placenta-Derived Mesenchymal Stromal Cells Conditioned Media on an LPS-Induced Mouse Model of Preeclampsia. *Int. J. Mol. Sci.* **2022**, *23*, 1674. https://doi.org/10.3390/ijms23031674

Academic Editor: Luisa Campagnolo

Received: 27 December 2021
Accepted: 28 January 2022
Published: 31 January 2022

Publisher's Note: MDPI stays neutral with regard to jurisdictional claims in published maps and institutional affiliations.

1. Introduction

The preeclamptic syndrome (PE), exclusive to human pregnancy, represents the main cause of fetal–maternal mortality and morbidity worldwide [1,2]. PE generally resolves at delivery with placenta removal, but it causes severe long-term complications for both the mother and the fetus, such as cardiovascular and neurological disorders, diabetes, and metabolic syndrome [3]. Despite almost three decades of intensive investigation, PE still remains an unsolved medical need.

Indeed, preeclampsia has a major social–economic impact due to the lack of effective therapies, except for a timely and often premature delivery. There were several unsuccessful attempts to find a resolutive cure for PE, ranging from new drug candidates to drug relocation. The main problem with preeclampsia is that, as a syndrome, it is multifactorial, a destructive mix of inflammation, endothelial damage, and immunological impairment [2,4]. Key features of PE are a maternal immune maladaptation towards the fetoplacental district with a shift towards Th1 immunity [5–7], increased placental release of proinflammatory cytokines (e.g., Tumor Necrosis Factor-α—TNF-α; Interleukin-6–IL-6), and anti-angiogenic factors (e.g., soluble FMS-like tyrosine kinase-1—sFlt-1) that promote aberrant placental angiogenesis and generalized endothelial cell activation and damage [8–11]. Therefore, an ideal PE therapeutic approach must be able to contemporarily target all preeclamptic culprits and not to just mitigate a single clinical symptom as hypertension or inflammation.

The human placenta has been identified as a source of mesenchymal stromal cells (placenta-derived mesenchymal stromal cells—PDMSCs). PDMSCs could be isolated from the chorionic villi, the amnion, and the decidua and possess an increased self-renewal potential. Moreover, PDMSCs express stem cell markers (e.g., OCT-4, NANOG) and could differentiate into condrogenic, adipogenic, and osteogenic lineages [12–14]. Importantly, PDMSCs are characterized by unique immunologic and immune-regulatory properties, thus exerting a powerful immunosuppressive effect on T-cells [15–18]. Placental MSCs have been shown to promote angiogenic growth and to possess anti-inflammatory, anti-fibrotic, and cytoprotective abilities mediated by both direct cell-to-cell contact and/or specific trophic mediators more than cell differentiation [15,19].

Thus, PDMSCs may be an attractive therapeutic candidate for PE treatment. Recently, decidual MSCs were injected in vivo in a Th1 cell-induced PE-like mouse model demonstrating the ability to ameliorate PE-like symptoms as blood pressure and proteinuria [20]. In line with these results, an endotoxin-induced PE rat model infused with umbilical cord blood-derived MSCs showed decreased blood pressure, proteinuria, and inflammation relative to untreated controls [21]. Finally, commercially available placental mesenchymal cells were administered to hypertensive TLRs-induced pregnant mice, decreasing blood pressure, placental injury, and inflammation [22]. MSC-based therapy definitely sounds an intriguing potential multitarget therapeutic tool for preeclampsia.

Nevertheless, it could be hazardous to hypothesize a cell therapy for such a sensitive and delicate environment as human pregnancy for both ethical and biosafety reasons. No long-term studies on MSCs oncogenic potentials are available, and data about MSCs ability to invade maternal organs are contrasting [21,23–27].

Since mesenchymal stromal cells exerts their beneficial effects mainly through the release of trophic mediators, in the present study we tested the hypothesis that PDMSCs' conditioned media (CM) could be used as an effective, ethical, and safe therapeutic approach for preeclampsia. Therefore, we evaluated the effects of PDMSCs-CM administration on maternal blood pressure, proteinuria, fetal outcome, and placental expression of sFlt-1, TNF-α, and IL-6 in an LPS-induced mouse model of preeclampsia.

2. Results

2.1. PDMSCs Presented Proper Mesenchymal Stromal Cell Profile

PDMSCs used for CM preparation presented proper mesenchymal stromal phenotype as assessed by flow cytometry. As previously published, cells were positive for CD105, CD166, CD90, and CD73, and negative for HLA-II, CD34 and CD45 (hematopoietic markers), and CD133 and CD31 (endothelial progenitor markers). PDMSCs were also negative for B cells, neutrophils, and macrophages markers CD20 and CD14 and for trophoblast and epithelial marker CD326, thus excluding any type of contamination [19,28]. RT-PCR detected the expression of typical stemness markers Oct-4 and Nanog in all PDMSCs cell lines [19,28].

2.2. Characteristics of the Study Population

Female pregnant mice from both control and CM group did not display significant differences in body weight after LPS injection and treatment with unconditioned/conditioned media (CM = median 27.4 g; controls: median 26.3 g). Mice from the control group showed adverse pregnancy outcomes, including fetal absorption (n = 5) and significant lower placental weight ($p < 0.01$), compared with PDMSCs-CM group (Table 1). Moreover, one case of miscarriage was observed in the control group. No significant differences were observed in fetal weight (Table 1).

Table 1. Effect of PDMSCs-CM treatment on clinical and biochemical parameters of mice in PDMSCs-CM and control groups.

	PDMSCs-CM (n = 5)	Control (n = 5)	*p*-Value
Number of fetuses	41	24	$p < 0.01$
Fetal reabsorption	0	5	$p = 0.02$
Fetal weight, grams (median and range)	0.82 (0.62–1.26)	0.75 (0.59–0.99)	ns
Placental weight, grams (median and range)	0.12 (0.07–0.25)	0.09 (0.05–0.14)	$p < 0.01$
Hematocrit (%)	11.1	10.8	ns
RBC	7.2	7.08	ns
WBC	1.1	1.2	ns
Plt	330	135	ns
Htc (%)	10.4	10.3	ns
Hb	10.1	9.9	ns
ALT (mg/dL)	39.7	39.7	ns
AST (mg/dL)	247.7	263.7	ns
Urea (mg/dL)	35.2	34.2	ns
Creatinine (mg/dL)	0.07	0.07	ns

Significant main effect of PDMSCs-CM treatment on number of fetus, fetal reabsorption, and placental weight. Data are expressed as means ± SEM. ns: not significant.

RBC (red blood cells) and WBC (white blood cells) count, Htc (hematocrit), and hemoglobin concentration did not change between groups but there was a trend of decrease in Plt (platelets) count in control relative to CM-treated mice (Table 1).

2.3. PDMSCs Conditioned Media Ameliorated Maternal Hypertension and Proteinuria in LPS-Induced PE Mouse Model

We first investigated whether LPS injection was able to induce maternal hypertension. Average maternal basal SBP at days 9–11 was 93.3 ± 1.3 mmHg and it significantly increased to 101.3 ± 1.4 mmHg at day 12 ($p = 0.04$), 24 h after LPS injection (Figure 1B). We next examined the effect of plain media or PDMSCs-CM injection in LPS-induced hypertensive pregnant mice. In control mice (LPS + plain media), maternal SBP continued to increase at day 13 (104.4 ± 1.9) and day 18 (113 ± 2.3), while in CM pregnant females (LPS + PDMSCs-CM), SBP significantly decreased at days 13 (95.6 ± 0.63 mmHg, $p < 0.01$), 15 (95.7 ± 1.2 mmHg, $p < 0.01$), 16 (95.7 ± 2.1 mmHg, $p = 0.03$), and 18 (101.2 ± 2.3 mmHg, $p < 0.01$) relative to control mice (Figure 1B).

On day 12, after LPS injection and before CM or plain media injection, mean mice urine protein concentration was 0.19 ± 0.04 μg/μL. In LPS pregnant females treated by plain media, proteinuria showed a trend of decrease from day 13 to 17 relative to day 12, even though it was not significant and less dramatic compared to CM-treated mice (Figure 1C). Proteinuria decreased on d13 (0.12 ± 0.04, $p > 0.05$), d15 (0.04 ± 0.02, $p = 0.016$), and d17 (0.06 ± 0.02, $p = 0.045$) relative to day 12 CM mice (Figure 1C).

Finally, in order to investigate the effects of plain media or PDMSCs-CM infusion on liver and renal functions in LPS-induced hypertensive pregnant mice, we tested serum levels of selected parameters. On day 19, no differences were found in CM group relative to control in AST (aspartate aminotransferase) (39.75 ± 7.31 mg/dL versus 39.75 ± 5.17, $p > 0.05$) and ALT (alanine transaminase) (247.75 ± 62.88 mg/dL versus 263.75 ± 34.47 mg/dL, $p > 0.05$) levels, used as markers of liver functionality, nor in creatinine (0.0175 ± 0.01 mg/dL versus 0.07 ± 0.01 mg/dL, $p > 0.05$) and urea levels (35.25 ± 5.72 versus 34.25 ± 2.28 mg/dL, $p > 0.05$), used as markers of kidney function.

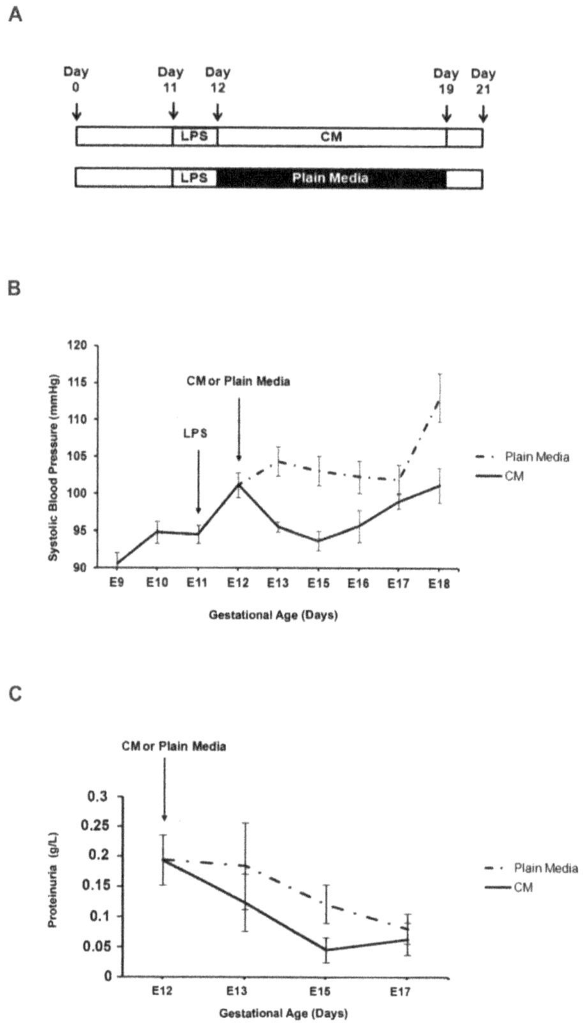

Figure 1. Effects of PDMSCs-CM treatment on maternal parameters during days 11 to 18 of gestation. (**A**) Study design. Blood pressure (**B**) and proteinuria (**C**) in LPS-induced PE mouse model injected with PDMSCs-CM or plain media.

2.4. Placental sFlt-1, TNF-α, and IL-6 Expression Were Inhibited by PDMSCs-CM in PE Mice

In order to determine if PDMSCs-CM treatment was effective also at the placental level, we evaluated placental expression of sFlt-1, TNF-α, and IL-6, key hallmarks of preeclampsia, in CM and control mice. We found a significant reduction of mRNA expression for the three molecular targets investigated, namely sFlt-1 (5.5-fold decrease, $p = 0.04$), TNF-α (6.2-fold decrease, $p = 0.03$), and IL-6 (5-fold decrease, $p = 0.03$) in CM mice compared to controls (Figure 2A). Decreased sFlt-1 ($p = 0.013$, 1.29-fold decrease) and IL-6 ($p = 0.034$, 1.14-fold decrease) expression in the placentae of CM mice, compared to controls, was confirmed also at the protein level (Figure 2B). No significant differences in TNF-α protein level were found in CM compared to control placentae (Figure 2B).

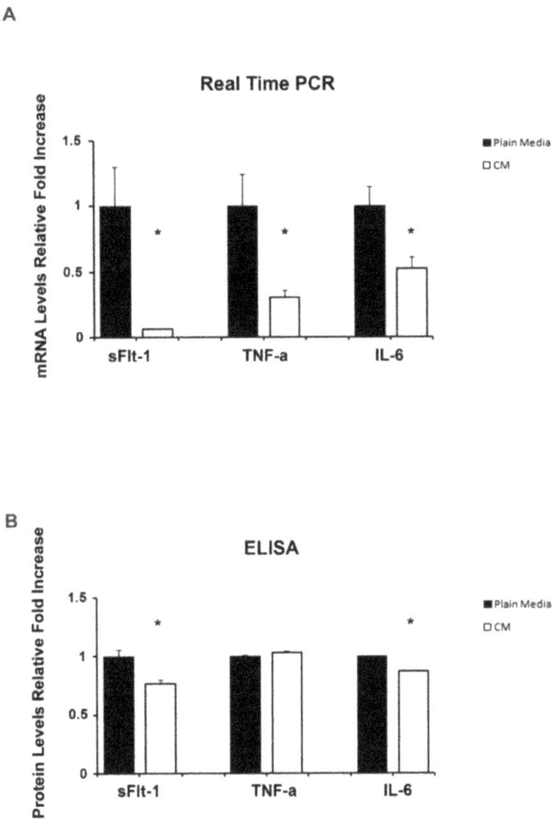

Figure 2. Effects of PDMSCs-CM treatment on placental sFlt-1, TNF-α, and IL-6 expression. Gene (**A**) and protein (**B**) sFlt-1, TNF-α, and IL-6 expression in placentae from LPS-induced PE mouse model injected with PDMSCs-CM or plain media. Statistical significance (*) has been considered as $p < 0.05$.

3. Discussion

Although the etiology of preeclampsia is still elusive, recent evidence has demonstrated that MSCs of different origins, when directly infused in animal models, are able to ameliorate PE-like symptoms, thus suggesting their potential role as therapeutic agents [20,21,24]. Nevertheless, systemic administration of living cells implies significant biosafety and ethical issues that must be considered when designing therapies for such a sensitive environment as pregnancy. Intravenously injected GFP-labeled umbilical cord MSCs were detected in the renal parenchyma and placenta of PE pregnant rats and in fetal kidneys, liver, lungs, and heart [21]. Moreover, locally administered cells often die before they significantly contribute to the healing response due to poor diffusion of nutrients and oxygen [29]. Alternative approaches are therefore mandatory.

The conditioned media obtained from MSCs consist of biologically active molecules whose function is to simultaneously modulate key biological mechanisms such as inflammation and immune response, angiogenesis, cell proliferation, apoptosis, and senescence [19,30–33]. Therefore, MSCs-derived CM can be investigated as an alternative approach to cell therapy.

MSCs have been successfully isolated from a variety of tissues, including adipose tissue, pancreas, and umbilical cord blood [34–36]. In the present study, we used physiological MSCs derived from placental chorionic villi because of their immunosuppressive, pro-angiogenic, anti-inflammatory, and cytoprotective activities potentially useful against the aberrant placentation typical of PE [19,31,35,37–39]. In our previous works we found that PDMSCs-CM is able to modulate in vitro the expression of inflammatory cytokines, angiogenic factors, senescence markers, and cell cycle modulators in the placental villi [19,31].

Herein, we performed intravenous PDMSCs-CM administration to investigate the paracrine effects of placental mesenchymal cells in an LPS-induced mouse model of PE. LPS was chosen to mimic PE symptoms since it induces generalized endothelial dysfunction via the activation of inflammatory pathways [40]. In accordance with previous reports, in our model, LPS exposure led to gestational hypertension and proteinuria [24,41,42], as observed in PE pregnancies. Importantly, we demonstrated that systemic administration of PDMSC-CM, and not of living cells, significantly reduced LPS-induced gestational hypertension and proteinuria.

In line with our data, other groups reported the efficacy of MSCs in reducing hypertension in vivo [22,24,43]. It was suggested that the mechanism by which MSCs may ameliorate hypertension is through the modulation of endothelium-derived factors that control vasodilatation, vasoconstriction, and microvascular density increase [44]. In two-kidney, one-clip rats, MSCs minimized hypertension at least in part by interfering with sympathetic nerve activity leading to reduction of sympathetic activity in the cardiovascular system [45].

As mentioned above, our results were obtained by injecting PDMSCs conditioned media and not living cells, thus opening the door to less invasive, more ethical, and safe therapeutic approaches for preeclampsia. PDMSCs-CM composition is complex and it includes free proteins, small molecules, and extracellular vesicles which can be further divided into apoptotic bodies, microparticles, and exosomes [46,47]. We previously published PDMSCs-CM partial characterization performed by cytokine array technology [19], demonstrating the presence of cytokines and chemokines in the conditioned media. In particular, we described Interleukin 8 (IL-8), Osteopontin, Tissue Inhibitor of Metalloproteinases-2 (TIMP-2), Neutrophil Activating Peptide 2 (NAP-2), Monocyte Chemotactic Protein-1 (MCP-1), Osteoprotegerin, Transforming Growth Factor-b2 (TGF-b2), Interferon-inducible protein-10 (IP-10), GRO, Vascular Endothelial Growth Factor (VEGF), Placental Growth Factor (PlGF) and Interleukin 10 (IL-10) as components of physiological PDMSCs-CM [19]. Despite that some of these molecules were classified as proinflammatory and Th1 mediators increased in PDMSCs and maternal blood from PE patients, they are also pivotal for physiological embryo implantation, placentation, and maternal–placental vascular remodeling [48–50]. For example, IL-8 is able to specifically counteract inflammation at the endothelial level [51], while TIMP-2 and IL-10 are powerful anti-inflammatory molecules, suggesting a multifactorial action directed against the exacerbated inflammation typical of PE. Moreover, since VEGF/sFlt-1 unbalance is widely accepted as the main trigger for the endothelial dysfunction, leading to hypertension in preeclampsia [52–54], we hypothesize that PDMSCs-CM counteracted endothelial dysfunction in our LPS-induced PE model via pro-angiogenic VEGF modulation and anti-angiogenic sFlt-1 inhibition. This hypothesis is in line with our previous findings showing that PDMSCs-CM promoted placental VEGF accumulation and sFlt-1 downregulation in physiological human villous explants [19].

Next, we described that PDMSCs-CM administration induced a significant reduction in LPS-induced urinary protein excretion. Similar effects were observed when human placental expanded (PLX-PAD) mesenchymal-like cells were injected in two models of innate immunity-induced PE, thus confirming PDMSCs potential to reduce PE-associated proteinuria [22]. Chatterjee et al. suggested that the release of paracrine factors led to the observed decrease in oxidative stress, angiogenesis, inflammation, and endothelial dysfunction, without the need of cell–cell contact. Since endothelial damage underlies many PE manifestations, including proteinuria [55], our data are in line with the above mentioned results, confirming the ability of human PDMSCs to counteract vascular dysfunction throughout the release of trophic mediators, as VEGF that it is able to stimulate endothelial repair and reduce circulating sFlt-1 levels. In our model, LPS-induced proteinuria slowly decreased with advancing gestation also in animals infused with plain media, even if to a significantly lesser extent relative to CM animals. This effect is likely due to spontaneous LPS clearance and consequent inflammation remission.

Moreover, we reported that LPS followed by plain media infusion in pregnant mice induced reduction in fetuses' number, fetal absorption, and decrease in fetal and placental weight, thus mimicking placenta development anomalies typical of PE. Similar results were reported by Rivera et al., showing that a 100 μg/kg day LPS for 7 days in pregnant rats significantly reduced fetal size and increased fetal demise [56]. In stark contrast, we demonstrated that PDMSCs-CM treatment resulted in a higher number of fetuses, increased fetal–placental weight, and no fetal absorption. These outcomes are likely due to decreased endothelial damage and placental inflammation as demonstrated by placental sFlt-1, TNF-a, and IL-6 downregulation, reduced maternal blood pressure and proteinuria derived from improved placental functionality, and fetal nutritional status as previously suggested [43,57]. Importantly, we specifically investigated TNF-α, Il-6, and sFlt-1 because they are key players in PE pathogenesis [58] and severe endothelial dysfunction [54,59].

To investigate the impact of PDMSCs-CM on liver and renal functions, serum AST-ALT and creatinine–urea levels were monitored. It was previously described that after MSCs injection (e.g., bone marrow MSC, hematopoietic stem cells, umbilical cords MSC, amniotic fluid MSCs), AST, ALT, creatinine, and urea were normalized to physiological levels restoring compromised liver and kidneys activities [60–62]. In our model, AST, ALT, creatinine, and urea serum levels were not modified by LPS injection nor by PDMSCs-CM/plain media infusion, in line with data reported by Oludare and colleagues that described no significant differences in liver enzymes' levels in LPS mice compared to controls [63]. Importantly, our results demonstrated that PDMSCs-CM did not affect liver and kidney functionalities, thus excluding adverse effects.

In conclusion, our findings demonstrated that human PDMSCs-CM administration significantly ameliorated PE-like symptoms and improved fetal–placental outcomes in pregnant LPS mice. Our data strongly suggested that PDMSCs-CM acted through the restoration of endothelial function and the suppression of the proinflammatory cascade, thus contrasting systemic and placental injury. A limitation of our study was that we have not yet completed PDMSCs conditioned media characterization; therefore, we could not indicate its exact mechanism of action. Due to CM complexity, different pathways were most likely promoted and/or suppressed by PDMSCs trophic mediators, thus explaining the multitarget therapeutic activity demonstrated by our results. Indeed, PDMSCs-CM-based therapy could be considered a promising tool due to its ability to simultaneously target different drivers of the preeclamptic syndrome. This approach could minimize the biological variability, biosafety, and ethical issues of cell-based therapies, thus leading to the development of a safe and ethical cell-free strategy against preeclampsia. Further investigations are required.

4. Materials and Methods

4.1. PDMSCs Conditioned Media Preparation

PDMSCs-CM was prepared as previously described [19,31]. Briefly, placentae from healthy women with a singleton physiological pregnancy were collected immediately after delivery. Physiological pregnancy was defined as term normotensive pregnancy and no signs of preeclampsia or FGR. Exclusion criteria were congenital malformations, chromosomal abnormalities (in number and/or structure), maternal and/or intrauterine infections, cardiovascular diseases, metabolic syndrome, diabetes, and immunological disorders.

PDMSCs were isolated by enzymatic digestion and gradient as previously described [19,28,31]. PDMSCs were next resuspended in Dulbecco's Modified Minimum Essential Medium (DMEM, Gibco, Life Technologies, Monza, Italy) supplemented with 10% fetal bovine serum (FBS Australian origin, Life Technologies, Monza, Italy) and maintained at 37 °C and 5% CO_2. At every passage, physiological PDMSCs were characterized by flow cytometry for the expression of the following antigens: HLA-I, HLA-DR, CD105, C166, CD90, CD34, CD73, CD133, CD20, CD326, CD31, CD45, and CD14 (Miltenyi Biotech, Bologna, Italy). PDMSCs were analyzed by semiquantitative PCR to assess gene expression levels of stem cell markers Oct-4 and Nanog. Primers were designed as previously described [19].

At passage three of culture, after obtaining a pure PDMSCs population, cells were plated and expanded in 1264 cm^2 EasyFill cell factories (Carlo Erba, Cornaredo (MI), Italy) at a concentration of 3 × 106 cells. When cells reached confluency, media was removed and replaced by 400 mL of DMEM LG without FBS. After 48 h of culture, CM was collected, filtered, and stored at −20 °C until use.

4.2. Preeclamptic Mouse Model Preparation and PDMSCs-CM Treatment

The preeclamptic mouse model was prepared following a modified protocol from Wang et al. [57]. Briefly, C57BL/6NCrl virgin mice females (n = 30) and males (n = 10) at 4 weeks of age were purchased from Charles River Laboratories (Calco (LC), Italy). All mice were maintained on a 12 h/12 h dark and light cycle with relative humidity of 50–70% at 18–22 °C. Tap water and standard laboratory pelleted formula were provided.

Female mice were mated with males at 9 weeks of age and plug discovery was considered as day 0 of pregnancy. Female mice lacking copulation plugs (n = 20) were returned to the breeding colony. At day 11 of pregnancy, all pregnant females (n = 10) were removed from breeding cages and received intravenous tail injection of 1 µg/kg LPS solution in order to induce inflammation-mediated endothelial damage and hypertension. At day 12, after blood pressure measurements, mice were randomized into two groups as follows: (1) animals that received a single intravenous tail injection of 300 µL of plain unconditioned media (control group, n = 5); (2) animals that received a single intravenous tail injection of 300 µL PDMSCs-CM (CM group, n = 5) (Figure 1A). LPS was purchased from Sigma Aldrich (Milan, Italy). PDMSCs-CM was prepared as described above.

Maternal systolic blood pressure (SBP) was monitored by tail cuff plethysmography by using BP-2000 Series II Blood Pressure Analysis System, 2 channels mouse platform (Visitech Systems, Napa Pl, Apex, United States) from day 9 to 18 of pregnancy. Urine samples were collected at days 12, 13, and 17 of pregnancy and analyzed for protein content by Bradford assay (Sigma Aldrich, Milan, Italy). Mice were sacrificed by cervical dislocation at day 19 of pregnancy and uteri were removed. Maternal blood samples were taken from mice carotid and collected in heparin tubes to determine the following hematological parameters: RBC count, WBC count, Plt count, Htc, and Hb. All analyses were performed using a veterinary hematology analyzer. Serum was separated and used for measuring ALT, AST, urea, and creatinine. Placental weights, number of fetuses, and fetal weights were recorded. Placentae were collected and stored at −80 °C until the next molecular analysis.

4.3. RNA Isolation and Real-Time PCR

Total RNA was isolated from PDMSCs-CM-treated and untreated PE placentae using TRIzol reagent (Life Technologies, Invitrogen, Monza, Italy) according to manufacturer instructions. Genomic DNA contamination was removed by DNAse I digestion before RT-PCR. CDNA was generated from 5 µg of total RNA using a random hexamers approach and RevertAid H Minus First Strand cDNA Synthesis kit (Life Technologies, Monza, Italy).

Gene expressions levels of sFlt-1, TNF-α, and IL-6 were determined by real-time PCR using specific TaqMan primers and probes (Life Technologies, Monza, Italy). MRNA levels were normalized using endogenous 18 s as internal reference (Life Technologies, Monza, Italy). Relative expression and fold change were calculated according to Livak and Schmittgen [64].

4.4. Enzyme-Linked Immunosorbent Assay (ELISA)

Total proteins were isolated from PDMSCs-CM-treated and untreated PE placentae using 1X radio immunoprecipitation assay (RIPA) buffer. Quantitative measurement of sFlt-1 (R&D System, Milan, Italy), TNF-α (RayBiotech, Prodotti Gianni, Milan, Italy), and IL-6 (Abcam, Milan, Italy) placental levels were determined using commercially available competitive ELISA kits according to manufacturer's instruction.

4.5. Statistical Analysis

All data are represented as mean $\pm$ standard error (SE). For comparison of data between multiple groups, one-way analysis of variance (ANOVA) with post hoc Dunnett's test was used. For comparison between two groups, Mann–Whitney U-test was used as appropriate. Fisher's exact test was used for small sample sizes. Statistical analysis was carried out using SPSS Version 23 statistical software, and significance was accepted at $p < 0.05$.

Author Contributions: A.R. and C.E. conceived and designed the study; A.M.N., L.M. and P.M. performed the experiments; A.R. and A.M.N. analyzed the data and performed statistical analysis; T.T. selected and recruited patients population for PDMSCs isolation; A.R., A.M.N. and C.E. wrote and edited the article; L.M., P.M. and T.T. reviewed the article. All authors have read and agreed to the published version of the manuscript.

Funding: A.R. was supported by Carlo Denegri Foundation. Corion Biotech S.R.L. (Turin, Italy) funded the study.

Institutional Review Board Statement: Placental collection and PDMSCs isolation was approved by the Institutional Ethical Committee of O.I.R.M. S. Anna Hospital and "Ordine Mauriziano di Torino" (n.209; protocol 39226/C.27.104/08/09) (Turin, Italy). All animal experiments were performed in accordance with the European Community Council Directive of 24 November 1986 86/609/EEC and 6106/10/EU and approved by the University of Turin Ethical Committee for animal research and by the Italian Ministry of Health (License No. 180/2006-B).

Informed Consent Statement: Written Informed consent was obtained from all subjects involved in the study.

Data Availability Statement: The raw data supporting the conclusions of this article will be made available by the authors, without undue reservation.

Acknowledgments: We thank Rossella Barrile (University of Turin, Italy) for her support in animal model establishment.

Conflicts of Interest: A.R. and T.T. are co-founders and shareholders of Corion Biotech S.R.L. The remaining authors declare that the research was conducted in the absence of any commercial or financial relationships that could be construed as a potential conflict of interest.

References

1. Ishihara, N.; Matsuo, H.; Murakoshi, H.; Laoag-Fernandez, J.B.; Samoto, T.; Maruo, T. Increased Apoptosis in the Syncytiotrophoblast in Human Term Placentas Complicated by Either Preeclampsia or Intrauterine Growth Retardation. *Am. J. Obstet. Gynecol.* **2002**, *186*, 158–166. [CrossRef] [PubMed]
2. Redman, C.W.; Sargent, I.L. Latest Advances in Understanding Preeclampsia. *Science* **2005**, *308*, 1592–1594. [CrossRef] [PubMed]
3. Cunningham, F.G.; Lindheimer, M.D. Hypertension in Pregnancy. *N. Engl. J. Med.* **1992**, *326*, 927–932. [CrossRef] [PubMed]
4. Pennington, K.A.; Schlitt, J.M.; Jackson, D.L.; Schulz, L.C.; Schust, D.J. Preeclampsia: Multiple Approaches for a Multifactorial Disease. *Dis. Model. Mech.* **2012**, *5*, 9–18. [CrossRef]
5. Marzi, M.; Vigano, A.; Trabattoni, D.; Villa, M.L.; Salvaggio, A.; Clerici, E.; Clerici, M. Characterization of Type 1 and Type 2 Cytokine Production Profile in Physiologic and Pathologic Human Pregnancy. *Clin. Exp. Immunol.* **1996**, *106*, 127–133. [CrossRef]
6. Saito, S.; Sakai, M.; Sasaki, Y.; Tanebe, K.; Tsuda, H.; Michimata, T. Quantitative Analysis of Peripheral Blood Th0, Th1, Th2 and the Th1:Th2 Cell Ratio during Normal Human Pregnancy and Preeclampsia. *Clin. Exp. Immunol.* **1999**, *117*, 550–555. [CrossRef]
7. Saito, S.; Umekage, H.; Sakamoto, Y.; Sakai, M.; Tanebe, K.; Sasaki, Y.; Morikawa, H. Increased T-Helper-1-Type Immunity and Decreased T-Helper-2-Type Immunity in Patients with Preeclampsia. *Am. J. Reprod. Immunol.* **1999**, *41*, 297–306. [CrossRef]
8. Pijnenborg, R.; Vercruysse, L.; Verbist, L.; Van Assche, F.A. Interaction of Interstitial Trophoblast with Placental Bed Capillaries and Venules of Normotensive and Pre-Eclamptic Pregnancies. *Placenta* **1998**, *19*, 569–575. [CrossRef]

9. Sibai, B.; Dekker, G.; Kupferminc, M. Pre-Eclampsia. *Lancet* **2005**, *365*, 785–799. [CrossRef]
10. Gammill, H.S.; Roberts, J.M. Emerging Concepts in Preeclampsia Investigation. *Front. Biosci.* **2007**, *12*, 2403–2411. [CrossRef]
11. Lockwood, C.J.; Yen, C.-F.; Basar, M.; Kayisli, U.A.; Martel, M.; Buhimschi, I.; Buhimschi, C.; Huang, S.J.; Krikun, G.; Schatz, F. Preeclampsia-Related Inflammatory Cytokines Regulate Interleukin-6 Expression in Human Decidual Cells. *Am. J. Pathol.* **2008**, *172*, 1571–1579. [CrossRef] [PubMed]
12. Mandò, C.; Razini, P.; Novielli, C.; Anelli, G.M.; Belicchi, M.; Erratico, S.; Banfi, S.; Meregalli, M.; Tavelli, A.; Baccarin, M.; et al. Impaired Angiogenic Potential of Human Placental Mesenchymal Stromal Cells in Intrauterine Growth Restriction. *Stem Cells Transl. Med.* **2016**, *5*, 451–463. [CrossRef] [PubMed]
13. Yen, B.L.; Huang, H.-I.; Chien, C.-C.; Jui, H.-Y.; Ko, B.-S.; Yao, M.; Shun, C.-T.; Yen, M.-L.; Lee, M.-C.; Chen, Y.-C. Isolation of Multipotent Cells from Human Term Placenta. *Stem Cells* **2005**, *23*, 3–9. [CrossRef] [PubMed]
14. Brooke, G.; Tong, H.; Levesque, J.-P.; Atkinson, K. Molecular Trafficking Mechanisms of Multipotent Mesenchymal Stem Cells Derived from Human Bone Marrow and Placenta. *Stem Cells Dev.* **2008**, *17*, 929–940. [CrossRef] [PubMed]
15. Parolini, O.; Alviano, F.; Bergwerf, I.; Boraschi, D.; De Bari, C.; De Waele, P.; Dominici, M.; Evangelista, M.; Falk, W.; Hennerbichler, S.; et al. Toward Cell Therapy Using Placenta-Derived Cells: Disease Mechanisms, Cell Biology, Preclinical Studies, and Regulatory Aspects at the Round Table. *Stem Cells Dev.* **2010**, *19*, 143–154. [CrossRef]
16. Fukuchi, Y.; Nakajima, H.; Sugiyama, D.; Hirose, I.; Kitamura, T.; Tsuji, K. Human Placenta-Derived Cells Have Mesenchymal Stem/Progenitor Cell Potential. *Stem Cells* **2004**, *22*, 649–658. [CrossRef]
17. Chang, C.-J.; Yen, M.-L.; Chen, Y.-C.; Chien, C.-C.; Huang, H.-I.; Bai, C.-H.; Yen, B.L. Placenta-Derived Multipotent Cells Exhibit Immunosuppressive Properties That Are Enhanced in the Presence of Interferon-Gamma. *Stem Cells* **2006**, *24*, 2466–2477. [CrossRef]
18. Li, C.; Zhang, W.; Jiang, X.; Mao, N. Human-Placenta-Derived Mesenchymal Stem Cells Inhibit Proliferation and Function of Allogeneic Immune Cells. *Cell Tissue Res.* **2007**, *330*, 437–446. [CrossRef]
19. Rolfo, A.; Giuffrida, D.; Nuzzo, A.M.; Pierobon, D.; Cardaropoli, S.; Piccoli, E.; Giovarelli, M.; Todros, T. Pro-Inflammatory Profile of Preeclamptic Placental Mesenchymal Stromal Cells: New Insights into the Etiopathogenesis of Preeclampsia. *PLoS ONE* **2013**, *8*, e59403. [CrossRef]
20. Liu, L.; Zhao, G.; Fan, H.; Zhao, X.; Li, P.; Wang, Z.; Hu, Y.; Hou, Y. Mesenchymal Stem Cells Ameliorate Th1-Induced Pre-Eclampsia-like Symptoms in Mice via the Suppression of TNF-α Expression. *PLoS ONE* **2014**, *9*, e88036. [CrossRef]
21. Fu, L.; Liu, Y.; Zhang, D.; Xie, J.; Guan, H.; Shang, T. Beneficial Effect of Human Umbilical Cord-Derived Mesenchymal Stem Cells on an Endotoxin-Induced Rat Model of Preeclampsia. *Exp. Ther. Med.* **2015**, *10*, 1851–1856. [CrossRef] [PubMed]
22. Chatterjee, P.; Chiasson, V.L.; Pinzur, L.; Raveh, S.; Abraham, E.; Jones, K.A.; Bounds, K.R.; Ofir, R.; Flaishon, L.; Chajut, A.; et al. Human Placenta-Derived Stromal Cells Decrease Inflammation, Placental Injury and Blood Pressure in Hypertensive Pregnant Mice. *Clin. Sci.* **2016**, *130*, 513–523. [CrossRef] [PubMed]
23. Dominina, A.P.; Fridliandskaia, I.I.; Zemel'ko, V.I.; Pugovkina, N.A.; Kovaleva, Z.V.; Zenin, V.V.; Grinchuk, T.M.; Nikol'skiĭ, N.N. Mesenchymal stem cells of human endometrium do not undergo spontaneous transformation during long-term cultivation. *Tsitologiia* **2013**, *55*, 69–74. [CrossRef] [PubMed]
24. Wang, S.; Qu, X.; Zhao, R.C. Clinical Applications of Mesenchymal Stem Cells. *J. Hematol. Oncol.* **2012**, *5*, 19. [CrossRef]
25. Wu, W.; He, Q.; Li, X.; Zhang, X.; Lu, A.; Ge, R.; Zhen, H.; Chang, A.E.; Li, Q.; Shen, L. Long-Term Cultured Human Neural Stem Cells Undergo Spontaneous Transformation to Tumor-Initiating Cells. *Int. J. Biol. Sci.* **2011**, *7*, 892–901. [CrossRef]
26. Bernardo, M.E.; Zaffaroni, N.; Novara, F.; Cometa, A.M.; Avanzini, M.A.; Moretta, A.; Montagna, D.; Maccario, R.; Villa, R.; Daidone, M.G.; et al. Human Bone Marrow Derived Mesenchymal Stem Cells Do Not Undergo Transformation after Long-Term in Vitro Culture and Do Not Exhibit Telomere Maintenance Mechanisms. *Cancer Res.* **2007**, *67*, 9142–9149. [CrossRef]
27. Popov, B.V.; Petrov, N.S.; Mikhailov, V.M.; Tomilin, A.N.; Alekseenko, L.L.; Grinchuk, T.M.; Zaichik, A.M. Spontaneous Transformation and Immortalization of Mesenchymal Stem Cells in Vitro. *Cell Tissue Biol.* **2009**, *3*, 110–120. [CrossRef]
28. Nuzzo, A.M.; Giuffrida, D.; Zenerino, C.; Piazzese, A.; Olearo, E.; Todros, T.; Rolfo, A. JunB/Cyclin-D1 Imbalance in Placental Mesenchymal Stromal Cells Derived from Preeclamptic Pregnancies with Fetal-Placental Compromise. *Placenta* **2014**, *35*, 483–490. [CrossRef]
29. Karp, J.M.; Leng Teo, G.S. Mesenchymal Stem Cell Homing: The Devil Is in the Details. *Cell Stem Cell* **2009**, *4*, 206–216. [CrossRef]
30. Asami, T.; Ishii, M.; Fujii, H.; Namkoong, H.; Tasaka, S.; Matsushita, K.; Ishii, K.; Yagi, K.; Fujiwara, H.; Funatsu, Y.; et al. Modulation of Murine Macrophage TLR7/8-Mediated Cytokine Expression by Mesenchymal Stem Cell-Conditioned Medium. *Mediat. Inflamm.* **2013**, *2013*, 264260. [CrossRef]
31. Nuzzo, A.M.; Giuffrida, D.; Masturzo, B.; Mele, P.; Piccoli, E.; Eva, C.; Todros, T.; Rolfo, A. Altered Expression of G1/S Phase Cell Cycle Regulators in Placental Mesenchymal Stromal Cells Derived from Preeclamptic Pregnancies with Fetal-Placental Compromise. *Cell Cycle* **2017**, *16*, 200–212. [CrossRef] [PubMed]
32. Peng, C.-K.; Wu, S.-Y.; Tang, S.-E.; Li, M.-H.; Lin, S.-S.; Chu, S.-J.; Huang, K.-L. Protective Effects of Neural Crest-Derived Stem Cell-Conditioned Media against Ischemia-Reperfusion-Induced Lung Injury in Rats. *Inflammation* **2017**, *40*, 1532–1542. [CrossRef] [PubMed]
33. Yang, C.; Lei, D.; Ouyang, W.; Ren, J.; Li, H.; Hu, J.; Huang, S. Conditioned Media from Human Adipose Tissue-Derived Mesenchymal Stem Cells and Umbilical Cord-Derived Mesenchymal Stem Cells Efficiently Induced the Apoptosis and Differentiation in Human Glioma Cell Lines in Vitro. *BioMed. Res. Int.* **2014**, *2014*, 109389. [CrossRef] [PubMed]

34. Zuk, P.A.; Zhu, M.; Ashjian, P.; De Ugarte, D.A.; Huang, J.I.; Mizuno, H.; Alfonso, Z.C.; Fraser, J.K.; Benhaim, P.; Hedrick, M.H. Human Adipose Tissue Is a Source of Multipotent Stem Cells. *Mol. Biol. Cell* **2002**, *13*, 4279–4295. [CrossRef]
35. Lee, O.K.; Kuo, T.K.; Chen, W.-M.; Lee, K.-D.; Hsieh, S.-L.; Chen, T.-H. Isolation of Multipotent Mesenchymal Stem Cells from Umbilical Cord Blood. *Blood* **2004**, *103*, 1669–1675. [CrossRef]
36. Hu, Y.; Liao, L.; Wang, Q.; Ma, L.; Ma, G.; Jiang, X.; Zhao, R.C. Isolation and Identification of Mesenchymal Stem Cells from Human Fetal Pancreas. *J. Lab. Clin. Med.* **2003**, *141*, 342–349. [CrossRef]
37. Abomaray, F.M.; Al Jumah, M.A.; Kalionis, B.; AlAskar, A.S.; Al Harthy, S.; Jawdat, D.; Al Khaldi, A.; Alkushi, A.; Knawy, B.A.; Abumaree, M.H. Human Chorionic Villous Mesenchymal Stem Cells Modify the Functions of Human Dendritic Cells, and Induce an Anti-Inflammatory Phenotype in CD1+ Dendritic Cells. *Stem Cell Rev. Rep.* **2015**, *11*, 423–441. [CrossRef]
38. Abumaree, M.H.; Abomaray, F.M.; Alshabibi, M.A.; AlAskar, A.S.; Kalionis, B. Immunomodulatory Properties of Human Placental Mesenchymal Stem/Stromal Cells. *Placenta* **2017**, *59*, 87–95. [CrossRef]
39. Du, W.; Li, X.; Chi, Y.; Ma, F.; Li, Z.; Yang, S.; Song, B.; Cui, J.; Ma, T.; Li, J.; et al. VCAM-1+ Placenta Chorionic Villi-Derived Mesenchymal Stem Cells Display Potent pro-Angiogenic Activity. *Stem Cell Res. Ther.* **2016**, *7*, 49. [CrossRef]
40. Grylls, A.; Seidler, K.; Neil, J. Link between Microbiota and Hypertension: Focus on LPS/TLR4 Pathway in Endothelial Dysfunction and Vascular Inflammation, and Therapeutic Implication of Probiotics. *Biomed. Pharmacother.* **2021**, *137*, 111334. [CrossRef]
41. Ding, X.; Yang, Z.; Han, Y.; Yu, H. Fatty Acid Oxidation Changes and the Correlation with Oxidative Stress in Different Preeclampsia-like Mouse Models. *PLoS ONE* **2014**, *9*, e109554. [CrossRef] [PubMed]
42. Lin, F.; Zeng, P.; Xu, Z.; Ye, D.; Yu, X.; Wang, N.; Tang, J.; Zhou, Y.; Huang, Y. Treatment of Lipoxin A(4) and Its Analogue on Low-Dose Endotoxin Induced Preeclampsia in Rat and Possible Mechanisms. *Reprod. Toxicol.* **2012**, *34*, 677–685. [CrossRef] [PubMed]
43. Zhang, D.; Fu, L.; Wang, L.; Lin, L.; Yu, L.; Zhang, L.; Shang, T. Therapeutic Benefit of Mesenchymal Stem Cells in Pregnant Rats with Angiotensin Receptor Agonistic Autoantibody-Induced Hypertension: Implications for Immunomodulation and Cytoprotection. *Hypertens. Pregnancy* **2017**, *36*, 247–258. [CrossRef] [PubMed]
44. De Oliveira, L.F.; Almeida, T.R.; Ribeiro Machado, M.P.; Cuba, M.B.; Alves, A.C.; da Silva, M.V.; Rodrigues Júnior, V.; Dias da Silva, V.J. Priming Mesenchymal Stem Cells with Endothelial Growth Medium Boosts Stem Cell Therapy for Systemic Arterial Hypertension. *Stem Cells Int.* **2015**, *2015*, 685383. [CrossRef]
45. Oliveira-Sales, E.B.; Maquigussa, E.; Semedo, P.; Pereira, L.G.; Ferreira, V.M.; Câmara, N.O.; Bergamaschi, C.T.; Campos, R.R.; Boim, M.A. Mesenchymal Stem Cells (MSC) Prevented the Progression of Renovascular Hypertension, Improved Renal Function and Architecture. *PLoS ONE* **2013**, *8*, e78464. [CrossRef]
46. Vizoso, F.J.; Eiro, N.; Cid, S.; Schneider, J.; Perez-Fernandez, R. Mesenchymal Stem Cell Secretome: Toward Cell-Free Therapeutic Strategies in Regenerative Medicine. *Int. J. Mol. Sci.* **2017**, *18*, 1852. [CrossRef]
47. Beer, L.; Mildner, M.; Ankersmit, H.J. Cell Secretome Based Drug Substances in Regenerative Medicine: When Regulatory Affairs Meet Basic Science. *Ann. Transl. Med.* **2017**, *5*, 170. [CrossRef]
48. Albonici, L.; Benvenuto, M.; Focaccetti, C.; Cifaldi, L.; Miele, M.T.; Limana, F.; Manzari, V.; Bei, R. PlGF Immunological Impact during Pregnancy. *Int. J. Mol. Sci.* **2020**, *21*, 8714. [CrossRef]
49. Jones, R.L.; Stoikos, C.; Findlay, J.K.; Salamonsen, L.A. TGF-Beta Superfamily Expression and Actions in the Endometrium and Placenta. *Reprod. Camb. Engl.* **2006**, *132*, 217–232. [CrossRef]
50. Wu, L.-Z.; Liu, X.-L.; Xie, Q.-Z. Osteopontin Facilitates Invasion in Human Trophoblastic Cells via Promoting Matrix Metalloproteinase-9 in Vitro. *Int. J. Clin. Exp. Pathol.* **2015**, *8*, 14121–14130.
51. Qazi, B.S.; Tang, K.; Qazi, A. Recent advances in underlying pathologies provide insight into interleukin-8 expression-mediated inflammation and angiogenesis. *Int. J. Inflam.* **2011**, *2011*, 908468. [CrossRef] [PubMed]
52. Tang, Y.; Ye, W.; Liu, X.; Lv, Y.; Yao, C.; Wei, J. VEGF and SFLT-1 in Serum of PIH Patients and Effects on the Foetus. *Exp. Ther. Med.* **2019**, *17*, 2123–2128. [CrossRef] [PubMed]
53. Nuzzo, A.M.; Giuffrida, D.; Moretti, L.; Re, P.; Grassi, G.; Menato, G.; Rolfo, A. Placental and Maternal SFlt1/PlGF Expression in Gestational Diabetes Mellitus. *Sci. Rep.* **2021**, *11*, 2312. [CrossRef] [PubMed]
54. Rolfo, A.; Attini, R.; Nuzzo, A.M.; Piazzese, A.; Parisi, S.; Ferraresi, M.; Todros, T.; Piccoli, G.B. Chronic Kidney Disease May Be Differentially Diagnosed from Preeclampsia by Serum Biomarkers. *Kidney Int.* **2013**, *83*, 177–181. [CrossRef] [PubMed]
55. Ali, S.M.J.; Khalil, R.A. Genetic, Immune and Vasoactive Factors in the Vascular Dysfunction Associated with Hypertension in Pregnancy. *Expert Opin. Ther. Targets* **2015**, *19*, 1495–1515. [CrossRef] [PubMed]
56. Rivera, D.L.; Olister, S.M.; Liu, X.; Thompson, J.H.; Zhang, X.J.; Pennline, K.; Azuero, R.; Clark, D.A.; Miller, M.J. Interleukin-10 Attenuates Experimental Fetal Growth Restriction and Demise. *FASEB J. Off. Publ. Fed. Am. Soc. Exp. Biol.* **1998**, *12*, 189–197. [CrossRef]
57. Vonlaufen, A.; Phillips, P.A.; Xu, Z.; Zhang, X.; Yang, L.; Pirola, R.C.; Wilson, J.S.; Apte, M.V. Withdrawal of Alcohol Promotes Regression While Continued Alcohol Intake Promotes Persistence of LPS-Induced Pancreatic Injury in Alcohol-Fed Rats. *Gut* **2011**, *60*, 238–246. [CrossRef]
58. Wang, L.-L.; Yu, Y.; Guan, H.-B.; Qiao, C. Effect of Human Umbilical Cord Mesenchymal Stem Cell Transplantation in a Rat Model of Preeclampsia. *Reprod. Sci.* **2016**, *23*, 1058–1070. [CrossRef]
59. Zhang, Z.; Dai, M. Effect of paeonol on adhesive function of rat vascular endothelial cells induced by lipopolysaccharide and co-cultured with smooth muscle cells. *Zhongguo Zhong Yao Za Zhi Zhongguo Zhongyao Zazhi/China J. Chin. Mater. Med.* **2014**, *39*, 1058–1063.

60. Zekri, A.-R.N.; Salama, H.; Medhat, E.; Musa, S.; Abdel-Haleem, H.; Ahmed, O.S.; Khedr, H.A.H.; Lotfy, M.M.; Zachariah, K.S.; Bahnassy, A.A. The Impact of Repeated Autologous Infusion of Haematopoietic Stem Cells in Patients with Liver Insufficiency. *Stem Cell Res. Ther.* **2015**, *6*, 118. [CrossRef]
61. Shi, M.; Liu, Z.; Wang, Y.; Xu, R.; Sun, Y.; Zhang, M.; Yu, X.; Wang, H.; Meng, L.; Su, H.; et al. A Pilot Study of Mesenchymal Stem Cell Therapy for Acute Liver Allograft Rejection. *Stem Cells Transl. Med.* **2017**, *6*, 2053–2061. [CrossRef]
62. Hauser, P.V.; De Fazio, R.; Bruno, S.; Sdei, S.; Grange, C.; Bussolati, B.; Benedetto, C.; Camussi, G. Stem Cells Derived from Human Amniotic Fluid Contribute to Acute Kidney Injury Recovery. *Am. J. Pathol.* **2010**, *177*, 2011–2021. [CrossRef] [PubMed]
63. Oludare, G.O.; Ilo, O.J.; Lamidi, B.A. Effects of Lipopolysaccharide and High Saline Intake on Blood Pressure, Angiogenic Factors and Liver Enzymes of Pregnant Rats. *Niger. J. Physiol. Sci. Off. Publ. Physiol. Soc. Niger.* **2017**, *32*, 129–136.
64. Livak, K.J.; Schmittgen, T.D. Analysis of Relative Gene Expression Data Using Real-Time Quantitative PCR and the 2(-Delta Delta C(T)) Method. *Methods San Diego Calif* **2001**, *25*, 402–408. [CrossRef] [PubMed]

International Journal of *Molecular Sciences*

MDPI

Review

PlGF Immunological Impact during Pregnancy

Loredana Albonici [1,*], Monica Benvenuto [1,2], Chiara Focaccetti [1,3], Loredana Cifaldi [1,4], Martino Tony Miele [5], Federica Limana [3,6], Vittorio Manzari [1] and Roberto Bei [1]

1 Department of Clinical Sciences and Translational Medicine, University of Rome "Tor Vergata", Via Montpellier 1, 00133 Rome, Italy; monica.benvenuto@unicamillus.org (M.B.); chiara.focaccetti@uniroma5.it (C.F.); cifaldi@med.uniroma2.it (L.C.); manzari@med.uniroma2.it (V.M.); bei@med.uniroma2.it (R.B.)

2 Saint Camillus International University of Health and Medical Sciences, Via di Sant' Alessandro 8, 00131 Rome, Italy

3 Department of Human Science and Promotion of the Quality of Life, San Raffaele Roma Open University, Via di Val Cannuta 247, 00166 Rome, Italy; federica.limana@uniroma5.it

4 Academic Department of Pediatrics (DPUO), Ospedale Pediatrico Bambino Gesù, IRCCS, Piazza Sant' Onofrio 4, 00165 Rome, Italy

5 Department of Experimental Medicine, University of Rome "Tor Vergata", Via Montpellier 1, 00133 Rome, Italy; miele@med.uniroma2.it

6 Laboratory of Cellular and Molecular Pathology, IRCCS San Raffaele Pisana, Via della Pisana 235, 00163 Rome, Italy

* Correspondence: albonici@med.uniroma2.it

Received: 20 October 2020; Accepted: 16 November 2020; Published: 18 November 2020

Abstract: During pregnancy, the mother's immune system has to tolerate the persistence of paternal alloantigens without affecting the anti-infectious immune response. Consequently, several mechanisms aimed at preventing allograft rejection, occur during a pregnancy. In fact, the early stages of pregnancy are characterized by the correct balance between inflammation and immune tolerance, in which proinflammatory cytokines contribute to both the remodeling of tissues and to neo-angiogenesis, thus, favoring the correct embryo implantation. In addition to the creation of a microenvironment able to support both immunological privilege and angiogenesis, the trophoblast invades normal tissues by sharing the same behavior of invasive tumors. Next, the activation of an immunosuppressive phase, characterized by an increase in the number of regulatory T (Treg) cells prevents excessive inflammation and avoids fetal immuno-mediated rejection. When these changes do not occur or occur incompletely, early pregnancy failure follows. All these events are characterized by an increase in different growth factors and cytokines, among which one of the most important is the angiogenic growth factor, namely placental growth factor (PlGF). PlGF is initially isolated from the human placenta. It is upregulated during both pregnancy and inflammation. In this review, we summarize current knowledge on the immunomodulatory effects of PlGF during pregnancy, warranting that both innate and adaptive immune cells properly support the early events of implantation and placental development. Furthermore, we highlight how an alteration of the immune response, associated with PlGF imbalance, can induce a hypertensive state and lead to the pre-eclampsia (PE).

Keywords: PlGF; Flt-1/VEGFR1; pregnancy; immune modulation

1. Introduction

1.1. The Immunological Paradox of Human Pregnancy

In the middle of the last century, Peter Medawar was the first to question the complexity of the immunological networks that allow maternal-fetal tolerance. Although Medwar's intuition was still far from the description of the real immune mechanisms, the immunological tolerance, especially during the early stages of pregnancy, represented a current issue that has only been partially clarified [1].

The uterine microenvironment can be considered to be a privileged immune site similar to other body districts (e.g., mucosae-associated lymphoid tissue), where the immune response is tightly controlled to prevent improper activation of immune cells. However, recent evidence supports the idea that the immune response, at the maternal-fetal interface, is not suppressed, and instead, is in a highly dynamic state [2–4]. The placenta develops from the outer cell layer of the blastocyst, which forms the primary trophoblastic cell mass. Then, the trophoblast, which represents the fetal part of placenta and contains the father's genetic material, deeply invades the decidua and usually its development is not counteracted by the maternal immune system [5–10]. Indeed, implantation implicates attachment and invasion of the blastocyst into the uterus and remodeling of the endometrial stroma, known as "decidualization", as well as modifications of the endometrial vascular bed. Thus, the implantation site is characterized by an inflammatory response, with an abundance of recruited immune cells and the induction of inflammatory genes [11–15]. In addition, the blastocyst needs to aggressively adhere to the endometrium to obtain oxygen and nutrients. During this process, uterine tissue remodeling and inflammatory mechanisms are required [4,16–18]. This proliferative and invasive behavior, associated with immune tolerance mechanisms, is similar to that activated by invasive cancer cells, for obtaining a nutrient supply and evading or editing a host immune response [19–22]. Therefore, in addition to the ability to invade normal tissues, both developing placenta cells and cancer cells are able to create a microenvironment supportive of angiogenesis and immunological privilege [13,21–25] (Figure 1). Although these two processes show overlapping mechanisms, pregnancy is still a physiological process unlike tumor invasion, which is a pathological process. Consequently, a fine balance between inflammatory and anti-inflammatory mediators is required to prevent fetal rejection [7,13,26]. Among the several inflammatory mediators present in the uterus, a predominant role is played by placental growth factor (PlGF) [12,27–32]. In addition to being a factor involved in the implantation and development of the placenta, PlGF also has an impact on the immune response by acting on both innate and adaptive immune cells [23,29,33–37].

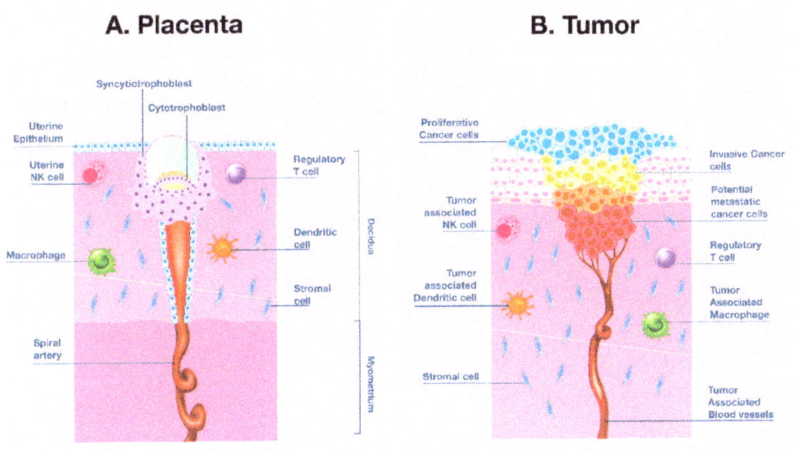

Figure 1. Human placentation and malignancy share proliferative and invasive features to establish a nutrient supply and evade or modify the host's immune response. (**A**) Human placenta development. In the early stage of implantation, the blastocyst displays an invasive phenotype that allows implanting inside the endometrial stroma. In this process cytokines, growth factors, hormones, extracellular matrix metalloproteinases (MMPs), and immune cells, all modulate cell invasion of maternal decidua and myometrium and their capacity to transform spiral arteries. Among growth factors, placental growth factor (PlGF) secreted by the decidua, trophoblast, and uterine natural killer (uNK) cells have a determinant role in regulating invasion, vascular development, and maternal immune tolerance mechanisms to semi-allogeneic fetus; (**B**) Tumor invasion and progression. Malignancy is able to create both a vascular network that warrants perfusion of tumor mass and an immunosuppressive microenvironment by recruiting specific immune cells. Molecules (cytokines, growth factors, extracellular MMPs), produced by tumor and inflammatory cells in the tumor microenvironment, recruit (tumor-associated) immune cells, thus, creating a tolerogenic milieu that inhibits the development of an efficient immune response against cancer cells that foster tumor growth and progression. PlGF blockade reduces both neoangiogenesis and lymphangiogenesis, inhibits the M2 macrophages polarization, hinders the recruitment of tumor-associated macrophages (TAM), and decreases the recruitment of myeloid suppressor cells [23]. (Illustration inspired by Holtan et al. [21] with kind permission of Elsevier, License Number: 4911830648497).

Although numerous authors have attributed the onset of pre-eclampsia (PE) to maternal cardiovascular system maladaptation [38–42], the idea that PE is a pathological condition associated with an alteration of the vascular development of the placenta [43] and it is also the result of an aberrant maternal immune response has become increasingly consistent [44–49]. This hypothesis stems from the evidence that immune-mediated mechanisms are able to regulate the cells response to angiogenic growth factors and, in turn, angiogenic growth factors can modulate the behavior of immune cells. In addition, angiogenic growth factor PlGF imbalance has long been associated with pathological pregnancies, such as implantation failure and the development of PE [48–60]. In this review, we discuss how an altered maternal immune response, modulated by PlGF imbalance, can induce a hypertensive state and lead to pathological pregnancy.

1.2. Placental Growth Factor (PlGF)

PlGF is a pleiotropic angiogenic growth factor originally isolated from the human placenta in which it displays a major role in vasculogenesis and angiogenesis [61,62]. It is encoded by a single *plgf* gene belonging to the vascular endothelial growth factor (VEGF) family and gives rise to four splice isoforms in humans, PlGF-1–4 [61]. Whereas PlGF-1 and PlGF-3 are diffusible isoforms, PlGF-2 and PlGF-4 have heparin-binding domains. Mice express only one isoform equivalent to human PlGF-2. Heparin can modulate PlGF-2-induced proliferation of extravillous trophoblast (EVT) cells, by sequestering PlGF-2 from its receptor or by modifying its structural conformation, without affecting migration or invasiveness of EVT cells [28].

PlGF expression is low to undetectable in most tissues in normal health; however, its angiogenic activity is mainly restricted to pathological conditions, such as inflammation or ischemia [12,63–65]. PlGF is upregulated by different stimuli such as hypoxia, inflammatory cytokines, growth factors, and hormones, all events present during implantation and placenta development. Hypoxia also upregulates the PlGF receptor fms-related tyrosine kinase (Flt)-1/vascular endothelial growth factor receptor (VEGFR)-1 and its co-receptor neuropilin-1 (NRP1) in disease conditions [66–69]. However, unlike tumor growth, during pregnancy, the transcriptional activity of PlGF is suppressed by hypoxia and increased by a normoxic environment in the trophoblast, indicating a specific regulatory mechanism in these cells. Furthermore, the regulation of PlGF transcription in trophoblast cells is not mediated by the functional activity of hypoxia-inducible factor (HIF)-1 [70]. Since PlGF plays an important role in the vascular development of the placenta, these results suggest the existence of a protective regulation mechanism of PlGF levels in the prevention of the oxidative damage caused by hypoxia. PlGF has been shown to play a negligible role in the development and in the physiological angiogenesis process, even if adult $plgf^{-/-}$ mice are apparently healthy, fertile, and without vascular defects [71]. Nevertheless, in mice with genetic deletion of *plgf*, uterine natural killer (uNK) cells were smaller, less granular, and binucleated than in control mice. However, uNK cell numbers were almost unchanged as compared with control mice, indicating that PlGF plays an important role in successful uNK cells proliferation or differentiation [72]. However, although decidual invasion was not influenced in $plgf^{-/-}$ mice, the implantation site showed abnormal placental vasculature with decreased branching of feto-placental vessels and increased lacunarity, indicating a lack of uniformity of vessel formation [73].

During pregnancy, the placental trophoblast is the main source of PlGF and its expression is significantly upregulated at an early gestational age after implantation [12,27,28,74]. PlGF is also produced by the human endometrium and released into the uterine lumen [29]. An additional source of PlGF during implantation is from the production mediated by uNK cells [30,75]. Accordingly, the abnormal expression of PlGF during pregnancy affects the trophoblast function as much as the vascularity in the placental bed [30,53,75–77]. Immunohistochemical analysis has shown that PlGF was significantly lower in the placentas of women with severe PE as compared with those with mild PE and placentas of normotensive women [52]. This result confirmed previous data from serum levels of PlGF. Indeed, serum PlGF levels were lower among women who developed PE (23 ± 24 pg/mL vs. 63 ± 145 pg/mL) or gestational hypertension (27 ± 19 pg/mL) as compared with the controls [60].

PlGF homodimers bind Flt-1/VEGFR-1 [67]. However, only PlGF-2 and PlGF-4 are able to bind the co-receptors NRP1 and NRP2, due to the insertion of 21 basic amino acids [78–80]. Although the downstream Flt-1/VEGFR-1 signaling is still elusive under physiological condition, Flt-1/VEGFR-1 plays a negative role by suppressing pro-angiogenic signals, as displayed by an early death in embryogenesis due to the uncontrolled growth of endothelial cells and disorganization of the vascular architecture in *vegfr1-null* mutant mice [81].

In addition to the direct effects on endothelial cells, the binding of PlGF to Flt-1/VEGFR-1 shows indirect effects on nonvascular cells by modulating the behavior of immune cells [23,71,82–87]. PlGF enhances macrophages proliferation, migration, and survival [33] and also shows a direct effect on the inflammatory reaction by triggering tumor necrosis factor (TNF)-α and interleukin (IL)-6 production in a calcineurin-dependent pathway [86]. PlGF significantly increases IL-8 secretion, cyclooxygenase (Cox)-2 expression, and consequent prostaglandin (PG)-E2 and PG-F2α release, matrix metalloproteinases (MMP)-9 gene expression, and enzyme production via Flt1/VEGFR1 on monocytes [35,36]. Overall, all these molecules play a role during the decidualization and tumor cell growth and progression (Figure 1). Moreover, TNF-α, by promoting PlGF synthesis, can regulate angiogenesis via PlGF/Flt-1/VEGFR-1 [87]. Flt-1/VEGFR-1 is a cell surface marker for the monocyte-macrophage lineage in humans [85], and it is also expressed on the surface of activated T cells in which it increases migration and IL-10 production [88], thus, indicating that Flt-1/VEGFR-1 is able to mediate a direct immunomodulatory effect.

The co-receptors NRP1 and NRP2 were initially characterized as receptors expressed by neuronal cells, where the natural ligand of NRP1 was semaphoring 3A (Sema3A) and, subsequently, endothelial cells [89]. Apart from vessels and axons, NRPs are also expressed by immune cells in which they display an inhibitory effect [90]. NRP1 is expressed primarily by dendritic cells (DCs) and regulatory T (Treg) cells [66,91] and exerts mainly inhibitory effects on the immune response. Indeed, NRP1 is constitutively expressed at a high level, independently of its activation status, only on the surface of $CD4^{+}CD25^{high}$ natural Treg cells (nTreg), which arise in the thymus, but not on inducible Treg cells (iTreg) generated in the periphery through the induction of Foxp3 [92]. NRP1 functions as a component of the immunological synapse and promotes prolonged interaction between Treg cells and immature dendritic cells (iDCs). This long contact results in higher nuclear factor kappa beta (NF-κB) transcriptional activity that might prevent an autoimmune response by inducing immunosuppression probably because of the delay of iDCs maturation [93]. These findings could envisage a possible competition between PlGF and Sema3A as they bind NRP1 in the same domain and, consequently, an immunosuppressive role played by PlGF. Although NRP1 downregulation has been rarely described in few studies involving PE or fetal growth restriction due to deficient vascular branching [94,95], a recent study by Moldenhauer et al. confirmed the role of NRP1 on the immune system during the preimplantation period. They reported that mating mice elicited a five-fold increase in uterine Treg cells, followed by an extensive Treg proliferation in the uterus-draining lymph nodes, comprising 70% $NRP1^{+}$ thymus-derived Treg cells and 30% $NRP1^{-}$ peripheral Treg cells, as compared with virgin mice. This increase was due to epigenetic modification of the transcription factor Foxp3, induced by the presence of the seminal fluid [96].

The alternative splicing of both Flt1/VEGFR-1 and NRP1 pre-mRNA produces soluble receptor isoforms (sFlt-1/sVEGFR-1 and sNRP1, respectively) that can bind to and inhibit the action of both PlGF and VEGF [97–99]. Excessive sFlt-1/sVEGFR-1 generated by the human placenta and released into the maternal circulation leads to hypertension and proteinuria in PE, thus, contributing to maternal vascular injury [50,51,53,55,56,59,77,100–104]. At least four different tissue-specific splice variants of sFlt-1/sVEGFR-1 have been described [101,105,106]. Among these variants, sFlt-1/sVEGFR-1 e15a is the main variant produced primarily in the placenta and it is believed to be responsible for PE, being significantly elevated in the placenta and circulation of women with PE as compared with normal pregnancies. sFlt-1/sVEGFR-1 e15a could be responsible for endothelial dysfunction and terminal organ dysfunction as observed in PE-like mouse models. These biochemical changes appear to precede the clinical features of disease [106].

2. Pregnancy and Inflammation

The inflammatory response of the PlGF-Flt-1/VEGFR-1 axis is mediated by the transcription factor NF-κB involved in genes related to inflammation and immune response regulation. NF-κB has regulatory binding sites in the promoter of the PlGF gene and can modulate the transcriptional activity of PlGF via Rel-A in hypoxic human monocytes [107,108]. PlGF, by increasing the degradation of IκB-α, can increase the DNA binding activity of p65, an NF-κB subunit, thus, generating a self-feeding mechanism [107]. NF-κB is involved in several cellular pathways including inflammation, hypoxia, and angiogenesis, all processes implicated in placental development, thus consequently, when dysregulated, it is considered to be one of the main factors responsible for PE [109–111]. In fact, NF-κB is involved in the production of both TNF-α and PlGF resulting in aberrant activation of innate immune cells and imbalanced differentiation of $CD4^{+}$ T lymphocyte subsets, which may account for high cytokine levels and the cytotoxic environment in utero [48,112–114].

In normal pregnancy, three different phases can be identified, each characterized by a specific correlation between NF-κB and PlGF. The first trimester is characterized by a proinflammatory profile in which NF-κB is activated by inflammation and hypoxia that occurs during the development of the placenta. The second trimester shows an anti-inflammatory state as the pregnancy progresses. During the third trimester, NF-κB returns to be expressed at high levels in the decidua in preparation for parturition [115]. In addition, NF-κB can also directly upregulate the expression of TNF-α by lymphocytes during hypoxic stress, thus, establishing a vicious circle that feeds inflammation [116,117]. Therefore, the expression of NF-κB is critical in maintaining an adequate level of cytokines/PlGF required during the different periods of pregnancy.

A further link between the immune response and PlGF results from the involvement of another transcription factor, namely nuclear factor of activated T cells (NFAT)-1, which was initially identified in activated T cells [118,119] and has also been shown to be involved in the control of innate immunity [120]. Cytoplasmic NFAT is activated through calcineurin-mediated dephosphorylation, then, NFAT translocates into the nucleus, where it upregulates the expression of IL-2 and stimulates the growth and differentiation of the T cells [118]. Ding et al. reported that TNF-α was upregulated by tumor-derived PlGF in myelomonocytic cells via NFAT-1, which in turn contributed to the recruitment of PlGF-induced myelomonocytic cells [34]. Moreover, a region of the Flt-1/VEGFR-1 promoter contains a binding site for the transcription factor NFAT-1, thus, providing evidence that Flt-1/VEGFR-1 represents a NFAT-1 target gene [121]. The definitive confirmation of the role played by the PlGF-Flt-1/VEGFR-1-NFAT-1 axis in the placenta derives from the evidence that the inhibition of NFAT reduces both Flt-1/VEGFR-1 and sFlt-1/sVEGFR-1 splice variant e15a transcript secretion from primary human cytotrophoblasts [122]. In addition to being involved in cell proliferation, invasive migration, and angiogenesis, NFAT-1 mediates both the induced anergy of $CD4^+$ T cells through the expression of different inflammatory cytokines and the Treg-mediated suppression of T-helper (Th) cells activation [123–125]. Thus, PlGF, in addition to mediating inflammation, could contribute to induce a state of tolerance via NFAT-1, by binding Flt-1/VEGFR-1. Overall, these results further confirm the effects of PlGF on the immune response.

2.1. Interplay between Immune Cells and PlGF during Pregnancy

During pregnancy, innate immune cells are the main leukocyte population in the uterus at the time of embryo implantation. Although most studies analyze the role played by a single immune cell type, it is clear that the creation of an adequate microenvironment, supporting the gestation in its various stages, is the result of the reciprocal interaction among mediators and immune cells present during the early stage of pregnancy. In addition to the uNK cells, which are the most abundant and the main protagonists of innate immunity, macrophages and DCs are also present. Importantly, depletion of any of these cell types modifies the uterine environment and hampers the implantation. Changes in the behavior of these cells lead to an imbalance of angiogenic factors and the proinflammatory cytokines and, at the same time, these altered levels of proinflammatory cytokines and angiogenic factors are able to affect the function of immune cells. An altered uterine environment causes defects in trophoblast invasion and placenta damage that trigger a systemic inflammatory response and widespread activation of the endothelium. Thus, the type and function of the immune cells involved in this response are critical and determine whether a viable pregnancy will occur [11,46,126,127].

2.2. Natural Killer (NK) Cells

During pregnancy, NK cells are the predominant population of lymphoid immune cells in the uterus at the maternal-fetal interface and are involved both in placental vascular remodeling and in a mother's immune tolerance [128–134].

NK cells are members of a rapidly expanding family of innate lymphoid cells (ILCs), derived from early common lymphoid progenitors, the $CD34^+$ hematopoietic stem cell [135,136]. Given their heterogeneous characteristics, a new classification that categorizes ILCs into five subsets based on transcription factors and cytokines production has been proposed [137]. These subsets include ILCs1, ILCs2, ILCs3, lymphoid tissue-inducer (LTi) cells, and conventional NK (cNK) cells [138,139]. ILCs interact with the tissue cells to ensure, in addition to the immune response, homeostasis and tissue repair [140,141]. Among these subsets, cNKs are very similar to Tbet$^+$ ILCs1 producing IFN-γ but differ in their higher cytotoxic potential. A common feature of cNK cells and tissue resident NK (trNK) cells is their IL-15-dependent signaling during early development, however, they differ in their ability to recirculate [142–144]. In fact, cNK cells circulate freely, while trNK cells are resident in the liver, skin, kidney, and virgin uterus [135,136,140,143–145].

On the basis of different receptors and transcription factors expression, it is possible to further distinguish cNK cells from trNK cells. In contrast to cNK cells, where CD56 and CD16 expression allows for discrimination of cell cytotoxic and regulatory subsets, uterine trNK cells are almost exclusively CD16$^-$CD56^{bright} [146]. In addition, whereas Tbet is required for the development of trNK cells in the liver and skin, uterine trNK develop independently of the Tbet transcription factor [147]. All these differences indicate that cNK and trNK cells in the uterus represent different lineages of NK cells rather than different differentiation states [130,140,145].

In mice, the onset of decidualization is characterized by a series of events such as the extracellular matrix (ECM) remodeling of the endometrial stroma, the induction of angiogenesis, and a significant increase in uNK cells, which originates mainly from local proliferating trNK cells, [73,143,144,148–150] while the recruitment of cNK cells takes place later [144,151].

Unlike cytotoxic T lymphocytes, NK cells can eliminate tumor cells, infected cells or nonself cells, by direct contact without a previous activation, due to their natural cytotoxic activity. Direct contact between NK and the target cells may engage activating or inhibitory receptors expressed on NK cells. Each NK cell can express simultaneously several different activating or inhibitory receptors, resulting in the potential for many specificities. The uNK cells cytotoxic ability is regulated by an education process where only those cells that recognize the "self" are promoted to have cytotoxic ability, become tolerant, and act when there is a "dangerous self" or "missing self" signals [152,153]. This ability of NK cells is tightly regulated by a complex of interactions among the target cell and activating or inhibitory receptors expressed on the NK cell surface. If the strength of the activating signals outweighs the inhibitory signals, the cell releases cytolytic granules directed against the target cell and produces cytokines [154,155]. Among these receptors, the expression of killer immunoglobulin-like receptors (KIRs) confers to uNK cells an important function, by inhibiting the production of cytotoxic cytokines and stimulating the production of angiogenic factors [75,131,156]. Thus, resting uNK cells have a low cytotoxic ability as compared with "primed" uNK cells. Additionally, uNK cells priming is regulated by the microenvironment in which the NK cell is present and can also be regulated by the proximity of other immune cell types such as monocytes, DCs, and T cells [157–161].

Genetic association studies have indicated that both uNK cell-activating receptors, KIR2DS4 and KIR2DS1, recognize fetal HLA-A ligands and protect from PE [157,162]. Conversely, pregnant women with a specific KIR haplotype and fetal HLA-C2 genotype combination have a significantly higher risk of PE [163]. Thus, uNK cells can respond to fetal MHC class I via their inhibitory and activating receptors to control appropriate placental vascularization and development.

During the first trimester of pregnancy, uNK cells represent as much as 50–70% of decidual infiltrating lymphocytes and are characterized by CD56^{bright}CD16$^-$KIR$^+$CD9$^+$ phenotype [146,164,165]. Unlike cNK cells, uNK cells are poorly cytolytic; they release cytokines/chemokines that regulate the immune environment and angiogenic growth factors, such as PlGF, for placentation [73,129,148–150,156,166,167]. One of the plausible mechanisms by which uNK cells are inhibited in their cytotoxic activity is through hypoxia, a condition that is able to stimulate angiogenesis and which is normally present during decidualization. Indeed, it was shown that the cytolytic capacities of uNK cells were markedly and significantly impaired under hypoxic conditions and this inhibition was associated with the activation of transcriptional factor STAT3 [168]. Of interest, as mentioned above, hypoxia upregulates the expression of PlGF, its receptor Flt-1/VEGFR-1 and co-receptor NRP1 [68,69,82]. A further mechanism by which uNK cells induce maternal tolerance is through the crosstalk between uNK and CD14$^+$ myelomonocytic cells. These cells are in close contact in the decidua and their interaction is mediated by IFN-γ. Following interaction with uNK cells, decidual CD14$^+$ cells express indoleamine 2,3-dioxygenase (IDO) resulting in the induction of Treg cells and immunosuppression [169]. In addition, decidual CD14$^+$ cells may also induce Treg cells through transforming growth factor-β (TGF-β) production or Cytotoxic T-Lymphocyte Antigen (CTLA)-4-mediated interactions. Notably, only the interaction between uNK and decidual CD14$^+$ cells results in Treg cell induction, whereas cNK or CD14$^+$ cells isolated from peripheral blood are ineffective [169]. Therefore, uNK cells, in addition to being involved in the regulation of invading trophoblastic cells and in providing immunity during pregnancy, play an essential role in modulating maternal tolerance [7,128,132,152].

Although the details of signaling pathway triggering cytokines production are still elusive, IFN-γ and angiogenic growth factors such as PlGF produced by uNK cells contribute to spiral arteriole remodeling by acting on endothelial cells and decidual stromal cells [73,129,148,159,170]. By contrast, genetic evidence in humans and mice suggests that excessive inhibition of uNK cells function impedes both decidual arterial remodeling and fetal growth [75,148,167,170–172] (Figure 2). However, it has been reported that uNK cells depletion did not reproduce a deficiency in uterine spiral artery remodeling but was associated with a marked maternal uterine vasculopathy at a later time point in gestation in a rat model of PE induced by antibody-based NK cells depletion [171]. This evidence could account for the onset of endothelial dysfunction, and subsequently, PE in late pregnancy.

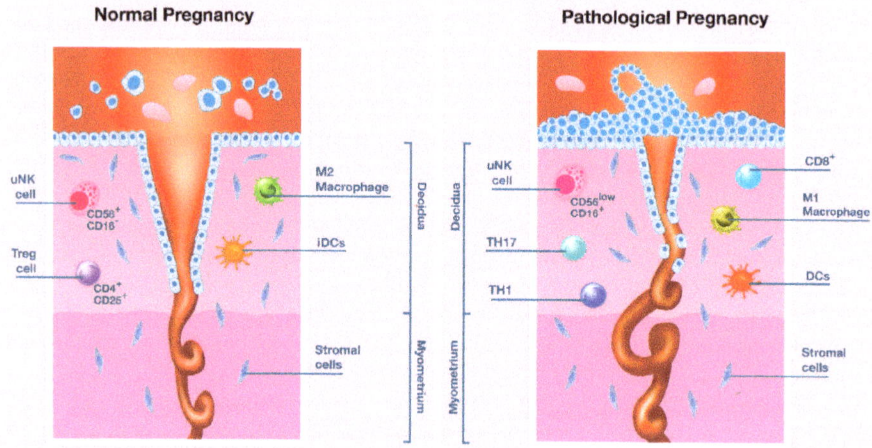

Figure 2. Reduced invasiveness and remodeling of maternal spiral arteries by trophoblastic cells is associated with a risk of pathological pregnancy.

During a normal pregnancy (Figure 2, left panel) with adequate PlGF levels, EVT cells migrate to the myometrium and infiltrate the endothelium of the maternal spiral arteries. This results in dilatation and increased flow of maternal blood at low pressure into the intervillous space. Uterine NK cells (CD56$^+$CD16$^-$) and M2 macrophages facilitate deep invasion of trophoblast cells into the myometrium. Moreover, tolerogenic immature dendritic cells (iDCs) promote Treg lymphocytes. Conversely, in pathological pregnancies (Figure 2, right panel) with unbalanced levels of PlGF, the immune cells in the uterine microenvironment (uNK CD56^{low}CD16$^+$ cells, M1 macrophages, mature DCs, Th1 and Th17 lymphocytes, and CD8$^+$ cells) fuel an excessive inflammatory response and reduce the invasion of the trophoblast with less remodeling of the spiral artery. Blood flows at higher pressure resulting in placental stress and reduced placental development.

Uterine NK cells have also been reported to contribute directly to fetal growth by producing growth-promoting factors essential for embryo development prior to the development of the placenta [173]. They also produce several cytokines that recruit and regulate cells of the adaptive immune system as well [161,174]. In addition to the production of PlGF and cytokines, an additional mechanism by which uNK could induce maternal-fetal tolerance appears to promote the induction of Treg cells and suppress Th17 cells [175,176]. Indeed, PE is associated with a reduction in Treg cells in the circulation and decidua and the severity of the disease is related to their reduced number [177]. It is plausible to assume that one of the mechanisms by which uNK cells induce maternal tolerance is through the production of an adequate level of PlGF. In fact, as previously mentioned, the involvement of Flt-1/VEGFR-1 and NRP1 receptors induces the secretion of IL-10 by T lymphocytes [88] and the delayed maturation of DCs [93], respectively, thus, favoring the development of Treg cells.

Of note, serum IL-17 levels have been shown to be significantly higher in PE patients than in healthy non-pregnant and pregnant women, and elevated IL-17 and sFlt-1/PlGF ratio serum level had an additive effect on the risk of PE [178]. Predictably, Th17 cells are higher in PE than in non-PE pregnancy, resulting in an imbalance between Treg/Th17 with a proinflammatory phenotype and an increased secretion of inflammatory cytokines [176]. This evidence has been recently confirmed by Yoo et al. Indeed, Th17 cells selectively secrete PlGF, and PlGF in turn, specifically induces the differentiation of inflammatory Th17 cells by activating the transcription factor STAT3 via binding to Flt-1/VEGFR-1. Moreover, PlGF replaces the activity of IL-6 in the production of IL-17, suppressing the generation of Treg cells [37]. Remarkably, as mentioned above, the hypoxic microenvironment reduces the killing capacity of uNK cells via STAT3 and, at the same time, upregulates PlGF production by uNK cells. By contrast, PlGF secreted by Th17 cells suppresses the generation of Treg cells via STAT3 and induces proinflammatory IL-17 secretion. Overall, these results suggest a refined regulation of the inflammatory response in the early stages of pregnancy, confirming the important role of PlGF in modulating the vascular development of the placenta, and also the behavior of the immune cells.

2.3. Macrophages

Macrophages are the second largest group of cells in the decidua and they comprise 20–30% of all leukocytes with $CD14^{+}CD206^{+}$ phenotype [179]. Macrophages are characterized by functional plasticity, and therefore their activation can be proinflammatory or anti-inflammatory. They can be classified as "classic" and "alternative" or M1 and M2, respectively. M1 macrophages, by triggering Th1 adaptive immune reactions, have inflammatory and antimicrobial properties and promote the destruction of tissue cells, while M2 macrophages have anti-inflammatory properties and contribute to tissue remodeling, angiogenesis, and wound healing [180–182]. The balance of polarization between M1 and M2 macrophages is important for different processes of normal pregnancy, such as trophoblast invasion, spiral artery remodeling, and apoptotic cell phagocytosis [183,184]. In normal pregnancy, many evidences have shown that uterine macrophages (uMφ) have an immunosuppressive phenotype, therefore, M2 immunosuppressive macrophages are necessary for normal pregnancy to also maintain fetal-maternal tolerance [185–187].

Although the ratio of M1/M2 macrophages changes during different gestation phases to protect the fetus from the maternal immune microenvironment, some studies have reported that PE could be associated with either a decrease or increase in the number of uMφ [8,185,188]. However, it appears that their phenotype is more important than their number and it is influenced by soluble factors in the microenvironment [182,187]. Moreover, the recruitment of macrophages is a receptor-dependent process and is largely regulated by chemotactic factors and by hypoxia [189–191]. In this regard, one of the main chemotactic factors capable of recruiting and polarizing macrophages is precisely the PlGF [82,189]. Additionally, as well as being a survival factor for macrophages, the binding of PlGF to VEGFR-1 promotes and stimulates activation (e.g., cytokine production) of macrophages [33,86].

Although much of the knowledge about PlGF-induced M2 macrophages polarization has been derived from evidence in tumor models, different in vitro and in vivo studies have proven that PlGF polarized macrophages to the M2 phenotype, which in turn, was characterized by PlGF upregulation [83,192]. In fact, in a laryngeal carcinoma model, PlGF-induced M2 polarization was associated with an increase in the expression level of MMP-9 through the activation of TGF-β, which in turn upregulated the PlGF [193]. Moreover, as reported in tumor models, hypoxia strongly increased macrophage-mediated T-cell suppression in vivo, in a HIF-1α macrophage expression-dependent manner [194,195]. Indeed, in addition to being involved in the process of ECM remodeling, angiogenesis, and invasiveness, MMP-9 has been reported to mediate immunosuppression of $CD8^+$ lymphocytes through a proteolytic process of IL-2Rα [196]. These results further confirm that human placental development is based on features of tumorigenesis, such as invasiveness, angiogenesis, and evasion of the immune response (Figure 1). Thus, uMφs regulate vascular remodeling by secreting PlGF, VEGF, MMP-9, and MMP-2, enabling trophoblast invasion, and their number is correlated with the expression levels of these angiogenic growth factors in the endometrium. Furthermore, the M2 macrophage phenotype has a higher angiogenic potential than other macrophage subsets, as has been shown in C57BL/6 J mice [197]. However, as macrophages migrate and accumulate in the most hypoxic regions, severe hypoxic (1–3% O_2) exposure significantly suppressed PlGF expression by M2c subset macrophages as compared with physiological hypoxia (5% O_2) [190,191]. Of note, M2c macrophages were the cells that produced the highest levels of PlGF [198].

A further link between macrophage polarization and PlGF has been attributed to histidine-rich glycoprotein (HRG). HRG is a heparin-binding plasma protein produced in the liver with anti-inflammatory effects and also synthesized by monocytes and macrophages [199]. HRG is transported as either a free protein or stored in α-granules of platelets and released after thrombin stimulation, and it modulates several functions, including coagulation, immune response, and vascularization, by binding to different cells such as endothelial cells, T cells, and macrophages. HRG shows both antiangiogenic and pro-angiogenic activity depending on the components of the microenvironment or on proteolytic cleavage of the antiangiogenic fragment of HRG [200]. For its features, HRG is also involved in the hypercoagulability and the angiogenic imbalance seen in early-onset PE [201,202]. Furthermore, specific HRG polymorphisms have been associated with recurrent miscarriage [203]. Given its role in angiogenesis, HRG could be responsible for an inadequate implantation and placentation [202,204].

In tumor models, HRG upregulation downregulated M2 markers such as IL-10, CCL22, and PlGF, while simultaneously increasing M1 markers such as IL-6 and CXCL9. Therefore, the reduced expression of IL-10 and CCL22 decreased the recruitment of Treg cells and, consequently, improved the function of DC and T cells and promoted the infiltration of $CD8^+$ T cells and NK cells. The mechanism by which HRG influenced M2 polarization was largely due to the downregulation of PlGF [192].

Whether HRG-mediated mechanisms in the tumor microenvironment could play a positive role in the immune response against cancer cells, how they influence a pregnancy remains to be elucidated. In this regard, the answer arises from several studies which have shown that HRG altered levels or polymorphisms were associated with the onset of PE [202,203]. Indeed, levels of HRG decreased during pregnancy in all women, but the levels were significantly lower in women who later developed PE than in normal pregnant women [202]. This finding has been partly explained by authors through the relative hypoxia during early pregnancy in women with PE due to defective placentation because of inappropriate trophoblast invasion of the maternal spiral arteries [202]. Alternatively, it could instead be explained by the onset of PE due to an inadequate amount of PlGF, therefore finely tuned PlGF levels in normal pregnancy are necessary to ensure adequate levels of HRG.

Reasonably, uMφ dysregulated polarization has been associated with inadequate remodeling of the uterine vessels and defective trophoblast invasion, and finally has led to spontaneous abortion, PE, and preterm birth [184–187,191]. In this regard, Li et al. reported that a pregnancy-induced hypertension (PIH) patient group exhibited a significantly higher percentage of $CD86^+$ cells (M1) and a significant lower percentage of CD163$^+$ cells (M2), representing a higher M1/M2 ratio than a control group. Moreover, the PIH group expressed higher concentrations of TNF-α and IL-1β, and expressed lower concentrations of IL-4, IL-10, and IL-13 than the control group, indicating a Th1 polarization [205]. It was clear that the functional maturation of macrophages was impaired in patients with PE and that a proinflammatory imbalance with a predominance of the M1 phenotype would be present. This finding was corroborated by an increase in proinflammatory cytokines (such as TNF-α, IL-6, and IL-8) and a decrease in anti-inflammatory cytokines (such as IL-10) in the placenta of preeclamptic patients [13,205–207].

uMφs also play a role in controlling local maternal immune responses because they are involved in a crosstalk with NK cells by secreting active TGF-β, which in turn inhibits NK cell effector functions [169,208]. Indeed, after complete placental development, uMφs shift toward a predominantly M2 phenotype, which promotes maternal immune tolerance and protects fetal growth until parturition [8]. Finally, Mφs also have an essential role in adaptive immunity through induction of T cells recruitment and activation and by B cells interaction. Therefore, placental-derived macrophage colony-stimulating factor (M-CSF) and IL-10 induce macrophages to produce IL-10 and CCL18, but not IL-12 or IL-23, thus, driving the expansion of CD25$^+$Foxp3$^+$ Treg cells in parallel with increased IL-10 production [8,169,209–211].

2.4. Dendritic Cells

DCs are the third cell group of innate immune cells present in the decidua. DCs have a key role in triggering the immune response by inducing the activation and differentiation of naïve T cells and simultaneously play a critical role for the development of tolerance [212]. The biological plasticity of DCs in promoting immunity or tolerance appears to be dependent on both their maturation state and the microenvironment in which they dwell. Thus, immature DCs are specifically efficient in antigen uptake and presentation, but the low expression levels of MHC gene products and T cell costimulatory molecules may contribute to peripheral tolerance under homeostatic conditions [213].

Concerning the uterine microenvironment, various stimuli such as soluble mediators (e.g., hormones, cytokines, and growth factors) or interactions with other immune cells modulate the tolerogenic activity of DCs, indicating that this behavior is context dependent. It has been reported that estrogen inhibited the ability of DCs to stimulate T cell proliferation and the production of both Th1/Th2 cytokines by upregulating IDO that supported the maturation of DCs with tolerogenic properties in rodents [214].

A recent review that focused on the pleiotropic nature of IFN-γ could help clarify the role played by DCs during early pregnancy, a phase characterized by inflammation [215]. Although IFN-γ exhibits extensive proinflammatory effects and has been associated with pathological pregnancies, it can also paradoxically exert an immunosuppressive role on both innate and adaptive immune cell types, promoting DC tolerance through the induction of IDO expression. IFN-γ induced IDO competence is not limited to immune cells but extends to other cell types such as epithelial cells, mesenchymal stem cells, and tumor cells. Therefore, the depletion of tryptophan leads to the inhibition of the effector T cells and promotes the differentiation of the FoxP3$^+$ Treg cells [215]. In addition, uMφ may also be an additional source of IFN-γ [216], although uMφ has been reported to suppress IFN-γ production by T cells costimulatory B7-H1/PD-1 signaling during early pregnancy. Therefore, B7-H1, expressed by uMφ, functions as a key regulator of local IFN-γ production, and thus contributes to the development of appropriate maternal immune responses [217]. Several studies have shown that when DCs were stimulated by IFN-γ in the absence of danger or pathogen-related signals, its effects induced predominantly tolerogenic features [218]. In this scenario, the crosstalk between decidua and immune cells contributes to the adequate vascular remodeling of the spiral arterioles and to the protection of the fetus from infections, and also guarantees the development of the tolerogenic microenvironment.

Uterine DCs (uDCs) appear to also have a role in implantation and decidua formation because the presence of CD11c$^+$ DCs is critical during early placentation, as the number of uDCs increases at the implantation period and the depletion of these cells leads to implantation failure in mouse models [219,220].

The close contact between uDCs and uNK cells in decidua suggests that there are important reciprocal interactions between them in shaping the decidualizing microenvironment during the early stages of pregnancy [169]. Indeed, uDCs appear to promote the differentiation of uNK cells, because uNK cell functions are impaired in association with decreased levels of IL-15 and IL-12 in implantation sites depleted of uDCs [159,221]. Therefore, the depletion of uDCs, as well as the altered uNK cells maturation, impair tissue remodeling and angiogenesis [221–224]. In addition, studies have shown that uterine cell proliferation was dependent on a synergistic effect between uDCs and uNK cells, confirming that uDCs-uNK cells crosstalk may be important for the endometrial stroma differentiation during implantation [160,166,169].

Implant failure was also observed in combinations of allogeneic and syngeneic mating after DCs depletion, as DCs depletion altered decidua response [219]. Again, the expression of sFlt-1/sVEGFR-1 in mature human monocyte-derived dendritic cells counteracted their angiogenic properties, and thus prevented adequate implantation [223]. However, since uDCs are generally in an immature state, their role in implantation, on the one hand, is likely redundant and potentially compensated by other immune cells. On the other hand, they likely are more important in playing an early key role in promoting tolerance to paternal antigens. Indeed, altered levels of Treg cells and DCs have been demonstrated in the peripheral blood of women with PE [222,224].

Concerning the PlGF role in inducing tolerance, it has been reported that PlGF inhibited the activation and maturation of human DCs differentiated from $CD14^+$ monocytes, effectively and rapidly through the NF-κB signaling pathway [225], confirming the immunosuppressive effect shown by angiogenic growth factors [226,227]. PlGF-treated DCs resulted in the downregulation of the expression of maturation markers CD80, CD86, CD83, CD40, and MHC-II expression, as well as the inhibition of IL-12, IL-8, and TNF-α production in response to LPS stimulation, with respect to untreated DCs. PlGF inhibited DCs maturation through the VEGFR-1, and this PlGF-induced DCs dysfunction was rescued by anti-human Flt-1/VEGFR-1 mAb. Additionally, the treatment of DCs with PlGF resulted in the suppression of naïve $CD4^+$ T cell proliferation in an allogeneic mixed lymphocyte reaction. The results from this study indicated that PlGF could downregulate Th1 immune responses by modulating the function of DCs [225].

Last but not least, PlGF is able to recruit bone marrow-derived myelomonocytic cells through its receptor Flt-1/VEGFR-1 via NFAT-1 transcriptional activation [34,228]. These myeloid-derived cells have been shown to exert immunomodulatory effects on the immune cells, especially DCs. Common features of bone marrow-derived myelomonocytic cells are the immature state and a remarkable ability to suppress T cell responses [226,227]. In addition to their suppressive effects on adaptive immune responses, they have also been reported to regulate innate immune responses by modulating cytokines production by macrophages such as IL-6, IL-10, CCL-22, and TGF-β. These cytokines, in turn, promote the differentiation of monocytes to mature macrophages but block their differentiation to DC, decrease the expression of DCs maturation markers, attenuate the Th1 immune response, as well as impair the activity of cytotoxic T lymphocytes and NK cells.

3. Bridging PlGF and Hypertension

In recent years, in addition to the known altered mechanisms involved in inducing hypertension (e.g., salt-water balance, cardiovascular function, and peripheral vascular resistance), several clinical and experimental evidences have supported the involvement of the immune system in the occurrence of hypertension [47–49,229–233]. This belief arises from observations on animal models, in which immune cells are crucial players in the onset of hypertension, infiltrating vessel walls and kidneys of hypertensive animals [234,235]. However, only a few studies have provided a mechanistic explanation for how immune cell functions promote blood pressure increases. Severe combined immunodeficiency mice or recombinase-activating gene-1 (*Rag-1*$^{-/-}$) knockout mice, which lacked both T and B lymphocytes, showed blunted hypertension responses and did not develop abnormalities of vascular function to angiotensin II (ANGII) treatment [236,237]. However, hypertension and vascular dysfunction were restored when *Rag-1*$^{-/-}$ mice received adoptive transfer of T cells, but not B cells. Furthermore, after the adoptive transfer of T cells, *Rag-1*$^{-/-}$ mice challenged with ANGII significantly increased T cell production of IFN-γ and TNF-α and treatment with a TNF-α antagonist prevented hypertension induced by ANGII, indicating the role of inflammation in inducing hypertension [237]. In addition, IL-10 knockout mice, unable to direct T cells to an anti-inflammatory phenotype, developed symptoms similar to PE when pregnant, confirming the role of T cells in the onset of this syndrome [238]. Conversely, IL-10 treatment reduced inflammation, endothelial dysfunction, and blood pressure in hypertensive pregnant rats [239]. Again, the evidence indicates that production of proinflammatory cytokines, including TNF-α, IL-17, and IL-6, contribute to hypertension, likely by promoting vasoconstriction, production of reactive oxygen species, and sodium reabsorption in the kidney [240,241]. Remarkably, each of these cytokines is reciprocally interconnected with PlGF because, by activating the same transcription factors, they create a self-feeding circuit [34,36]. Overall, these observations support the role of PlGF in the immune system and also the development of PE.

Experimental evidence in pregnant animals indicates that PlGF is a potent arterial vasodilator and may participate in the mechanisms regulating the maternal vascular tone during pregnancy [242]. Therefore, an inadequate level of PlGF during pregnancy may be accountable for the onset of PE [51,59,77,104,243]. A recent study revealed that PlGF had biological effects on samples of uterine arteries, obtained from normotensive women undergoing cesarean hysterectomy, when challenged with ANGII. This study showed that PlGF contributed to the blunted vascular response to ANGII during normotensive pregnancies and sFlt-1/sVEGFR-1 appeared to attenuate this effect contributing to the regulation of vascular tone by altering the vascular response to ANGII [244].

Although the role of PlGF in inducing hypertension during pregnancy has been associated with several mechanisms involving immune cells and cytokines dysregulation, a definitive link between PlGF-induced hypertension and the immune system, even in non-pregnant animal models, was reported in a study by Carnevale et al. The authors demonstrated that PlGF, through mediating a neuroimmune interaction, played a key role in the spleen's immune system by organizing the T-cell response to hypertensive challenge in ANGII-treated mice. Whereas the chronic ANGII infusion produced a progressive increase in blood pressure in wild-type (wt) mice, due to an accumulation of T lymphocytes in the aortic wall and kidney, the hypertensive response was completely abolished in *plgf*$^{-/-}$ mice. Mechanistically, PlGF mediated the hypertension response to ANGII challenge by repressing tissue inhibitor of metalloproteinases (Timp)-3 protein expression in macrophages, through the transcriptional Sirtuin (Sirt)1-p53 axis. In turn, Timp3 repression allowed the costimulation of T cells and their deployment toward classical organs involved in hypertension [245]. Overall, these data are definitive evidence of the link between PlGF and hypertension through immune-mediated mechanisms.

4. Conclusions

The most common conditions encountered during pathological pregnancy are hypertensive complications, including PE, PIH, and maternal chronic hypertension. PE is a systemic syndrome affecting about 5–8% of all pregnancies. It is characterized by new onset of hypertension and proteinuria after 20 weeks of gestation and is a major cause of maternal mortality (15–20% in developed countries) and one of the major causes of infant morbidity and mortality, perinatal deaths, preterm birth, and intrauterine growth restriction [246–248].

PE may be the result of abnormal activation of the maternal immune system characterized by endothelial dysfunction and excessive inflammation [57,112,113,206,239,241]. Knowledge regarding the involvement of the immune system in both hypertension and in pregnancies complicated by PE has evolved significantly. During pregnancy, it was believed that alterations in the levels of angiogenic/anti-angiogenic factors such as PlGF and sFlt-1/sVEGFR-1 were essentially responsible for impaired vascular placental development, increased vascular resistance of spiral arteries, and endothelial dysfunction. Today, we know that these factors also have a decisive role in modulating the maternal immune response throughout the pregnancy. They are not only accountable for alterations of immune cell behavior but also for the altered release of proinflammatory mediators and systemic involvement of the inflammatory response.

PlGF showed an important effect on the development of hypertension by modulating egress of T cells from the spleen and their accumulation in vessel walls and kidneys, even in non-pregnant mouse model [245]. It is not surprisingly that its influence is modulated by the involvement of typical pathways implicated in tumor growth. Indeed, some of the effects attributed to PlGF inhibition in cancer might rely on epigenetic modulation of the p53-Timp3 axis, which is well known to also play a crucial role in tumor growth [249], again confirming the presence of overlapping mechanisms between pregnancy and malignancy.

Presently, the diagnostic criteria of PE remain uncertain because no known specific biomarkers are available, thus, women at risk are identified based on epidemiological and clinical risk factors [250]. Many studies on PE diagnosis, which have analyzed the sFlt-1/sVEGFR-1/PlGF ratio as a predictor for poor pregnancy, have demonstrated its reliability [100,102–104]. However, PlGF imbalance during pregnancy can be due to several mechanisms, which preclude the precise identification of a specific marker useful for diagnosis. In this regard, different scenarios can be expected. On the one hand, PlGF upregulation can be accountable for the activation of an inflammatory state associated with the release of mediators such as TNF-α, IL-6, IL-17, and COX-2. These cytokines, by modulating the cells of the immune system (e.g., M1 polarized Mφ, mature DCs, Th1 lymphocytes, reduction of Treg cells) could prevent maternal tolerance, fuel inflammation, and, at the same time, interfere with adequate vascular development of the placenta. Both of these events contribute to mechanisms that promote hypertension. On the other hand, reduced levels of PlGF, due either to inadequate production by trophoblast cells, uterine cells and uNK cells, or to its excessive sequestration by sFlt1/sVEGFR-1, can be responsible for the inadequate trophoblast invasion, the deficient vascular development and, at the same time, for a loss of maternal tolerance. Yet again, these events contribute to mechanisms that promote hypertension. Therefore, to date, a single reliable diagnostic marker is still difficult to identify. In the future, the definitive understanding of the immunological mechanisms involved in the onset of different PE phenotypes, such as placental PE associated with growth restriction, PE associated with a chronic maternal inflammatory condition, and maternal dysmetabolism associated with normal fetal growth, in which PlGF could be involved, may offer new specific diagnostic markers and therapeutic tools for managing pathological pregnancies.

Author Contributions: Conceptualization, L.A. and R.B.; Funding acquisition, R.B.; Supervision, L.A. and R.B.; writing—original draft, L.A. and R.B.; writing—review and editing, M.B., L.A., C.F., L.C., M.T.M., F.L., and V.M. All authors gave a significant intellectual contribution to the article. All authors have read and agreed to the published version of the manuscript.

Funding: The present article was supported by a grant from the University of Rome "Tor Vergata" BeiREPRVanillina (CUP: E88D19000980005 to R.B.).

Acknowledgments: The authors wish to thank Elena Bove for the graphic design of Figures 1 and 2 and for the creation of the graphic abstract.

Conflicts of Interest: The authors declare no conflict of interest.

Footnote: Figure 1 was modified by Holtan et al. [19] with kind permission by Publisher, License Number: 4911830648497.

References

1. Billingham, R.E.; Brent, L.; Medawar, P.B. Actively acquired tolerance of foreign cells. *Nature* **1953**, *172*, 603–606. [CrossRef] [PubMed]
2. Erlebacher, A. Immunology of the maternal-fetal interface. *Annu. Rev. Immunol.* **2013**, *31*, 387–411. [CrossRef] [PubMed]

3. Mor, G.; Aldo, P.; Alvero, A.B. The unique immunological and microbial aspects of pregnancy. *Nat. Rev. Immunol.* **2017**, *17*, 469–482. [CrossRef] [PubMed]
4. Schumacher, A.; Sharkey, D.J.; Robertson, S.A.; Zenclussen, A.C. Immune Cells at the Fetomaternal Interface: How the Microenvironment Modulates Immune Cells To Foster Fetal Development. *J. Immunol.* **2018**, *201*, 325–334. [CrossRef] [PubMed]
5. Aluvihare, V.R.; Kallikourdis, M.; Betz, A.G. Regulatory T cells mediate maternal tolerance to the fetus. *Nat. Immunol.* **2004**, *5*, 266–271. [CrossRef]
6. Bonney, E.A. Alternative theories: Pregnancy and immune tolerance. *J. Reprod. Immunol.* **2017**, *123*, 65–71. [CrossRef]
7. Colucci, F. The immunological code of pregnancy. *Science* **2019**, *365*, 862–863. [CrossRef]
8. Svensson-Arvelund, J.; Mehta, R.B.; Lindau, R.; Mirrasekhian, E.; Rodriguez-Martinez, H.; Berg, G.; Lash, G.E.; Jenmalm, M.C.; Ernerudh, J. The human fetal placenta promotes tolerance against the semiallogeneic fetus by inducing regulatory T cells and homeostatic M2 macrophages. *J. Immunol.* **2015**, *194*, 1534–1544. [CrossRef]
9. Trowsdale, J.; Betz, A.G. Mother's little helpers: Mechanisms of maternal-fetal tolerance. *Nat. Immunol.* **2006**, *7*, 241–246. [CrossRef]
10. Valencia-Ortega, J.; Saucedo, R.; Pena-Cano, M.I.; Hernandez-Valencia, M.; Cruz-Duran, J.G. Immune tolerance at the maternal-placental interface in healthy pregnancy and pre-eclampsia. *J. Obstet. Gynaecol. Res.* **2020**, *46*, 1067–1076. [CrossRef]
11. Brien, M.E.; Boufaied, I.; Bernard, N.; Forest, J.C.; Giguere, Y.; Girard, S. Specific inflammatory profile in each pregnancy complication: A comparative study. *Am. J. Reprod. Immunol.* **2020**, e13316. [CrossRef]
12. Griffith, O.W.; Chavan, A.R.; Protopapas, S.; Maziarz, J.; Romero, R.; Wagner, G.P. Embryo implantation evolved from an ancestral inflammatory attachment reaction. *Proc. Natl. Acad. Sci. USA* **2017**, *114*, E6566–E6575. [CrossRef] [PubMed]
13. Nadeau-Vallee, M.; Obari, D.; Palacios, J.; Brien, M.E.; Duval, C.; Chemtob, S.; Girard, S. Sterile inflammation and pregnancy complications: A review. *Reproduction* **2016**, *152*, R277–R292. [CrossRef] [PubMed]
14. Nejabati, H.R.; Latifi, Z.; Ghasemnejad, T.; Fattahi, A.; Nouri, M. Placental growth factor (PlGF) as an angiogenic/inflammatory switcher: Lesson from early pregnancy losses. *Gynecol. Endocrinol.* **2017**, *33*, 668–674. [CrossRef]
15. Zenclussen, A.C.; Hammerling, G.J. Cellular Regulation of the Uterine Microenvironment That Enables Embryo Implantation. *Front. Immunol.* **2015**, *6*, 321. [CrossRef] [PubMed]
16. Highet, A.R.; Khoda, S.M.; Buckberry, S.; Leemaqz, S.; Bianco-Miotto, T.; Harrington, E.; Ricciardelli, C.; Roberts, C.T. Hypoxia induced HIF-1/HIF-2 activity alters trophoblast transcriptional regulation and promotes invasion. *Eur. J. Cell Biol.* **2015**, *94*, 589–602. [CrossRef] [PubMed]
17. Huppertz, B. Traditional and New Routes of Trophoblast Invasion and Their Implications for Pregnancy Diseases. *Int. J. Mol. Sci.* **2019**, *21*, 289. [CrossRef] [PubMed]
18. Imakawa, K.; Bai, R.; Fujiwara, H.; Ideta, A.; Aoyagi, Y.; Kusama, K. Continuous model of conceptus implantation to the maternal endometrium. *J. Endocrinol.* **2017**, *233*, R53–R65. [CrossRef]
19. Burton, G.J.; Jauniaux, E.; Murray, A.J. Oxygen and placental development; parallels and differences with tumour biology. *Placenta* **2017**, *56*, 14–18. [CrossRef]
20. DaSilva-Arnold, S.; James, J.L.; Al-Khan, A.; Zamudio, S.; Illsley, N.P. Differentiation of first trimester cytotrophoblast to extravillous trophoblast involves an epithelial-mesenchymal transition. *Placenta* **2015**, *36*, 1412–1418. [CrossRef]
21. Holtan, S.G.; Creedon, D.J.; Haluska, P.; Markovic, S.N. Cancer and pregnancy: Parallels in growth, invasion, and immune modulation and implications for cancer therapeutic agents. *Mayo Clin. Proc.* **2009**, *84*, 985–1000. [CrossRef]
22. Kareva, I. Immune Suppression in Pregnancy and Cancer: Parallels and Insights. *Transl. Oncol.* **2020**, *13*, 100759. [CrossRef] [PubMed]
23. Albonici, L.; Giganti, M.G.; Modesti, A.; Manzari, V.; Bei, R. Multifaceted Role of the Placental Growth Factor (PlGF) in the Antitumor Immune Response and Cancer Progression. *Int. J. Mol. Sci.* **2019**, *20*, 2970. [CrossRef] [PubMed]
24. Costanzo, V.; Bardelli, A.; Siena, S.; Abrignani, S. Exploring the links between cancer and placenta development. *Open Biol.* **2018**, *8*, 180081. [CrossRef]

25. Ferroni, P.; Palmirotta, R.; Spila, A.; Martini, F.; Formica, V.; Portarena, I.; Del Monte, G.; Buonomo, O.; Roselli, M.; Guadagni, F. Prognostic value of carcinoembryonic antigen and vascular endothelial growth factor tumor tissue content in colorectal cancer. *Oncology* **2006**, *71*, 176–184. [CrossRef]
26. PrabhuDas, M.; Bonney, E.; Caron, K.; Dey, S.; Erlebacher, A.; Fazleabas, A.; Fisher, S.; Golos, T.; Matzuk, M.; McCune, J.M.; et al. Immune mechanisms at the maternal-fetal interface: Perspectives and challenges. *Nat. Immunol.* **2015**, *16*, 328–334. [CrossRef]
27. Achen, M.G.; Gad, J.M.; Stacker, S.A.; Wilks, A.F. Placenta growth factor and vascular endothelial growth factor are co-expressed during early embryonic development. *Growth Factors* **1997**, *15*, 69–80. [CrossRef]
28. Athanassiades, A.; Lala, P.K. Role of placenta growth factor (PIGF) in human extravillous trophoblast proliferation, migration and invasiveness. *Placenta* **1998**, *19*, 465–473. [CrossRef]
29. Binder, N.K.; Evans, J.; Salamonsen, L.A.; Gardner, D.K.; Kaitu'u-Lino, T.J.; Hannan, N.J. Placental Growth Factor Is Secreted by the Human Endometrium and Has Potential Important Functions during Embryo Development and Implantation. *PLoS ONE* **2016**, *11*, e0163096. [CrossRef]
30. Holme, A.M.; Roland, M.C.; Henriksen, T.; Michelsen, T.M. In vivo uteroplacental release of placental growth factor and soluble Fms-like tyrosine kinase-1 in normal and preeclamptic pregnancies. *Am. J. Obstet. Gynecol.* **2016**, *215*, 782e1–782e9. [CrossRef]
31. Knuth, A.; Liu, L.; Nielsen, H.; Merril, D.; Torry, D.S.; Arroyo, J.A. Placenta growth factor induces invasion and activates p70 during rapamycin treatment in trophoblast cells. *Am. J. Reprod. Immunol.* **2015**, *73*, 330–340. [CrossRef] [PubMed]
32. Vrachnis, N.; Kalampokas, E.; Sifakis, S.; Vitoratos, N.; Kalampokas, T.; Botsis, D.; Iliodromiti, Z. Placental growth factor (PlGF): A key to optimizing fetal growth. *J. Matern. Fetal Neonatal Med.* **2013**, *26*, 995–1002. [CrossRef] [PubMed]
33. Adini, A.; Kornaga, T.; Firoozbakht, F.; Benjamin, L.E. Placental growth factor is a survival factor for tumor endothelial cells and macrophages. *Cancer Res.* **2002**, *62*, 2749–2752. [PubMed]
34. Ding, Y.; Huang, Y.; Song, N.; Gao, X.; Yuan, S.; Wang, X.; Cai, H.; Fu, Y.; Luo, Y. NFAT1 mediates placental growth factor-induced myelomonocytic cell recruitment via the induction of TNF-alpha. *J. Immunol.* **2010**, *184*, 2593–2601. [CrossRef] [PubMed]
35. Newell, L.F.; Holtan, S.G.; Yates, J.E.; Pereira, L.; Tyner, J.W.; Burd, I.; Bagby, G.C. PlGF enhances TLR-dependent inflammatory responses in human mononuclear phagocytes. *Am. J. Reprod. Immunol.* **2017**, *78*, e12709. [CrossRef]
36. Wu, M.Y.; Yang, R.S.; Lin, T.H.; Tang, C.H.; Chiu, Y.C.; Liou, H.C.; Fu, W.M. Enhancement of PLGF production by 15-(S)-HETE via PI3K-Akt, NF-kappaB and COX-2 pathways in rheumatoid arthritis synovial fibroblast. *Eur. J. Pharmacol.* **2013**, *714*, 388–396. [CrossRef]
37. Yoo, S.A.; Kim, M.; Kang, M.C.; Kong, J.S.; Kim, K.M.; Lee, S.; Hong, B.K.; Jeong, G.H.; Lee, J.; Shin, M.G.; et al. Placental growth factor regulates the generation of TH17 cells to link angiogenesis with autoimmunity. *Nat. Immunol.* **2019**, *20*, 1348–1359. [CrossRef]
38. Clapp, J.F., 3rd; Capeless, E. Cardiovascular function before, during, and after the first and subsequent pregnancies. *Am. J. Cardiol.* **1997**, *80*, 1469–1473. [CrossRef]
39. Ridder, A.; Giorgione, V.; Khalil, A.; Thilaganathan, B. Preeclampsia: The Relationship between Uterine Artery Blood Flow and Trophoblast Function. *Int. J. Mol. Sci.* **2019**, *20*, 3263. [CrossRef]
40. Tay, J.; Masini, G.; McEniery, C.M.; Giussani, D.A.; Shaw, C.J.; Wilkinson, I.B.; Bennett, P.R.; Lees, C.C. Uterine and fetal placental Doppler indices are associated with maternal cardiovascular function. *Am. J. Obstet. Gynecol.* **2019**, *220*, 96e1–96e8. [CrossRef]
41. Thilaganathan, B. Pre-eclampsia and the cardiovascular-placental axis. *Ultrasound Obstet. Gynecol.* **2018**, *51*, 714–717. [CrossRef] [PubMed]
42. Thilaganathan, B.; Kalafat, E. Cardiovascular System in Preeclampsia and Beyond. *Hypertension* **2019**, *73*, 522–531. [CrossRef] [PubMed]
43. Yagel, S.; Verlohren, S. The Role of the Placenta in the Development of Preeclampsia: Revisited. *Ultrasound Obstet. Gynecol.* **2020**. [CrossRef] [PubMed]
44. Borzychowski, A.M.; Croy, B.A.; Chan, W.L.; Redman, C.W.; Sargent, I.L. Changes in systemic type 1 and type 2 immunity in normal pregnancy and pre-eclampsia may be mediated by natural killer cells. *Eur. J. Immunol.* **2005**, *35*, 3054–3063. [CrossRef]

45. Care, A.S.; Bourque, S.L.; Morton, J.S.; Hjartarson, E.P.; Robertson, S.A.; Davidge, S.T. Reduction in Regulatory T Cells in Early Pregnancy Causes Uterine Artery Dysfunction in Mice. *Hypertension* **2018**, *72*, 177–187. [CrossRef]
46. Faas, M.M.; De Vos, P. Innate immune cells in the placental bed in healthy pregnancy and preeclampsia. *Placenta* **2018**, *69*, 125–133. [CrossRef]
47. Lu, H.Q.; Hu, R. The role of immunity in the pathogenesis and development of pre-eclampsia. *Scand. J. Immunol.* **2019**, *90*, e12756. [CrossRef]
48. Norlander, A.E.; Madhur, M.S.; Harrison, D.G. The immunology of hypertension. *J. Exp. Med.* **2018**, *215*, 21–33. [CrossRef]
49. Rambaldi, M.P.; Weiner, E.; Mecacci, F.; Bar, J.; Petraglia, F. Immunomodulation and preeclampsia. *Best Pract. Res. Clin. Obstet. Gynaecol.* **2019**, *60*, 87–96. [CrossRef]
50. Agarwal, I.; Karumanchi, S.A. Preeclampsia and the Anti-Angiogenic State. *Pregnancy Hypertens.* **2011**, *1*, 17–21. [CrossRef]
51. Agrawal, S.; Cerdeira, A.S.; Redman, C.; Vatish, M. Meta-Analysis and Systematic Review to Assess the Role of Soluble FMS-Like Tyrosine Kinase-1 and Placenta Growth Factor Ratio in Prediction of Preeclampsia: The SaPPPhirE Study. *Hypertension* **2018**, *71*, 306–316. [CrossRef] [PubMed]
52. Ali, L.E.; Salih, M.M.; Elhassan, E.M.; Mohmmed, A.A.; Adam, I. Placental growth factor, vascular endothelial growth factor, and hypoxia-inducible factor-1alpha in the placentas of women with pre-eclampsia. *J. Matern. Fetal Neonatal Med.* **2019**, *32*, 2628–2632. [CrossRef] [PubMed]
53. Chau, K.; Hennessy, A.; Makris, A. Placental growth factor and pre-eclampsia. *J. Hum. Hypertens.* **2017**, *31*, 782–786. [CrossRef] [PubMed]
54. Hurrell, A.; Beardmore-Gray, A.; Duhig, K.; Webster, L.; Chappell, L.C.; Shennan, A.H. Placental growth factor in suspected preterm pre-eclampsia: A review of the evidence and practicalities of implementation. *BJOG* **2020**, *127*, 1590–1597. [CrossRef] [PubMed]
55. Levine, R.J.; Maynard, S.E.; Qian, C.; Lim, K.H.; England, L.J.; Yu, K.F.; Schisterman, E.F.; Thadhani, R.; Sachs, B.P.; Epstein, F.H.; et al. Circulating angiogenic factors and the risk of preeclampsia. *N. Engl. J. Med.* **2004**, *350*, 672–683. [CrossRef]
56. Maynard, S.E.; Min, J.Y.; Merchan, J.; Lim, K.H.; Li, J.; Mondal, S.; Libermann, T.A.; Morgan, J.P.; Sellke, F.W.; Stillman, I.E.; et al. Excess placental soluble fms-like tyrosine kinase 1 (sFlt1) may contribute to endothelial dysfunction, hypertension, and proteinuria in preeclampsia. *J. Clin. Invest.* **2003**, *111*, 649–658. [CrossRef]
57. Sarween, N.; Drayson, M.T.; Hodson, J.; Knox, E.M.; Plant, T.; Day, C.J.; Lipkin, G.W. Humoral immunity in late-onset Pre-eclampsia and linkage with angiogenic and inflammatory markers. *Am. J. Reprod. Immunol.* **2018**, *80*, e13041. [CrossRef]
58. Sezer, S.D.; Kucuk, M.; Doger, F.K.; Yuksel, H.; Odabasi, A.R.; Turkmen, M.K.; Cakmak, B.C.; Omurlu, I.K.; Kinas, M.G. VEGF, PIGF and HIF-1alpha in placentas of early- and late-onset pre-eclamptic patients. *Gynecol. Endocrinol.* **2013**, *29*, 797–800. [CrossRef]
59. Zeisler, H.; Llurba, E.; Chantraine, F.J.; Vatish, M.; Staff, A.C.; Sennstrom, M.; Olovsson, M.; Brennecke, S.P.; Stepan, H.; Allegranza, D.; et al. Soluble fms-like tyrosine kinase-1 to placental growth factor ratio: Ruling out pre-eclampsia for up to 4 weeks and value of retesting. *Ultrasound Obstet. Gynecol.* **2019**, *53*, 367–375. [CrossRef]
60. Thadhani, R.; Mutter, W.P.; Wolf, M.; Levine, R.J.; Taylor, R.N.; Sukhatme, V.P.; Ecker, J.; Karumanchi, S.A. First trimester placental growth factor and soluble fms-like tyrosine kinase 1 and risk for preeclampsia. *J. Clin. Endocrinol. Metab.* **2004**, *89*, 770–775. [CrossRef]
61. De Falco, S. The discovery of placenta growth factor and its biological activity. *Exp. Mol. Med.* **2012**, *44*, 1–9. [CrossRef] [PubMed]
62. Maglione, D.; Guerriero, V.; Viglietto, G.; Delli-Bovi, P.; Persico, M.G. Isolation of a human placenta cDNA coding for a protein related to the vascular permeability factor. *Proc. Natl. Acad. Sci. USA* **1991**, *88*, 9267–9271. [CrossRef]
63. Autiero, M.; Luttun, A.; Tjwa, M.; Carmeliet, P. Placental growth factor and its receptor, vascular endothelial growth factor receptor-1: Novel targets for stimulation of ischemic tissue revascularization and inhibition of angiogenic and inflammatory disorders. *J. Thromb. Haemost.* **2003**, *1*, 1356–1370. [CrossRef] [PubMed]
64. Dewerchin, M.; Carmeliet, P. PlGF: A multitasking cytokine with disease-restricted activity. *Cold Spring Harb. Perspect. Med.* **2012**, *2*, a011056. [CrossRef] [PubMed]

65. Oura, H.; Bertoncini, J.; Velasco, P.; Brown, L.F.; Carmeliet, P.; Detmar, M. A critical role of placental growth factor in the induction of inflammation and edema formation. *Blood* **2003**, *101*, 560–567. [CrossRef] [PubMed]
66. Roy, S.; Bag, A.K.; Singh, R.K.; Talmadge, J.E.; Batra, S.K.; Datta, K. Multifaceted Role of Neuropilins in the Immune System: Potential Targets for Immunotherapy. *Front. Immunol.* **2017**, *8*, 1228. [CrossRef]
67. Simons, M.; Gordon, E.; Claesson-Welsh, L. Mechanisms and regulation of endothelial VEGF receptor signalling. *Nat. Rev. Mol. Cell. Biol.* **2016**, *17*, 611–625. [CrossRef]
68. Torry, R.J.; Tomanek, R.J.; Zheng, W.; Miller, S.J.; Labarrere, C.A.; Torry, D.S. Hypoxia increases placenta growth factor expression in human myocardium and cultured neonatal rat cardiomyocytes. *J. Heart Lung Transpl.* **2009**, *28*, 183–190. [CrossRef]
69. Tudisco, L.; Orlandi, A.; Tarallo, V.; De Falco, S. Hypoxia activates placental growth factor expression in lymphatic endothelial cells. *Oncotarget* **2017**, *8*, 32873–32883. [CrossRef]
70. Gobble, R.M.; Groesch, K.A.; Chang, M.; Torry, R.J.; Torry, D.S. Differential regulation of human PlGF gene expression in trophoblast and nontrophoblast cells by oxygen tension. *Placenta* **2009**, *30*, 869–875. [CrossRef]
71. Luttun, A.; Brusselmans, K.; Fukao, H.; Tjwa, M.; Ueshima, S.; Herbert, J.M.; Matsuo, O.; Collen, D.; Carmeliet, P.; Moons, L. Loss of placental growth factor protects mice against vascular permeability in pathological conditions. *Biochem. Biophys. Res. Commun.* **2002**, *295*, 428–434. [CrossRef]
72. Tayade, C.; Hilchie, D.; He, H.; Fang, Y.; Moons, L.; Carmeliet, P.; Foster, R.A.; Croy, B.A. Genetic deletion of placenta growth factor in mice alters uterine NK cells. *J. Immunol.* **2007**, *178*, 4267–4275. [CrossRef] [PubMed]
73. Ratsep, M.T.; Felker, A.M.; Kay, V.R.; Tolusso, L.; Hofmann, A.P.; Croy, B.A. Uterine natural killer cells: Supervisors of vasculature construction in early decidua basalis. *Reproduction* **2015**, *149*, R91–R102. [CrossRef] [PubMed]
74. Kang, M.C.; Park, S.J.; Kim, H.J.; Lee, J.; Yu, D.H.; Bae, K.B.; Ji, Y.R.; Park, S.J.; Jeong, J.; Jang, W.Y.; et al. Gestational loss and growth restriction by angiogenic defects in placental growth factor transgenic mice. *Arterioscler. Thromb. Vasc. Biol.* **2014**, *34*, 2276–2282. [CrossRef]
75. Kopcow, H.D.; Karumanchi, S.A. Angiogenic factors and natural killer (NK) cells in the pathogenesis of preeclampsia. *J. Reprod. Immunol.* **2007**, *76*, 23–29. [CrossRef]
76. Ratsep, M.T.; Carmeliet, P.; Adams, M.A.; Croy, B.A. Impact of placental growth factor deficiency on early mouse implant site angiogenesis. *Placenta* **2014**, *35*, 772–775. [CrossRef]
77. Yonekura Collier, A.R.; Zsengeller, Z.; Pernicone, E.; Salahuddin, S.; Khankin, E.V.; Karumanchi, S.A. Placental sFLT1 is associated with complement activation and syncytiotrophoblast damage in preeclampsia. *Hypertens. Pregnancy* **2019**, *38*, 193–199. [CrossRef]
78. Mamluk, R.; Gechtman, Z.; Kutcher, M.E.; Gasiunas, N.; Gallagher, J.; Klagsbrun, M. Neuropilin-1 binds vascular endothelial growth factor 165, placenta growth factor-2, and heparin via its b1b2 domain. *J. Biol. Chem.* **2002**, *277*, 24818–24825. [CrossRef]
79. Migdal, M.; Huppertz, B.; Tessler, S.; Comforti, A.; Shibuya, M.; Reich, R.; Baumann, H.; Neufeld, G. Neuropilin-1 is a placenta growth factor-2 receptor. *J. Biol. Chem.* **1998**, *273*, 22272–22278. [CrossRef]
80. Neufeld, G.; Kessler, O.; Herzog, Y. The interaction of Neuropilin-1 and Neuropilin-2 with tyrosine-kinase receptors for VEGF. *Adv. Exp. Med. Biol.* **2002**, *515*, 81–90.
81. Anisimov, A.; Leppanen, V.M.; Tvorogov, D.; Zarkada, G.; Jeltsch, M.; Holopainen, T.; Kaijalainen, S.; Alitalo, K. The basis for the distinct biological activities of vascular endothelial growth factor receptor-1 ligands. *Sci. Signal* **2013**, *6*, ra52. [CrossRef] [PubMed]
82. Eltzschig, H.K.; Carmeliet, P. Hypoxia and inflammation. *N. Engl. J. Med.* **2011**, *364*, 656–665. [CrossRef] [PubMed]
83. Incio, J.; Tam, J.; Rahbari, N.N.; Suboj, P.; McManus, D.T.; Chin, S.M.; Vardam, T.D.; Batista, A.; Babykutty, S.; Jung, K.; et al. PlGF/VEGFR-1 Signaling Promotes Macrophage Polarization and Accelerated Tumor Progression in Obesity. *Clin. Cancer Res.* **2016**, *22*, 2993–3004. [CrossRef] [PubMed]
84. Perrotta, M.; Lori, A.; Carnevale, L.; Fardella, S.; Cifelli, G.; Iacobucci, R.; Mastroiacovo, F.; Iodice, D.; Pallante, F.; Storto, M.; et al. Deoxycorticosterone acetate-salt hypertension activates placental growth factor in the spleen to couple sympathetic drive and immune system activation. *Cardiovasc. Res.* **2018**, *114*, 456–467. [CrossRef] [PubMed]
85. Sawano, A.; Iwai, S.; Sakurai, Y.; Ito, M.; Shitara, K.; Nakahata, T.; Shibuya, M. Flt-1, vascular endothelial growth factor receptor 1, is a novel cell surface marker for the lineage of monocyte-macrophages in humans. *Blood* **2001**, *97*, 785–791. [CrossRef] [PubMed]

86. Selvaraj, S.K.; Giri, R.K.; Perelman, N.; Johnson, C.; Malik, P.; Kalra, V.K. Mechanism of monocyte activation and expression of proinflammatory cytochemokines by placenta growth factor. *Blood* **2003**, *102*, 1515–1524. [CrossRef]
87. Tanaka, K.; Watanabe, M.; Tanigaki, S.; Iwashita, M.; Kobayashi, Y. Tumor necrosis factor-alpha regulates angiogenesis of BeWo cells via synergy of PlGF/VEGFR1 and VEGF-A/VEGFR2 axes. *Placenta* **2018**, *74*, 20–27. [CrossRef]
88. Shin, J.Y.; Yoon, I.H.; Kim, J.S.; Kim, B.; Park, C.G. Vascular endothelial growth factor-induced chemotaxis and IL-10 from T cells. *Cell Immunol.* **2009**, *256*, 72–78. [CrossRef]
89. Zachary, I. Neuropilins: Role in signalling, angiogenesis and disease. *Chem. Immunol. Allergy* **2014**, *99*, 37–70.
90. Lepelletier, Y.; Moura, I.C.; Hadj-Slimane, R.; Renand, A.; Fiorentino, S.; Baude, C.; Shirvan, A.; Barzilai, A.; Hermine, O. Immunosuppressive role of semaphorin-3A on T cell proliferation is mediated by inhibition of actin cytoskeleton reorganization. *Eur. J. Immunol.* **2006**, *36*, 1782–1793. [CrossRef]
91. Kalekar, L.A.; Schmiel, S.E.; Nandiwada, S.L.; Lam, W.Y.; Barsness, L.O.; Zhang, N.; Stritesky, G.L.; Malhotra, D.; Pauken, K.E.; Linehan, J.L.; et al. CD4(+) T cell anergy prevents autoimmunity and generates regulatory T cell precursors. *Nat. Immunol.* **2016**, *17*, 304–314. [CrossRef] [PubMed]
92. Yadav, M.; Louvet, C.; Davini, D.; Gardner, J.M.; Martinez-Llordella, M.; Bailey-Bucktrout, S.; Anthony, B.A.; Sverdrup, F.M.; Head, R.; Kuster, D.J.; et al. Neuropilin-1 distinguishes natural and inducible regulatory T cells among regulatory T cell subsets in vivo. *J. Exp. Med.* **2012**, *209*, 1713–1722. [CrossRef] [PubMed]
93. Sarris, M.; Andersen, K.G.; Randow, F.; Mayr, L.; Betz, A.G. Neuropilin-1 expression on regulatory T cells enhances their interactions with dendritic cells during antigen recognition. *Immunity* **2008**, *28*, 402–413. [CrossRef] [PubMed]
94. Arad, A.; Nammouz, S.; Nov, Y.; Ohel, G.; Bejar, J.; Vadasz, Z. The Expression of Neuropilin-1 in Human Placentas from Normal and Preeclamptic Pregnancies. *Int. J. Gynecol. Pathol.* **2017**, *36*, 42–49. [CrossRef] [PubMed]
95. Maulik, D.; De, A.; Ragolia, L.; Evans, J.; Grigoryev, D.; Lankachandra, K.; Mundy, D.; Muscat, J.; Gerkovich, M.M.; Ye, S.Q. Down-regulation of placental neuropilin-1 in fetal growth restriction. *Am. J. Obstet. Gynecol.* **2016**, *214*, 279e1–279e9. [CrossRef] [PubMed]
96. Moldenhauer, L.M.; Schjenken, J.E.; Hope, C.M.; Green, E.S.; Zhang, B.; Eldi, P.; Hayball, J.D.; Barry, S.C.; Robertson, S.A. Thymus-Derived Regulatory T Cells Exhibit Foxp3 Epigenetic Modification and Phenotype Attenuation after Mating in Mice. *J. Immunol.* **2019**, *203*, 647–657. [CrossRef]
97. Failla, C.M.; Carbo, M.; Morea, V. Positive and Negative Regulation of Angiogenesis by Soluble Vascular Endothelial Growth Factor Receptor-1. *Int. J. Mol. Sci.* **2018**, *19*, 1306. [CrossRef]
98. Panigrahy, D.; Adini, I.; Mamluk, R.; Levonyak, N.; Bruns, C.J.; D'Amore, P.A.; Klagsbrun, M.; Bielenberg, D.R. Regulation of soluble neuropilin 1, an endogenous angiogenesis inhibitor, in liver development and regeneration. *Pathology* **2014**, *46*, 416–423. [CrossRef]
99. Yang, F.; Jin, C.; Jiang, Y.J.; Li, J.; Di, Y.; Fu, D.L. Potential role of soluble VEGFR-1 in antiangiogenesis therapy for cancer. *Expert Rev. Anticancer Ther.* **2011**, *11*, 541–549. [CrossRef]
100. Chang, Y.S.; Chen, C.N.; Jeng, S.F.; Su, Y.N.; Chen, C.Y.; Chou, H.C.; Tsao, P.N.; Hsieh, W.S. The sFlt-1/PlGF ratio as a predictor for poor pregnancy and neonatal outcomes. *Pediatr Neonatol* **2017**, *58*, 529–533. [CrossRef]
101. Jebbink, J.; Keijser, R.; Veenboer, G.; van der Post, J.; Ris-Stalpers, C.; Afink, G. Expression of placental FLT1 transcript variants relates to both gestational hypertensive disease and fetal growth. *Hypertension* **2011**, *58*, 70–76. [CrossRef] [PubMed]
102. Kumazaki, K.; Nakayama, M.; Suehara, N.; Wada, Y. Expression of vascular endothelial growth factor, placental growth factor, and their receptors Flt-1 and KDR in human placenta under pathologic conditions. *Hum. Pathol.* **2002**, *33*, 1069–1077. [CrossRef] [PubMed]
103. McKeeman, G.C.; Ardill, J.E.; Caldwell, C.M.; Hunter, A.J.; McClure, N. Soluble vascular endothelial growth factor receptor-1 (sFlt-1) is increased throughout gestation in patients who have preeclampsia develop. *Am. J. Obstet. Gynecol.* **2004**, *191*, 1240–1246. [CrossRef] [PubMed]
104. Yusuf, A.M.; Kahane, A.; Ray, J.G. First and Second Trimester Serum sFlt-1/PlGF Ratio and Subsequent Preeclampsia: A Systematic Review. *J. Obstet. Gynaecol. Can.* **2018**, *40*, 618–626. [CrossRef]

105. Palmer, K.R.; Kaitu'u-Lino, T.J.; Hastie, R.; Hannan, N.J.; Ye, L.; Binder, N.; Cannon, P.; Tuohey, L.; Johns, T.G.; Shub, A.; et al. Placental-Specific sFLT-1 e15a Protein Is Increased in Preeclampsia, Antagonizes Vascular Endothelial Growth Factor Signaling, and Has Antiangiogenic Activity. *Hypertension* **2015**, *66*, 1251–1259. [CrossRef] [PubMed]
106. Palmer, K.R.; Tong, S.; Kaitu'u-Lino, T.J. Placental-specific sFLT-1: Role in pre-eclamptic pathophysiology and its translational possibilities for clinical prediction and diagnosis. *Mol. Hum. Reprod.* **2017**, *23*, 69–78. [CrossRef] [PubMed]
107. Cramer, M.; Nagy, I.; Murphy, B.J.; Gassmann, M.; Hottiger, M.O.; Georgiev, O.; Schaffner, W. NF-kappaB contributes to transcription of placenta growth factor and interacts with metal responsive transcription factor-1 in hypoxic human cells. *Biol. Chem.* **2005**, *386*, 865–872. [CrossRef]
108. Liu, T.; Zhang, L.; Joo, D.; Sun, S.C. NF-kappaB signaling in inflammation. *Signal Transduct. Target. Ther.* **2017**, *2*. [CrossRef]
109. Armistead, B.; Kadam, L.; Drewlo, S.; Kohan-Ghadr, H.R. The Role of NFkappaB in Healthy and Preeclamptic Placenta: Trophoblasts in the Spotlight. *Int. J. Mol. Sci.* **2020**, *21*, 1775. [CrossRef]
110. Dorrington, M.G.; Fraser, I.D.C. NF-kappaB Signaling in Macrophages: Dynamics, Crosstalk, and Signal Integration. *Front. Immunol.* **2019**, *10*, 705. [CrossRef]
111. Taylor, C.T.; Cummins, E.P. The role of NF-kappaB in hypoxia-induced gene expression. *Ann. N.Y. Acad. Sci.* **2009**, *1177*, 178–184. [CrossRef] [PubMed]
112. Guney, G.; Taskin, M.I.; Tokmak, A. Increase of circulating inflammatory molecules in preeclampsia, an update. *Eur. Cytokine Netw.* **2020**, *31*, 18–31. [PubMed]
113. Harmon, A.C.; Cornelius, D.C.; Amaral, L.M.; Faulkner, J.L.; Cunningham, M.W., Jr.; Wallace, K.; LaMarca, B. The role of inflammation in the pathology of preeclampsia. *Clin. Sci.* **2016**, *130*, 409–419. [CrossRef] [PubMed]
114. Lappas, M. Nuclear factor-kappaB mediates placental growth factor induced pro-labour mediators in human placenta. *Mol. Hum. Reprod.* **2012**, *18*, 354–361. [CrossRef] [PubMed]
115. Sakowicz, A. The role of NFkappaB in the three stages of pregnancy—Implantation, maintenance, and labour: A review article. *BJOG* **2018**, *125*, 1379–1387. [CrossRef]
116. D'Ignazio, L.; Rocha, S. Hypoxia Induced NF-kappaB. *Cells* **2016**, *5*, 10. [CrossRef]
117. Hayden, M.S.; Ghosh, S. Regulation of NF-kappaB by TNF family cytokines. *Semin. Immunol.* **2014**, *26*, 253–266. [CrossRef]
118. Hogan, P.G. Calcium-NFAT transcriptional signalling in T cell activation and T cell exhaustion. *Cell Calcium* **2017**, *63*, 66–69. [CrossRef]
119. Macian, F. NFAT proteins: Key regulators of T-cell development and function. *Nat. Rev. Immunol.* **2005**, *5*, 472–484. [CrossRef]
120. Fric, J.; Zelante, T.; Wong, A.Y.; Mertes, A.; Yu, H.B.; Ricciardi-Castagnoli, P. NFAT control of innate immunity. *Blood* **2012**, *120*, 1380–1389. [CrossRef]
121. Jinnin, M.; Medici, D.; Park, L.; Limaye, N.; Liu, Y.; Boscolo, E.; Bischoff, J.; Vikkula, M.; Boye, E.; Olsen, B.R. Suppressed NFAT-dependent VEGFR1 expression and constitutive VEGFR2 signaling in infantile hemangioma. *Nat. Med.* **2008**, *14*, 1236–1246. [CrossRef] [PubMed]
122. Ye, L.; Gratton, A.; Hannan, N.J.; Cannon, P.; Deo, M.; Palmer, K.R.; Tong, S.; Kaitu'u-Lino, T.J.; Brownfoot, F.C. Nuclear factor of activated T-cells (NFAT) regulates soluble fms-like tyrosine kinase-1 secretion (sFlt-1) from human placenta. *Placenta* **2016**, *48*, 110–118. [CrossRef] [PubMed]
123. Abe, B.T.; Shin, D.S.; Mocholi, E.; Macian, F. NFAT1 supports tumor-induced anergy of CD4(+) T cells. *Cancer Res.* **2012**, *72*, 4642–4651. [CrossRef] [PubMed]
124. Ono, M. Control of regulatory T-cell differentiation and function by T-cell receptor signalling and Foxp3 transcription factor complexes. *Immunology* **2020**, *160*, 24–37. [CrossRef]
125. Shin, D.S.; Jordan, A.; Basu, S.; Thomas, R.M.; Bandyopadhyay, S.; de Zoeten, E.F.; Wells, A.D.; Macian, F. Regulatory T cells suppress CD4+ T cells through NFAT-dependent transcriptional mechanisms. *EMBO Rep.* **2014**, *15*, 991–999. [CrossRef]
126. Lima, P.D.; Zhang, J.; Dunk, C.; Lye, S.J.; Croy, B.A. Leukocyte driven-decidual angiogenesis in early pregnancy. *Cell Mol. Immunol.* **2014**, *11*, 522–537. [CrossRef]
127. Racicot, K.; Kwon, J.Y.; Aldo, P.; Silasi, M.; Mor, G. Understanding the complexity of the immune system during pregnancy. *Am. J. Reprod. Immunol.* **2014**, *72*, 107–116. [CrossRef]

128. Gaynor, L.M.; Colucci, F. Uterine Natural Killer Cells: Functional Distinctions and Influence on Pregnancy in Humans and Mice. *Front. Immunol.* **2017**, *8*, 467. [CrossRef]
129. Hanna, J.; Goldman-Wohl, D.; Hamani, Y.; Avraham, I.; Greenfield, C.; Natanson-Yaron, S.; Prus, D.; Cohen-Daniel, L.; Arnon, T.I.; Manaster, I.; et al. Decidual NK cells regulate key developmental processes at the human fetal-maternal interface. *Nat. Med.* **2006**, *12*, 1065–1074. [CrossRef]
130. Jabrane-Ferrat, N.; Siewiera, J. The up side of decidual natural killer cells: New developments in immunology of pregnancy. *Immunology* **2014**, *141*, 490–497. [CrossRef]
131. Le Bouteiller, P.; Tabiasco, J. Killers become builders during pregnancy. *Nat. Med.* **2006**, *12*, 991–992. [CrossRef] [PubMed]
132. Moffett, A.; Colucci, F. Uterine NK cells: Active regulators at the maternal-fetal interface. *J. Clin. Invest.* **2014**, *124*, 1872–1879. [CrossRef] [PubMed]
133. Moffett-King, A. Natural killer cells and pregnancy. *Nat. Rev. Immunol.* **2002**, *2*, 656–663. [CrossRef] [PubMed]
134. Sojka, D.K.; Yang, L.; Yokoyama, W.M. Uterine Natural Killer Cells. *Front. Immunol.* **2019**, *10*, 960. [CrossRef] [PubMed]
135. Huang, Q.; Seillet, C.; Belz, G.T. Shaping Innate Lymphoid Cell Diversity. *Front. Immunol.* **2017**, *8*, 1569. [CrossRef] [PubMed]
136. Miller, D.; Motomura, K.; Garcia-Flores, V.; Romero, R.; Gomez-Lopez, N. Innate Lymphoid Cells in the Maternal and Fetal Compartments. *Front. Immunol.* **2018**, *9*, 2396. [CrossRef] [PubMed]
137. Spits, H.; Artis, D.; Colonna, M.; Diefenbach, A.; Di Santo, J.P.; Eberl, G.; Koyasu, S.; Locksley, R.M.; McKenzie, A.N.; Mebius, R.E.; et al. Innate lymphoid cells–a proposal for uniform nomenclature. *Nat. Rev. Immunol.* **2013**, *13*, 145–149. [CrossRef]
138. Colonna, M. Innate Lymphoid Cells: Diversity, Plasticity, and Unique Functions in Immunity. *Immunity* **2018**, *48*, 1104–1117. [CrossRef]
139. Vacca, P.; Chiossone, L.; Mingari, M.C.; Moretta, L. Heterogeneity of NK Cells and Other Innate Lymphoid Cells in Human and Murine Decidua. *Front. Immunol.* **2019**, *10*, 170. [CrossRef]
140. Bonanni, V.; Sciume, G.; Santoni, A.; Bernardini, G. Bone Marrow NK Cells: Origin, Distinctive Features, and Requirements for Tissue Localization. *Front. Immunol.* **2019**, *10*, 1569. [CrossRef]
141. Klose, C.S.; Artis, D. Innate lymphoid cells as regulators of immunity, inflammation and tissue homeostasis. *Nat. Immunol.* **2016**, *17*, 765–774. [CrossRef] [PubMed]
142. Croy, B.A.; Esadeg, S.; Chantakru, S.; van den Heuvel, M.; Paffaro, V.A.; He, H.; Black, G.P.; Ashkar, A.A.; Kiso, Y.; Zhang, J. Update on pathways regulating the activation of uterine Natural Killer cells, their interactions with decidual spiral arteries and homing of their precursors to the uterus. *J. Reprod. Immunol.* **2003**, *59*, 175–191. [CrossRef]
143. Sojka, D.K.; Yang, L.; Plougastel-Douglas, B.; Higuchi, D.A.; Croy, B.A.; Yokoyama, W.M. Cutting Edge: Local Proliferation of Uterine Tissue-Resident NK Cells during Decidualization in Mice. *J. Immunol.* **2018**, *201*, 2551–2556. [CrossRef] [PubMed]
144. van den Heuvel, M.J.; Chantakru, S.; Xuemei, X.; Evans, S.S.; Tekpetey, F.; Mote, P.A.; Clarke, C.L.; Croy, B.A. Trafficking of circulating pro-NK cells to the decidualizing uterus: Regulatory mechanisms in the mouse and human. *Immunol. Invest.* **2005**, *34*, 273–293. [CrossRef]
145. Bjorkstrom, N.K.; Ljunggren, H.G.; Michaelsson, J. Emerging insights into natural killer cells in human peripheral tissues. *Nat. Rev. Immunol.* **2016**, *16*, 310–320. [CrossRef]
146. Tao, Y.; Li, Y.H.; Piao, H.L.; Zhou, W.J.; Zhang, D.; Fu, Q.; Wang, S.C.; Li, D.J.; Du, M.R. CD56(bright)CD25+ NK cells are preferentially recruited to the maternal/fetal interface in early human pregnancy. *Cell Mol. Immunol.* **2015**, *12*, 77–86. [CrossRef]
147. Tayade, C.; Fang, Y.; Black, G.P.; Paffaro, V.A., Jr.; Erlebacher, A.; Croy, B.A. Differential transcription of Eomes and T-bet during maturation of mouse uterine natural killer cells. *J. Leukoc. Biol.* **2005**, *78*, 1347–1355. [CrossRef]
148. Fraser, R.; Whitley, G.S.; Thilaganathan, B.; Cartwright, J.E. Decidual natural killer cells regulate vessel stability: Implications for impaired spiral artery remodelling. *J. Reprod. Immunol.* **2015**, *110*, 54–60. [CrossRef]
149. Lash, G.E.; Schiessl, B.; Kirkley, M.; Innes, B.A.; Cooper, A.; Searle, R.F.; Robson, S.C.; Bulmer, J.N. Expression of angiogenic growth factors by uterine natural killer cells during early pregnancy. *J. Leukoc. Biol.* **2006**, *80*, 572–580. [CrossRef]

150. Robson, A.; Harris, L.K.; Innes, B.A.; Lash, G.E.; Aljunaidy, M.M.; Aplin, J.D.; Baker, P.N.; Robson, S.C.; Bulmer, J.N. Uterine natural killer cells initiate spiral artery remodeling in human pregnancy. *FASEB J.* **2012**, *26*, 4876–4885. [CrossRef]
151. Lockwood, C.J.; Huang, S.J.; Chen, C.P.; Huang, Y.; Xu, J.; Faramarzi, S.; Kayisli, O.; Kayisli, U.; Koopman, L.; Smedts, D.; et al. Decidual cell regulation of natural killer cell-recruiting chemokines: Implications for the pathogenesis and prediction of preeclampsia. *Am. J. Pathol* **2013**, *183*, 841–856. [CrossRef] [PubMed]
152. Orr, M.T.; Lanier, L.L. Natural killer cell education and tolerance. *Cell* **2010**, *142*, 847–856. [CrossRef] [PubMed]
153. Sharkey, A.M.; Xiong, S.; Kennedy, P.R.; Gardner, L.; Farrell, L.E.; Chazara, O.; Ivarsson, M.A.; Hiby, S.E.; Colucci, F.; Moffett, A. Tissue-Specific Education of Decidual NK Cells. *J. Immunol.* **2015**, *195*, 3026–3032. [CrossRef] [PubMed]
154. Fukui, A.; Funamizu, A.; Fukuhara, R.; Shibahara, H. Expression of natural cytotoxicity receptors and cytokine production on endometrial natural killer cells in women with recurrent pregnancy loss or implantation failure, and the expression of natural cytotoxicity receptors on peripheral blood natural killer cells in pregnant women with a history of recurrent pregnancy loss. *J. Obstet. Gynaecol. Res.* **2017**, *43*, 1678–1686.
155. Lanier, L.L. NK cell recognition. *Annu. Rev. Immunol.* **2005**, *23*, 225–274. [CrossRef]
156. Ashkar, A.A.; Di Santo, J.P.; Croy, B.A. Interferon gamma contributes to initiation of uterine vascular modification, decidual integrity, and uterine natural killer cell maturation during normal murine pregnancy. *J. Exp. Med.* **2000**, *192*, 259–270. [CrossRef]
157. Kennedy, P.R.; Chazara, O.; Gardner, L.; Ivarsson, M.A.; Farrell, L.E.; Xiong, S.; Hiby, S.E.; Colucci, F.; Sharkey, A.M.; Moffett, A. Activating KIR2DS4 Is Expressed by Uterine NK Cells and Contributes to Successful Pregnancy. *J. Immunol.* **2016**, *197*, 4292–4300. [CrossRef]
158. Kerdiles, Y.; Ugolini, S.; Vivier, E. T cell regulation of natural killer cells. *J. Exp. Med.* **2013**, *210*, 1065–1068. [CrossRef]
159. Knorr, M.; Munzel, T.; Wenzel, P. Interplay of NK cells and monocytes in vascular inflammation and myocardial infarction. *Front. Physiol* **2014**, *5*, 295. [CrossRef]
160. Lucas, M.; Schachterle, W.; Oberle, K.; Aichele, P.; Diefenbach, A. Dendritic cells prime natural killer cells by trans-presenting interleukin 15. *Immunity* **2007**, *26*, 503–517. [CrossRef]
161. Zingoni, A.; Sornasse, T.; Cocks, B.G.; Tanaka, Y.; Santoni, A.; Lanier, L.L. NK cell regulation of T cell-mediated responses. *Mol. Immunol.* **2005**, *42*, 451–454. [CrossRef] [PubMed]
162. Xiong, S.; Sharkey, A.M.; Kennedy, P.R.; Gardner, L.; Farrell, L.E.; Chazara, O.; Bauer, J.; Hiby, S.E.; Colucci, F.; Moffett, A. Maternal uterine NK cell-activating receptor KIR2DS1 enhances placentation. *J. Clin. Invest.* **2013**, *123*, 4264–4272. [CrossRef] [PubMed]
163. Hiby, S.E.; Apps, R.; Sharkey, A.M.; Farrell, L.E.; Gardner, L.; Mulder, A.; Claas, F.H.; Walker, J.J.; Redman, C.W.; Morgan, L.; et al. Maternal activating KIRs protect against human reproductive failure mediated by fetal HLA-C2. *J. Clin. Invest.* **2010**, *120*, 4102–4110. [CrossRef] [PubMed]
164. Drake, P.M.; Gunn, M.D.; Charo, I.F.; Tsou, C.L.; Zhou, Y.; Huang, L.; Fisher, S.J. Human placental cytotrophoblasts attract monocytes and CD56(bright) natural killer cells via the actions of monocyte inflammatory protein 1alpha. *J. Exp. Med.* **2001**, *193*, 1199–1212. [CrossRef] [PubMed]
165. Le Gars, M.; Seiler, C.; Kay, A.W.; Bayless, N.L.; Starosvetsky, E.; Moore, L.; Shen-Orr, S.S.; Aziz, N.; Khatri, P.; Dekker, C.L.; et al. Pregnancy-Induced Alterations in NK Cell Phenotype and Function. *Front. Immunol.* **2019**, *10*, 2469. [CrossRef]
166. Blois, S.M.; Klapp, B.F.; Barrientos, G. Decidualization and angiogenesis in early pregnancy: Unravelling the functions of DC and NK cells. *J. Reprod. Immunol.* **2011**, *88*, 86–92. [CrossRef]
167. Cartwright, J.E.; James-Allan, L.; Buckley, R.J.; Wallace, A.E. The role of decidual NK cells in pregnancies with impaired vascular remodelling. *J. Reprod. Immunol.* **2017**, *119*, 81–84. [CrossRef]
168. Solocinski, K.; Padget, M.R.; Fabian, K.P.; Wolfson, B.; Cecchi, F.; Hembrough, T.; Benz, S.C.; Rabizadeh, S.; Soon-Shiong, P.; Schlom, J.; et al. Overcoming hypoxia-induced functional suppression of NK cells. *J. Immunother Cancer* **2020**, *8*. [CrossRef]
169. Vacca, P.; Cantoni, C.; Vitale, M.; Prato, C.; Canegallo, F.; Fenoglio, D.; Ragni, N.; Moretta, L.; Mingari, M.C. Crosstalk between decidual NK and CD14+ myelomonocytic cells results in induction of Tregs and immunosuppression. *Proc. Natl. Acad. Sci. USA* **2010**, *107*, 11918–11923. [CrossRef]

170. Fukui, A.; Yokota, M.; Funamizu, A.; Nakamua, R.; Fukuhara, R.; Yamada, K.; Kimura, H.; Fukuyama, A.; Kamoi, M.; Tanaka, K.; et al. Changes of NK cells in preeclampsia. *Am. J. Reprod. Immunol.* **2012**, *67*, 278–286. [CrossRef]
171. Golic, M.; Haase, N.; Herse, F.; Wehner, A.; Vercruysse, L.; Pijnenborg, R.; Balogh, A.; Saether, P.C.; Dissen, E.; Luft, F.C.; et al. Natural Killer Cell Reduction and Uteroplacental Vasculopathy. *Hypertension* **2016**, *68*, 964–973. [CrossRef] [PubMed]
172. Kuon, R.J.; Vomstein, K.; Weber, M.; Muller, F.; Seitz, C.; Wallwiener, S.; Strowitzki, T.; Schleussner, E.; Markert, U.R.; Daniel, V.; et al. The "killer cell story" in recurrent miscarriage: Association between activated peripheral lymphocytes and uterine natural killer cells. *J. Reprod. Immunol.* **2017**, *119*, 9–14. [CrossRef] [PubMed]
173. Fu, B.; Zhou, Y.; Ni, X.; Tong, X.; Xu, X.; Dong, Z.; Sun, R.; Tian, Z.; Wei, H. Natural Killer Cells Promote Fetal Development through the Secretion of Growth-Promoting Factors. *Immunity* **2017**, *47*, 1100–1113. [CrossRef] [PubMed]
174. Moretta, A.; Marcenaro, E.; Parolini, S.; Ferlazzo, G.; Moretta, L. NK cells at the interface between innate and adaptive immunity. *Cell Death Differ.* **2008**, *15*, 226–233. [CrossRef] [PubMed]
175. Kofod, L.; Lindhard, A.; Hviid, T.V.F. Implications of uterine NK cells and regulatory T cells in the endometrium of infertile women. *Hum. Immunol.* **2018**, *79*, 693–701. [CrossRef]
176. Santner-Nanan, B.; Peek, M.J.; Khanam, R.; Richarts, L.; Zhu, E.; Fazekas de St Groth, B.; Nanan, R. Systemic increase in the ratio between Foxp3+ and IL-17-producing CD4+ T cells in healthy pregnancy but not in preeclampsia. *J. Immunol.* **2009**, *183*, 7023–7030. [CrossRef]
177. Sasaki, Y.; Darmochwal-Kolarz, D.; Suzuki, D.; Sakai, M.; Ito, M.; Shima, T.; Shiozaki, A.; Rolinski, J.; Saito, S. Proportion of peripheral blood and decidual CD4(+) CD25(bright) regulatory T cells in pre-eclampsia. *Clin. Exp. Immunol.* **2007**, *149*, 139–145. [CrossRef]
178. Molvarec, A.; Czegle, I.; Szijarto, J.; Rigo, J., Jr. Increased circulating interleukin-17 levels in preeclampsia. *J. Reprod. Immunol.* **2015**, *112*, 53–57. [CrossRef]
179. Vishnyakova, P.; Elchaninov, A.; Fatkhudinov, T.; Sukhikh, G. Role of the Monocyte-Macrophage System in Normal Pregnancy and Preeclampsia. *Int. J. Mol. Sci.* **2019**, *20*, 3695. [CrossRef]
180. Murray, P.J. Macrophage Polarization. *Annu. Rev. Physiol.* **2017**, *79*, 541–566. [CrossRef]
181. Porta, C.; Riboldi, E.; Ippolito, A.; Sica, A. Molecular and epigenetic basis of macrophage polarized activation. *Semin. Immunol.* **2015**, *27*, 237–248. [CrossRef] [PubMed]
182. Shapouri-Moghaddam, A.; Mohammadian, S.; Vazini, H.; Taghadosi, M.; Esmaeili, S.A.; Mardani, F.; Seifi, B.; Mohammadi, A.; Afshari, J.T.; Sahebkar, A. Macrophage plasticity, polarization, and function in health and disease. *J. Cell. Physiol.* **2018**, *233*, 6425–6440. [CrossRef] [PubMed]
183. Faas, M.M.; Spaans, F.; De Vos, P. Monocytes and macrophages in pregnancy and pre-eclampsia. *Front. Immunol.* **2014**, *5*, 298. [CrossRef]
184. Lash, G.E.; Pitman, H.; Morgan, H.L.; Innes, B.A.; Agwu, C.N.; Bulmer, J.N. Decidual macrophages: Key regulators of vascular remodeling in human pregnancy. *J. Leukoc. Biol.* **2016**, *100*, 315–325. [CrossRef] [PubMed]
185. Ning, F.; Liu, H.; Lash, G.E. The Role of Decidual Macrophages during Normal and Pathological Pregnancy. *Am. J. Reprod. Immunol.* **2016**, *75*, 298–309. [CrossRef] [PubMed]
186. Yao, Y.; Xu, X.H.; Jin, L. Macrophage Polarization in Physiological and Pathological Pregnancy. *Front. Immunol.* **2019**, *10*, 792. [CrossRef]
187. Zhang, Y.H.; He, M.; Wang, Y.; Liao, A.H. Modulators of the Balance between M1 and M2 Macrophages during Pregnancy. *Front. Immunol.* **2017**, *8*, 120. [CrossRef]
188. Nagamatsu, T.; Schust, D.J. The contribution of macrophages to normal and pathological pregnancies. *Am. J. Reprod. Immunol.* **2010**, *63*, 460–471. [CrossRef]
189. Muramatsu, M.; Yamamoto, S.; Osawa, T.; Shibuya, M. Vascular endothelial growth factor receptor-1 signaling promotes mobilization of macrophage lineage cells from bone marrow and stimulates solid tumor growth. *Cancer Res.* **2010**, *70*, 8211–8221. [CrossRef]
190. Riboldi, E.; Porta, C.; Morlacchi, S.; Viola, A.; Mantovani, A.; Sica, A. Hypoxia-mediated regulation of macrophage functions in pathophysiology. *Int. Immunol.* **2013**, *25*, 67–75. [CrossRef]
191. Zhao, H.; Kalish, F.S.; Wong, R.J.; Stevenson, D.K. Hypoxia regulates placental angiogenesis via alternatively activated macrophages. *Am. J. Reprod. Immunol.* **2018**, *80*, e12989. [CrossRef] [PubMed]

192. Rolny, C.; Mazzone, M.; Tugues, S.; Laoui, D.; Johansson, I.; Coulon, C.; Squadrito, M.L.; Segura, I.; Li, X.; Knevels, E.; et al. HRG inhibits tumor growth and metastasis by inducing macrophage polarization and vessel normalization through downregulation of PlGF. *Cancer Cell* **2011**, *19*, 31–44. [CrossRef] [PubMed]
193. Zhou, X.; Qi, Y. Larynx carcinoma regulates tumor-associated macrophages through PLGF signaling. *Sci. Rep.* **2015**, *5*, 10071. [CrossRef] [PubMed]
194. Doedens, A.L.; Stockmann, C.; Rubinstein, M.P.; Liao, D.; Zhang, N.; DeNardo, D.G.; Coussens, L.M.; Karin, M.; Goldrath, A.W.; Johnson, R.S. Macrophage expression of hypoxia-inducible factor-1 alpha suppresses T-cell function and promotes tumor progression. *Cancer Res.* **2010**, *70*, 7465–7475. [CrossRef]
195. Li, Y.; Patel, S.P.; Roszik, J.; Qin, Y. Hypoxia-Driven Immunosuppressive Metabolites in the Tumor Microenvironment: New Approaches for Combinational Immunotherapy. *Front. Immunol.* **2018**, *9*, 1591. [CrossRef]
196. Wang, B.Q.; Zhang, C.M.; Gao, W.; Wang, X.F.; Zhang, H.L.; Yang, P.C. Cancer-derived matrix metalloproteinase-9 contributes to tumor tolerance. *J. Cancer Res. Clin. Oncol.* **2011**, *137*, 1525–1533. [CrossRef]
197. Jetten, N.; Verbruggen, S.; Gijbels, M.J.; Post, M.J.; De Winther, M.P.; Donners, M.M. Anti-inflammatory M2, but not pro-inflammatory M1 macrophages promote angiogenesis in vivo. *Angiogenesis* **2014**, *17*, 109–118. [CrossRef]
198. Mantovani, A.; Biswas, S.K.; Galdiero, M.R.; Sica, A.; Locati, M. Macrophage plasticity and polarization in tissue repair and remodelling. *J. Pathol.* **2013**, *229*, 176–185. [CrossRef]
199. Wakabayashi, S. New insights into the functions of histidine-rich glycoprotein. *Int. Rev. Cell. Mol. Biol.* **2013**, *304*, 467–493.
200. Blank, M.; Shoenfeld, Y. Histidine-rich glycoprotein modulation of immune/autoimmune, vascular, and coagulation systems. *Clin. Rev. Allergy Immunol.* **2008**, *34*, 307–312. [CrossRef]
201. Aksornphusitaphong, A.; Phupong, V. Combination of serum histidine-rich glycoprotein and uterine artery Doppler to predict preeclampsia. *Hypertens. Res.* **2018**, *41*, 275–281. [CrossRef] [PubMed]
202. Bolin, M.; Akerud, P.; Hansson, A.; Akerud, H. Histidine-rich glycoprotein as an early biomarker of preeclampsia. *Am. J. Hypertens.* **2011**, *24*, 496–501. [CrossRef] [PubMed]
203. Lindgren, K.E.; Karehed, K.; Karypidis, H.; Hosseini, F.; Bremme, K.; Landgren, B.M.; Skjoldebrand-Sparre, L.; Stavreus-Evers, A.; Sundstrom-Poromaa, I.; Akerud, H. Histidine-rich glycoprotein gene polymorphism in patients with recurrent miscarriage. *Acta Obstet. Gynecol. Scand.* **2013**, *92*, 974–977. [CrossRef] [PubMed]
204. Lindgren, K.E.; Hreinsson, J.; Helmestam, M.; Wanggren, K.; Poromaa, I.S.; Karehed, K.; Akerud, H. Histidine-rich glycoprotein derived peptides affect endometrial angiogenesis in vitro but has no effect on embryo development. *Syst. Biol. Reprod. Med.* **2016**, *62*, 192–200. [CrossRef]
205. Li, Y.; Xie, Z.; Wang, Y.; Hu, H. Macrophage M1/M2 polarization in patients with pregnancy-induced hypertension. *Can. J. Physiol. Pharmacol.* **2018**, *96*, 922–928. [CrossRef]
206. Cornelius, D.C.; Cottrell, J.; Amaral, L.M.; LaMarca, B. Inflammatory mediators: A causal link to hypertension during preeclampsia. *Br. J. Pharmacol.* **2019**, *176*, 1914–1921. [CrossRef]
207. Nunes, P.R.; Romao-Veiga, M.; Peracoli, J.C.; Araujo Costa, R.A.; de Oliveira, L.G.; Borges, V.T.M.; Peracoli, M.T. Downregulation of CD163 in monocytes and its soluble form in the plasma is associated with a pro-inflammatory profile in pregnant women with preeclampsia. *Immunol. Res.* **2019**, *67*, 194–201. [CrossRef]
208. Co, E.C.; Gormley, M.; Kapidzic, M.; Rosen, D.B.; Scott, M.A.; Stolp, H.A.; McMaster, M.; Lanier, L.L.; Barcena, A.; Fisher, S.J. Maternal decidual macrophages inhibit NK cell killing of invasive cytotrophoblasts during human pregnancy. *Biol. Reprod.* **2013**, *88*, 155. [CrossRef]
209. Ruytinx, P.; Proost, P.; Van Damme, J.; Struyf, S. Chemokine-Induced Macrophage Polarization in Inflammatory Conditions. *Front. Immunol.* **2018**, *9*, 1930. [CrossRef]
210. Svensson, J.; Jenmalm, M.C.; Matussek, A.; Geffers, R.; Berg, G.; Ernerudh, J. Macrophages at the fetal-maternal interface express markers of alternative activation and are induced by M-CSF and IL-10. *J. Immunol.* **2011**, *187*, 3671–3682. [CrossRef]
211. Wang, N.; Liang, H.; Zen, K. Molecular mechanisms that influence the macrophage m1-m2 polarization balance. *Front. Immunol.* **2014**, *5*, 614. [CrossRef] [PubMed]
212. Tagliani, E.; Erlebacher, A. Dendritic cell function at the maternal-fetal interface. *Expert Rev. Clin. Immunol.* **2011**, *7*, 593–602. [CrossRef] [PubMed]
213. Iberg, C.A.; Jones, A.; Hawiger, D. Dendritic Cells as Inducers of Peripheral Tolerance. *Trends Immunol.* **2017**, *38*, 793–804. [CrossRef] [PubMed]

214. Xiao, B.G.; Liu, X.; Link, H. Antigen-specific T cell functions are suppressed over the estrogen-dendritic cell-indoleamine 2,3-dioxygenase axis. *Steroids* **2004**, *69*, 653–659. [CrossRef] [PubMed]
215. Rozman, P.; Svajger, U. The tolerogenic role of IFN-gamma. *Cytokine Growth Factor Rev.* **2018**, *41*, 40–53. [CrossRef]
216. Mezouar, S.; Mege, J.L. Changing the paradigm of IFN-gamma at the interface between innate and adaptive immunity: Macrophage-derived IFN-gamma. *J. Leukoc. Biol.* **2020**, *108*, 419–426. [CrossRef]
217. Sayama, S.; Nagamatsu, T.; Schust, D.J.; Itaoka, N.; Ichikawa, M.; Kawana, K.; Yamashita, T.; Kozuma, S.; Fujii, T. Human decidual macrophages suppress IFN-gamma production by T cells through costimulatory B7-H1:PD-1 signaling in early pregnancy. *J. Reprod. Immunol.* **2013**, *100*, 109–117. [CrossRef]
218. Svajger, U.; Rozman, P. Induction of Tolerogenic Dendritic Cells by Endogenous Biomolecules: An Update. *Front. Immunol.* **2018**, *9*, 2482. [CrossRef]
219. Krey, G.; Frank, P.; Shaikly, V.; Barrientos, G.; Cordo-Russo, R.; Ringel, F.; Moschansky, P.; Chernukhin, I.V.; Metodiev, M.; Fernandez, N.; et al. In vivo dendritic cell depletion reduces breeding efficiency, affecting implantation and early placental development in mice. *J. Mol. Med.* **2008**, *86*, 999–1011. [CrossRef]
220. Plaks, V.; Birnberg, T.; Berkutzki, T.; Sela, S.; BenYashar, A.; Kalchenko, V.; Mor, G.; Keshet, E.; Dekel, N.; Neeman, M.; et al. Uterine DCs are crucial for decidua formation during embryo implantation in mice. *J. Clin. Invest.* **2008**, *118*, 3954–3965. [CrossRef]
221. Karsten, C.M.; Behrends, J.; Wagner, A.K.; Fuchs, F.; Figge, J.; Schmudde, I.; Hellberg, L.; Kruse, A. DC within the pregnant mouse uterus influence growth and functional properties of uterine NK cells. *Eur. J. Immunol.* **2009**, *39*, 2203–2214. [CrossRef] [PubMed]
222. Hsu, P.; Santner-Nanan, B.; Dahlstrom, J.E.; Fadia, M.; Chandra, A.; Peek, M.; Nanan, R. Altered decidual DC-SIGN+ antigen-presenting cells and impaired regulatory T-cell induction in preeclampsia. *Am. J. Pathol.* **2012**, *181*, 2149–2160. [CrossRef] [PubMed]
223. Kishuku, M.; Nishioka, Y.; Abe, S.; Kishi, J.; Ogino, H.; Aono, Y.; Azuma, M.; Kinoshita, K.; Batmunkh, R.; Makino, H.; et al. Expression of soluble vascular endothelial growth factor receptor-1 in human monocyte-derived mature dendritic cells contributes to their antiangiogenic property. *J. Immunol.* **2009**, *183*, 8176–8185. [CrossRef] [PubMed]
224. Li, J.; Huang, L.; Wang, S.; Zhang, Z. The prevalence of regulatory T and dendritic cells is altered in peripheral blood of women with pre-eclampsia. *Pregnancy Hypertens.* **2019**, *17*, 233–240. [CrossRef] [PubMed]
225. Lin, Y.L.; Liang, Y.C.; Chiang, B.L. Placental growth factor down-regulates type 1 T helper immune response by modulating the function of dendritic cells. *J. Leukoc. Biol.* **2007**, *82*, 1473–1480. [CrossRef] [PubMed]
226. Dikov, M.M.; Ohm, J.E.; Ray, N.; Tchekneva, E.E.; Burlison, J.; Moghanaki, D.; Nadaf, S.; Carbone, D.P. Differential roles of vascular endothelial growth factor receptors 1 and 2 in dendritic cell differentiation. *J. Immunol.* **2005**, *174*, 215–222. [CrossRef]
227. Gabrilovich, D.I.; Chen, H.L.; Girgis, K.R.; Cunningham, H.T.; Meny, G.M.; Nadaf, S.; Kavanaugh, D.; Carbone, D.P. Production of vascular endothelial growth factor by human tumors inhibits the functional maturation of dendritic cells. *Nat. Med.* **1996**, *2*, 1096–1103. [CrossRef]
228. Laurent, J.; Hull, E.F.; Touvrey, C.; Kuonen, F.; Lan, Q.; Lorusso, G.; Doucey, M.A.; Ciarloni, L.; Imaizumi, N.; Alghisi, G.C.; et al. Proangiogenic factor PlGF programs CD11b(+) myelomonocytes in breast cancer during differentiation of their hematopoietic progenitors. *Cancer Res.* **2011**, *71*, 3781–3791. [CrossRef]
229. Drummond, G.R.; Vinh, A.; Guzik, T.J.; Sobey, C.G. Immune mechanisms of hypertension. *Nat. Rev. Immunol.* **2019**, *19*, 517–532. [CrossRef]
230. Harrison, D.G. The immune system in hypertension. *Trans. Am. Clin. Climatol. Assoc.* **2014**, *125*, 130–138, discussion 130–138.
231. Laresgoiti-Servitje, E. A leading role for the immune system in the pathophysiology of preeclampsia. *J. Leukoc. Biol.* **2013**, *94*, 247–257. [CrossRef] [PubMed]
232. Mikolajczyk, T.P.; Guzik, T.J. Adaptive Immunity in Hypertension. *Curr. Hypertens. Rep.* **2019**, *21*, 68. [CrossRef] [PubMed]
233. Rodriguez-Iturbe, B.; Pons, H.; Johnson, R.J. Role of the Immune System in Hypertension. *Physiol. Rev.* **2017**, *97*, 1127–1164. [CrossRef] [PubMed]
234. Trott, D.W.; Thabet, S.R.; Kirabo, A.; Saleh, M.A.; Itani, H.; Norlander, A.E.; Wu, J.; Goldstein, A.; Arendshorst, W.J.; Madhur, M.S.; et al. Oligoclonal CD8+ T cells play a critical role in the development of hypertension. *Hypertension* **2014**, *64*, 1108–1115. [CrossRef]

235. Vinh, A.; Chen, W.; Blinder, Y.; Weiss, D.; Taylor, W.R.; Goronzy, J.J.; Weyand, C.M.; Harrison, D.G.; Guzik, T.J. Inhibition and genetic ablation of the B7/CD28 T-cell costimulation axis prevents experimental hypertension. *Circulation* **2010**, *122*, 2529–2537. [CrossRef]
236. Crowley, S.D.; Song, Y.S.; Lin, E.E.; Griffiths, R.; Kim, H.S.; Ruiz, P. Lymphocyte responses exacerbate angiotensin II-dependent hypertension. *Am. J. Physiol. Regul. Integr. Comp. Physiol.* **2010**, *298*, R1089–R1097. [CrossRef]
237. Guzik, T.J.; Hoch, N.E.; Brown, K.A.; McCann, L.A.; Rahman, A.; Dikalov, S.; Goronzy, J.; Weyand, C.; Harrison, D.G. Role of the T cell in the genesis of angiotensin II induced hypertension and vascular dysfunction. *J. Exp. Med.* **2007**, *204*, 2449–2460. [CrossRef]
238. Chatterjee, P.; Chiasson, V.L.; Kopriva, S.E.; Young, K.J.; Chatterjee, V.; Jones, K.A.; Mitchell, B.M. Interleukin 10 deficiency exacerbates toll-like receptor 3-induced preeclampsia-like symptoms in mice. *Hypertension* **2011**, *58*, 489–496. [CrossRef]
239. Tinsley, J.H.; South, S.; Chiasson, V.L.; Mitchell, B.M. Interleukin-10 reduces inflammation, endothelial dysfunction, and blood pressure in hypertensive pregnant rats. *Am. J. Physiol. Regul. Integr. Comp. Physiol.* **2010**, *298*, R713–R719. [CrossRef]
240. Loperena, R.; Van Beusecum, J.P.; Itani, H.A.; Engel, N.; Laroumanie, F.; Xiao, L.; Elijovich, F.; Laffer, C.L.; Gnecco, J.S.; Noonan, J.; et al. Hypertension and increased endothelial mechanical stretch promote monocyte differentiation and activation: Roles of STAT3, interleukin 6 and hydrogen peroxide. *Cardiovasc. Res.* **2018**, *114*, 1547–1563. [CrossRef]
241. Sharma, A.; Satyam, A.; Sharma, J.B. Leptin, IL-10 and inflammatory markers (TNF-alpha, IL-6 and IL-8) in pre-eclamptic, normotensive pregnant and healthy non-pregnant women. *Am. J. Reprod. Immunol.* **2007**, *58*, 21–30. [CrossRef] [PubMed]
242. Osol, G.; Celia, G.; Gokina, N.; Barron, C.; Chien, E.; Mandala, M.; Luksha, L.; Kublickiene, K. Placental growth factor is a potent vasodilator of rat and human resistance arteries. *Am. J. Physiol. Heart Circ. Physiol.* **2008**, *294*, H1381–H1387. [CrossRef] [PubMed]
243. Saleh, L.; Vergouwe, Y.; van den Meiracker, A.H.; Verdonk, K.; Russcher, H.; Bremer, H.A.; Versendaal, H.J.; Steegers, E.A.P.; Danser, A.H.J.; Visser, W. Angiogenic Markers Predict Pregnancy Complications and Prolongation in Preeclampsia: Continuous Versus Cutoff Values. *Hypertension* **2017**, *70*, 1025–1033. [CrossRef] [PubMed]
244. Espinoza, J.; Betancourt, A.; Belfort, M.A.; Shamshirsaz, A.A.; Fox, K.A.; Yallampalli, C. Placental growth factor blunts uterine artery responses to angiotensin II. *BJOG* **2019**, *126*, 1058–1064. [CrossRef]
245. Carnevale, D.; Pallante, F.; Fardella, V.; Fardella, S.; Iacobucci, R.; Federici, M.; Cifelli, G.; De Lucia, M.; Lembo, G. The angiogenic factor PlGF mediates a neuroimmune interaction in the spleen to allow the onset of hypertension. *Immunity* **2014**, *41*, 737–752. [CrossRef]
246. Phipps, E.A.; Thadhani, R.; Benzing, T.; Karumanchi, S.A. Pre-eclampsia: Pathogenesis, novel diagnostics and therapies. *Nat. Rev. Nephrol.* **2019**, *15*, 275–289. [CrossRef]
247. Rana, S.; Lemoine, E.; Granger, J.P.; Karumanchi, S.A. Preeclampsia: Pathophysiology, Challenges, and Perspectives. *Circ. Res.* **2019**, *124*, 1094–1112. [CrossRef]
248. Steegers, E.A.; von Dadelszen, P.; Duvekot, J.J.; Pijnenborg, R. Pre-eclampsia. *Lancet* **2010**, *376*, 631–644. [CrossRef]
249. Carnevale, D.; Lembo, G. PlGF, immune system and hypertension. *Oncotarget* **2015**, *6*, 18246–18247. [CrossRef]
250. Magee, L.A.; Khalil, A.; Kametas, N.; von Dadelszen, P. Toward personalized management of chronic hypertension in pregnancy. *Am. J. Obstet. Gynecol.* **2020**, in press. [CrossRef]

International Journal of
Molecular Sciences

Article

Protease Inhibitor Anti-HIV, Lopinavir, Impairs Placental Endocrine Function

Camille Fraichard [1], Fidéline Bonnet-Serrano [2], Christelle Laguillier-Morizot [1,2], Marylise Hebert-Schuster [3], René Lai-Kuen [4], Jeanne Sibiude [5], Thierry Fournier [1], Marie Cohen [3] and Jean Guibourdenche [1,2,*]

1 INSERM UMR-S 1139, Faculté de Pharmacie, Université de Paris, 75006 Paris, France; camille.fraichard@gmail.com (C.F.); christelle.laguillier@aphp.fr (C.L.-M.); thierry.fournier@parisdescartes.fr (T.F.)
2 Service d'Hormonologie, CHU Cochin, HUPC, AP-HP, 75014 Paris, France; fideline.bonnet@aphp.fr
3 Service de Gynécologie-Obstétrique, Faculté de Médecine, Université de Genève, 1206 Genève, Suisse; maryliseschuster@gmail.com (M.H.-S.); marie.cohen@hcuge.ch (M.C.)
4 INSERM UMS 025—CNRS UMS 3612, Faculté de Pharmacie, Université de Paris, 75006 Paris, France; rene.lai-kuen@parisdescartes.fr
5 Service de Gynécologie-Obstétrique, CHU Louis Mourier, HUPN, AP-HP, 92700 Colombes, France; jeanne.sibiude@aphp.fr
* Correspondence: jean.guibourdenche@aphp.fr

Abstract: Protease Inhibitors (PI e.g., ritonavir (RTV) and lopinavir (LPV)) used to treat pregnant mothers infected by HIV induce prematurity and endocrine dysfunctions. The maintenance of pregnancy relies on placental hormone production (human Chorionic Gonadotrophin (hCG) and progesterone (P4)). Those functions are ensured by the villous trophoblast and are mainly regulated by the Unfolded Protein Response (UPR) pathway and mitochondria. We investigated, in vitro, if PI impair hCG and P4 production and the potential intracellular mechanisms involved. Term villous cytotrophoblast (VCT) were cultured with or without RTV or LPV from 6 to 48 h. VCT differentiation into syncytiotrophoblast (ST) was followed measuring hCG and P4 secretion. We evaluated the expression of P4 synthesis partners (Metastatic Lymph Node 64 (MLN64), cholesterol side-chain cleavage (P450SCC), Hydroxy-delta-5-Steroid Dehydrogenase and 3 Beta-and steroid delta-isomerase 1 (HSD3B1)), of mitochondrial pro-fusion factors (Mitofusin 2 (Mfn2), Optic Atrophy 1 (OPA1)) and of UPR factors (Glucose-Regulated Protein 78 (GRP78), Activating Transcription Factor 4 (ATF4), Activating Transcription Factor 6 (ATF6), spliced X-box Binding Protein 1 (sXBP1)). RTV had no significant effect on hCG and P4 secretion, whereas lopinavir significantly decreased both secretions. LPV also decreased P450SCC and HSD3B1 expression, whereas it increased Mfn2, GRP78 and sXBP1 expression in ST. RTV has no effect on the endocrine placenta. LPV impairs both villous trophoblast differentiation and P4 production. It is likely to act via mitochondrial fusion and UPR pathway activation. These trophoblastic alterations may end in decreased P4 levels in maternal circulation, inducing prematurity.

Keywords: human placenta; lopinavir; progesterone; mitochondria; Mfn2; UPR; IRE1α

Citation: Fraichard, C.; Bonnet-Serrano, F.; Laguillier-Morizot, C.; Hebert-Schuster, M.; Lai-Kuen, R.; Sibiude, J.; Fournier, T.; Cohen, M.; Guibourdenche, J. Protease Inhibitor Anti-HIV, Lopinavir, Impairs Placental Endocrine Function. *Int. J. Mol. Sci.* **2021**, *22*, 683. https://doi.org/10.3390/ijms22020683

Received: 18 December 2020
Accepted: 8 January 2021
Published: 12 January 2021

1. Introduction

To prevent mother-to-fetus transmission of Human Immunodeficiency Virus (HIV), World Organization of Health (WHO) recommends to continue or initiate antiretroviral therapy (ART) during pregnancy regardless of the clinical stage or CD4 cell count. ART consists in the association of two NRTI (Nucleoside Reverse Transcriptase Inhibitor) with a Protease Inhibitor (PI) such as Lopinavir (LPV) or Ritonavir (RTV) [1]. However, the use of PI during pregnancy increases the risk of preterm birth and obstetric complications (e.g., pre-eclampsia, diabetes or intra uterine growth restriction) [2–4]. Those treatments have been shown to alter both adrenal and placental steroidogenesis. Indeed, neonates exposed in utero to PI exhibit adrenal dysfunction with an increase in 17-OH progesterone [5]. Anti-HIV treatment during pregnancy also induces a decrease in maternal serum progesterone (P4), especially when using PI [6].

From the end of the first month of pregnancy, P4 hormone, like human chorionic gonadotrophin (hCG), is produced by the villous trophoblast in the chorionic villi, mainly the syncytiotrophoblast (ST) [7,8]. We recently showed that the villous cytotrophoblast (VCT), which will differentiate into ST, is also able to produce placental glycoproteic hormones such as hCG and steroids such as P4 in vitro [7,9–11]. In placenta, P4 is synthesized from maternal serum cholesterol, which is captured by the trophoblast and enters the mitochondria via Metastatic Lymph Node 64 (MLN64) protein. The cholesterol molecule is then converted by cholesterol side-chain cleavage (P450SCC) enzyme in pregnenolone (P5), which is further converted in P4 by Hydroxy-delta-5-Steroid Dehydrogenase and 3 Beta-and steroid delta-isomerase 1 (HSD3B1) [8,12].

The production of hCG and P4 relies on the good functionality of the trophoblast. This functionality is regulated by numerous factors including cyclic Adenosine MonoPhosphate (cAMP)/Protein Kinase A (PKA) pathway, oxidative stress and stress of the Endoplasmic Reticulum (ER). ER stress involves different organelles (mitochondria, ER) and pathways such as the Unfolded Protein Response (UPR) pathway [13–18].

Mitochondria are essential to ensure energy regulation and steroid hormones production. We previously confirmed a change in mitochondrial function associated with structural modifications during VCT differentiation [11]. Different studies demonstrated that the structural modifications observed between VCT and ST mitochondria are related to mitochondria dynamics, relying on fusion/fission process [19–23]. The fusion is regulated by different factors such as Mitofusin 2 (Mfn2) and Optic Atrophy 1 (OPA1) [24–27]. In the placenta, mitochondrial dynamics are known to change with trophoblast differentiation but the mechanisms and factors involved remain controversial [18,27,28]. Any disruption in the fusion/fission process may lead to mitochondrial dysfunction, particularly steroidogenesis alteration [29,30].

The ER and Golgi apparatus are key organelles involved in the production of peptide hormones such as hCG [31]. In case of ER stress, an accumulation of unfolded proteins in ER lumen is observed. In response, Glucose-Regulated Protein 78 (GRP78) dissociates from the ER membrane activating the UPR pathway. This pathway involves (i) Inositol-Requiring Enzyme 1α (IRE1α), which induces the splice of X-box Binding Protein 1(XBP1) transcription factor controlling the expression of IRE1α target genes; (ii) Activating Transcription Factor 6 (ATF6), which is cleaved in its active form to control the transcription of its target genes; and (iii) Protein kinase RNA-like ER protein Kinase (PERK), which induces activation of Activating Transcription Factor 4 (ATF4) transcription factor to control the expression of PERK target genes [32]. These UPR pathways are known to regulate trophoblast differentiation, hCG secretion and the steroidogenesis [17,33].

It has been established that anti-HIV treatment by PI alters adrenal steroidogenesis both in the mothers and in their neonates exposed in utero [5,6,34]. As little is known about the effect of PI on the human placenta, we aimed to investigate the effect of two widely used PI (RTV and LPV) on the villous trophoblast differentiation in vitro, its endocrine function, and to identify their potential targets focusing on the mitochondria and the UPR pathway.

2. Results

2.1. *Effect of RTV on the Villous Trophoblast*

hCG and P4 levels were measured in supernatants of trophoblast cells incubated with RTV or control dimethylsulfoxide (DMSO). Neither hCG nor P4 secretion was disrupted during differentiation of VCT into ST whatever the incubation time (6 to 48 h) or RTV concentration (5 to 20 μM) (Figure 1A). On Western blot, the expression of P450SCC and HSD3B1, two enzymes involved in P4 synthesis, was not affected during RTV exposition (Figure 1B). RTV had no effect on villous trophoblast differentiation (data not shown).

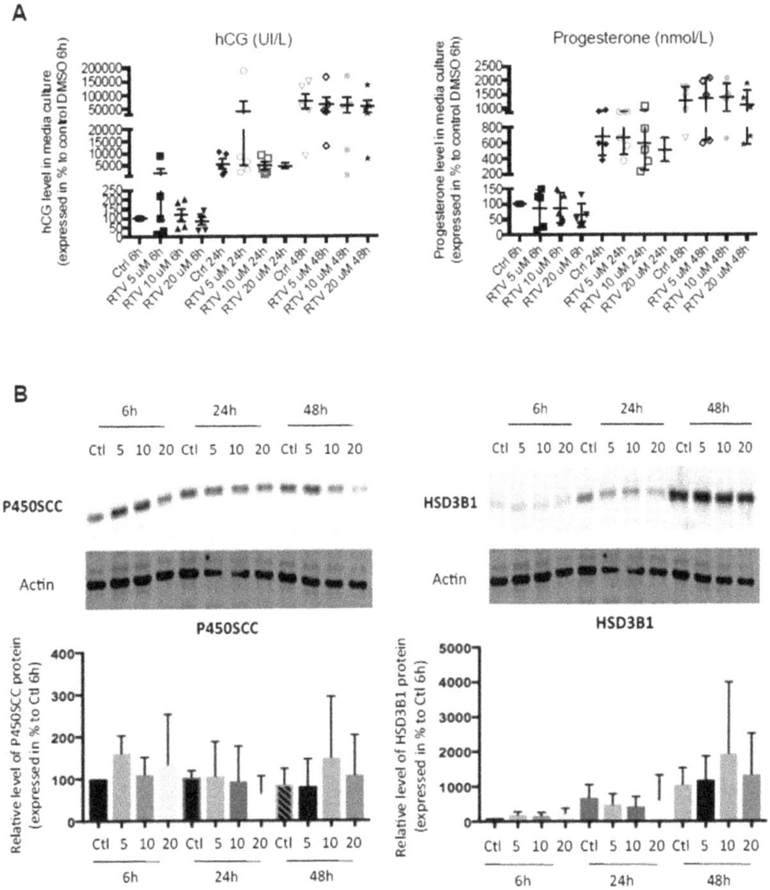

Figure 1. Ritonavir has no effect on hCG and P4 production during trophoblast differentiation. Cytotrophoblasts isolated and purified from term placenta were cultured for 15 h and then incubated with RTV at 5, 10 or 20 μM or DMSO for 6 h, 24 h or 48 h to allow fusion process. (**A**) hCG concentration and P4 concentration were measured by immuno-analysis in culture supernatant. (**B**) The protein expression of P450SCC and HSD3B1 were evaluated using immunoblotting with anti-P450SCC and anti-HSD3B1 antibodies. The protein expression of actin was determined with anti-actin antibody, used as a loading control. The lanes intensity was measured with ImageJ program. Results are expressed as the mean +/− SEM of $n = 6$ independent experiments. Two-tailed paired no parametric student t-tests were performed to compare RTV to DMSO exposition at the same incubation time.

2.2. Effect of LPV on the Villous Trophoblast

In controls, VCT spontaneously fuses to form a ST at 72 h of culture. After 48 h of incubation with LPV at 10 μM, ST formation was decreased as demonstrated with desmoplakin staining distribution (Figure 2A). The fusion index calculation points out a significant decrease ($p < 0.05$) of 20% in VCT fusion into ST after 48 h of incubation with LPV at 10 μM (Figure 2B).

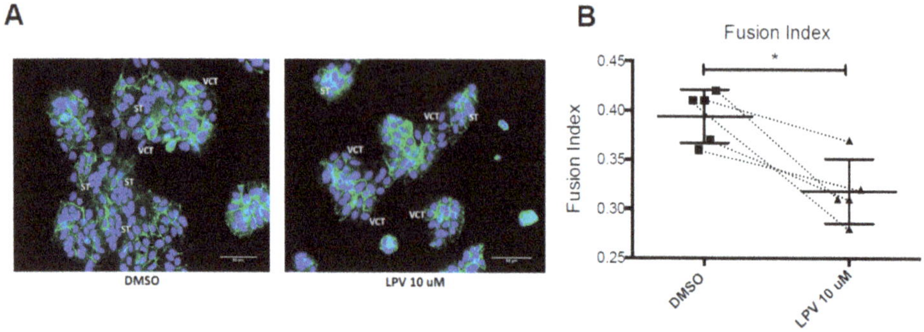

Figure 2. Lopinavir decreases VCTs fusion. VCT isolated and purified from term placenta were cultured for 15 h and then incubated with LPV 10 μM or DMSO for 48 h to allow fusion process. (**A**) Picture of cytotrophoblast fusion process by fixing and immunostaining of cells for the distribution of desmoplakin (green) and nuclei (4′,6-diamidino-2-phenylindole [DAPI] staining). 400 × magnification. Scale bar: 50 μm. (**B**) Representation of syncytium formation as a fusion index graph. Results are expressed as the mean +/− SD of $n = 5$ independent experiments. * $p < 0.05$ vs. DMSO, Mann-Whitney t-test.

As expected, VCT fusion into ST was associated with an increase in hCG (by 1,000-fold) and P4 (by 10-fold) secretion in controls. LPV at 10 μM significantly ($p < 0.001$) decreased hCG secretion by 35% in average after 6 h of incubation, reaching 84% of decrease at 48 h of incubation (Figure 3A). LPV also induced an early significant ($p < 0.01$) decrease in P4 secretion by 41% in average that tended to disappear thereafter (Figure 3B).

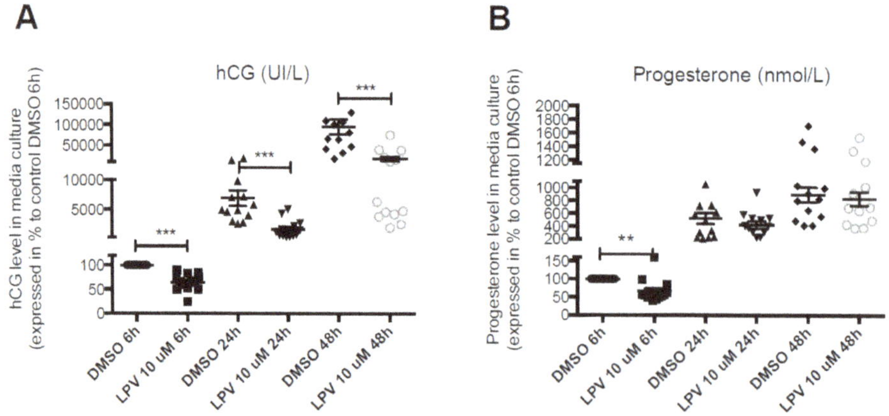

Figure 3. Lopinavir decreases hCG and progesterone secretion during differentiation of VCT into ST. Cytotrophoblasts isolated and purified from term placenta were cultured for 15 h and then incubated with LPV 10 μM or DMSO for 6 h, 24 h or 48 h to allow fusion process. hCG concentrations (**A**) and progesterone concentrations (**B**) were measured by immuno-analysis in culture supernatant. Results are expressed as the mean +/− SEM of $n = 11$ independent experiments. ** $p < 0.01$; *** $p < 0.001$ vs. DMSO at the same incubation time, two-tailed paired no parametric student t-test.

2.3. *Expression of Trophoblastic Enzymes Involved in P4 Synthesis during LPV Exposition*

As only LPV decreases both ST formation and P4 secretion, P4 synthesis partners were further investigated. On Western blots, LPV significantly ($p < 0.05$) decreased expression of P450SCC enzyme by 58% in average and HSD3B1 enzyme by 62% in average after 48 h of incubation (Figure 4B,C). However, LPV did not affect expression of mitochondrial cholesterol transporter MLN64 whatever the incubation time (Figure 4A).

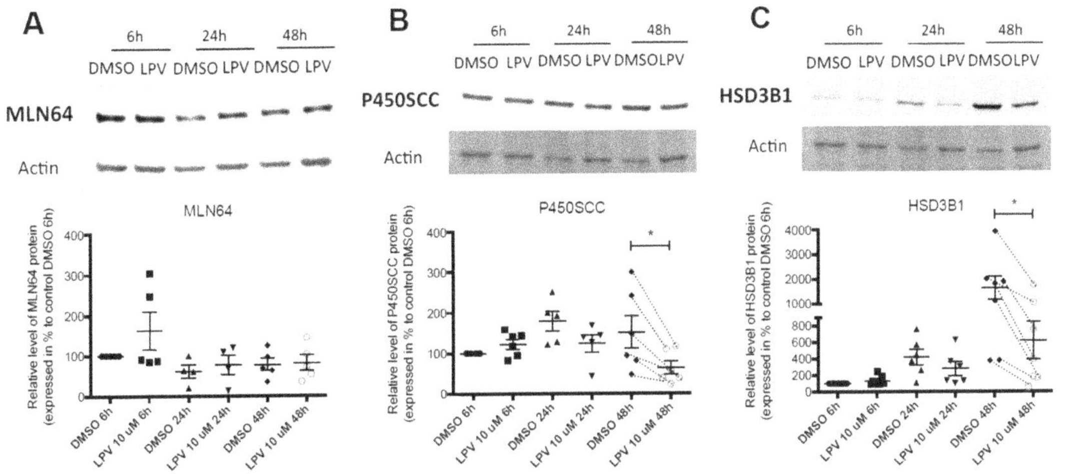

Figure 4. Lopinavir decreases protein expression of enzymes P450SCC and HSD3B1 involved in progesterone synthesis during differentiation of VCT into ST. Human VCT isolated and purified from term placenta were cultured for 15 h and then incubated with LPV 10 μM or DMSO for 6 h, 24 h or 48 h. MLN64 (**A**), P450SCC (**B**) and HSD3B1 (**C**) protein expression was determined using immunoblotting with anti-MLN64, anti-P450SCC and anti-HSD3B1 antibodies. Actin protein was determined with anti-actin antibody, used as a loading control. The lanes intensity was measured with ImageJ program. Results are expressed as a percentage of the control DMSO 6 h conditions and are shown as mean −/+ SEM from six independent experiments. * $p < 0.05$ vs. DMSO at the same incubation time, two-tailed paired no parametric paired *t*-test.

2.4. Trophoblastic Nuclei, Mitochondria and Endoplasmic Reticulum under LPV Treatment

We then analyzed by electron microscopy two main organelles involved in P4 and hCG synthesis, respectively. In controls, at 6 h, i.e., when VCT are still predominant and ST not yet formed, mitochondria present a few dense matrix with clearly defined cristae (Figure 5). On the contrary, at 48 h, i.e., when the vast majority of VCT has differentiated into ST, mitochondria present a clearly denser matrix with a less defined and more atypical cristae structure than in VCT (Figure 5). Moreover, we observed an increase in nuclei chromatin condensation with differentiation of VCT into ST (Figure 5). In VCT cells (i.e., 6 h of incubation), ER is thin, while in ST (i.e., 48 h of incubation), ER is larger (Figure 5). These physiological changes were not modified under LPV treatment in VCT (6 h). On the contrary, in ST (48 h), chromatin was less condensed than in controls and ER was thinner and rather empty. Mitochondria presented a less dense matrix with clearly defined cristae (Figure 5). In cells incubated with LPV for 24 h, the results were less significant as some VCT had not already started their differentiation.

2.5. Mitochondrial Dynamics in Villous Trophoblast under LPV

The mitochondrial structure relies on the fusion and fission process regulated by several proteins (OPA1, Mfn2). LPV induced a significant decrease ($p < 0.05$) in Mfn2 protein expression after 48 h incubation (Figure 6A), while OPA1 protein expression was not affected (Figure 6B).

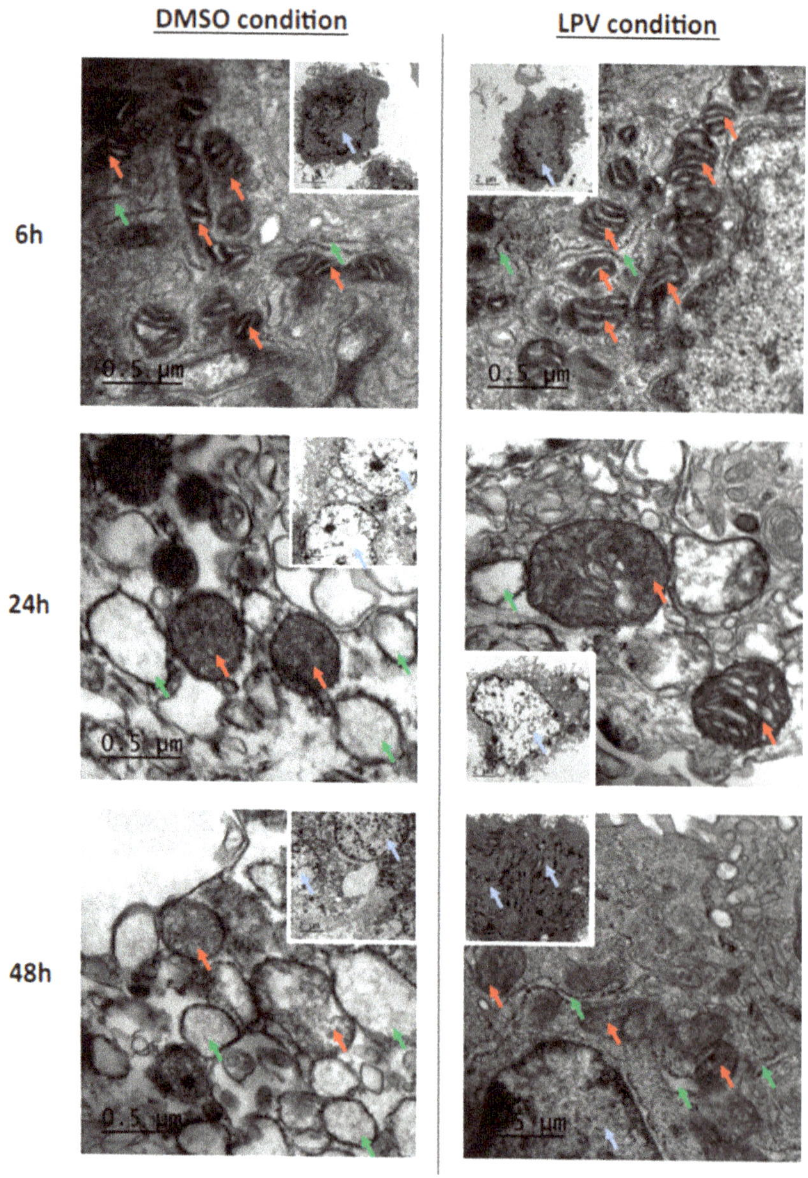

Figure 5. Lopinavir alters mitochondria and endoplasmic reticulum structure during trophoblast differentiation. Cytotrophoblasts isolated and purified from term placenta were cultured for 15 h and then incubated with LPV 10 µM or DMSO for 6 h, 24 h or 48 h. Cells have been fixed and prepared for transmission electronic microscopy. The blue, red and green arrows indicate respectively the nuclei, mitochondria and endoplasmic reticulum (ER). 1000× magnification was used for the small pictures; scale bar: 2 µm. Zoom in 5000× magnification was used for the principal pictures. Scale bar: 0.5 µm. These pictures are representative of 3 independent experiments.

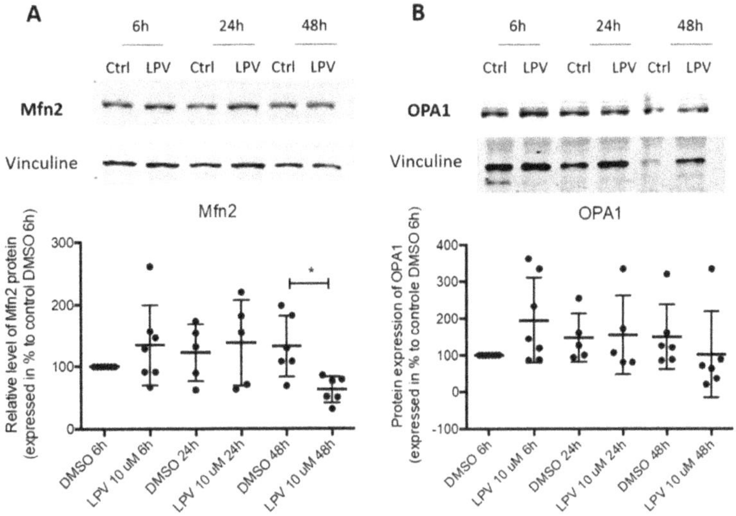

Figure 6. Impairment in mitochondrial dynamics during trophoblast differentiation under LPV exposition. Cytotrophoblast cells isolated from human term placentas were cultured for 15 h before to the incubation with LPV 10 μM or DMSO for 6 h, 24 h or 48 h. Mfn2 (**A**) or OPA1 (**B**) protein expression were evaluated by immunoblotting with anti-Mfn2 and anti-OPA1 antibodies. Vinculin protein expression determined with anti-vinculin antibody was used as loading control. Results are expressed as the mean +/− SD of $n = 7$ independent experiments. * $p < 0.05$ vs. DMSO at the same incubation time, two-tailed paired no parametric student *t*-test.

2.6. ER Stress in Villous Trophoblast under LPV

In collaboration with Dr Marie Cohen, we first checked whether UPR pathways are involved in steroidogenesis regulation in our in vitro model. We evaluated mRNA expression of *P450SCC* and *HSD3B1* by RT-qPCR on VCT cells transfected with 3 siRNA against *IRE1α*, *PERK* and *ATF6*, respectively. Inhibition of UPR pathway tended to induce an increase in *HSD3B1* and *P450SCC* expression in the same proportion (50% and 55%, respectively) (Figure 7).

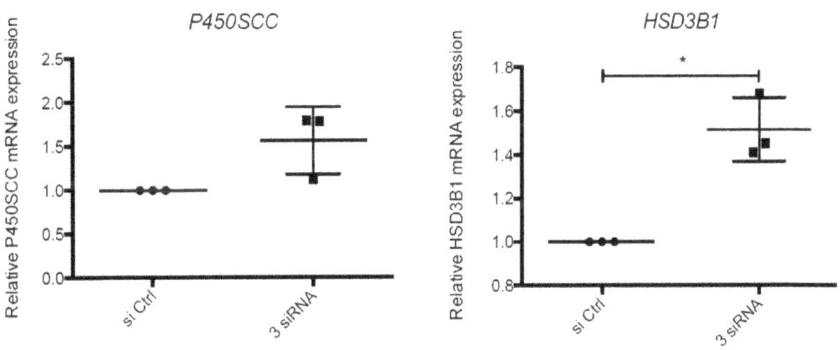

Figure 7. UPR pathways are involved in transcriptional regulation of enzymes involved in P4 synthesis. VCT cells isolated from human term placenta were transfected with siRNA against *IRE1α*, *PERK* and *ATF6* (3 siRNA) or siRNA control (si Ctrl). Total RNA was extracted and *P450SCC* and *HSD3B1* expression were evaluated by RT-qPCR. The results are presented as mean +/− SD of 3 independent experiments. * $p < 0.05$ vs. si Ctrl, two-tailed paired no parametric *t*-test.

We then investigated whether LPV activates UPR pathways measuring mRNA expression of UPR markers: *GRP78, ATF6, ATF4* and *sXBP1* by qPCR. In VCT (i.e., 6 h incubation with LPV), *GRP78* and *sXBP1* expression significantly increased 1.5- and 2-fold ($p < 0.05$) (Figure 8A). During trophoblast differentiation, GRP78 protein expression under LPV increased 1.5-fold in VCT cells, after 6 h of incubation but decreased 0.65-fold in ST, after 48 h of incubation (Figure 8B).

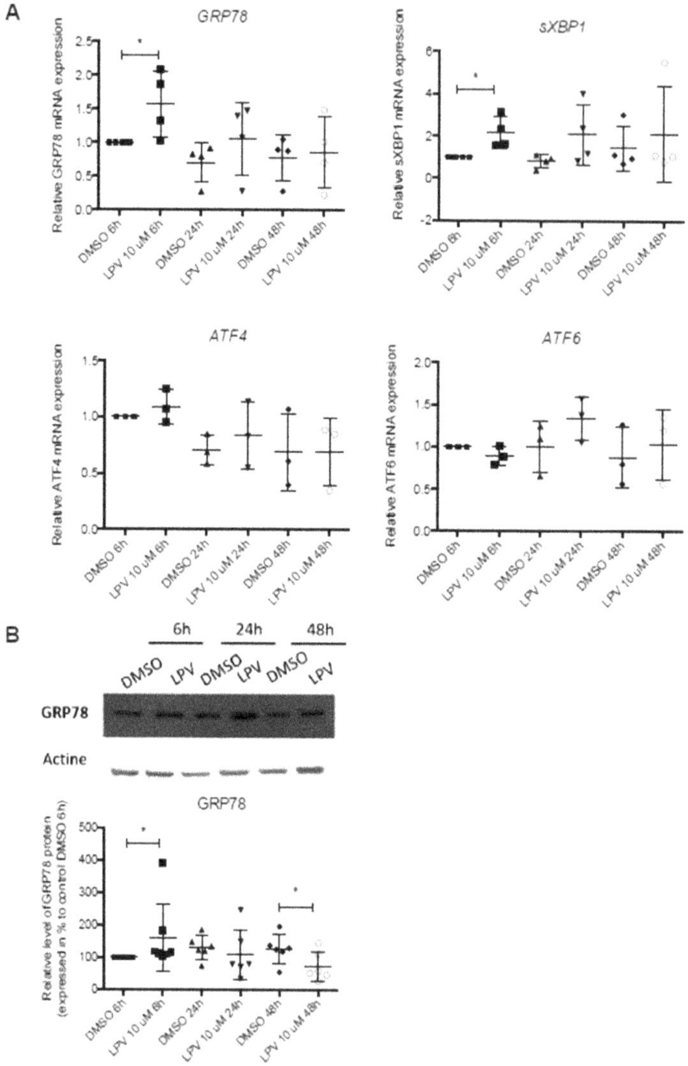

Figure 8. LPV induces activation of IRE1α pathway from 6 h of incubation in trophoblasts cells. VCT cells were isolated from human term placenta. After 15 h of culture, cells were incubated for 6 h, 24 h or 48 h with 10 μM LPV. (**A**) Transcriptional expression of *GRP78*, *sXBP1*, *AFT4* and *AFT6* were measured by RT-qPCR. (**B**) Protein expression of GRP78 was evaluated by immunoblotting with anti-GRP78 antibody. Actin protein expression determined with anti-actin antibody was used as loading control. The results are expressed as mean +/− SD of $n = 4$ independent experiments. * $p < 0.05$ vs. DMSO at the same incubation time, two-tailed paired no parametric t-test.

2.7. Effects of LPV on Preformed ST

As ST is trophoblastic tissue in direct contact with maternal blood, we analyzed the impact of LPV directly on ST (i.e., when VCT differentiation is already performed). VCT were cultured in complete Dulbecco's Modified Eagle Medium (DMEM) for 72 h forming the ST. The resulting ST was further exposed to LPV at 10 μM for 6 h. Exposition to LPV induced a significant decrease in hCG and P4 secretion by 20% and 40%, respectively ($p < 0.05$) (Figure 9).

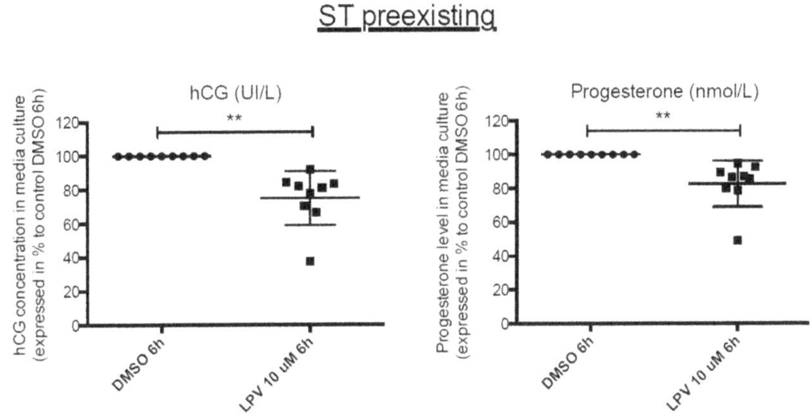

Figure 9. Brief incubation with Lopinavir decreases hCG and progesterone secretion both in VCT and ST cells. Cytotrophoblasts isolated and purified from term placenta were cultured for 72 h in complete DMEM to allow ST formation before incubation for 6 h with LPV 10 μM or DMSO. hCG and progesterone concentrations were measured by immuno-analysis in culture supernatant. Results are expressed as the mean +/− SEM of $n = 11$ independent experiments. ** $p < 0.01$ vs. DMSO, two-tailed paired no parametric student *t*-test.

P450SCC and HSD3B1 protein expression was also significantly decreased by 20% in ST exposed to LPV for 6 h ($p < 0.05$) (Figure 10).

Exposition of ST to LPV for 6 h also induced a significant increase by 20% in Mfn2 mitochondrial protein expression ($p < 0.05$) (Figure 11A), while OPA1 mitochondrial protein expression was unchanged (Figure 11B).

Exposition of ST to LPV tended to activate UPR pathway as attested by the increase in *GRP78* and *sXBP1* transcripts (Figure 12A).

We investigated whether the IRE1α-pathway was targeted by LPV. Using STF-083010 (STF), an IRE1α inhibitor, we checked that IRE1α inhibition led to a decrease in LPV effects on UPR pathway, quantifying *sXBP1* and *GRP78* gene expression by RT-qPCR (Figure 13).

We then evaluated the effect of IRE1α inhibition on hCG and P4 production by ST under LPV exposition. Pre-incubation with STF-083010 (STF), the IRE1α inhibitor, did not restore hCG and P4 secretion, decreased by LPV in ST (Figure 14).

In addition, this pre-incubation did not modify the expression of either P450SCC or HSD3B1 (Figure 15) at both mRNA and protein levels.

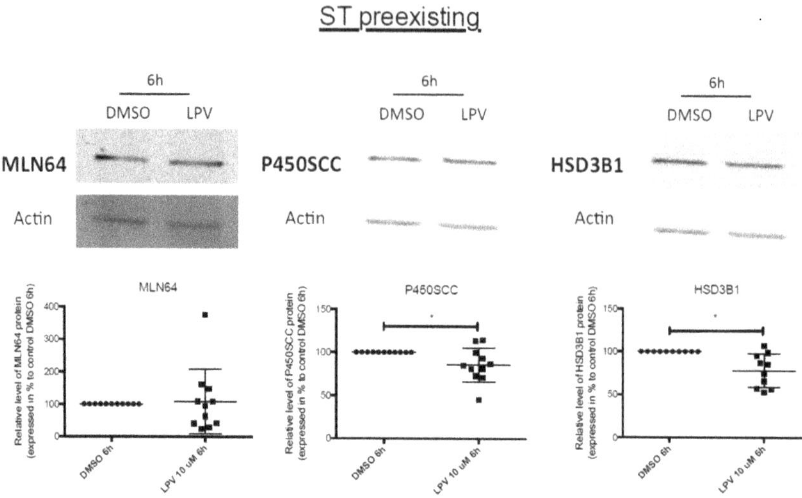

Figure 10. Decrease in enzymes protein expression in ST cells under brief LPV exposition. Cytotrophoblast cells isolated from human term placentas were cultured for 72 h before incubation with LPV 10 μM or DMSO for 6 h. MLN64, P450SCC and HSD3B1 protein expression was evaluated by immunoblotting with anti-MLN64, anti-P450SCC and anti-HSD3B1 antibodies. Actin protein expression determined with anti-actin antibody was used as loading control. Results are expressed as the mean +/− SD of $n = 8$ independent experiments. * $p < 0.05$ vs. DMSO, two-tailed paired no parametric student t-test.

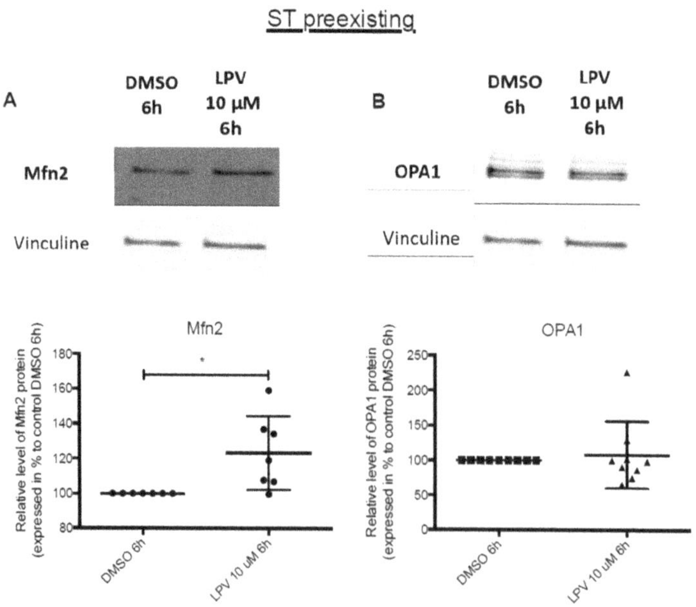

Figure 11. Impairment in mitochondrial dynamics in ST cells under LPV exposition. Cytotrophoblast cells isolated from human term placentas were cultured for 72 h before incubation with LPV 10 μM or DMSO for 6 h. Mfn2 (**A**) and OPA1 (**B**) protein expression was evaluated by immunoblotting with anti-Mfn2 and anti-OPA1 antibodies. Vinculin protein expression determined with anti-vinculin antibody was used as loading control. Results are expressed as the mean +/− SD of $n = 7$ independent experiments. * $p < 0.05$ vs. DMSO, two-tailed paired no parametric student t-test.

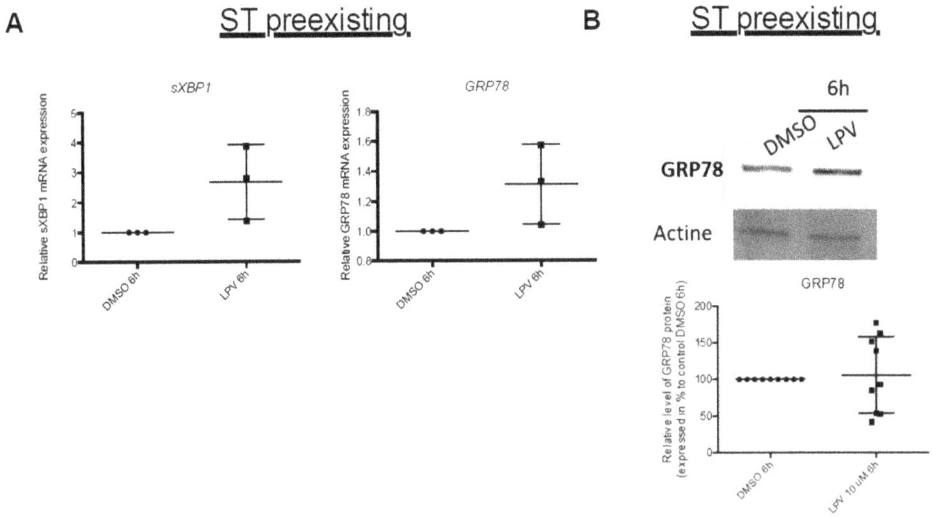

Figure 12. Activation of UPR pathway in VCT cells under LPV exposition. Cytotrophoblast cells isolated from human term placentas were cultured for 72 h before incubation with LPV 10 μM or DMSO for 6 h. (**A**) mRNA expression of *sXBP1* and *GRP78* were measured by RT-qPCR. The results are expressed as mean +/− SD of n = 3 independent experiments. Two-tailed paired no parametric t-test were realized. (**B**) GRP78 protein expression was evaluated by immunoblotting with anti-GRP78 antibody. Actin protein determined with anti-actin antibody was used as loading control. The results are expressed as mean +/− SD of n = 6 independent experiments. Two-tailed paired no parametric t-test were realized.

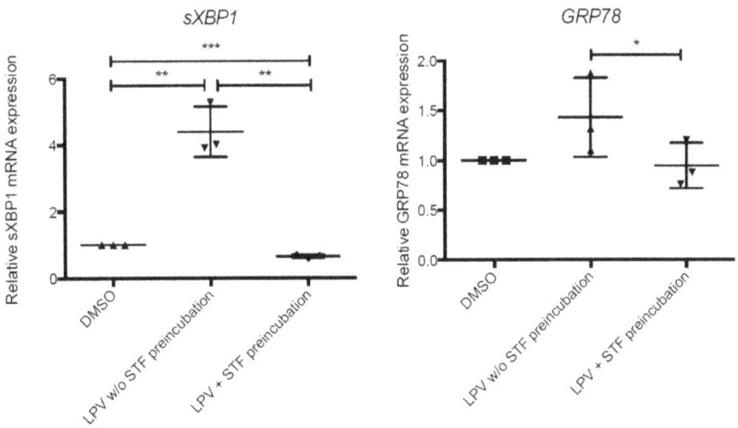

Figure 13. Inhibition of IRE1α pathway partially prevents the UPR pathway activation. Cytotrophoblast cells isolated from human term placentas were cultured for 72 h. Cells were then incubated for 2 h with STF (100 μM) (LPV with STF preincubation) or DMSO control (LPV without (*w/o*) STF preincubation) before incubation for 4 h with LPV (10 μM). (A) mRNA expression of *sXBP1* and *GRP78* were measured by RT-qPCR. The results are expressed as mean +/− SD of n = 3 independent experiments. * $p < 0.05$; ** $p < 0.01$; *** $p < 0.001$ vs. DMSO, two-tailed paired no parametric t-test.

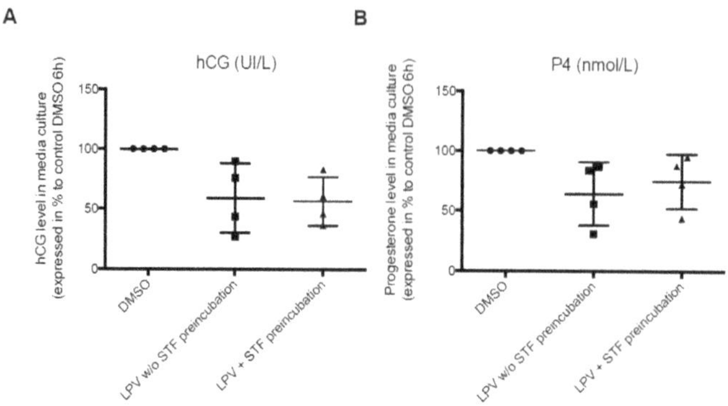

Figure 14. Inhibition of IRE1α pathway does not prevent the inhibitory effect of LPV on hCG and P4 secretions. VCT cells isolated from human term placenta were cultured for 72 h. Cells were further incubated for 2 h with STF (100 μM) (LPV + STF preincubation) or DMSO control (LPV without (*w/o*) STF preincubation) before incubation for 4 h with LPV (10 μM). hCG (**A**) and P4 (**B**) levels were measured in supernatant. The results are presented as the mean +/− SD from four independent experiments.

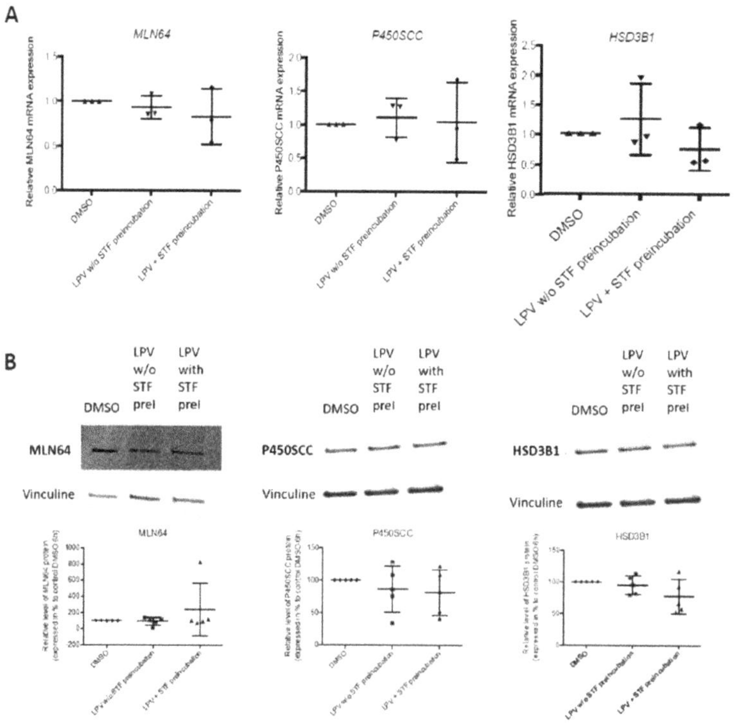

Figure 15. Inhibition of IRE1α pathway does not prevent the inhibitory effect of LPV on P4 synthesis partners expression. VCT cells isolated from human term placenta were cultured for 72 h. Cells were further incubated for 2 h with STF (100 μM) (LPV + STF preincubation) or DMSO control (LPV without (*w/o*) STF preincubation) before incubation for 4 h with LPV (10 μM). MLN64, P450SCC and HSD3B1 expression was evaluated by RT-qPCR (**A**) and Western blot using anti-MLN64, anti-P450SCC and anti-HSD3B1 antibodies (**B**). Vinculin protein expression determined by Western blot using anti-vinculin antibody was used as loading control. The results are presented as the mean +/− SD from n = 3 independent experiments for RT-qPCR and n = 5 independent experiments for Western blot.

3. Discussion

During pregnancy, HIV-infected mothers are treated with two NRTI and one PI to prevent the viral transmission to the fetus [1]. According to several studies those treatment, notably the PI, have secondary effects such as pre-term birth, pre-eclampsia, and intra uterine growth restriction [3,4]. Moreover, PI impair steroidogenesis, inducing a decreased P4 level in maternal blood and impaired adrenal function in neonates exposed in utero [3,5,6]. However, very little is known about the placenta, which produces P4, a steroid hormone mandatory for the maintenance of pregnancy [7,8,35]. Our aim was to investigate in vitro whether PI such as RTV and LPV disturb human placental steroidogenesis, focusing on the P4, mitochondria and main intracellular pathways potentially targeted.

For many years, the syncytiotrophoblast (ST) has been considered as the main endocrine tissue of the placenta as it is in direct contact with maternal blood in the intervilli chamber [36]. As in the chorionic villi, the ST arises from the differentiation of the villous cytotrophoblast (VCT), we used our in vitro model to extensively characterize P4 production in human villous placenta. In our previous study [11], we established that placental steroidogenesis is not restricted to ST only but starts early in the VCT, which expresses the cholesterol transporter MLN64 and the key enzymes P450SCC and HSD3B1 required for P4 synthesis and secretes significant levels of P4. We confirmed these findings in this work, pointing out that the whole trophoblast is able to produce placental steroid and protein hormones. This is in agreement with previous studies showing that VCT and also extravillous cytotrophoblast (EVCT) produce significant amounts of hCG and its subunits [37]. We confirmed in this in vitro study that the endocrine production increases with the morphological differentiation of VCT into ST, hCG secretion increase being associated with ST formation. We also established that this increase in hCG secretion is correlated with an increase in P4 synthesis and secretion [11].

The mitochondria is a key organelle involved both in steroidogenesis and in trophoblast differentiation [16,27,30]. We have previously demonstrated that the differentiation of VCT into ST is associated with a decrease in mitochondrial transmembrane potential. This functional change in mitochondria could explain the observed changes in P4 synthesis [11]. Indeed, a previous study on placental tissue demonstrated that a decrease in transmembrane potential in mitochondrial fraction is associated with a high steroidogenesis activity [38]. In addition, we showed that in vitro differentiation of VCT into ST is associated with morphological changes in mitochondria using transmission electronic microscopy. Indeed, in our in vitro model, VCT are predominant at 24 h of culture, whereas ST is formed at 72 h of culture. In VCT, mitochondria are larger than in ST. Moreover, mitochondria in VCT present cristae with typical structure, whereas they present in ST an atypical structure of cristae and a denser matrix than in VCT, in agreement with previous histological studies [20]. These structural modifications are known to be related to fusion/fission dynamics of mitochondria [21]. However, in our model, we observed no modification in the expression of two main fusion proteins Mfn2 and OPA1 during differentiation of VCT into ST. These results are not consistent with a previous study in BeWo cells that demonstrated a decrease in Mfn2 and OPA1 expression but, under forskolin, induced differentiation [27]. This suggests that the observed changes in mitochondria depend on the model used, i.e., primary trophoblast culture in our model compared to choriocarcinoma cell lines and on experimental conditions, i.e., basal or stimulated by forskolin. Mitochondrial dynamics are known to be involved in the regulation of steroidogenesis. Some studies demonstrated that fission process is necessary for steroidogenesis [22,27], while others reported on the contrary that steroidogenesis relies on fusion process [30,39]. In our model, it is likely that fusion process is not necessary for P4 synthesis as we found no change in Mfn2 and OPA1 expression during the ST formation. It would thus be of interest to evaluate the expression of other factors such as fission factors Drp1 and Fiss1. An increase in Peroxisome proliferator-activated receptor Gamma Coactivator 1-α (PGC1-α) expression induced by P4 could lead to an increase in mitochondria biogenesis allowing an increase in P4 synthesis itself [40].

We also found morphological changes in nuclei and ER during VCT differentiation into ST. In ST, we observed that the nuclei present a denser chromatin compared to VCT, confirming previous placental histological findings [36]. The ER is also thinner in VCT than in ST cells. The presence of large ER in ST is associated with an intense protein synthesis [41,42], consistent with the observed increase in hCG secretion. It is also in agreement with the UPR activation during VCT differentiation into ST [17]. To conclude, the use of transmission electronic microscopy allowed us to fully characterize the differentiation of VCT into ST at cellular level, pointing out a membrane fusion, a nuclear high-condensed chromatin, small and dense mitochondria and large ER. This is in agreement with in vivo findings showing a release of syncytial knots containing ST DNA and UPR activation and an increased hormonal production.

We tested the impact of PI, notably the LPV, at a concentration of 10 μM, which corresponds to maternal blood concentration. Higher concentrations are likely to induce apoptosis as observed in HEK293 cells [43]. LPV at 10 μM induced an early decrease in P4 production in VCT that could be in part responsible for the decrease in P4 level in treated mother serum [3,6]. However, it would be of interest to investigate the effect of LPV/RTV as RTV is often associated to LPV as a booster [3]. Surprisingly, the decrease in P4 secretion under LPV was not associated with a decrease in expression of P450SCC and HSD3B1 enzymes in VCT. LPV could act rather indirectly by modulating P4 catabolism or transport. Indeed, LPV could activate Cytochrome P450 (CYP) family enzymes [44] and could modulate Adenosine-Tri-Phosphate (ATP)-Binding Cassette transporter activity [45]. Moreover, in BeWo cells, Papp et al. have demonstrated that PI could act on 20-α Hydroxysteroid Dehydrogenase, an enzyme involved in progesterone catabolism [46]. Consequently, in short incubation times, PI could decrease progesterone secretion in inhibiting enzymes involved in its synthesis and in activating enzymes involved in its catabolism. In ST, after 48 h of incubation with LPV, we also noted a discrepancy between the decreased expression of P450SCC and HSD3B1 enzymes, and the level of P4 secretion. The decrease in enzyme expression could be due to a global alteration in protein synthesis involving the eucaryotic Elongation Factor-2 (eEF2) translation factor. Indeed, it has previously been shown in the myocyte that LPV induces a decrease in protein synthesis [47]. The absence of P4 decrease after 48 h of incubation with lopinavir, while P450SCC and HSD3B1 expression is reduced, could be explained by the high stability of steroid hormones in our media culture. Consequently, we are not able to detect significant changes in P4 concentration after 48 h of incubation. So, it would be interesting to measure P4 concentration at shorter time or by changing our media culture.

LPV did not only induce an early decrease in P4 but also in hCG secretion in VCT. However, the decrease in hCG also persisted in ST after 48 h of incubation. This could be due to the fact that LPV affects rather ER and protein pathway than mitochondria and steroidogenesis. During ST formation, dynamic changes in some mitochondria could overlap the initial effect of LPV observed on VCT. On the contrary, as far as hormone production is concerned, we observed that LPV leads to a decrease in plasma membrane fusion as demonstrated by the decrease in fusion index associated with a decreased nuclei chromatin condensation and a thinner ER. Consequently, the 48 h incubation with LPV may prevent the morphological changes physiologically associated with functional differentiation. However, it is possible that the early decrease in P4 and hCG secretion induced by LPV from 6 h of incubation in VCT is responsible for the reduced morphological differentiation as both hormones are involved in autocrine an paracrine regulation of the trophoblast differentiation. As a matter of fact, hCG receptors are more expressed in VCT cells than in ST [13] and could be regulated by P4 [48] in order to control plasma membrane fusion [14]. Consequently, the decrease in P4 observed in VCT cells could impair hCG receptor expression. The decrease in hCG receptors expression and hCG secretion would lead to a decrease in plasma membrane fusion, preventing the global differentiation of VCT into ST.

Several intracellular mechanisms are involved in trophoblast differentiation [13,14] such as mitochondria dynamics and UPR pathways [16,17]. It is known that PI, notably LPV, alter both mitochondria DNA content and dynamics [49]. We analyzed the protein expression of two mitochondrial pro-fusion factors Mfn2 and OPA1. We demonstrated that LPV induced a decrease in Mfn2 expression after 48 h of incubation while OPA1 expression remained unchanged. The decrease in at least one fusion factor could result in a decrease in fusion process. These results are consistent with previous studies demonstrating that mitochondrial fusion is necessary for steroidogenesis [30,39]. However, they are inconsistent with the observed P4 secretion in ST after 48 h of incubation. This discrepancy could be explained by a global decrease in protein synthesis including Mfn2. It could be induced by a global cellular stress targeting rather the ER and the proteins than the mitochondria and the steroidogenesis [47].

Marie Cohen et al. also previously demonstrated that trophoblast differentiation and hCG secretion are under the control of UPR pathway [17]. Interestingly, several studies showed an activation of UPR, especially the IRE1α pathways under LPV exposition [43,50–53]. In collaboration with Dr Marie Cohen, we firstly demonstrated that UPR pathway regulates the expression of enzymes involved in P4 synthesis in placenta. We established in our model that LPV induces an early increase in sXBP1 and GRP78 expression in VCT pointing out an activation of IRE1α pathway in agreement with previous studies [43,50–53]. The activation of IRE1α is not maintained in ST, where GRP78 expression is on the contrary decreased. All these findings under LPV exposition, i.e., UPR activation, and decrease in GRP78, P450SCC, HSD3B1 and Mfn2, are likely to reflect an uncontrolled global cellular stress. Physiologically, UPR activation allows the maintenance of cellular homeostasis in case of stress. During a stress which cannot be controlled by UPR pathways activation, other processes take place, including particularly a general decrease in protein synthesis [32]. Thus, LPV could induce a stress activating IRE1α pathways. This stress could become uncontrollable after 48 h incubation with LPV ending in a global cell alteration and decreased protein synthesis as previously demonstrated in myocytes cells [47]. In summary, LPV have dual effects, an early effect (after 6 h of incubation) on enzyme and transporter activities to reduce P4 secretion and a later effect (after 48 h of incubation) on protein synthesis due to an ER stress.

In comparison to other PI, we demonstrated that RTV has no effect either on the ST formation or on hCG and P4 secretion in vitro, whatever the used concentration.

LPV exposition could not only disrupt VCT and its functional differentiation into ST, but it could also directly damage the ST covering the villi in direct contact with maternal blood and thus with PI. Indeed, on preexisting ST, we showed that LPV induces a decrease in both hCG and P4 secretion. In ST, on the contrary of VCT, this decreased secretion is associated with a decrease in P450SCC and HSD3B1 expression. LPV induces an increase in Mfn2 expression in ST. Consequently, the mitochondria in ST are more fused under LPV exposition. The increase in mitochondrial fusion induced by LPV could explain the decrease in P4 secretion by ST. We also investigated the effect on UPR pathways. In the preexisting ST, LPV induces an increase in sXBP1 expression demonstrating the activation of IRE1α pathway. To investigate whether this activation is responsible for the disruption in hCG and P4 secretion, we used an IRE1α inhibitor. We showed that the IRE1α inhibitor does not prevent either the decrease in hCG and P4 secretion or the decrease in P450SCC and HSD3B1 expression induced by LPV. The absence of effect of the IRE1α inhibitor may be explained by compensatory mechanisms of other UPR pathways. Indeed, the use of chemical UPR inhibitors is known to lead to compensatory mechanisms via the activation of the other UPR pathways [54]. It will thus be of interest to perform additional experiments combining different chemical inhibitors or siRNA targeting the 3 UPR pathways. Further investigation would also be necessary to check whether LPV directly activates UPR pathways or not. As a matter of fact, in HEK293 cells, Taura et al. [43] demonstrated that LPV is able to produce Reactive Oxygen Species (ROS), leading to an activation of JUNK kinase pathway, further activating UPR pathways. Consequently, ROS production could be involved in the indirect effect of LPV on UPR activation in trophoblast.

The treatment of HIV infection uses a combination of two NRTI combined with a PI. In our study, we have demonstrated for the first time that RTV has no significant effects on endocrine function in trophoblast cells. On the contrary, LPV induces alteration of trophoblast morphological differentiation related to a decrease in plasma membrane fusion and in nuclei chromatin condensation, larger mitochondria and a thinner ER. These morphological alterations are associated with a decrease in P4 and hCG secretions. LPV also impairs mitochondrial dynamics as attested by an increase in Mfn2 expression and an induction of ER stress. ER stress leads to IRE1α activation marked by an increase in sXBP1 and GRP78 expression. These findings give the beginning of an explanation for the PI in vivo toxicity as they are known to induce preterm birth especially when they are "boosted" with RTV.

4. Materials and Methods

4.1. Placental Tissue Collection

Placental tissues from patients delivering by cesarean section at full term (mean gestational age: 39 $\pm$ 1 weeks of gestation) were obtained from Antoine Béclère Hospital (Clamart, France), Antony Hospital (Antony, France), Port-Royal Maternity (Paris, France) and Montsouris Mutual Institute (Paris, France). Indications for cesarean sections were: maternal uterus abnormalities (uterus scar, myomectomy), maternal wish, breech presentation, narrow maternal pelvis. Placentas were collected following informed patient written consent and approval from our local ethics committee (CPP 2015-mai-13909). All the collected placentas resulted from monofetal non-complicated pregnancies, i.e., with no fetal abnormalities, no maternal diseases or treatment (i.e., diabetes, thyroid disorder, hypertension, pre-eclampsia). Placentas included were all macroscopically normal (weight and macroscopic examination).

4.2. Isolation and In Vitro Culture of VCT

Chorionic villi were obtained by manual dissection of placental tissues from term placentas as previously described about one hour after delivery [11,55]. Villous tissue was dissected free of membranes, rinsed and minced in Hank's Balanced Salt Solution (HBSS) 1X. The villous sample was then subjected to sequential enzymatic digestion in HBSS 1X containing 0.2% trypsin (*w/v*), 25 IU/mL DNAse I, 0.1 mM $MgSO_4$, 0.1 mM $CaCl_2$ and 4% milk (*v/v*). Cell dissociation was monitored by light microscopy. The first three digests were discarded to eliminate residual ST fragments and erythrocytes. Cell suspensions resulting from the following four or five sequential digestions were pooled. Cells were then purified on a discontinuous Percoll gradient (5% to 70% in 14 steps) and their viability was determined in *v/v* solution with trypan blue.

Isolated cells were seeded in DMEM containing 10% Fetal Calf Serum (FCS), 1% Penicillin-Streptomycin and 1% L-Glutamin (complete DMEM) at 1000 cells/mm^2 during 15 h at 37 °C with 5% CO_2. After plating, cells were incubated with RTV, LPV or DMSO control during 6 h, 24 h or 48 h (time necessary to allow fusion process). RTV concentrations used ranged from 5 to 20 μM. LPV chosen concentration was 10 μM (=6.288 μg/mL or 6288 ng/mL) corresponding to the mean LPV blood concentration measured in mothers treated for HIV infection. RTV (SML-0491, Sigma-Aldrich® Inc, St-Louis, MO, USA) or LPV (SML-1222, Sigma-Aldrich® Inc, St-Louis, MO, USA) were dissolved in DMSO. Consequently, control DMSO was equivalent to the percentage of DMSO necessary to dissolve LPV.

Cell culture media were collected at different times of culture, supernatants were kept frozen at −80 °C as part of the Equipex 10-PhC/SC-11/243 project "Perinatcollection" until use. Cells were either fixed or snap-frozen for RNA/Protein extraction and stored at −80 °C until use.

4.3. Immunofluorescence Microscopy

Immunocytofluorescence staining was performed after 48 h of incubation with LPV or control DMSO, which corresponds to the time required for ST formation, as previously described [11]. Briefly, cells were fixed and permeabilized in methanol and blocked in Phosphate Buffered Saline (PBS) 1× Bovine Serum Albumine (BSA) 1%-Tween 0.1%. Cells were first incubated overnight at 4 °C with primary polyclonal rabbit antibody to desmoplakin (5 µg/mL, ab16434 abcam®, Cambridge, UK). They were then incubated for 1 h at room temperature and protected from light, with secondary goat anti-rabbit antibody conjugated with Alexa Fluor 488 (1/500, A-11008 LifeTechnologies, Carlsbad, CA, USA). Nuclei were stained with 4′,6-diamidino-2-phenylindole (DAPI) and samples were conserved in mounted-medium. Pictures were taken using a BX60 epifluorescence microscope (Olympus) equipped with a 40× oil objective (Olympus 1.00), an ultrahigh-vacuum mercury lamp and a Hamamatsu camera (C4742-95) and analyzed with VisionStage Orca software (v 1.6). The "control immunoglobulin without specific epitope" conditions were used to evaluate the background signal and to set up the acquisition and colorization of pictures. Resulting pictures allowed us to calculate fusion index i.e., (nuclei number in ST-ST number)/total nuclei number. ST was considered when at least two nuclei were not separated by plasmic membrane, observed thanks to desmoplakin staining in green.

4.4. Transmission Electronic Microscopy

Trophoblastic cells were seeded for 15 h post isolation and incubated with LPV or control DMSO for 6, 24 or 48 h. Cells were collected after each time by trypsin-EDTA, centrifuged 5 min at 1400× g, and washed twice in PB (0.05 M PIPES [piperazine-N,N′-bis(2-ethanesulfonic acid)] buffer, 5 mM $CaCl_2$, pH 7.3) for 10 min, then centrifuged 5 min at 1400× g. Each sample was fixed during 45 min at room temperature protected from light in PB containing 2.5% glutaraldehyde and 2% paraformaldehyde. After 5 min of centrifugation at 1400× g, samples were washed twice for 10 min in PB and then post-fixed first in PB–1% osmium tetroxide (45 min at 4 °C) and then in 1% aqueous uranyl acetate solution for 2 h at room temperature. Samples were then dehydrated in increasing concentrations of ethanol (30%, 50%, 70%, 95% and 100%) followed by ethanol/propylene oxide (1/1 (vol/vol)) and propylene oxide and were finally embedded in Epon epoxy resin. Ultrathin sections (80 nm of thickness) were performed with a Leica ultracut S microtome fitted with a diamond knife (Diatome histoknife Jumbo or Diathome ultrathin). Theses sections were stained with lead citrate and placed on copper grids. The sections were analyzed at 80 kV with a Jeol electron transmission microscope (JEM-100S transmission electron microscope, Croisy sur Seine, France). Acquisitions were made with Gatan software (Gatan Microscopy Suite ®, Gatan Inc., AMETEK, Pleasanton, CA, USA).

4.5. Hormones Immuno Assays

As previously described [11], total hCG and P4 concentrations were determined in supernatants of culture using ECLIA immuno-assay (Liaison®, DiaSorin, Sallugia, Italy). Measuring ranges of both assays were respectively of 1.5–10,000 UI/L and of 0.7–190 nmol/L. Between assay precision expressed by the coefficient of variation (CV %) were <5% and <11% and detection limits were 0.3 UI/L and 0.4 nmol/L respectively.

4.6. Western Blots

Total proteins were isolated from cells incubated with RTV, LPV or DMSO using Lysis Buffer (NP40 Cell Lysis Buffer; Invitrogen™, Carlsbad, CA, USA) combined with a protease inhibitor cocktail 100× (1× final; Protease Inhibitor Cocktail Set I, Calbiochem®, EMD Chemicals Inc., Merck KGaA, Darmstadt, Germany) and a phosphatase inhibitor cocktail 50× (1× final; Phosphatase Inhibitor Cocktail 50× Set V, Calbiochem®, EMD Chemicals Inc., Merck KGaA, Darmstadt, Germany). Proteins (20 µg) were loaded on 4–15% gradient gel (Mini-PROTEAN® TGX™ Precast Gels, BIORAD®, Hercules, CA, USA), and were then transferred on nitrocellulose membrane which was blocked with TBS 1×-

Milk 5%-Tween 0.1% solution. The resulting proteins blots were probed with anti-MLN64, anti-P450SCC, anti-HSD3B1, anti-OPA1, anti-Mfn2, anti-GRP78, monoclonal mouse anti-actine or anti-vinculine antibodies (references and concentration in Table 1) [11,17]. Actine or vinculine were used as loading control. Addition of secondary goat anti-mouse antibody conjugated with DyLight 680 (1/15,000, #35518 Thermo Fisher Scientific, Waltham, MA, USA) or secondary goat anti-rabbit antibody conjugated with DyLight800 4× PEG (1/15,000, SA5-35571 Thermo Fischer Scientific, Waltham, MA, USA) allowed blots revelation using Odyssey infrared fluorescent system (LI-COR).

Table 1. References and concentrations of primary antibodies.

Target Protein	Concentration	Species	Clonality	Reference
MLN64	1 μg/mL	Rabbit	Polyclonal	ab3478 abcam®
P450SCC	1 μg/mL	Rabbit	Polyclonal	ab75497 abcam®
HSD3B1	0.18 μg/mL	Rabbit	Monoclonal	ab167417 abcam®
Mfn2	1 μg/mL	Mouse	Monoclonal	ab56889 abcam®
OPA1	1 μg/mL	Rabbit	Polyclonal	ab42364 abcam®
GRP78	1 μg/mL	Rabbit	Polyclonal	G8918 Sigma-Aldrich
Actine	0.2 μg/mL	Mouse	Monoclonal	A5441 Sigma-Aldrich
Vinculine	1 μg/mL	Mouse	Monoclonal	V9131 Sigma-Aldrich

4.7. Reverse Transcription-Quantitative Polymerase Chain Reaction

Total RNA from human term trophoblast previously transfected with siRNA against IRE1α, ATF6 and PERK were used [17]. Briefly, VCT cells were transfected with 16.6 nM siATF6 (SantaCruz Biotechnology, Labforce, Muttenz, Switzerland), 16,6 nM of siIRE1α (SantaCruz Biotechnology, Labforce, Muttenz, Switzerland) and 16,6 nM of siPERK (SantaCruz Biotechnology, Labforce, Muttenz, Switzerland) or 50 nM control siRNA (SantaCruz Biotechnology, Labforce, Muttenz, Switzerland) using Interferin transfection reagent (Polyplus transfection SA, Illkirch-Graffenstaden, France) and following the manufacturer's protocol [17]. RNA extracted from trophoblast cells incubated with LPV or DMSO control with or without IRE1α inhibitor (STF-083010) was also analyzed. An amount of 500 ng of total RNA were reversed transcript with SuperScript® III Reverse Transcriptase Kit (Invitrogen™, Carlsbad, CA, USA). The qPCR was performed using cDNA diluted 1/5 in RNase-DNase free water using the the Takyon™ ROX SYBR® MasterMix blue dTTP (Eurogentec, Kaneka, Liège, Belgium). Data were normalized using SDHA, 18S and HPRT as endogenous controls. Used primers are described in Table 2.

Table 2. Sequences of primers used for RT-qPCR.

Target mRNA	Gene Name	Sequences
MLN64	*STARD3*	Forward GAGCGATGGTATCTTGCCGC Reverse CTGCAAAGGATTCTGGGGGT
P450SCC	*CYP11A1*	Forward TTTTTGCCCCTGTTGGATGCA Reverse CCCTGGCGCTCCCCAAAAAT
HSD3B1	*HSD3B1*	Forward AGTACGTCCACTCTTCTGTCCA Reverse TTCTCCTTCACCAAGAGGCG
GRP78	*GRP78*	Forward CGTGGAGATCATCGCCAAC Reverse ACATAGGACGGCGTGATGC
sXBP1	*sXBP1*	Forward CTGAGTCCGAATCAGGTGCAG Reverse ATCCATGGGGAGATGTTCTGG
ATF4	*ATF4*	Forward GTTCTCCAGCGACAAGGCTA Reverse ATCCTGCTTGCTGTTGTTGG
ATF6	*ATF6*	Forward GAGTATTTTGTCCGCCTGCC Reverse CGGGCTAAAAGGTGACTCCA

4.8. Statistical Analysis

Statistical analysis was performed using GraphPad Prism software package® (San Diego, CA, USA). Results were expressed as raw values or mean ± SD. Significant differences ($p < 0.05$) were identified using paired non-parametric student t test.

Author Contributions: Conceptualization: M.C. and J.G.; methodology: C.F., M.H.-S., M.C. and J.G.; formal analysis: C.F.; investigation: C.F., C.L.-M. and R.L.-K.; resources: T.F.; writing—original draft preparation: C.F. and J.G.; writing—review and editing: F.B.-S.; visualization: C.F.; supervision: M.C. and J.G.; project administration: J.G.; funding acquisition, M.H.-S., J.S. and J.G. All authors have read and agreed to the published version of the manuscript.

Funding: This research was funded by ANRS, PhD Grant number 17198.

Institutional Review Board Statement: The study was conducted according to the guidelines of the Declaration of Helsinki, and approved by the Ethics Committee (CPP 2015-mai-13909) of INSERM.

Informed Consent Statement: Informed consent was obtained from all subjects involved in the study.

Data Availability Statement: Data available on request.

Acknowledgments: The authors thank the consenting patients and the clinical staff midwives of Cochin Port-Royal, Antony, Béclère, Beaujon and Montsouris hospitals for providing placental tissues.

Conflicts of Interest: The authors declare no conflict of interest.

References

1. World Health Organization. *Consolidated Guidelines on the Use of Antiretroviral Drugs for Treating and Preventing HIV Infection: Recommendations for a Public Health Approach*; World Health Organization: Geneva, Switzerland, 2016.
2. Powis, K.M.; Kitch, D.; Ogwu, A.; Hughes, M.D.; Lockman, S.; Leidner, J.; van Widenfelt, E.; Moffat, C.; Moyo, S.; Makhema, J.; et al. Increased Risk of Preterm Delivery among HIV-Infected Women Randomized to Protease Versus Nucleoside Reverse Transcriptase Inhibitor-Based HAART During Pregnancy. *J. Infect. Dis.* **2011**, *204*, 506–514. [CrossRef]
3. Sibiude, J.; Warszawski, J.; Tubiana, R.; Dollfus, C.; Faye, A.; Rouzioux, C.; Teglas, J.-P.; Ekoukou, D.; Blanche, S.; Mandelbrot, L. Premature Delivery in HIV-Infected Women Starting Protease Inhibitor Therapy During Pregnancy: Role of the Ritonavir Boost? *Clin. Infect. Dis.* **2012**, *54*, 1348–1360. [CrossRef] [PubMed]
4. Short, C.-E.S.; Taylor, G.P. Antiretroviral Therapy and Preterm Birth in HIV-Infected Women. *Expert Rev. Anti Infect. Ther.* **2014**, *12*, 293–306. [CrossRef] [PubMed]
5. Kariyawasam, D.; Simon, A.; Laborde, K.; Parat, S.; Souchon, P.-F.; Frange, P.; Blanche, S.; Polak, M. Adrenal Enzyme Impairment in Neonates and Adolescents Treated with Ritonavir and Protease Inhibitors for HIV Exposure or Infection. *Horm. Res. Paediatr.* **2014**, *81*, 226–231. [CrossRef] [PubMed]
6. McDonald, C.R.; Conroy, A.L.; Gamble, J.L.; Papp, E.; Hawkes, M.; Olwoch, P.; Natureeba, P.; Kamya, M.; Silverman, M.; Cohan, D. Estradiol Levels Are Altered in Human Immunodeficiency Virus-Infected Pregnant Women Randomized to Efavirenz-Versus Lopinavir/Ritonavir-Based Antiretroviral Therapy. *Clin. Infect. Dis.* **2018**, *66*, 428. [CrossRef]
7. Rodway, M.; Zhou, F.; Benoit, J.; Yuen, B.H.; Leung, P.C.K. Differential Effects of 8-Bromo-Cyclic AMP on Human Chorionic Gonadotropin (HCG), Progesterone and Estrogen Production by Term Placental Cells. *Life Sci.* **1988**, *43*, 1451–1458. [CrossRef]
8. Tuckey, R.C. Progesterone Synthesis by the Human Placenta. *Placenta* **2005**, *26*, 273–281. [CrossRef] [PubMed]
9. Lacroix, M.C.; Guibourdenche, J.; Frendo, J.L.; Muller, F.; Evain-Brion, D. Human Placental Growth Hormone—A Review. *Placenta* **2002**, *23*, S87–S94. [CrossRef] [PubMed]
10. Fournier, T.; Guibourdenche, J.; Evain-Brion, D. Review: HCGs: Different Sources of Production, Different Glycoforms and Functions. *Placenta* **2015**, *36*, S60–S65. [CrossRef]
11. Fraichard, C.; Bonnet, F.; Garnier, A.; Hébert-Schuster, M.; Bouzerara, A.; Gerbaud, P.; Ferecatu, I.; Fournier, T.; Hernandez, I.; Trabado, S.; et al. Placental Production of Progestins Is Fully Effective in Villous Cytotrophoblasts and Increases with the Syncytiotrophoblast Formation. *Mol. Cell. Endocrinol.* **2020**, *499*, 110586. [CrossRef]
12. Pasqualini, J.R. Enzymes Involved in the Formation and Transformation of Steroid Hormones in the Fetal and Placental Compartments. *J. Steroid Biochem. Mol. Biol.* **2005**, *97*, 401–415. [CrossRef] [PubMed]
13. Pidoux, G.; Gerbaud, P.; Tsatsaris, V.; Marpeau, O.; Ferreira, F.; Meduri, G.; Guibourdenche, J.; Badet, J.; Evain-Brion, D.; Frendo, J.-L. Biochemical Characterization and Modulation of LH/CG-Receptor during Human Trophoblast Differentiation. *J. Cell. Physiol.* **2007**, *212*, 26–35. [CrossRef] [PubMed]
14. Pidoux, G.; Gerbaud, P.; Dompierre, J.; Lygren, B.; Solstad, T.; Evain-Brion, D.; Taskén, K. A PKA–Ezrin–Cx43 Signaling Complex Controls Gap Junction Communication and Thereby Trophoblast Cell Fusion. *J. Cell Sci.* **2014**, *127*, 4172–4185. [CrossRef] [PubMed]

15. Gerbaud, P.; Taskén, K.; Pidoux, G. Spatiotemporal Regulation of CAMP Signaling Controls the Human Trophoblast Fusion. *Front. Pharmacol.* **2015**, *6*. [CrossRef] [PubMed]
16. Poidatz, D.; Dos Santos, E.; Gronier, H.; Vialard, F.; Maury, B.; De Mazancourt, P.; Dieudonné, M.-N. Trophoblast Syncytialisation Necessitates Mitochondrial Function through Estrogen-Related Receptor-γ Activation. *Mol. Hum. Reprod.* **2015**, *21*, 206–216. [CrossRef] [PubMed]
17. Bastida-Ruiz, D.; Yart, L.; Wuillemin, C.; Ribaux, P.; Morris, N.; Epiney, M.; Martinez de Tejada, B.; Cohen, M. The Fine-Tuning of Endoplasmic Reticulum Stress Response and Autophagy Activation during Trophoblast Syncytialization. *Cell Death Dis.* **2019**, *10*, 651. [CrossRef] [PubMed]
18. Walker, O.S.; Ragos, R.; Wong, M.K.; Adam, M.; Cheung, A.; Raha, S. Reactive Oxygen Species from Mitochondria Impacts Trophoblast Fusion and the Production of Endocrine Hormones by Syncytiotrophoblasts. *PLoS ONE* **2020**, *15*, e0229332. [CrossRef]
19. Martinez, F.; Kiriakidou, M.; Strauss, J.F. Structural and Functional Changes in Mitochondria Associated with Trophoblast Differentiation: Methods to Isolate Enriched Preparations of Syncytiotrophoblast Mitochondria. *Endocrinology* **1997**, *138*, 12. [CrossRef]
20. De Los Rios Castillo, D.; Zarco-Zavala, M.; Olvera-Sanchez, S.; Pardo, J.P.; Juarez, O.; Martinez, F.; Mendoza-Hernandez, G.; García-Trejo, J.J.; Flores-Herrera, O. Atypical Cristae Morphology of Human Syncytiotrophoblast Mitochondria. *J. Biol. Chem.* **2011**, *286*, 23911–23919. [CrossRef]
21. Meyer, J.N.; Leuthner, T.C.; Luz, A.L. Mitochondrial Fusion, Fission, and Mitochondrial Toxicity. *Toxicology* **2017**, *391*, 42–53. [CrossRef]
22. Holland, O.; Dekker Nitert, M.; Gallo, L.A.; Vejzovic, M.; Fisher, J.J.; Perkins, A.V. Review: Placental Mitochondrial Function and Structure in Gestational Disorders. *Placenta* **2017**, *54*, 2–9. [CrossRef] [PubMed]
23. Fisher, J.J.; McKeating, D.R.; Cuffe, J.S.; Bianco-Miotto, T.; Holland, O.J.; Perkins, A.V. Proteomic Analysis of Placental Mitochondria Following Trophoblast Differentiation. *Front. Physiol.* **2019**, *10*, 1536. [CrossRef] [PubMed]
24. Frank, S.; Gaume, B.; Bergmann-Leitner, E.S.; Leitner, W.W.; Robert, E.G.; Catez, F.; Smith, C.L.; Youle, R.J. The Role of Dynamin-Related Protein 1, a Mediator of Mitochondrial Fission, in Apoptosis. *Dev. Cell* **2001**, *1*, 515–525. [CrossRef]
25. Rojo, M.; Legros, F.; Chateau, D.; Lombès, A. Characterization of Mammalian Mitofusins. *J. Cell Sci.* **2002**, *115*, 163–1674.
26. Chen, H.; Detmer, S.A.; Ewald, A.J.; Griffin, E.E.; Fraser, S.E.; Chan, D.C. Mitofusins Mfn1 and Mfn2 Coordinately Regulate Mitochondrial Fusion and Are Essential for Embryonic Development. *J. Cell Biol.* **2003**, *160*, 189–200. [CrossRef]
27. Wasilewski, M.; Semenzato, M.; Rafelski, S.M.; Robbins, J.; Bakardjiev, A.I.; Scorrano, L. Optic Atrophy 1-Dependent Mitochondrial Remodeling Controls Steroidogenesis in Trophoblasts. *Curr. Biol.* **2012**, *22*, 1228–1234. [CrossRef]
28. Cai, H.; Chen, L.; Zhang, M.; Xiang, W.; Su, P. Low Expression of MFN2 Is Associated with Early Unexplained Miscarriage by Regulating Autophagy of Trophoblast Cells. *Placenta* **2018**, *70*, 34–40. [CrossRef]
29. Dal Yontem, F.; Kim, S.; Ding, Z.; Grimm, E.; Ekmekcioglu, S.; Akcakaya, H. Mitochondrial Dynamic Alterations Regulate Melanoma Cell Progression. *J. Cell. Biochem.* **2019**, *120*, 2098–2108. [CrossRef]
30. Shan, A.; Li, M.; Li, X.; Li, Y.; Yan, M.; Xian, P.; Chang, Y.; Chen, X.; Tang, N. BDE-47 Decreases Progesterone Levels in BeWo Cells by Interfering with Mitochondrial Functions and Genes Related to Cholesterol Transport. *Chem. Res. Toxicol.* **2019**, *32*, 621–628. [CrossRef]
31. Bielinska, M.; Boime, I. The Glycoprotein Hormone Family: Structure and Function of the Carbohydrate Chains. In *New Comprehensive Biochemistry*; Elsevier: Amsterdam, The Netherlands, 1995.
32. Bastida-Ruiz, D.; Aguilar, E.; Ditisheim, A.; Yart, L.; Cohen, M. Endoplasmic Reticulum Stress Responses in Placentation—A True Balancing Act. *Placenta* **2017**, *57*, 163–169. [CrossRef]
33. Park, H.-J.; Park, S.-J.; Koo, D.-B.; Lee, S.-R.; Kong, I.-K.; Ryoo, J.-W.; Park, Y.-I.; Chang, K.-T.; Lee, D.-S. Progesterone Production Is Affected by Unfolded Protein Response (UPR) Signaling during the Luteal Phase in Mice. *Life Sci.* **2014**, *113*, 60–67. [CrossRef] [PubMed]
34. Papp, E.; Mohammadi, H.; Loutfy, M.R.; Yudin, M.H.; Murphy, K.E.; Walmsley, S.L.; Shah, R.; MacGillivray, J.; Silverman, M.; Serghides, L. HIV Protease Inhibitor Use during Pregnancy Is Associated with Decreased Progesterone Levels, Suggesting a Potential Mechanism Contributing to Fetal Growth Restriction. *J. Infect. Dis.* **2015**, *211*, 10–18. [CrossRef] [PubMed]
35. Wilson, R.A.; Mesiano, S.A. Progesterone Signaling in Myometrial Cells: Role in Human Pregnancy and Parturition. *Curr. Opin. Physiol.* **2020**, *13*, 117–122. [CrossRef]
36. Benirschke, K.; Burton, G.J.; Baergen, R.N. *Pathology of the Human Placenta*, 6th ed.; Springer: Berlin/Heidelberg, Germany, 2013.
37. Handschuh, K.; Guibourdenche, J.; Tsatsaris, V.; Guesnon, M.; Laurendeau, I.; Evain-Brion, D.; Fournier, T. Human Chorionic Gonadotropin Produced by the Invasive Trophoblast but Not the Villous Trophoblast Promotes Cell Invasion and Is Down-Regulated by Peroxisome Proliferator-Activated Receptor-Gamma. *Endocrinology* **2007**, *148*, 5011–5019. [CrossRef]
38. Bustamante, J.; Ramírez-Vélez, R.; Czerniczyniec, A.; Cicerchia, D.; Aguilar de Plata, A.C.; Lores-Arnaiz, S. Oxygen Metabolism in Human Placenta Mitochondria. *J. Bioenerg. Biomembr.* **2014**, *46*, 459–469. [CrossRef]
39. Duarte, A.; Poderoso, C.; Cooke, M.; Soria, G.; Cornejo Maciel, F.; Gottifredi, V.; Podestá, E.J. Mitochondrial Fusion Is Essential for Steroid Biosynthesis. *PLoS ONE* **2012**, *7*, e45829. [CrossRef]
40. Ruiz, M.; Courilleau, D.; Jullian, J.-C.; Fortin, D.; Ventura-Clapier, R.; Blondeau, J.-P.; Garnier, A. A Cardiac-Specific Robotized Cellular Assay Identified Families of Human Ligands as Inducers of PGC-1α Expression and Mitochondrial Biogenesis. *PLoS ONE* **2012**, *7*, e46753. [CrossRef]

41. Guzel, E.; Arlier, S.; Guzeloglu-Kayisli, O.; Tabak, M.S.; Ekiz, T.; Semerci, N.; Larsen, K.; Schatz, F.; Lockwood, C.J.; Kayisli, U.A. Endoplasmic Reticulum Stress and Homeostasis in Reproductive Physiology and Pathology. *Int. J. Mol. Sci.* **2017**, *18*, 792. [CrossRef]
42. Burton, G.J.; Yung, H.W.; Murray, A.J. Mitochondrial—Endoplasmic Reticulum Interactions in the Trophoblast: Stress and Senescence. *Placenta* **2017**, *52*, 146–155. [CrossRef]
43. Taura, M.; Kariya, R.; Kudo, E.; Goto, H.; Iwawaki, T.; Amano, M.; Suico, M.A.; Kai, H.; Mitsuya, H.; Okada, S. Comparative Analysis of ER Stress Response into HIV Protease Inhibitors: Lopinavir but Not Darunavir Induces Potent ER Stress Response via ROS/JNK Pathway. *Free Radic. Biol. Med.* **2013**, *65*, 778–788. [CrossRef]
44. Yeh, R.F.; Gaver, V.E.; Patterson, K.B.; Rezk, N.L.; Baxter-Meheux, F.; Blake, M.J.; Eron, J.J.J.; Klein, C.E.; Rublein, J.C.; Kashuba, A.D.M. Lopinavir/Ritonavir Induces the Hepatic Activity of Cytochrome P450 Enzymes CYP2C9, CYP2C19, and CYP1A2 But Inhibits the Hepatic and Intestinal Activity of CYP3A as Measured by a Phenotyping Drug Cocktail in Healthy Volunteers. *JAIDS J. Acquir. Immune Defic. Syndr.* **2006**, *42*, 52–60. [CrossRef] [PubMed]
45. Besse, A.; Stolze, S.C.; Rasche, L.; Weinhold, N.; Morgan, G.J.; Kraus, M.; Bader, J.; Overkleeft, H.S.; Besse, L.; Driessen, C. Carfilzomib Resistance Due to ABCB1/MDR1 Overexpression Is Overcome by Nelfinavir and Lopinavir in Multiple Myeloma. *Leukemia* **2018**, *32*, 391–401. [CrossRef] [PubMed]
46. Papp, E.; Balogun, K.; Banko, N.; Mohammadi, H.; Loutfy, M.; Yudin, M.H.; Shah, R.; MacGillivray, J.; Murphy, K.E.; Walmsley, S.L.; et al. Low Prolactin and High 20-α-Hydroxysteroid Dehydrogenase Levels Contribute to Lower Progesterone Levels in HIV-Infected Pregnant Women Exposed to Protease Inhibitor–Based Combination Antiretroviral Therapy. *J. Infect. Dis.* **2016**, *213*, 1532–1540. [CrossRef] [PubMed]
47. Hong-Brown, L.Q.; Brown, C.R.; Huber, D.S.; Lang, C.H. Lopinavir Impairs Protein Synthesis and Induces EEF2 Phosphorylation via the Activation of AMP-Activated Protein Kinase. *J. Cell. Biochem.* **2008**, *105*, 814–823. [CrossRef]
48. Yung, Y.; Maman, E.; Ophir, L.; Rubinstein, N.; Barzilay, E.; Yerushalmi, G.M.; Hourvitz, A. Progesterone Antagonist, RU486, Represses LHCGR Expression and LH/HCG Signaling in Cultured Luteinized Human Mural Granulosa Cells. *Gynecol. Endocrinol.* **2014**, *30*, 42–47. [CrossRef]
49. Guitart-Mampel, M.; Hernandez, A.S.; Moren, C.; Catalan-Garcia, M.; Tobias, E.; Gonzalez-Casacuberta, I.; Juarez-Flores, D.L.; Gatell, J.M.; Cardellach, F.; Milisenda, J.C.; et al. Imbalance in Mitochondrial Dynamics and Apoptosis in Pregnancies among HIV-Infected Women on HAART with Obstetric Complications. *J. Antimicrob. Chemother.* **2017**. [CrossRef]
50. Ganta, K.K.; Chaubey, B. Endoplasmic Reticulum Stress Leads to Mitochondria-Mediated Apoptosis in Cells Treated with Anti-HIV Protease Inhibitor Ritonavir. *Cell Biol. Toxicol.* **2019**, *35*, 189–204. [CrossRef]
51. Chen, L.; Jarujaron, S.; Wu, X.; Sun, L.; Zha, W.; Liang, G.; Gurley, E.C.; Studer, E.J.; Hylemon, P.B.; Pandak, W.M.; et al. HIV Protease Inhibitor Lopinavir-Induced TNF-a and IL-6 Expression Is Coupled to the Unfolded Protein Response and ERK Signaling Pathways in Macrophages. *Biochem. Pharmacol.* **2009**, *78*, 70–77. [CrossRef]
52. Zha, B.S.; Wan, X.; Zhang, X.; Zha, W.; Zhou, J.; Wabitsch, M.; Wang, G.; Lyall, V.; Hylemon, P.B.; Zhou, H. HIV Protease Inhibitors Disrupt Lipid Metabolism by Activating Endoplasmic Reticulum Stress and Inhibiting Autophagy Activity in Adipocytes. *PLoS ONE* **2013**, *8*, e59514. [CrossRef]
53. Brüning, A.; Kimmich, T.; Brem, G.J.; Buchholtz, M.L.; Mylonas, I.; Kost, B.; Weizsäcker, K.; Gingelmaier, A. Analysis of Endoplasmic Reticulum Stress in Placentas of HIV-Infected Women Treated with Protease Inhibitors. *Reprod. Toxicol.* **2014**, *50*, 122–128. [CrossRef]
54. Gallagher, C.M.; Garri, C.; Cain, E.L.; Ang, K.K.-H.; Wilson, C.G.; Chen, S.; Hearn, B.R.; Jaishankar, P.; Aranda-Diaz, A.; Arkin, M.R.; et al. Ceapins Are a New Class of Unfolded Protein Response Inhibitors, Selectively Targeting the ATF6α Branch. *eLife* **2016**, *5*, e11878. [CrossRef] [PubMed]
55. Alsat, E.; Haziza, J.; Evain-Brion, D. Increase in Epidermal Growth Factor Receptor and Its Messenger Ribonucleic Acid Levels with Differentiation of Human Trophoblast Cells in Culture. *J. Cell. Physiol.* **1993**, *154*, 122–128. [CrossRef] [PubMed]

International Journal of
Molecular Sciences

MDPI

Article

Persistent Human KIT Receptor Signaling Disposes Murine Placenta to Premature Differentiation Resulting in Severely Disrupted Placental Structure and Functionality

Franziska Kaiser [1,†], Julia Hartweg [1,2,†], Selina Jansky [1,3,4], Natalie Pelusi [1,5], Caroline Kubaczka [1,6], Neha Sharma [1,7], Dominik Nitsche [1,8], Jan Langkabel [1] and Hubert Schorle [1,*]

1 Department of Developmental Pathology, Institute of Pathology, University Hospital Bonn, 53127 Bonn, Germany; Franziska.Kaiser@ukbonn.de (F.K.); E_Hartweg_J@ukw.de (J.H.); s.jansky@kitz-heidelberg.de (S.J.); Natalie.Pelusi@ukbonn.de (N.P.); Caroline.Schuster-Kubaczka@childrens.harvard.edu (C.K.); ns.nehasharma01@gmail.com (N.S.); s6donits@uni-bonn.de (D.N.); Jan.Langkabel@ukbonn.de (J.L.)

2 Department of Medicine II and IZKF Research Laboratory, Würzburg University Hospital, 97080 Würzburg, Germany

3 Hopp Children's Cancer Center (KiTZ), 69120 Heidelberg, Germany

4 Division of Neuroblastoma Genomics, German Cancer Research Center (DKFZ), 69120 Heidelberg, Germany

5 Department of Molecular Pathology, Institute of Pathology, University Hospital Bonn, 53127 Bonn, Germany

6 Division of Pediatric Hematology/Oncology, Children's Hospital Boston, Harvard Medical School, Boston, MA 02115, USA

7 Department of Obstetrics and Gynaecology, Yong Loo Lin School of Medicine, National University of Singapore, Singapore 119077, Singapore

8 Life & Medical Sciences-Institute (LIMES), University of Bonn, 53115 Bonn, Germany

* Correspondence: Hubert.Schorle@ukbonn.de; Tel.: +49-228-287-16342; Fax: +49-228-287-19757

† These authors contributed equally to this work.

Received: 8 July 2020; Accepted: 30 July 2020; Published: 31 July 2020

Abstract: Activating mutations in the human KIT receptor is known to drive severe hematopoietic disorders and tumor formation spanning various entities. The most common mutation is the substitution of aspartic acid at position 816 to valine (D816V), rendering the receptor constitutively active independent of ligand binding. As the role of the KIT receptor in placental signaling cascades is poorly understood, we analyzed the impact of KIT^{D816V} expression on placental development using a humanized mouse model. Placentas from KIT^{D816V} animals present with a grossly changed morphology, displaying a reduction in labyrinth and spongiotrophoblast layer and an increase in the Parietal Trophoblast Giant Cell (P-TGC) layer. Elevated differentiation to P-TGCs was accompanied with reduced differentiation to other Trophoblast Giant Cell (TGC) subtypes and by severe decrease in proliferation. The embryos display growth retardation and die in utero. KIT^{D816V}-trophoblast stem cells (TSC) differentiate much faster compared to wild type (WT) controls. In undifferentiated KIT^{D816V}-TSCs, levels of Phosphorylated Extracellular-signal Regulated Kinase (P-ERK) and Phosphorylated Protein Kinase B (P-AKT) are comparable to wildtype cultures differentiating for 3–6 days. Accordingly, P-TGC markers Placental Lactogen 1 (PL1) and Proliferin (PLF) are upregulated as well. The results reveal that KIT signaling orchestrates the fine-tuned differentiation of the placenta, with special emphasis on P-TGC differentiation. Appropriate control of KIT receptor action is therefore essential for placental development and nourishment of the embryo.

Keywords: KIT receptor; KITD816V; placental development; premature differentiation; trophoblast stem cell; trophoblast giant cell; spongiotrophoblast; invasion; embryonic growth retardation

1. Introduction

Proper function of the placenta is essential for the development of the embryo as it is responsible for exchanging gases, nutrients, and waste products between the mother and the fetus. Placental insufficiency can result in adverse effects for the embryo including intrauterine growth retardation, embryonic defects, or even fatal outcome. Therefore, placenta development and differentiation are closely regulated [1–3].

The membrane-bound tyrosine receptor kinase KIT is activated by its ligand stem cell factor (SCF), which causes dimerization of the receptor followed by phosphorylation of its tyrosine residues [4–6]. Activation of KIT results in induction of various downstream signaling cascades such as the Mitogen-Activated Protein Kinase (MAPK)/Extracellular-Signal Regulated Kinase (ERK) and Janus Kinase (JAK)/Signal Transducers and Activators of Transcription (STAT) pathways which orchestrate cell proliferation, angiogenesis, cell migration, and cell cycle control [7–10]. During placental development, the KIT protein is detected in the uterine epithelium as well as the maternal and fetal part of the placenta. Importantly, KIT expression has not been detected in uteri of nonpregnant females, indicating a pregnancy-related role of KIT signaling [11–13]. In detail, starting at E9, KIT and SCF are detectable in trophoblast-chorion, ectoplacental cone, and decidua [12]. Placental hematopoietic activity begins around mid-gestation. Hematopoietic stem cells (HSC) are detected in the placenta at E11, peak in numbers at E12 to E13, and disappear thereafter [13,14]. These HSC were shown to be KIT positive, whereas endothelial cells surrounding the HSC niches are SCF positive [13,15]. At E12, KIT can also be found in mesenchymal cells of the chorionic plate and of the labyrinth, in endothelial cells lining the vessels, as well as in few trophoblast giant cells (TGC) [13]. At E14.5, KIT is mainly expressed in maternal decidua cells as well as the labyrinth whereas SCF is found only in labyrinth [11,12]. KIT is restricted to labyrinthine trophoblast cells exposed to maternal blood at E19 and is no longer detected in endothelial or mesenchymal cells [11].

The KIT receptor has been implicated with several disorders such as tumor formation, hematopoietic disorders, and systemic mastocytosis [16–20]. The most common activating mutation in the human KIT receptor encodes for the substitution of the aspartic acid at position 816 with a valine (D816V) causing constitutive phosphorylation of the protein [17,21,22]. Depletion of KIT receptor results in various defects spanning from hematopoietic failure, macrocytic anemia, pigmentation deficiency, and sterility to intestinal dysfunction [23–26]. Mice carrying homozygous $KIT^{W/W}$ mutation die pre- or perinatally; however, can be rescued by microinjection of wildtype fetal liver hematopoietic cells into placentas of affected fetuses [23,24,26]. Introduction of the viable c-Kit-deficient mouse line allowed for studying the loss of KIT expression in the adult mouse and showed that KIT is essential for adult lymphopoiesis in bone marrow and thymus [26]

Previously, we reported the generation of a humanized KIT^{D816V} mouse model [27]. The transgenic KIT receptor consists of murine extracellular and transmembrane domains and the human intracellular domain carrying the D816V mutation. The chimeric KIT receptor is fused to a green fluorescent protein (GFP) and integrated in the ROSA26 locus, allowing for Cre-mediated conditional expression of the protein under control of the endogenous ROSA26 promoter [27,28]. We then examined KIT signaling in fetal liver erythropoiesis and adult hematopoiesis [27,29]. Constitutively active KIT resulted in inhibition of terminal differentiation of erythroid precursors and embryonic death after embryonic day (E) 13.5 [27]. In adult mice, expression of KIT^{D816V} led to a polycythemia vera-like myeloproliferative neoplasm with highly increased red blood cells and splenomegaly [29].

While the role of KIT signaling in other tissues is well understood, its role in placental development, however, has so far not been described. Here, we examine the effect of the constitutively active KIT receptor using the KIT^{D816V} mouse model. Interestingly, expression of KIT^{D816V} resulted in decreased proliferation of trophoblast cells. As a consequence, placental structure was affected. The labyrinth layer was decreased accompanied by increased differentiation to parietal (P-)TGCs, whereas other TGC subtypes remained underrepresented. Trophoblast Stem Cells (TSC) derived from KIT^{D816V} blastocysts showed altered activation of signaling pathways upon differentiation as well as an increased invasive capacity. Due to these effects, the placenta is not able to sustain regular development and the embryos show severe growth retardation and die in utero.

2. Results

2.1. Embryos and Placentas Carrying KIT^{D816V} Mutation Suffer from Severe Growth Retardation

Previously, we generated the R26-LSL-KIT^{D816V} mouse line, which allows for conditional Cre-induced expression of chimeric KIT^{D816V} receptor and a GFP reporter driven by the ROSA26 promoter. The KIT^{D816V} cDNA is linked to the GFP cDNA via the coding sequence for a viral 2A peptide [27]. Here, mice carrying the ROSA26-KIT^{D816V}-GFP transgene were mated with Deleter-Cre mice, inducing ubiquitous expression of the transgenic receptor in the embryonic as well as the extra-embryonic tissue. Presence of ROSA26-KIT^{D816V} and Cre transgenes was verified by genotyping PCR using genomic DNA obtained from yolk sac or embryo (Supplementary Materials Figure S1A). Animals/placentas harboring both the Cre- and the ROSA26-KIT^{D816V} allele are further referred to as KIT^{D816V} animals/placentas. Presence of transgenes in KIT^{D816V} animals was further validated by analyzing RNA expression and protein levels in KIT^{D816V} placentas (Figure 1A–C). Of note, in KIT^{D816V} animals expressing human *KIT^{D816V}*, the expression of murine *Kit* remained unchanged, showing that endogenous *Kit* expression is not affected by transgene induction (Figure 1A,B). As expected, KIT as well as 2A-peptide proteins were present in KIT^{D816V} placentas only (Figure 1C). For further analyses, embryos and placentas were obtained on E9.5–E11.5 from KIT^{D816V} and wildtype (WT) animals. KIT^{D816V} embryos and placentas dissected on E11.5 showed growth retardation and developmental delay (Figure 1D). We previously reported that KIT^{D816V} expression restricted to the embryo proper leads to disturbance of the hematopoietic system and that such animals die at E14.5. Therefore, we hypothesize that severe growth retardation observed in KIT^{D816V} animals at 11.5 is an effect of placental insufficiency. Of note, KIT^{D816V} embryo and placentas showed GFP positivity (Figure 1D).

2.2. KIT^{D816V} Placentas Show Fewer Proliferating Cells

Next, we examined whether reduced sizes in KIT^{D816V} placentas occurred due to diminished proliferation. Proliferation in placental tissue was assessed by immunohistochemical staining for KI-67—a protein that is present in proliferating cells as well as cells undergoing endoreduplication [30,31]. At both timepoints analyzed, KI-67-positive cells were mainly detectable in the labyrinth and spongiotrophoblast layers of KIT^{D816V} and WT placentas. On day E9.5, the labyrinth layer and overall amount of KI-67-positive cells were markedly diminished in KIT^{D816V} in comparison to WT (Figure 2A,B). While the labyrinth and spongiotrophoblast layers were enlarged on day E10.5 in comparison to E9.5 in both samples, less KI-67-positive cells were detected in KIT^{D816V} placentas than in WT placentas (Figure 2A,B). KI-67-positive P-TGCs can be detected in both KIT^{D816V} and WT placentas. These results suggest that the smaller size of KIT^{D816V} placentas results from decreased proliferation in the labyrinth and spongiotrophoblast.

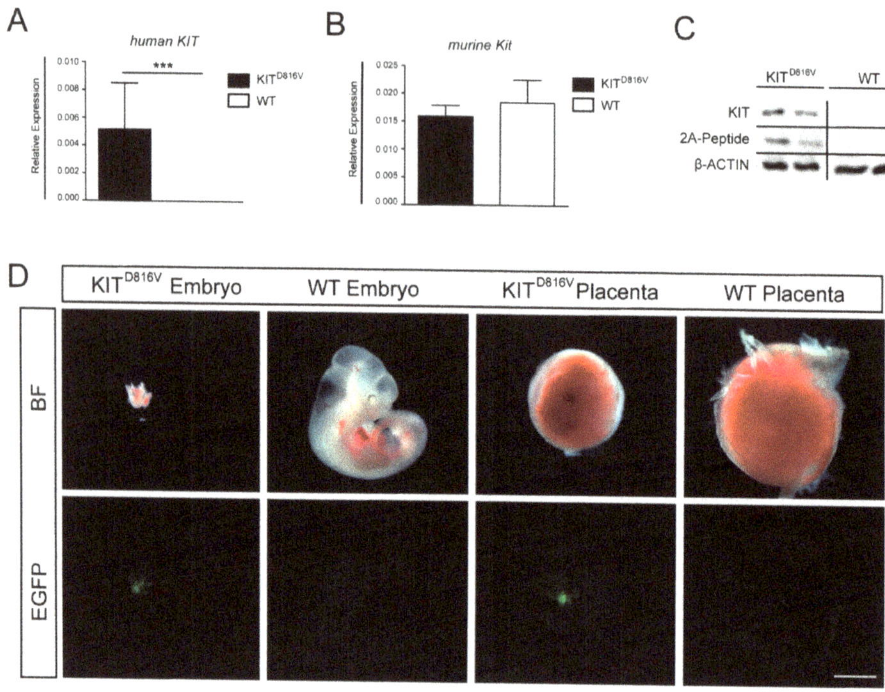

Figure 1. Detection of KIT^{D816V}-positive embryos and placentas: (**A**) qRT-PCR for transgene expression of human *KIT* in KIT^{D816V} and wildtype (WT) placentas obtained on E10.5 and (**B**) qRT-PCR for endogenous expression of murine *Kit* in KIT^{D816V} and WT placentas obtained on E10.5. RNA was obtained from at least three biological repeats. Expression is normalized to the housekeeping gene *Gapdh*. Bars display mean value ± SD. Significance was determined by unpaired t-test and indicated with *** $p < 0.001$. (**C**) Western Blot detecting KIT and 2A-peptide in placentas at E11.5 in comparison to β-ACTIN and (**D**) photographs showing KIT^{D816V} and WT expressing embryos and placentas at E11.5 in brightfield (BF) and GFP fluorescence (scale bar: 1 mm): Experiments were performed in at least tree biological repeats.

2.3. *KIT^{D816V} Placentas Show Reduced Labyrinth Layer and Disrupted Formation of Vasculature*

To assess the placental structure, hematoxylin and eosin (H&E) staining was performed on paraffin sections of KIT^{D816V} and WT placentas. Starting from day E9.5 in KIT^{D816V} placentas, H&E stained sections and quantification of the area showed a reduction in labyrinth size (Figure 3A,B). Also, the spongiotrophoblast layer was slightly reduced (Figure 3A,B). Interestingly, at E10.5, the layer of P-TGCs was increased in KIT^{D816V} placentas (Figure 3A,B). Quantification of P-TGCs within this area confirmed a significantly higher total number of P-TGCs in KIT^{D816V} sample (Figure 3C). Next, RNA was isolated from KIT^{D816V} and WT placentas on days E10.5 and E11.5. Expression of labyrinth TGC markers Placental Lactogen 2 (*Pl2*), Cathepsin q (*Ctsq*), and Glial Cells Missing Homolog 1 (*Gcm1*) is diminished in comparison to WT placentas (Figure 3D). This data indicates that the number of *Pl2*$^{+}$ and *Ctsq*$^{+}$ sinusoidal (S-)TGC and the number of *Gcm 1*$^{+}$ labyrinthine cells is reduced.

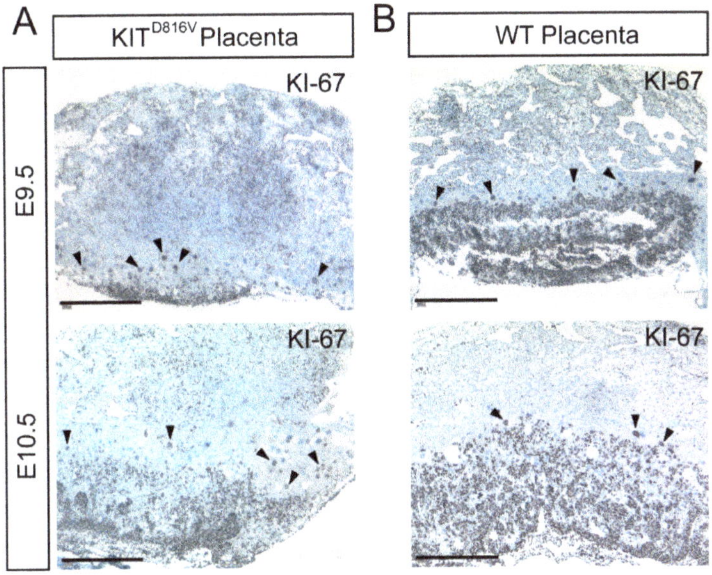

Figure 2. KIT^{D816V} shows a reduced number of KI-67-positive cells. (**A**,**B**) Immunohistochemical staining of paraffin sections of KIT^{D816V} and WT placentas for KI-67 at E9.5 and E10.5: Black arrowheads mark Parietal Trophoblast Giant Cells (P-TGCs) undergoing endoreduplication (biological replicates = 4; scale bar represents 500 μm).

Labyrinthine architecture was further examined by performing anti-CD31 staining. CD31 is expressed in fetal endothelial cells [32]. The staining revealed that at E9.5 there are fewer CD31-positive cells present in KIT^{D816V} placentas in contrast to WT placentas (Figure 3E). CD31-positive cells are indicated by red arrowheads. While the vasculature was established from E9.5 to E10.5 in KIT^{D816V} as well as WT placentas, the capillary network in KIT^{D816V} remained less prominent than in WT tissue. S-TGCs, indicated by black arrowheads, were lining maternal sinusoids (red asterisks) in both conditions. Surprisingly, maternal blood was detected in between P-TGCs of KIT^{D816V} placentas at E10.5, suggesting a disruption of the developing vascular structure (Figure 3E).

2.4. KIT^{D816V} Placentas Show Prominent Differentiation into P-TGCs

Since the number of P-TGCs is significantly increased in KIT^{D816V} placentas (Figure 3C), we next investigated this TGC subtype in more detail by detecting the P-TGC markers PL1 and PLF using in situ hybridization. At E9.5, the number and distribution of cells positive for PL1 and PLF appeared unaltered in KIT^{D816V} compared to WT placentas (Figure 4A). At E10.5, however, the number of PL1$^+$/PLF$^+$ cells was visibly higher in KIT^{D816V} placentas than WT placentas (Figure 4B). Further, when comparing PL1 and PLF staining in WT placentas, we observed areas that are PL1 negative but PLF positive (Figure 4B). Such cells are not P-TGCs but rather invading Spiral Artery (SpA-)TGCs (black arrowhead). We assume that the cell clusters indicated by red arrowheads are canal (C-) TGCs (Figure 4B). By contrast, PL1$^-$/PLF$^+$ cells are not detected in KIT^{D816V} placentas, further suggesting a severe reduction in specific TGC subtypes such as SpA-TGC and C-TGC.

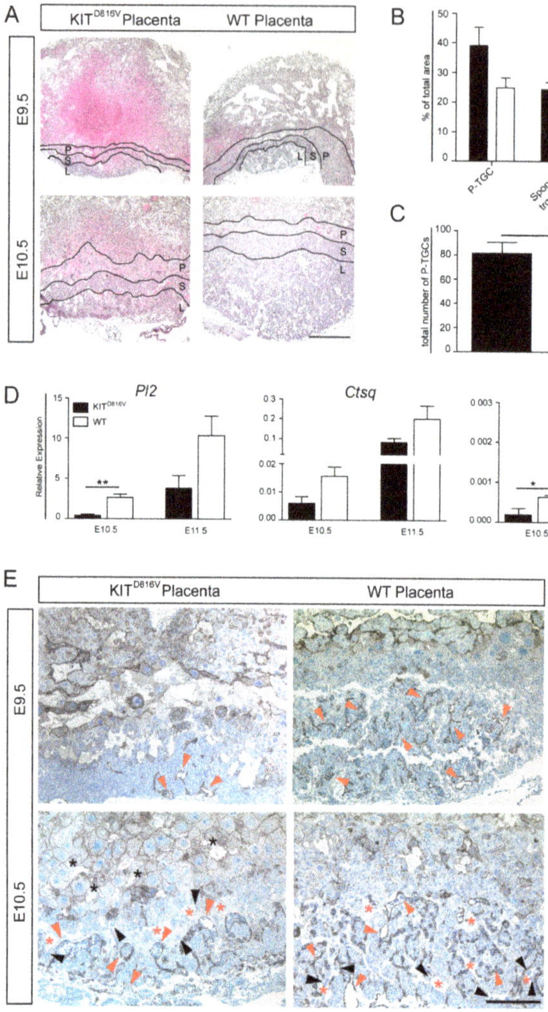

Figure 3. Placental structure is grossly affected by KIT^{D816V}. (**A**) Photomicrographs of hematoxylin and eosin (H&E) staining of paraffin sections of KIT^{D816V} and WT placentas on E9.5 (biological replicates = 2) and E10.5 (biological replicates = 3): Black lines mark areas of the P-TGCs (P), spongiotrophoblast (S), and labyrinth layer (L). Scale bar represents 500 µm. (**B**) Quantification of P-TGC, spongiotrophoblast, and labyrinth areas at E10.5 compared to the total area of respective placenta (biological replicates = 3) and (**C**) quantification of number of P-TGCs in KIT^{D816V} and WT placenta at E10.5 using ImageJ in three biological replicates: Significance was determined by unpaired t-test and indicated with * $p < 0.05$. (**D**) qRT-PCR analysis of labyrinth-specific TGC marker *Pl2*, *Ctsq*, and *Gcm* in KIT^{D816V} and WT placentas at E10.5 and E11.5. RNA was obtained from at least three biological replicates. Expression is normalized to the housekeeping gene *Gapdh*. Bars display mean value ± SD. Significance was determined by unpaired t-test and indicated with * $p < 0.05$ and ** $p < 0.01$; (**E**) Immunohistochemical staining of CD31 on paraffin sections of KIT^{D816V} and WT placentas at E9.5 (biological replicates = 2) and E10.5 (biological replicates = 2): Red arrowheads indicate fetal endothelial cells enclosing fetal blood spaces, and black arrowheads indicate S-TGCs. Maternal sinusoids are indicated by red asterisks, and maternal blood in the P-TGC layer is indicated by black asterisks. Scale bar represents 200 µm.

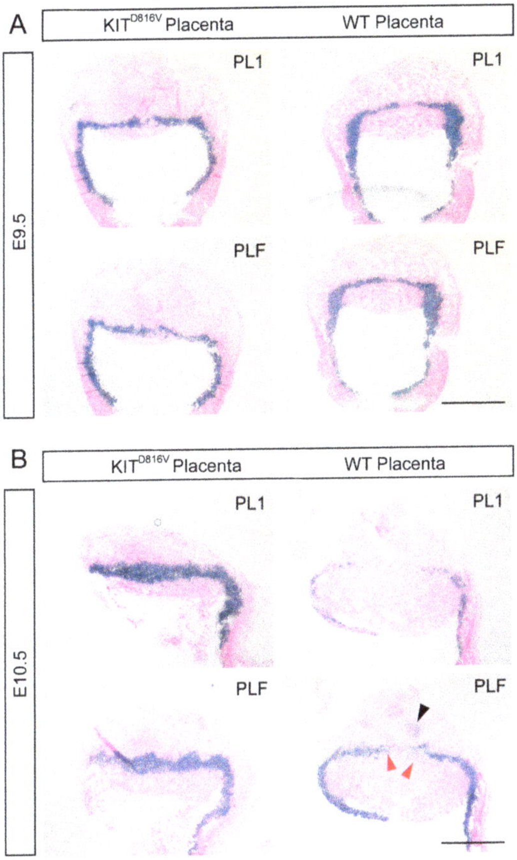

Figure 4. Increased levels of PL1 and PLF in KIT^{D816V} placentas: (**A**) In situ hybridization of PL1 and PLF on paraffin sections of KIT^{D816V} and control placentas at E9.5 using specific probe for PL1 and PLF (biological replicates = 2). Scale bar represents 1 mm. (**B**) In situ hybridization of PL1 and PLF on paraffin sections of KIT^{D816V} and control placentas at E10.5 (biological replicates = 2): Red arrowheads indicate PL1$^-$/PLF$^+$ cells. Black arrowhead indicates invading Spiral Artery (SpA-)TGCs. Scale bar represents 1 mm.

2.5. Spongiotrophoblast Cells and Glycogen Trophoblasts are Reduced in KIT^{D816V} Placentas

Next, we analyzed the development of the spongiotrophoblast layer in more detail. In situ hybridization revealed that the spongiotrophoblast marker Trophoblast Specific Protein Alpha (TPBPA) is hardly present in KIT^{D816V} placentas compared to WT placentas at both timepoints analyzed (Supplementary Materials Figure S1B). Also, expression of *Tpbpa* was significantly reduced at E10.5 and E11.5. Further, we investigated the expression of Gap Junction Beta 3-protein (*Gjb3)* and Procadherin 12 (*Pcdh12)*, both markers for glycogen trophoblasts (GlyT). They were significantly decreased in KIT^{D816V} placentas on days E10.5 and E11.5 (Supplementary Materials Figure S1C–E). Thus, these results suggest that expression of KIT^{D816V} not only blocks cellular proliferation but also heavily impinges on the fine-balanced differentiation of the various subtypes essential for placental proper development. The differentiation to P-TGCs seems increased at the expense of other TGC subtypes such as C-TGCs and SpA-TGCs as well as spongiotrophoblast and GlyT cells.

2.6. KIT^{D816V}-TSC Show Signs of Premature Differentiation

In order to analyze the molecular effect of KIT^{D816V} on a cellular level in more detail, TSCs were derived from E3.5 blastocysts obtained from R26-LSL-KIT^{D816V} and Deleter-Cre mice according to published procedures [27,33]. Genotyping at passage five was used to distinguish between KIT^{D816V}- and WT-TSC lines (Supplementary Materials Figure S2A). Ultimately, we had established two lines of KIT^{D816V}-TSC which were named KIT^{D816V} #3 and KIT^{D816V} #4. In these two lines, we were able to detect a GFP-signal using fluorescence-activated cell sorting (FACS) (Supplementary Materials Figure S2B) and expression of the human *KIT* transgene using qRT-PCR (Supplementary Materials Figure S2C), demonstrating that the ROSA26-GFP-2A-KIT^{D816} allele is functional in TSC culture in vitro. As in placental tissues, the level of endogenous murine *Kit* expression in KIT^{D816V}-TSC is not affected and remains comparable to that of WT TSC (Supplementary Materials Figure S2D).

Analysis of TSC-markers Transcription Factor AP-2 Gamma (Tfap2c), Caudal Type Homeobox 2 (Cdx2), and Eomesodermin homolog (Eomes) revealed that neither expression (Supplementary Materials Figure S2E–G) nor protein levels and distribution of TFAP2C, CDX2, and EOMES (Supplementary Materials Figure S2H) are affected in the KIT^{D816V}-TSC lines. Also, morphology as well as splitting ratio and frequency of KIT^{D816V}-TSC did not differ from WT TSC. Hence, we conclude that establishment, maintenance, and self-renewal of TSC is not altered by expression of the KIT^{D816V} transgene.

To evaluate whether KIT^{D816V}-TSC displays alterations in differentiation in vitro, WT and KIT^{D816V} lines were kept under differentiating conditions in trophoblast stem (TS) cell medium without conditioned medium (CM), Fibroblast Growth Factor (FGF) 4, and heparin for 9 days. Morphological analyses show no difference between studied lines. In both KIT^{D816V}-TSC and WT TSC, TGCs appeared in culture after three days (Figure 5A). RNA was isolated on days 0 and 6 and was analyzed for expression of P-TGC markers *Pl1*, *Pl2*, and *Plf*. Interestingly, in KIT^{D816V}-TSCs, all markers showed an increased level already at day 0, an effect which persisted to day 6 (Figure 5B). Further, spongiotrophoblast marker *Tpbpa* as well as labyrinth and S-TGC marker *Ctsq* and GlyT markers *Gjb3* and *Pcdh12* were expressed at lower levels in KIT^{D816V} TSC than in WT TSC (Supplementary Materials Figure S2I). Upregulation of *Tfap2c* which is detected during TSC differentiation was higher in KIT^{D816V}-TSC than in WT TSC [34] (Supplementary Materials Figure S2I). Of note, *Tfap2c* expression was already increased under stem cell culture conditions in KIT^{D816V}-TSC. *Hand1* is involved in mediation of TGC differentiation and is expressed in the ectoplacental cone [35]. In KIT^{D816V}-TSC it is significantly upregulated in comparison to WT-TSC after 6 days of differentiation (Supplementary Materials Figure S2I). Finally, we and others had previously shown that *Gata2* expression was upregulated upon KIT signaling in cells of the hematopoietic system [27,36]. Here, in KIT^{D816V} TSC, we detected a significant increase of *Gata2* expression after six days of differentiation whereas *Gata2* expression levels remained constant in WT TSC (Supplementary Materials Figure S2I). These results demonstrate that expression of spongiotrophoblast, labyrinth, and glycogen trophoblasts is underrepresented in KIT^{D816V} TSC upon differentiation consistent with data obtained from in vivo samples. Also, expression of KIT^{D816V} leads to an upregulation of P-TGC markers and *Tfap2c* already existing in TSC culture, which leads to premature and skewed induction of differentiation of TSCs, which is not morphologically apparent.

2.7. Signaling Cascades and Invasion Capability are Affected in KIT^{D816V}-TSC

Previously, we had shown that, in the hematopoietic system, expression of KIT^{D816V} leads to a block of differentiation of erythroblasts and a continuation of proliferation of precursor cells. There, we had demonstrated that KIT^{D816V} leads to induction of MAPK signaling as well as diminished AKT activation [27]. Hence, we used KIT^{D816V}-TSC to examine the activation of key signaling pathways during 9 days of TSC differentiation (Figure 5C). On day 0, levels of phosphorylated (P-) ERK and P-AKT in KIT^{D816V} TSC appeared reduced and comparable to day 3/6 of differentiating WT-TSC. Of note, on day 0, levels of P-STAT3 are comparable but become upregulated over the course of differentiation in KIT^{D816V}-TSC (Figure 5C). This suggests that chronic activation of the KIT signaling cascade in KIT^{D816V}-TSC results in diminished levels of P-ERK and P-AKT, priming the cells for rapid and premature differentiation.

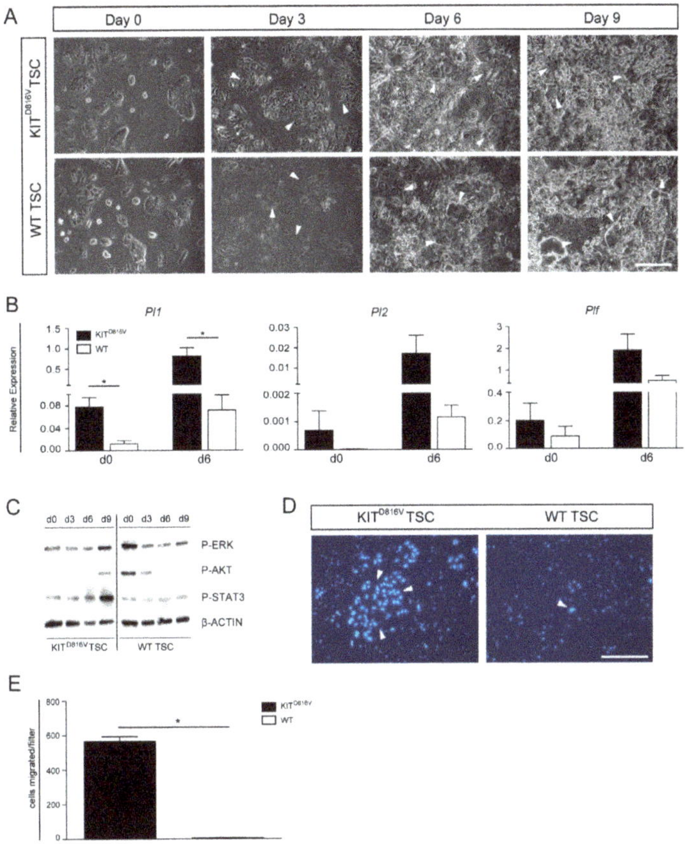

Figure 5. Differentiation of KIT^{D816V}-TSC: (**A**) Photomicrographs depicting in vitro differentiation of KIT^{D816V}-TSC line #3 and WT-TSC line 2.1 cultured for 0, 3, 6, and 9 days under differentiation conditions. White arrows indicate TGCs. Scale bar represents 250 μm. (**B**) qRT-PCR analysis of endogenous expression of *Pl1*, *Pl2*, and *Plf* in KIT^{D816V}-TSC line #3 and WT-TSC line 2.1 in undifferentiated states and after culture under differentiation conditions for 6 days. RNA was obtained from three biological replicates; expression was normalized to the housekeeping gene *Gapdh*. Data is represented by mean value ± SD. Significance was determined by unpaired t-test and indicated with * $p < 0.05$. (**C**) Western Blot detected phosphorylated Extracellular-Signal Regulated Kinase (ERK), Protein Kinase B (AKT), and Signal Transducers and Activators of Transcription 3 (STAT3) during differentiation of KIT^{D816V}-TSC line #3 and WT-TSC line 2.4 in comparison to β-ACTIN. (**D**) Representative photomicrographs of Hoechst-stained nuclei of cells that invaded through Matrigel coated filter membranes after five days under differentiating conditions: White arrows indicate nuclei of TGCs. Scale bar represents 250 μm. (**E**) Quantification of invaded cells per filter in KIT^{D816V}-TSC line and WT-TSC line (biological replicates = 2): Data is represented by mean value ± SD. Significance was determined by unpaired t-test and indicated with * $p < 0.05$.

Among their various functions such as guiding the attachment of the blastocyst and secretion of essential hormones and proteins, TGCs are also capable of invading into uterine tissue to establish the vital connection to the maternal blood vessels [31,37]. Finally, we analyzed the invasive capacity of KIT^{D816V}-TGCs in comparison to WT-TGCs by using a Transwell migration assay. Previously, it was shown that higher Matrigel concentrations result in preselection of TGCs as non-TGCs do not invade through thicker Matrigel layers [38]. Also, lower cell densities led to increased invasion [38]. Hence, we seeded 2×10^4 cells on a layer of 0.8 mg/mL Matrigel and omitted FGF4, heparin, and CM from culture medium to induce differentiation. After five days, Hoechst staining showed the presence of nuclei resembling TGCs in size and structure, indicating successful invasion (Figure 5D). Quantification of two independent experiments revealed that a significantly higher number of KIT^{D816V}-TGCs had migrated through the Matrigel and the pores (Figure 5E). Thus, KIT^{D816V} signaling in trophoblast cells seems to result in a higher portion of invasive cells. This might be due to the fact that KIT^{D816V}-TSC are much faster in inducing differentiation and gain migratory capabilities earlier than WT-TSC.

3. Discussion

Here, we show several effects of KIT^{D816V} on murine placental development. These findings comprise aberrant placental structure and increased differentiation into P-TGC subtypes, whereas other TGC subtypes such as SpA-TGCs, S-TGC, and C-TGCs remain underrepresented. Since TGC variety is essential for placental development, diminished levels of SpA-TGCs may lead to impaired establishment of blood flow at the implantation site by not properly formed blood vessels, lack of dilating spiral arteries, and insufficient secretion of vasodilators and other angiogenic factors [31,37]. C-TGCs contribute to formation of the vessels delivering maternal blood to the labyrinth [39]. Loss of these highly specialized cell types affects invasion, exchange of nutrients and gases, as well as hormone secretion, resulting in decreased functionality of the placenta. Moreover, proliferation in KIT^{D816V} placentas ceases at E10.5 when KI-67-positive cells cannot be detected anymore. At that timepoint, the embryo already shows severe growth retardation. Although we cannot exclude additive embryonic effects due to KIT^{D816V} expression, we suspect this phenotype to result from malnourishment due to placental failure.

While in other cell types activation of KIT resulted in induction of proliferation and impairment of differentiation [40], trophoblast and placenta development are hallmarked by reduced proliferation in combination with increased and skewed differentiation. In cells of the hematopoietic system, it had previously been shown that KIT signaling results in upregulation of c-Myc, c-Myb, and Gata2 [27,36]. Interestingly, in trophoblast cells, Gata2 binds to and transactivates expression of Pl1. Further, Gata2 level was also shown to be correlated with Plf expression [41,42]. In this study, KIT^{D816V}-TSC showed a significant increase in *Gata2* expression upon differentiation. Therefore, we speculate that, also in trophoblast cells, KIT signaling induces Gata2 which in turn transactivates Pl1 and Plf. This would lead to a rapid induction of differentiation, which is accompanied by exiting from the cell cycle. It also explains the high number of $PL1^+/PLF^+$ cells observed in KIT^{D816V} placentas. In the context of premature differentiation, it is interesting to note that we were able to generate self-renewing TSC lines. The established TSC did not show any aberrant overall growth parameters and ease of handling. However, the fact that, at day 0 of differentiation, P-ERK and P-AKT levels were comparable to day 3 of regular differentiation together with the already upregulated markers for TGC-differentiation strongly suggests that the growth conditions (FGF4, Heparin, and CM) are able to override the KIT-mediated signals leading to differentiation. This helps to explain the phenotype observed in KIT^{D816V}-placentas. During pre- and early post-implantation development, trophectoderm and ectoplacental cone cells rely on FGF4-induced signaling cascades leading to expression of trophectoderm and TSC master regulators Cdx2, Tfap2c, and Elf5 [43–45]. This protects the cells from differentiation inducing the effect of KIT signaling. Post implantation, when the different layers of the placenta are laid down, FGF signaling and expression of the marker genes decline. Since the KIT signaling cascade is already in place and active (as hallmarked by upregulation of *Pl1* and *Plf* at day 0 of TSC differentiation), a premature differentiation is induced, resulting in a smaller and disorganized placenta.

Interestingly, the placental alterations reported here are phenocopied in mouse models of Suppressor of Cytokine Signaling (SOCS) 3 deficiency as well as administration of retinoic acid (RA) to the mother. Administration of RA was demonstrated to result in loss of proliferation, differentiation to TGCs, and a reduced spongiotrophoblast layer [46]. Also, TPBPA levels were reduced whereas PL1 levels were increased [46]. Interestingly, in another study, RA resulted in an increase of invading TGCs [38]. In our model, chronic KIT activation results in significantly more invasive cells. Of note, RA was shown to promote KIT expression and translation in spermatogonia [47–49]. Hence, we speculate that, in trophoblast, KIT acts downstream of RA. In KIT^{D816V} mice, RA signaling is active, independent of RA presence (Figure 6).

In Leukemia Inhibitory Factor (LIF)/JAK/STAT3 signaling, Leukemia Inhibitory Factor (LIF) binds to its receptor and Janus Kinases (JAK) are activated. JAK then phosphorylates Signal Transducers and Activators of Transcription (STAT) 3 in the cytoplasm. Activated STAT3 induces the expression of SOCS3, which then represses LIF in a negative feedback loop [50]. Augmentation in LIF levels due to SOCS3 deficiency increases TGC differentiation [51]. SOCS3 deficiency was also reported to result in erythrocytosis and reduced spongiotrophoblast layer. Further, constitutively active STAT3 was observed [52]. We find increased levels of phosphorylated STAT3 in KIT^{D816V}-TSC upon differentiation for 9 days in comparison to WT-TSC. STAT3 also plays a role in cell migration and invasion [53], both of which we demonstrate to be affected in the KIT^{D816V} mouse model. Taken together, we conclude that KIT is also involved in LIF/JAK/STAT3 signaling (Figure 6).

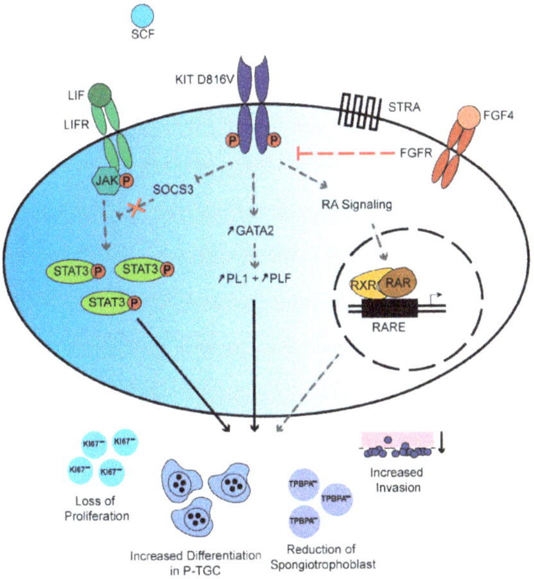

Figure 6. Proposed mechanism of KIT^{D816V} signaling in TSC: Schematic summary depicting possible signaling mechanisms that lead to KIT^{D816V} placental phenotype observed in this study. Independent of SCF binding, the KIT^{D816V} receptor is active. KIT inhibits SOCS3, thereby interrupting the negative feedback loop of Leukemia Inhibitory Factor (LIF) signaling resulting in accumulation of P-STAT3. Further, KIT signaling induces the expression of GATA2, which then transactivates PL1 and PLF. Independent of retinoic acid (RA) presence, KIT impinges on RA-controlled gene expression. Taken together, chronic KIT^{D816V} signaling results in severely diminished placental proliferation and reduction of spongiotrophoblast cells but increase in differentiation towards P-TGC and elevated invasion. In the presence of FGF4, Fibroblast Growth Factor Receptor (FGFR) signaling can override KIT-mediated signals. The black arrows indicate relations found in this study, whereas the gray and dashed arrows present hypothetical interactions, some of which have only been demonstrated in other cell types and T bars indicate repression.

Here, we demonstrate that KIT^{D816V} placentas are severely affected by constitutively active KIT signaling. The results show that, in placenta development, KIT signaling is required for the fine-tuned induction of differentiation. Moreover, we speculate that KIT signaling is essential for P-TGC differentiation since this is the major TGC-type observed in the studies.

4. Materials and Methods

4.1. Generation of Transgenic Animals

All experiments were conducted according to the German law of animal protection and in agreement with the approval of the local institutional animal care committees (Landesamt für Natur, Umwelt und Verbraucherschutz, North Rhine-Westphalia, approval number: 84-02.03.2013/A428 approved date: 31 January 2014). R26-LSL-KIT^{D816V} mice described by us were mated with Deleter-Cre mice carrying human cytomegalovirus minimal promoter (CMV) controlled Cre recombinase, thereby allowing for ubiquitous expression of the mutant human KIT receptor [27,54]. Both mouse strains were kept on a 129sv/S2 and C57BL/6 genetic background. The morning after mating, female mice were checked for vaginal plugs. In the case of plug detection, noon of that day was considered E0.5. Accordingly, mice were sacrificed 9 to 11 days later around noon.

4.2. DNA Isolation and Genotyping PCR

Embryos and placentas were lysed in tissue lysis buffer (50 mM Tris-HCl (Roth, Karlsruhe, Baden-Wuerttemberg, Germany; #9090.1), 100 mM ethylenediaminetetraacetic acid (EDTA) (AppliChem, Darmstadt, Hessen, Germany; #A4975), 100 mM NaCl (AppliChem, Darmstadt, Hessen, Germany; #A1149), 1% (*w*/*v*) SDS (Merck, Darmstadt, Hessen, Germany; #817039), and 10 mg/mL proteinase K (Merck, Darmstadt, Hessen, Germany; #1245680500)), while DNA from adherent cells was obtained using cell lysis buffer (10 mM NaCl (AppliChem, Darmstadt, Hessen, Germany; #A1149), 10 mM Tris, 10 mM EDTA, 0.5% sarcosyl, and 1 mg/mL proteinase K (Merck, Darmstadt, Hessen, Germany)). DNA was precipitated using isopropanol, pelleted by centrifugation, and washed twice with ethanol (80%). The pellet was then resuspended in H_2O (dd) at 37 °C. Using NanoDrop 1000 (Thermo Fisher Scientific, Waltham, MA, USA), concentration and purity of obtained DNA was measured. Genotyping was performed by polymerase chain reaction (PCR) using PCR primers. Primer sequences are listed in Supplementary Materials Table S1.

4.3. RNA Analysis

RNA isolation from tissue and cells was carried out using TRIzol™ reagent (Invitrogen, Thermo Scientific, Waltham, MA, USA; #15596026). RNA was precipitated using isopropanol and pelleted by centrifugation. After washing with ethanol (75%), pellet was resuspended in diethyl pyrocarbonate (DEPC)-treated water. Using NanoDrop 1000 (Thermo Fisher Scientific, Waltham, MA, USA), concentration and purity of obtained RNA was measured. DNase digestion (Thermo Scientific, Waltham, Massachusetts, USA; #EN0525) and cDNA synthesis (RevertAid Premium, Fermentas, Thermo Scientific, Waltham, MA, USA; #EP0441) were carried out on 1 µg RNA. Quantitative real time PCR (qPCR) was performed using SYBR Green Master Mix (Fermentas, Thermo Scientific, Waltham, MA, USA; #4309155) and ViiA7 (AppliedBiosystems, Life Technologies, Foster City, CA, USA). Primer sequences are listed in Supplementary Materials Table S3. Target gene expression was normalized to the housekeeping reference gene *Gapdh*, reactions were performed in triplicate, and *p*-values were calculated using unpaired t-test.

4.4. Hematoxylin and Eosin Staining

Paraffin sections were rehydrated and incubated in hematoxylin (Roth, Karlsruhe, Baden-Wuerttemberg, Germany; #T865) for 3 min. Counterstaining with eosin (Roth, Karlsruhe, Baden-Wuerttemberg, Germany; #X883) was performed for 1 min. Sections were then dehydrated, incubated in xylene (VMP Chemie Kontor GmbH, #1000649172), and embedded with Entellan® (Merck, Darmstadt, Hessen, Germany; #107960). Analysis of stained sections was performed with a DM LB microscope (Leica, Wetzlar, Hessen, Germany).

4.5. Western Blotting

Protein was obtained by lysing tissue or cells in radioimmunoprecipitation assay (RIPA) buffer (NEB, #9806). Protein concentration was measured using Pierce™ BCA Protein Assay Kit (Thermo Scientific, Waltham, MA, USA; #23225) on an iMarkTM Microplate Reader (BioRad, Hercules, CA, USA). Proteins were electrophoresed in a polyacrylamide (10%) gel (Roth, Karlsruhe, Baden-Wuerttemberg, Germany; #3029.1) and transferred onto a Roti® polyvinylidene fluoride (PVDF) membrane (Roth, Karlsruhe, Baden-Wuerttemberg, Germany; #T830.1). Membranes were blocked, stained with primary antibodies, and horseradish peroxidase (HRP)-coupled secondary antibodies. Antibodies and concentrations are given in Supplementary Materials Table S2. Pierce™ ECL Western Blotting Substrate (Thermo Scientific, Waltham, MA, USA; #32106) and ChemiDoc™MP (Biorad, Hercules, CA, USA) were used for detection of stained proteins.

4.6. Immunohistochemical/Immunofluorescence Staining

For CD31 and KI-67 immunohistochemical staining, paraffin sections were incubated with antigen retrieval buffer (Medac GmbH, Wedel, Schleswig-Holstein, Germany; #PMB1-250) and blocked with peroxide block (Medac GmbH, Wedel, Schleswig-Holstein, Germany; #925B-05). After primary antibody staining, sections were incubated with HRP-coupled secondary antibody, followed by a 3,3′-diaminobenzidine (DAB, Medac GmbH, Wedel, Schleswig-Holstein, Germany; #957D-50) staining for 8 min. Nuclei were stained with hematoxylin (Roth, Karlsruhe, Baden-Wuerttemberg, Germany; #T865) For immunofluorescence staining, cells were fixed in formalin (4%, Merck, Darmstadt, Hessen, Germany; #100496), permeabilized using 0.5% (*v*/*v*) Triton X-100 (AppliChem, Darmstadt, Hessen, Germany; #142314), and blocked in 2% bovine serum albumin (BSA, Sigma-Aldrich, St. Louis, MO, USA) and 0.1% Triton X-100 in phosphate-buffered saline (PBS). Then, cells were stained with primary antibody and Alexa-Fluor-conjugated secondary antibodies. Nuclei were stained with Hoechst (Sigma-Aldrich, St. Louis, MO, USA; #H6024). Antibodies and concentrations are given in Supplementary Materials Table S2.

4.7. In Situ Hybridization

In situ hybridization was performed using protocol and digoxigenin (DIG)-labeled probes established by Simmons et al. [37]. In short, cryo- or paraffin sections were thawed or deparaffinized and rehydrated, respectively. Sections were fixed using 4% paraformaldehyde (Merck, Darmstadt, Hessen, Germany; #818715) and treated with proteinase K (20 µg/mL, Merck, Darmstadt, Hessen, Germany; #1245680500). After another fixation step, sections were blocked. Probes (2 ng/µL) for PL1, PLF, and TPBPA were denatured and incubated with sections overnight, followed by a RNase A (AppliChem, Darmstadt, Hessen, Germany; #A2760) treatment. Sections were blocked, and anti-digoxigenin-AP (1:1000, Sigma-Aldrich, St. Louis, MO, USA; #11093274910) was added for probe detection. Incubation with BM-purple followed overnight; then, sections were counterstained with nuclear fast red (Sigma-Aldrich, St. Louis, MO, USA; #N3020). Sections were dehydrated, treated with xylene (VMP Chemie Kontor GmbH, Siegburg, North Rhine-Westfalia, Germany; #1000649172), and mounted in Entellan® (Merck, Darmstadt, Hessen, Germany; #107960). Leica DM LB microscope (Wetzlar, Hessen, Germany) was used for analysis of sections.

4.8. Cell Culture

KIT^{D816V}-TSCs were derived from E3.5 blastocysts, which were obtained after mating of R26-LSL-KIT^{D816V} mice with Deleter-Cre mice [33]. Cre recombinase expression is controlled by human CMV minimal promoter [54]. Cells were cultured on tissue culture plastic in humidified incubators (Heracell 240i, Thermo Scientific, Waltham, MA, USA) with 7.5% CO_2 at 37 °C. TSC were grown in a trophoblast stem (TS) cell medium supplemented with 70% mouse embryonic fibroblast (MEF) conditioned medium (CM) [45]. TS medium was supplemented with 25 ng/mL human recombinant FGF4 (Reliatech, Wolfenbüttel, Lower Saxony, Germany; #100-017L) and 1 μg/mL heparin (Sigma-Aldrich, St. Louis, MO, USA; #H3149-10KU).

4.9. Flow Cytometric Analysis

Cells were harvested with 0.05% trypsin/EDTA (Gibco, Thermo Scientific, Waltham, MA, USA; #11580626) and filtered through a 40-μm cell strainer (Becton Dickinson, Franklin Lakes, NJ, USA; #352235). After washing with fluorescence-activated cell sorting (FACS) buffer (2% fetal bovine serum (FBS) in PBS (Gibco, Thermo Scientific, Waltham, MA, USA; #11503387), cells were resuspended in FACS buffer containing 1% 7-amino-actinomycin D (AppliChem, Darmstadt, Hessen, Germany; #A7850). Detection of GFP positivity was performed using a FACSCanto (Becton Dickinson, Franklin Lakes, NJ, USA) and analyzed with FlowJo Software (TreeStar, Becton Dickinson, Franklin Lakes, NJ, USA).

4.10. Invasion Assay

FluoroBlok™ Cell Culture Inserts with 8 μm pore size (Corning, New York, NY, USA; #351157) were coated using 0.8 mg/mL Matrigel (Sigma-Aldrich, St. Louis, Missouri, USA) and dried for 2 h at 37 °C and then at room temperature (RT) overnight. Matrigel was rehydrated using supplemented RPMI (Gibco, Thermo Scientific, Waltham, MA, USA; #11530586) [38]. Cells were harvested using 0.05 trypsin/EDTA (Gibco, Thermo Scientific, Waltham, MA, USA; #11580626), counted and reseeded at 2×10^4 cells per well, and kept under differentiating conditions (TS-Medium w/o CM, FGF4 and heparin) for five days. Medium in the lower chamber was changed every day. For analysis, Matrigel in upper chamber as well as medium in lower chamber were discarded and filters were washed several times with PBS. Filters were fixed and stained with 1% Hoechst (Sigma-Aldrich, St. Louis, MO, USA; #H6024) in methanol (VWR, Radnor, PA, USA; #20.847.307) for 10 min. After 3 washing steps with PBS, detection of invaded cells was carried out using a Leica DM LB microscope (Wetzlar, Hessen, Germany).

Supplementary Materials: Supplementary materials can be found at http://www.mdpi.com/1422-0067/21/15/5503/s1. Figure S1: Spongiotrophoblast marker TPBPA reduced in KIT^{D816V} placentas, Figure S2: Validation of KIT^{D816V}-TSC, Table S1: Genotyping Primers, Table S2: Antibodies, Table S3: qRT-PCR Primers.

Author Contributions: Conceptualization, C.K., N.S., and H.S.; formal analysis, F.K., S.J., and D.N.; funding acquisition, H.S.; investigation, F.K., J.H., S.J., and D.N.; methodology, F.K., J.H., S.J., J.L., and H.S.; project administration, F.K., C.K., and N.S.; resources, N.P.; supervision, F.K., N.P., C.K., N.S., and H.S.; validation, F.K., J.K., D.N., and J.L.; writing—original draft, F.K. and H.S.; writing—review and editing, F.K., N.P., C.K., and H.S. All authors have read and agreed to the published version of the manuscript.

Funding: This study was financially supported by DFG grant Scho 503 20/1 to H.S.

Acknowledgments: We kindly thank Angela Egert, Andrea Jäger, Gaby Beine, and Susanne Steiner for technical assistance. We also thank Marieta Toma and Laura Eßer for assisting with performance of Transwell invasion assays.

Conflicts of Interest: The authors declare no conflict of interest.

Abbreviations

AKT	Protein Kinase B
C-TGC	Canal TGC
CM	Conditioned Medium
ERK	Extracellular-Signal Regulated Kinase
FGF4	Fibroblast Growth Factor 4
FGFR	Fibroblast Growth Factor Receptor
GlyT	Glycogen Trophoblast
HSC	Hematopoietic Stem Cell

JAK	Janus Kinase
LIF	Leukemia Inhibitory Factor
MAPK	Mitogen-Activated Protein Kinase
P-TGC	Parietal TGC
RA	Retinoic Acid
RAR	Retinoic Acid Receptor
RXR	Retinoic X Receptor
S-TGC	Sinusoidal TGC
SOCS3	Suppressor of Cytokine Signaling 3
SpA	Spiral Artery
STAT3	Signal Transducers and Activators of Transcription
STRA	Stimulated by Retinoic Acid
TGC	Trophoblast Giant Cell
TSC	Trophoblast Stem Celle
WT	Wildtype

References

1. Bazer, F.W.; Spencer, T.E.; Johnson, G.A.; Burghardt, R.C.; Wu, G. Comparative aspects of implantation. *Reproduction* **2009**, *138*, 195–209. [CrossRef]
2. Cockburn, K.; Rossant, J. Making the blastocyst: Lessons from the mouse. *J. Clin. Investig.* **2010**, *120*, 995–1003. [CrossRef]
3. Woods, L.; Perez-Garcia, V.; Hemberger, M. Regulation of Placental Development and Its Impact on Fetal Growth—New Insights From Mouse Models. *Front. Endocrinol.* **2018**, *9*, 570. [CrossRef]
4. Ashman, L.K. The biology of stem cell factor and its receptor C-kit. *Int. J. Biochem. Cell Biol.* **1999**, *31*, 1037–1051. [CrossRef]
5. Chabot, B.; Stephenson, D.A.; Chapman, V.M.; Besmer, P.; Bernstein, A. The proto-oncogene c-kit encoding a transmembrane tyrosine kinase receptor maps to the mouse W locus. *Nature* **1988**, *335*, 88–89. [CrossRef]
6. Copeland, N.G.; Gilbert, D.J.; Cho, B.C.; Donovan, P.J.; Jenkins, N.A.; Cosman, D.; Anderson, D.; Lyman, S.D.; Williams, D.E. Mast cell growth factor maps near the steel locus on mouse chromosome 10 and is deleted in a number of steel alleles. *Cell* **1990**, *63*, 175–183. [CrossRef]
7. Liang, J.; Wu, Y.-L.; Chen, B.-J.; Zhang, W.; Tanaka, Y.; Sugiyama, H. The C-kit receptor-mediated signal transduction and tumor-related diseases. *Int. J. Biol. Sci.* **2013**, *9*, 435–443. [CrossRef] [PubMed]
8. Linnekin, D. Early signaling pathways activated by c-Kit in hematopoietic cells. *Int. J. Biochem. Cell Biol.* **1999**, *31*, 1053–1074. [CrossRef]
9. Roskoski, R. Signaling by Kit protein-tyrosine kinase-The stem cell factor receptor. *Biochem. Biophys. Res. Commun.* **2005**, *337*, 1–13. [CrossRef] [PubMed]
10. Yasuda, T.; Kurosaki, T. Regulation of lymphocyte fate by Ras/ERK signals. *Cell Cycle* **2008**, *7*, 3634–3640. [CrossRef]
11. Horie, K.; Fujita, J.; Takakura, K.; Kanzaki, H.; Kaneko, Y.; Iwai, M.; Nakayama, H.; Mori, T. Expression of C-Kit Protein during Placental Development1. *Biol. Reprod.* **1992**, *47*, 614–620. [CrossRef] [PubMed]
12. Motro, B.; van der Kooy, D.; Rossant, J.; Reith, A.; Bernstein, A. Contiguous patterns of c-kit and steel expression: Analysis of mutations at the W and Sl loci. *Development* **1991**, *113*, 1207–1221. [PubMed]
13. Ottersbach, K.; Dzierzak, E. The Murine Placenta Contains Hematopoietic Stem Cells within the Vascular Labyrinth Region. *Dev. Cell* **2005**, *8*, 377–387. [CrossRef] [PubMed]
14. Azevedo Portilho, N.; Pelajo-Machado, M. Mechanism of hematopoiesis and vasculogenesis in mouse placenta. *Placenta* **2018**, *69*, 140–145. [CrossRef]
15. Sasaki, T.; Mizuochi, C.; Horio, Y.; Nakao, K.; Akashi, K.; Sugiyama, D. Regulation of hematopoietic cell clusters in the placental niche through SCF/Kit signaling in embryonic mouse. *Development* **2010**, *137*, 3941 LP–3952. [CrossRef]
16. Bodemer, C.; Hermine, O.; Palmérini, F.; Yang, Y.; Grandpeix-Guyodo, C.; Leventhal, P.S.; Hadj-Rabia, S.; Nasca, L.; Georgin-Lavialle, S.; Cohen-Akenine, A.; et al. Pediatric mastocytosis is a clonal disease associated with D 816 v and other activating c-KIT mutations. *J. Invest. Dermatol.* **2010**, *130*, 804–815. [CrossRef]

17. Kemmer, K.; Corless, C.L.; Fletcher, J.A.; McGreevey, L.; Haley, A.; Griffith, D.; Cummings, O.W.; Wait, C.; Town, A.; Heinrich, M.C. KIT mutations are common in testicular seminomas. *Am. J. Pathol.* **2004**, *164*, 305–313. [CrossRef]
18. Lasota, J.; Jasinski, M.; Sarlomo-Rikala, M.; Miettinen, M. Mutations in exon 11 of c-Kit occur preferentially in malignant versus benign gastrointestinal stromal tumors and do not occur in leiomyomas or leiomyosarcomas. *Am. J. Pathol.* **1999**, *154*, 53–60. [CrossRef]
19. Malaise, M.; Steinbach, D.; Corbacioglu, S. Clinical implications of c-Kit mutations in acute myelogenous leukemia. *Curr. Hematol. Malig. Rep.* **2009**, *4*, 77–82. [CrossRef]
20. Miettinen, M.; Lasota, J. KIT (CD117): A Review on Expression in Normal and Neoplastic Tissues, and Mutations and Their Clinicopathologic Correlation. *Appl. Immunohistochem. Mol. Morphol.* **2005**, *13*, 205–220. [CrossRef]
21. Sun, J.; Pedersen, M.; Rönnstrand, L. The D816V mutation of c-Kit circumvents a requirement for Src family kinases in c-Kit signal transduction. *J. Biol. Chem.* **2009**, *284*, 11039–11047. [CrossRef] [PubMed]
22. Tian, Q.; Frierson, H.F.; Krystal, G.W.; Moskaluk, C.A. Activating c-kit gene mutations in human germ cell tumors. *Am. J. Pathol.* **1999**, *154*, 1643–1647. [CrossRef]
23. Fleischman, R.A.; Mintz, B. Prevention of genetic anemias in mice by microinjection of normal hematopoietic stem cells into the fetal placenta. *Proc. Natl. Acad. Sci. USA* **1979**, *76*, 5736–5740. [CrossRef] [PubMed]
24. Broudy, V.C. Stem Cell Factor and Hematopoiesis. *Blood* **1997**, *90*, 1345–1364. [CrossRef]
25. Lotinun, S.; Krishnamra, N. Disruption of c-Kit Signaling in KitW-sh/W-sh Growing Mice Increases Bone Turnover. *Sci. Rep.* **2016**, *6*, 31515. [CrossRef]
26. Waskow, C.; Paul, S.; Haller, C.; Gassmann, M.; Rodewald, H.-R. Viable c-KitW/W Mutants Reveal Pivotal Role for c-Kit in the Maintenance of Lymphopoiesis. *Immunity* **2002**, *17*, 277–288. [CrossRef]
27. Haas, N.; Riedt, T.; Labbaf, Z.; Baßler, K.; Gergis, D.; Fröhlich, H.; Gütgemann, I.; Janzen, V.; Schorle, H. Kit transduced signals counteract erythroid maturation by MAPK-dependent modulation of erythropoietin signaling and apoptosis induction in mouse fetal liver. *Cell Death Differ.* **2015**, *22*, 790–800. [CrossRef]
28. Xiang, Z.; Kreisel, F.; Cain, J.; Colson, A.; Tomasson, M.H. Neoplasia driven by mutant c-KIT is mediated by intracellular, not plasma membrane, receptor signaling. *Mol. Cell. Biol.* **2007**, *27*, 267–282. [CrossRef]
29. Pelusi, N.; Kosanke, M.; Riedt, T.; Rösseler, C.; Seré, K.; Li, J.; Gütgemann, I.; Zenke, M.; Janzen, V.; Schorle, H. The spleen microenvironment influences disease transformation in a mouse model of KITD816V-dependent myeloproliferative neoplasm. *Sci. Rep.* **2017**, *7*, 41427. [CrossRef]
30. Winking, H.; Gerdes, J.; Traut, W. Expression of the proliferation marker Ki-67 during early mouse development. *Cytogenet. Genome Res.* **2004**, *105*, 251–256. [CrossRef]
31. Hu, D.; Cross, J.C. Development and function of trophoblast giant cells in the rodent placenta. *Int. J. Dev. Biol.* **2010**, *54*, 341–354. [CrossRef] [PubMed]
32. Aoki, M.; Mieda, M.; Ikeda, T.; Hamada, Y.; Nakamura, H.; Okamoto, H. R-spondin3 is required for mouse placental development. *Dev. Biol.* **2007**, *301*, 218–226. [CrossRef] [PubMed]
33. Himeno, E.; Tanaka, S.; Kunath, T. Isolation and Manipulation of Mouse Trophoblast Stem Cells. *Curr. Protoc. Stem Cell Biol.* **2008**, *7*, 1E.4.1–1E.4.27. [CrossRef] [PubMed]
34. Kuckenberg, P.; Buhl, S.; Woynecki, T.; van Fürden, B.; Tolkunova, E.; Seiffe, F.; Moser, M.; Tomilin, A.; Winterhager, E.; Schorle, H. The Transcription Factor TCFAP2C/AP-2γ Cooperates with CDX2 To Maintain Trophectoderm Formation. *Mol. Cell. Biol.* **2010**, *30*, 3310–3320. [CrossRef] [PubMed]
35. Scott, I.C.; Anson-Cartwright, L.; Riley, P.; Reda, D.; Cross, J.C. The HAND1 Basic Helix-Loop-Helix Transcription Factor Regulates Trophoblast Differentiation via Multiple Mechanisms. *Mol. Cell. Biol.* **2000**, *20*, 530 LP–541. [CrossRef] [PubMed]
36. Zeuner, A.; Francescangeli, F.; Signore, M.; Venneri, M.A.; Pedini, F.; Felli, N.; Pagliuca, A.; Conticello, C.; De Maria, R. The Notch2-Jagged1 interaction mediates stem cell factor signaling in erythropoiesis. *Cell Death Differ.* **2011**, *18*, 371–380. [CrossRef]
37. Simmons, D.G.; Fortier, A.L.; Cross, J.C. Diverse subtypes and developmental origins of trophoblast giant cells in the mouse placenta. *Dev. Biol.* **2007**, *304*, 567–578. [CrossRef]
38. Hemberger, M.; Hughes, M.; Cross, J.C. Trophoblast stem cells differentiate in vitro into invasive trophoblast giant cells. *Dev. Biol.* **2004**, *271*, 362–371. [CrossRef]
39. Rai, A.; Cross, J.C. Development of the hemochorial maternal vascular spaces in the placenta through endothelial and vasculogenic mimicry. *Dev. Biol.* **2014**, *387*, 131–141. [CrossRef]

40. Muta, K.; Krantz, S.B.; Bondurant, M.C.; Dai, C.H. Stem cell factor retards differentiation of normal human erythroid progenitor cells while stimulating proliferation. *Blood* **1995**, *86*, 572–580. [CrossRef]
41. Ma, G.T.; Roth, M.E.; Groskopf, J.C.; Tsai, F.Y.; Orkin, S.H.; Grosveld, F.; Engel, J.D.; Linzer, D.I. GATA-2 and GATA-3 regulate trophoblast-specific gene expression in vivo. *Development* **1997**, *124*, 907. [PubMed]
42. Ng, Y.K.; George, K.M.; Engel, J.D.; Linzer, D.I. GATA factor activity is required for the trophoblast-specific transcriptional regulation of the mouse placental lactogen I gene. *Development* **1994**, *120*, 3257. [PubMed]
43. Latos, P.A.; Hemberger, M. From the stem of the placental tree: Trophoblast stem cells and their progeny. *Development* **2016**, *143*, 3650 LP–3660. [CrossRef] [PubMed]
44. Simmons, D.G.; Cross, J.C. Determinants of trophoblast lineage and cell subtype specification in the mouse placenta. *Dev. Biol.* **2005**, *284*, 12–24. [CrossRef] [PubMed]
45. Tanaka, S.; Kunath, T.; Hadjantonakis, A.-K.; Nagy, A.; Rossant, J. Promotion of Trophoblast Stem Cell Proliferation by FGF4. *Science (80-)* **1998**, *282*, 2072 LP–2075. [CrossRef]
46. Yan, J.; Tanaka, S.; Oda, M.; Makino, T.; Ohgane, J.; Shiota, K. Retinoic acid promotes differentiation of trophoblast stem cells to a giant cell fate. *Dev. Biol.* **2001**, *235*, 422–432. [CrossRef] [PubMed]
47. Busada, J.T.; Chappell, V.A.; Niedenberger, B.A.; Kaye, E.P.; Keiper, B.D.; Hogarth, C.A.; Geyer, C.B. Retinoic acid regulates Kit translation during spermatogonial differentiation in the mouse. *Dev. Biol.* **2015**, *397*, 140–149. [CrossRef]
48. Gely-Pernot, A.; Raverdeau, M.; Teletin, M.; Vernet, N.; Féret, B.; Klopfenstein, M.; Dennefeld, C.; Davidson, I.; Benoit, G.; Mark, M.; et al. Retinoic Acid Receptors Control Spermatogonia Cell-Fate and Induce Expression of the SALL4A Transcription Factor. *PLoS Genet.* **2015**, *11*, e1005501. [CrossRef]
49. Koli, S.; Mukherjee, A.; Reddy, K.V.R. Retinoic acid triggers *c-kit* gene expression in spermatogonial stem cells through an enhanceosome constituted between transcription factor binding sites for retinoic acid response element (RARE), spleen focus forming virus proviral integration onco. *Reprod. Fertil. Dev.* **2017**, *29*, 521–543. [CrossRef]
50. Suman, P.; Malhotra, S.S.; Gupta, S.K. LIF-STAT signaling and trophoblast biology. *JAK-STAT* **2013**, *2*, e25155. [CrossRef]
51. Takahashi, Y.; Takahashi, M.; Carpino, N.; Jou, S.-T.; Chao, J.-R.; Tanaka, S.; Shigeyoshi, Y.; Parganas, E.; Ihle, J.N. Leukemia Inhibitory Factor Regulates Trophoblast Giant Cell Differentiation via Janus Kinase 1-Signal Transducer and Activator of Transcription 3-Suppressor of Cytokine Signaling 3 Pathway. *Mol. Endocrinol.* **2008**, *22*, 1673–1681. [CrossRef] [PubMed]
52. Takahashi, Y.; Carpino, N.; Cross, J.C.; Torres, M.; Parganas, E.; Ihle, J.N. SOCS3: An essential regulator of LIF receptor signaling in trophoblast giant cell differentiation. *EMBO J.* **2003**, *22*, 372–384. [CrossRef] [PubMed]
53. Huang, S. Regulation of Metastases by Signal Transducer and Activator of Transcription 3 Signaling Pathway: Clinical Implications. *Clin. Cancer Res.* **2007**, *13*, 1362. [CrossRef] [PubMed]
54. Schwenk, F.; Baron, U.; Rajewsky, K. A cre -transgenic mouse strain for the ubiquitous deletion of loxP -flanked gene segments including deletion in germ cells. *Nucleic Acids Res.* **1995**, *23*, 5080–5081. [CrossRef] [PubMed]

International Journal of
Molecular Sciences

MDPI

Article

Transcription Factor PLAGL1 Is Associated with Angiogenic Gene Expression in the Placenta

Rebekah R. Starks [1,2], Rabab Abu Alhasan [1], Haninder Kaur [1], Kathleen A. Pennington [3], Laura C. Schulz [4] and Geetu Tuteja [1,2,*]

1 Genetics, Development, and Cell Biology, Iowa State University, Ames, IA 50011, USA; rstarks1@iastate.edu (R.R.S.); rabab@iastate.edu (R.A.A.); hkaur@iastate.edu (H.K.)
2 Bioinformatics and Computational Biology, Iowa State University, Ames, IA 50011, USA
3 Obstetrics and Gynecology, Baylor College of Medicine, Houston, TX 77030, USA; Kathleen.Pennington@bcm.edu
4 Obstetrics, Gynecology and Women's Health, University of Missouri, Columba, MO 65212, USA; schulzl@health.missouri.edu
* Correspondence: geetu@iastate.edu

Received: 7 October 2020; Accepted: 2 November 2020; Published: 6 November 2020

Abstract: During pregnancy, the placenta is important for transporting nutrients and waste between the maternal and fetal blood supply, secreting hormones, and serving as a protective barrier. To better understand placental development, we must understand how placental gene expression is regulated. We used RNA-seq data and ChIP-seq data for the enhancer associated mark, H3k27ac, to study gene regulation in the mouse placenta at embryonic day (e) 9.5, when the placenta is developing a complex network of blood vessels. We identified several upregulated transcription factors with enriched binding sites in e9.5-specific enhancers. The most enriched transcription factor, PLAGL1 had a predicted motif in 233 regions that were significantly associated with vasculature development and response to insulin stimulus genes. We then performed several experiments using mouse placenta and a human trophoblast cell line to understand the role of PLAGL1 in placental development. In the mouse placenta, *Plagl1* is expressed in endothelial cells of the labyrinth layer and is differentially expressed in placentas from mice with gestational diabetes compared to placentas from control mice in a sex-specific manner. In human trophoblast cells, siRNA knockdown significantly decreased expression of genes associated with placental vasculature development terms. In a tube assay, decreased *PLAGL1* expression led to reduced cord formation. These results suggest that *Plagl1* regulates overlapping gene networks in placental trophoblast and endothelial cells, and may play a critical role in placental development in normal and complicated pregnancies.

Keywords: RNA-seq; ChIP-seq; enhancers; transcription factors; placenta; PLAGL1; gestational diabetes; tube formation; blood vessel development

---

1. Introduction

The placenta is a specialized organ playing a crucial role in the health of the fetus and mother during pregnancy, regulating the exchange of nutrients, gases, hormones and waste [1]. Inefficient transfer of materials from the mother to the fetus is frequently associated with adverse health outcomes for the fetus [2]. The placenta also plays a protective role, regulating the immune response of cells and preventing toxic elements from reaching the fetus [3,4]. Placental insufficiency, as well as many other placental abnormalities, have been attributed to misregulated gene expression [5,6]. Despite its importance, many aspects of placental development and the genetic mechanisms involved in its function remain unknown.

To elucidate these mechanisms, many researchers utilize the mouse model, which has contributed greatly to our understanding of the development of the placenta. Similar to humans, rodents have a single disc-shaped, hemochorial placenta [7] to sustain the fetus during development. In the human placenta, trophoblast cells invade deeply into the myometrium, and partially replace the endothelial wall of maternal spiral arteries, thereby channeling maternal blood into the intervillous spaces. Trophoblast lining the villi can then exchange factors between the maternal blood and fetal vessels running through the villous core [1]. Similarly, in the mouse, trophoblast cells replace the endothelial wall of maternal spiral arteries in the near decidua, channeling maternal blood through trophoblast-lined canals to spaces in the labyrinth, which are surrounded by trophoblast that exchange material between the maternal blood and the adjacent fetal vessels [8]. Labyrinth formation begins around embryonic day (e) 8.5 with the fusing of the chorion and allantois. Subsequently, the chorion begins to fold and form branches that become the trophoblast-lined maternal blood spaces, while allantoic offshoots form the interdigitating fetal vessels [9]. Thus, coordinated interactions between trophoblast and endothelial cells, with trophoblasts assuming key endothelial properties, are critical for placental development and function.

Regulation of these complex interactions requires transcriptional machinery to interact with the appropriate regulatory elements and cofactors in a spatiotemporal manner. A common cause of genetic misregulation is disruption of cis-regulatory elements, specifically in enhancers [10,11]. Despite the crucial role enhancers play in gene regulation, there remains a large deficit in our understanding of the gene-enhancer regulatory networks that are important for placental function [12], including trophoblast-endothelial cell interactions.

A previous study compared the enhancer landscape before (e7.5) and after (e9.5) chorioallantoic fusion in the mouse placenta, identifying an e7.5-specific cell migration network of transcription factors (TFs) and enhancers [13]. The goal of the present study is to identify e9.5-specific TFs and enhancers critical to placental development. In order to identify novel TFs that could be important for regulating processes in the mouse placenta after chorioallantoic fusion, we utilized publicly available H3k27ac ChIP-seq data, defining putative active enhancers in the e7.5 and e9.5 placenta, as well as RNA-seq data generated at the same timepoints [13]. We identified several highly expressed TFs that were upregulated in the e9.5 placenta. One of these TFs, pleiomorphic adenoma gene-like 1 (PLAGL1), has a motif highly enriched in e9.5-specific enhancers compared to e7.5-specific enhancers. Overexpression of *PLAGL1* has been associated with transient neonatal diabetes mellitus [14], while reduced expression has been found in female intrauterine growth restricted placentas, suggesting it plays an important role [15].

To further understand the functions of PLAGL1, we carried out additional experiments in mouse placenta and a human trophoblast cell line, HTR-8/SVneo, due to the high expression of *Plagl1* in both systems. By combining computational predictions with experimental validations, we were able to identify novel functions for PLAGL1 in two models frequently used to study human placenta, which could contribute to insights into placental development and certain pregnancy disorders.

2. Results

2.1. E9.5-Specific Genes and Enhancers Are Associated with Vasculature Development

RNA-seq and H3k27ac ChIP-seq were previously carried out on mouse placenta at e7.5 and e9.5 to identify genes upregulated and enhancers specific to the e7.5 timepoint [13]. However, the genes upregulated at e9.5, as well as the enhancers more active at this timepoint were not thoroughly investigated. We first analyzed the 583 genes that were highly expressed and upregulated at e9.5 (fragments per kilobase of transcript per million of mapped reads (FPKM) ≥ 10, fold ≥ 2, and FDR ≤ 0.05). As expected, we see higher e9.5 expression of genes related to processes important in the e9.5 placenta including, *Apoa1* (lipid metabolism [16]); *Notch1*, *Col1a1*, and *Dlc1* (blood vessel development [17–19]); *Gcm1* (hormone production [20]); *Igfbp2*, *Pappa2*, *Ada* (fetal growth and development [21–23]); and *Syna* (trophoblast differentiation [24]) (Figure 1a). We then used the Genomic Regions Enrichment of Annotations Tool (GREAT) [25] to identify the biological processes associated with the highly expressed and upregulated e9.5 genes using the single nearest gene option. In general, we found that terms related to metabolism and vasculature development are more enriched in genes upregulated at e9.5 than genes upregulated at e7.5 (Figure 1b).

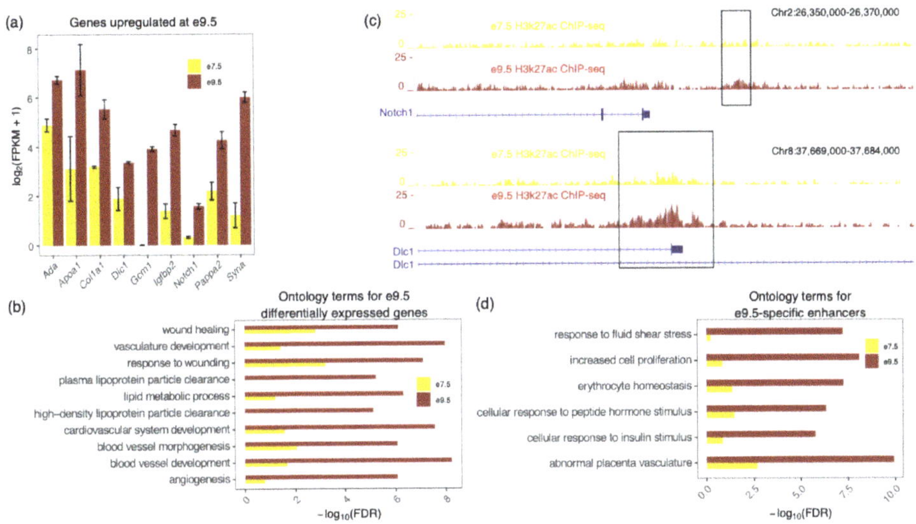

Figure 1. Genes and enhancers enriched in the placenta at e9.5 are associated with vasculature development. (**a**) Examples of genes significantly upregulated at e9.5 (FPKM $\geq$ 10, fold $\geq$ 2, and FDR $\leq$ 0.05) that are also associated with placental functions. (**b**) Top 10 GO biological process terms, according to GREAT, using the hypergeometric fold ($\geq$2), FDR ($\leq$0.05), and gene association ($\geq$5 associated genes) cutoffs. Genes significantly upregulated at e9.5 are related to lipid metabolism, and vasculature development. (**c**) Examples of e9.5-specific enhancers and their corresponding H3k27ac activity at e9.5 and e7.5, shown using the UCSC genome browser. Boxes correspond to regions identified as e9.5-specific. (**d**) E9.5-specific enhancers are associated with genes involved in proliferation, response to hormones, and placenta vasculature development.

Next, we analyzed e9.5-specific enhancers. We observed that e9.5-specific enhancers were located near genes with known roles in the midgestation placenta (Figure 1c). For example, *Notch1* plays a role in promoting trophoblast differentiation, and regulates angiogenesis and placental branching [17,26]. *Dlc1*, a tumor suppressing gene, also contributes to the development of placental vasculature [27]. Ontology analysis of the e9.5-specific enhancers showed that many terms, such as 'response to insulin stimulus' and 'abnormal placental vasculature', are more significantly enriched in e9.5-specific enhancers compared to e7.5-specific enhancers (Figure 1d; Supplemental Table S1).

Since both upregulated genes and enhancers specific to e9.5 were associated with placental development terms, such as vasculature and labyrinth morphology, we next investigated which TFs could be regulating these processes.

2.2. Plagl1 Is Highly Expressed in the e9.5 Placenta and the PLAGL1 Binding Motif Is Enriched in e9.5-Specific Enhancer Regions

To identify TFs that could be regulating e9.5-specific enhancers, we first determined which ones were upregulated at e9.5. Based on expression thresholds (FPKM $\geq$ 10), fold ($\geq$2), and q-value ($\leq$0.05), we identified 37 TFs upregulated at e9.5 (Figure 2a; Supplemental Figure S1a; Supplemental Table S2). We then used a phylofootprinting approach [28] to determine which of these TFs had motifs enriched in e9.5-specific enhancers. We ensured that the binding site predictions were conserved between the human and mouse genome since conserved binding sites are more likely to be functionally important [29]. Four TFs passed our motif fold (≥1.5) and *p*-value (≤0.05; Bonferonni correction) cutoffs: PLAGL1, GCM1, PPARγ, and BHLHB2 (Figure 2b). Interestingly, each of these TFs has a known role in the placenta, though some TFs are better studied. GCM1 is a well-known, important transcription factor expressed within labyrinthine trophoblast [30] that contributes to syncytiotrophoblast differentiation, chorionic branching [31], and hormone production [20]. PPARγ is also well-studied, known to play a variety of roles in the placenta including fatty acid uptake, differentiation, and vascularization [32–34]. BHLHB2 is expressed in cytotrophoblast and has been found to be upregulated in preeclampsia [35].

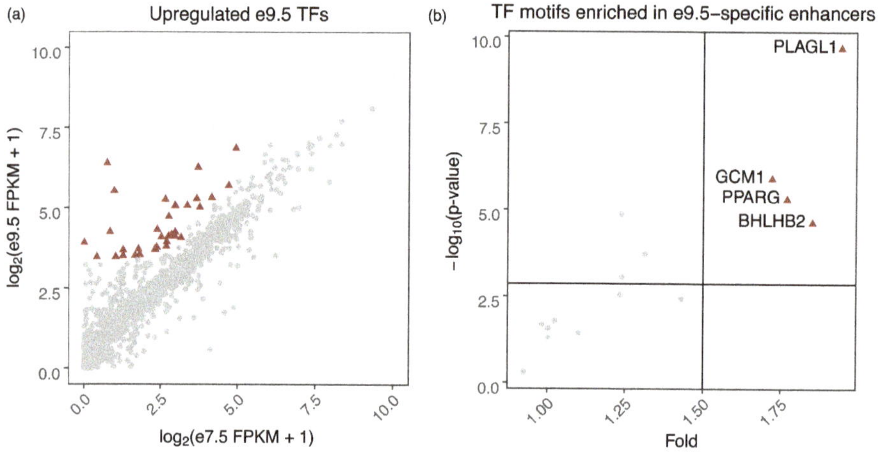

Figure 2. The PLAGL1 motif is enriched in e9.5-specific enhancers and is associated with blood vessel development genes. (**a**) Scatterplot showing the expression of all TFs at e7.5 and e9.5. Upregulated TFs at e9.5 are indicated by maroon triangles (fold ≥ 2, and *p*-value ≤ 0.05). (**b**) Of the 37 transcription factors upregulated at e9.5, four TFs (maroon triangles) pass the fold and Bonferroni corrected *p*-value thresholds (black lines).

PLAGL1 had a high expression fold change (fold change: 46.93; e7.5 FPKM: 0.99; e9.5 FPKM 45.26) as well as the highest, most significant fold change among binding sites of the four TFs (Figure 2b). To confirm expression differences, we performed qPCR using e7.5 and e9.5 mouse placenta and found *Plagl1* was indeed significantly more highly expressed at e9.5 (*p*-value = 0.015) (Supplemental Figure S1b). PLAGL1 is an imprinted zinc-finger transcription factor known to play roles in regulating glucose uptake [36], apoptosis [37], and proliferation [38] and peaks in expression in the midgestation human placenta [39]. Although several functions of PLAGL1 have been identified, its role in placenta has not been thoroughly studied. Therefore, we performed additional experiments in both the mouse placenta as well as human trophoblast cells.

2.3. *Plagl1 Is Expressed in Endothelial Cells of the Labyrinthine Layer*

To determine where within the mouse placenta *Plagl1* is expressed we performed RNAscope with an e9.5 implantation site. *Plagl1* expression was observed within the developing labyrinth; specifically in the endothelial cells forming the fetal blood vessels (Figure 3a, Supplemental Figure S2a). Positive (*PPIB*) and negative (*DapB*) control probes were used for RNAscope to ensure *Plagl1* signal was not due to noise or unspecific binding (Figure 3b–c, Supplemental Figure S2b–c). Immunohistochemistry for CD34, a marker for vascular endothelial cells [40,41], was also performed and showed a similar staining within the labyrinth (Figure 3d; Supplemental Figure S2d). *Plagl1* is also expressed throughout the allantois, where fetal vasculature begins developing, eventually forming the endothelia in the labyrinth layer of the placenta [42], suggesting a role for PLAGL1 in vasculogenesis and labyrinth layer morphology.

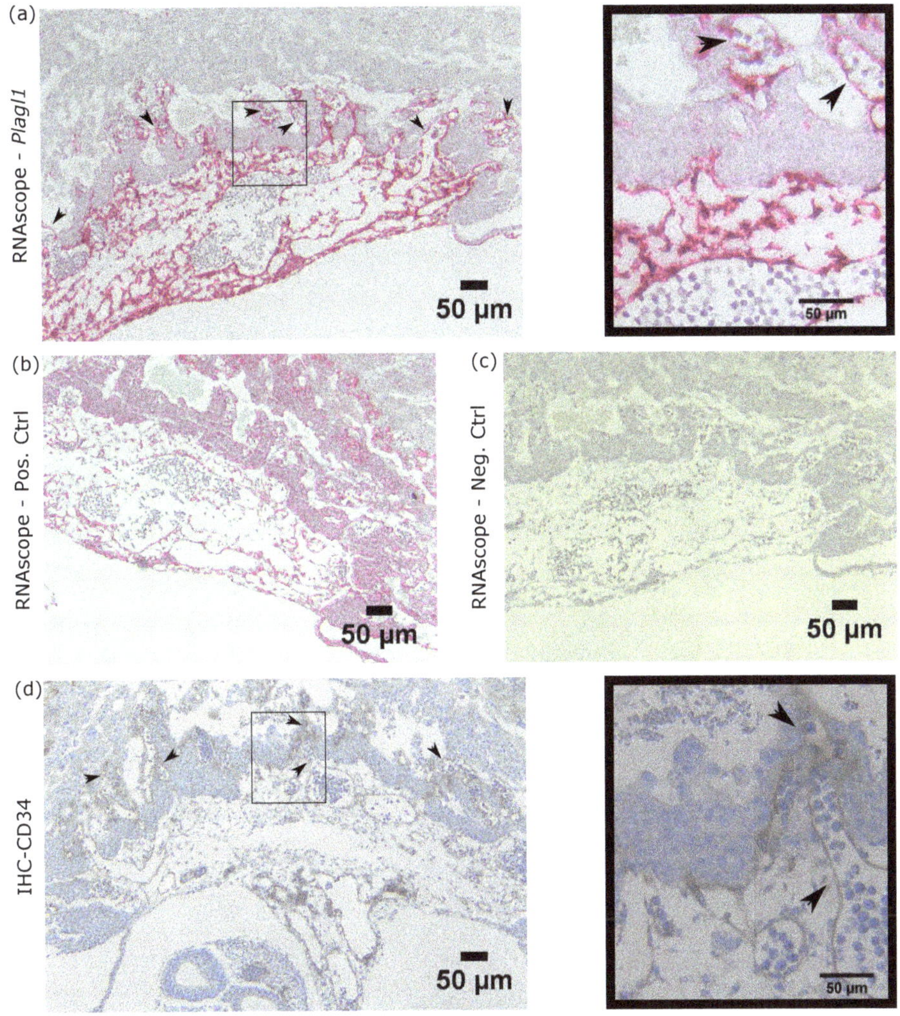

Figure 3. *Plagl1* is expressed within the developing labyrinth layer and allantois of the e9.5 mouse placenta. (**a**) RNAscope shows *Plagl1* RNA in the developing labyrinth as well as throughout the allantois (dark red) in the e9.5 placenta. Box shows zoomed in region (right). Arrowheads indicate *Plagl1* staining of endothelial cells forming fetal blood vessels. (**b**) RNAscope positive control staining *PPIB* within the placenta to determine RNA viability. (**c**) RNAscope negative control staining *DapB* within the placenta to determine nonspecific staining. (**d**) Immunohistochemistry shows CD34 staining of the vascular endothelial cells, displaying a similar pattern to *Plagl1* within the developing labyrinth. Box shows zoomed in region (right). Arrowheads indicate CD34 staining.

2.4. Plagl1 Is Associated with Blood Vessel Development and Insulin Response

The PLAGL1 motif was identified in 233 e9.5-specific enhancers that were predicted to associate with genes involved in fetal growth, placental labyrinth morphology, and insulin response (Figure 4a). These enhancers were also predicted to associate with genes involved in obesity and overnutrition

(Figure 4a). Interestingly, maternal gestational diabetes mellitus (GDM) has been associated with defective insulin signaling within the placenta [43] and increased vascularization [44], leading us to hypothesize that *Plagl1* could be misexpressed in GDM placentas.

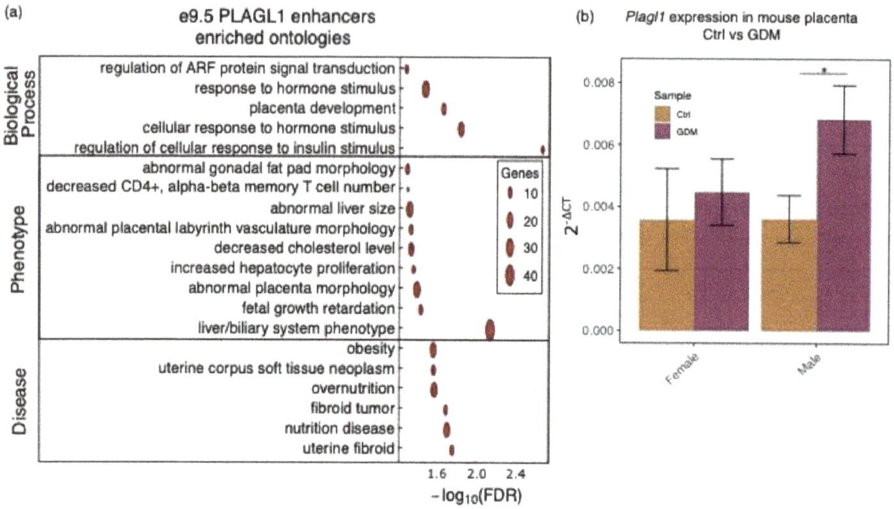

Figure 4. PLAGL1 is associated with placental morphology and *Plagl1* shows sex-specific differences in GDM mouse placentas. (**a**) The 233 e9.5-specific enhancers containing a PLAGL1 motif are associated with fetal growth, vasculature development, and other processes important in the placenta. (**b**) *Plagl1* is significantly overexpressed in the male placenta from mothers modeling GDM, but not in female placentas (p-value ≤ 0.05 (*)).

To determine if *Plagl1* expression differs between control placentas and those from GDM mothers, we measured its gene expression in placentas from a mouse model for GDM [45]. Mice were fed a high-fat, high-sucrose diet a week prior to mating and throughout pregnancy resulting in glucose intolerance during pregnancy [45]. When comparing placentas from control mice and placentas from GDM mice at e17.5, we observed no significant difference in *Plagl1* expression (p-value = 0.07). Since the placental environment has been shown to affect males and females differently [46,47], and *PLAGL1* has also been shown to have sex-specific differences in fetuses [15], we analyzed the sexes separately. First, we compared *Plagl1* gene expression in control female placentas and control male placentas and found no difference (p-value = 0.98). Then, we compared *Plagl1* expression in GDM placentas to control placentas for each sex, and found significant upregulation of *Plagl1* in the GDM placentas in males only (p-value = 0.031) (Figure 4b; Supplemental Table S3). These findings indicate that GDM affects *Plagl1* in the placenta in a sex-specific manner in a murine model of GDM.

2.5. PLAGL1 Knockdown in the HTR-8/SVneo Cell Line Predicts a Role in Blood Vessel Remodeling

We next investigated a potential role for PLAGL1 in human placenta. Using data from the human protein atlas [48] and the TissueEnrich tool [49], we found that *PLAGL1* has placenta-enriched gene expression (Figure 5a). The human protein atlas further showed that the placenta sections used for analysis were comprised primarily of trophoblast cells (Figure 5b). Therefore, we sought to investigate the role of *PLAGL1* in human trophoblast cells. First, we evaluated RNA-seq data from several human cell lines [50] for *PLAGL1* expression, including: choriocarcinomas representing villous trophoblast (BeWo [51] and JEG3 [52]); syncytiotrophoblast (PHTd_Syncytio [53]) differentiated from term placenta cytotrophoblast (PHTu_Cyto [53]); BAP treated human embryonic stem cells (ESCd [53]); and a cell line derived from chorionic villi explants of first trimester placenta (HTR-8/SVneo [54]). We proceeded with our experiments in HTR-8/SVneo cells, since they had the highest expression of *PLAGL1* (Figure 5c). When grown on Matrigel, HTR-8/SVneo cells express genes associated with blood vessel development, can form endothelial tube-like structures, and are commonly used to model endovascular differentiation of extravillous trophoblast [55–59].

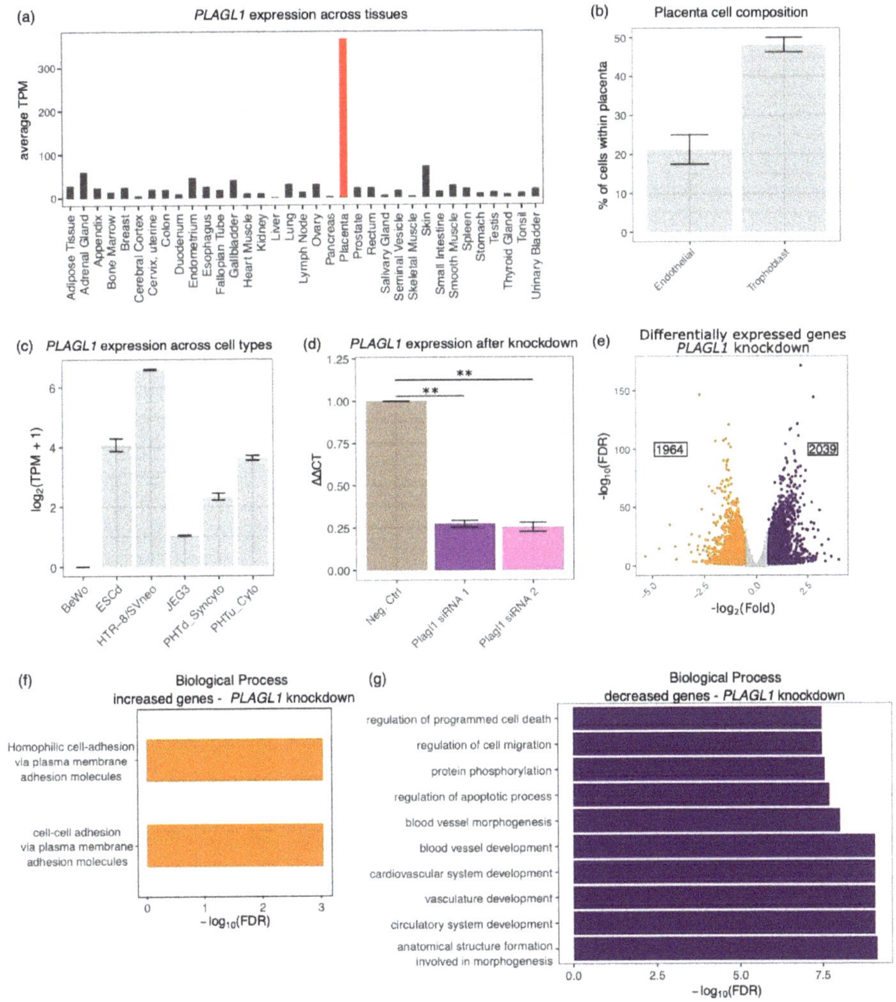

Figure 5. PLAGL1 regulates genes involved in blood vessel development. (**a**) Bar graph of PLAGL1 expression in human tissues, generated using TissueEnrich. Expression reported using transcripts per million (TPM). (**b**) Percentage of trophoblast and endothelial cells in human placental samples. Data is from the Human Protein Atlas. (**c**) Bar graph of PLAGL1 expression in multiple human cell lines. (**d**) *PLAGL1* expression is significantly reduced by two siRNAs (p-value ≤ 0.01 (**)) compared to a negative control. Values are normalized to the negative control siRNA. (**e**) 4003 genes are differentially expressed after *PLAGL1* knockdown in HTR-8/SVneo cells. Genes which increase (1964) in expression are indicated as orange dots on the volcano plot and those that decrease (2039) are purple. (**f**) Cell-adhesion terms are associated with genes that are upregulated when *PLAGL1* is knocked down. (**g**) Vasculature development terms are associated with genes that are downregulated when *PLAGL1* is knocked down.

To understand the global impacts PLAGL1 has on gene expression, we performed an siRNA knockdown of *PLAGL1* followed by RNA-seq in the HTR-8/SVneo cells. We first tested two siRNAs, and found that both siRNAs knocked down *PLAGL1* by 73–75% on average (Figure 5d). Since both siRNAs showed similar knockdown efficiencies, we proceeded with RNA-seq using siRNA 1. We identified 4003 genes that were differentially expressed (fold $\geq$ 1.5, adjusted p-value $\leq$ 0.05) between the *PLAGL1* and negative control knockdown samples using DESeq2 [60] (Figure 5e; Supplemental Table S4). The 1964 genes that were increased upon *PLAGL1* knockdown included several protocadherins such as *PCDH1*, *PCDH10*, and *PCDH7*. The only terms enriched in this group were related to cell-cell adhesion (Figure 5f). On the other hand, the 2039 genes that decreased as a result of *PLAGL1* knockdown were enriched for terms related to blood vessel development, cell migration, and both type 1 (insulin dependent) and type 2 (non-insulin dependent) diabetes (Figure 5g; Supplemental Figure S3).

To determine if *PLAGL1* could play a role in the ability of HTR-8/SVneo cells to mimic endothelial cell tube-like structure formation, we performed a tube formation assay. Ten hours after plating the cells transfected with the *PLAGL1* siRNA, we see a significant decrease in cord formation, as determined by the total branching length (p-value = 0.0093) and number of enclosed regions (meshes; p-value = 0.0024) compared to cells transfected with a negative control (Figure 6).

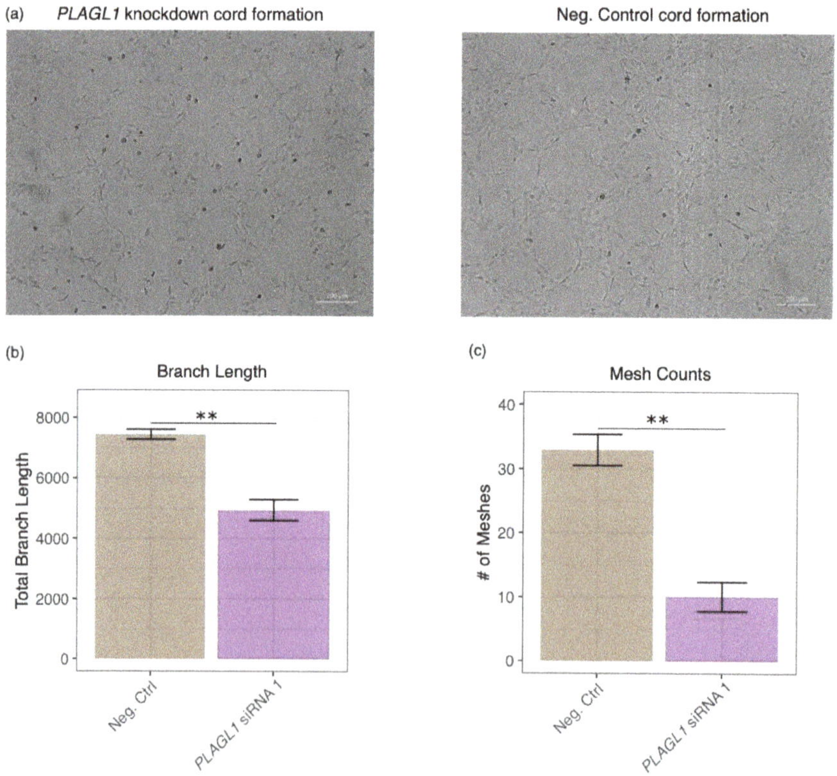

Figure 6. *PLAGL1* knockdown decreases cord formation ability in HTR-8/SVneo cells. (**a**) Representative images showing cord formation is reduced after PLAGL1 is knocked down (left) compared to a control (right). (**b**) Branch length was significantly reduced after PLAGL1 knockdown (p-value $\leq$ 0.01 (**)). (**c**) Enclosed regions, or meshes, were significantly reduced after PLAGL1 knockdown (p-value $\leq$ 0.01 (**)).

2.6. Relationship between Mouse and Human Plagl1 Results

The gene networks regulated by PLAGL1 in mouse placenta, where it is expressed in endothelial cells, and in HTR-8/SVneo human trophoblasts are at least partially conserved. For example, the GO terms enriched for genes downregulated upon *PLAGL1* knockdown in HTR-8/SVneo cells were similar to the terms associated with the genes we initially predicted to contain PLAGL1-binding motifs in their enhancer regions in the mouse genome (Figures 4a and 5g). Therefore, we tested whether the specific genes associated with terms enriched for predicted PLAGL1 enhancers decreased in expression after *PLAGL1* was knocked down in HTR-8/SVneo cells. To do this, we focused on the terms that were associated with *PLAGL1* enhancers—'abnormal placental labyrinth vasculature morphology', 'regulation of cellular response to insulin stimulus', and 'placental development'. Of the eight target genes associated with 'abnormal placental labyrinth vasculature morphology', four were downregulated upon *PLAGL1* knockdown in HTR-8/SVneo cells (p-value of overlap = 0.00188). Of the five predicted PLAGL1 target genes associated with 'regulation of cellular response to insulin stimulus', three were downregulated upon *PLAGL1* knockdown in HTR-8/SVneo cells (p-value of overlap = 0.00161). Of the 11 predicted PLAGL1 target genes associated with the more general term, 'placental development', five were downregulated upon *PLAGL1* knockdown (p-value of overlap = 0.00159) (Supplemental Figure S4a).

We further determined if the enhancers identified in mouse could drive gene activity in the HTR-8/SVneo cells. We performed a dual-glow luciferase assay using five enhancers predicted to be bound by PLAGL1 and target genes with varying functions, including blood vessel development (COL1A1 [18]), migration (DLC1 [61]), signal transduction (ARHGEF3 [62]), insulin regulation (IRS1 [63]) and fetal growth (IGF2BP1 [64]). We found that all of these regions were indeed acting as enhancers in the HTR-8/SVneo cells (relative luciferase activity ≥ 2) (Supplemental Figure S4b). To confirm PLAGL1 regulates the activity of these enhancers, we tested the activity with and without an siRNA-mediated knockdown. After *PLAGL1* knockdown, the enhancer activity significantly decreased in three out of five of these enhancers (Supplemental Figure S4c).

3. Discussions

By combining RNA-seq and ChIP-seq data from the mouse placenta, we identified *Plagl1* as an upregulated transcription factor that has its motif enriched within e9.5-specific enhancers. In mouse, gene ontology analysis showed that PLAGL1 was predicted to associate with genes involved in labyrinth layer development and fetal growth. *Plagl1* was expressed throughout the allantois and within vascular endothelial cells of the labyrinth layer. We also found that *Plagl1* has sex-specific gene expression differences between normal placentas and those from the GDM mouse model. Next, we performed an siRNA-mediated knockdown of *PLAGL1* in HTR-8/SVneo cells to determine if the gene could be important in human placental trophoblast cells. Ontology analysis of genes which decreased in expression upon *PLAGL1* knockdown showed enrichment of terms related to blood vessel remodeling. *PLAGL1* was further implicated in this role by a tube formation assay, where we observed a decrease in cord formation in *PLAGL1* knockdown cells.

In mouse, the importance of PLAGL1 in fetal growth has been established, as pups from global PLAGL1 knockout mice are smaller in size compared to wild type pups. The authors found a slight decrease in placental weight by gestational day 16.5. However, they did not report a difference in the histology of the placenta or its ability to transport glucose [65]. Our analysis identified a PLAGL1 motif in e9.5 placenta enhancers predicted to target *Pparγ* and *Soc3*. Both genes are associated with 'regulation of cellular response to insulin stimulus' and 'abnormal placental labyrinth vasculature morphology' (Supplemental Figure S4a). These genes were also found to significantly decrease in HTR-8/SVneo cells after *PLAGL1* knockdown and are a part of terms like 'blood vessel development'. Interestingly, PLAGL1 has been associated with the upregulation of each of these genes in beta cell proliferation, and insulin secretion or signaling, and both genes have been suggested as targets of study to better understand the role of PLAGL1 in transient neonatal diabetes [66]. However, such a relationship has not been established in placental vasculature although PPARγ null placentas, and SOCS3 null placentas both show defects in labyrinth formation and maternal blood sinuses [67,68]. Although we predict other targets of PLAGL1 that may be involved in blood vessel formation, future work, including PLAGL1 ChIP-seq, is necessary to confirm predicted, and identify novel, PLAGL1 binding sites. Given the association between PLAGL1 and angiogenic gene expression in the mouse placenta prior to major endovascular invasion, a role in blood vessel formation within the labyrinth layer is also possible. To test this, placenta-specific knockout of PLAGL1 would need to be generated, since *Plagl1* is highly expressed in other mouse embryonic tissues, including the liver and limb [69].

Our findings also revealed that, amongst male placentas, *Plagl1* is more highly expressed in a mouse model of GDM compared to controls. This adds to the evidence that *Plagl1* methylation and expression are sensitive to the maternal environment. *Plagl1* is hypomethylated and its expression is upregulated in mouse offspring generated by assisted reproductive technologies, and its methylation is associated with maternal folate concentrations, a Mediterranean diet, alcohol, and vitamin B2 consumption in women [70–73]. These *Plagl1* alterations are associated with birthweight and childhood obesity [70,74], and similarly, offspring of the mouse model of GDM utilized here have greater adiposity, and are more sensitive to metabolic disruption of their reproductive systems [75–77]. However, it is not clear whether placental *Plagl1* plays a functional role in these offspring outcomes, or is simply a marker for them; it is a putative metastable epiallele, meaning that environmental alterations in placental gene regulation may be maintained in offspring tissues and affect their adult functions [74]. *Plagl1* knockdown in the human trophoblast cells shows altered expression of *Oas1* and *Polr2g*, which are misexpressed in placentas from women with GDM, in the direction that would be predicted by PLAGL1 overexpression (downregulated and upregulated, respectively), suggesting a functional role for PLAGL1 in GDM-induced placental dysfunction [78]. Moreover, the findings here suggest that *Plagl1* misexpression could contribute to offspring outcomes in GDM by directly regulating placental function. For example, placental capillary density is increased, and branching decreased, in placentas from GDM pregnancies [44], consistent with the potential role of *Plagl1* in angiogenesis that we have uncovered. However, further experiments are needed to directly test whether placental angiogenesis is altered in this murine model of GDM, and whether *Plagl1* is responsible.

Notably, the effect of GDM on placental *Plagl1* expression in this model is sex-specific occurring in males only. In contrast, low birthweight is only associated with *Plagl1* methylation changes in female placentas [74]. Other genes have also been shown to be differentially expressed between male and female placentas associated with GDM, including lipolipase, which is involved in fatty acid transport and uptake [79]. Interestingly, GDM, like PLAGL1, is known to have sex-specific attributes. GDM is more common in pregnancies with a male fetus [80] and is also a risk factor for later type 2 diabetes of the fetus, with males developing it more than females [81]. GDM has also been shown to have different effects on thyroid hormone receptors within the placenta for male and female placentas [82], and affect glucose utilization differently between the sexes [83]. The mouse model of GDM used here affects the metabolism of both male and female offspring, but more dramatically impacts adipose tissue gene expression and substrate utilization in males [76].

The HTR-8/SVneo cells used for knockdown and subsequent experiments were of female origin [84]. Due to the sex-specific responses and expression of *Plagl1*, it would be advantageous to check the results in a male cell line to determine if PLAGL1 regulates the same genes in both genders. Much work remains in understanding the impact of sex on PLAGL1 functions and regulation.

Not only can PLAGL1 have sex-specific roles, but PLAGL1 is known to have different, or even opposite, functions depending on the type of cells being studied and on the co-factors that are expressed with it. For example, PLAGL1 induces activation of the PAC1 receptor promoter, unless co-transfected with ERA, which then leads it to repress promoter activity [37]. PLAGL1 expression has also been found to be increased in some tumor tissues [85,86], and decreased in others [87,88]. Since PLAGL1 can have different effects in different environments, it would be interesting to compare the effects of *PLAGL1* knockdown in specific subsets of trophoblast cells. Recently, human placental cells were cultured and differentiated into extravillous and syncytiotrophoblast cells [89], which have varying degrees of *PLAGL1* expression. Knockdown and functional assays on multiple trophoblast subtypes will give us a greater insight into the diverse functions and cell-specific roles of PLAGL1. Other cell lines that could also be more appropriate for studying aspects of blood vessel formation, and that could help better understand the results we observe in mouse, include human placental vascular endothelial cells (HPVEC) or human umbilical vein endothelial cells (HUVEC).

In HTR-8/SVneo cells, we found that genes downregulated upon *PLAGL1* knockdown were strongly enriched for blood vessel development terms. Interestingly, both HTR-8/SVneo cells and primary first trimester trophoblast, but not third trimester trophoblast, are capable of cord formation [90], indicating that this property is characteristic of trophoblast cells capable of endovascular invasion. We found that by knocking down *PLAGL1*, cord formation of HTR-8/SVneo cells was hindered. However, to determine if PLAGL1 has a role in tube formation, other methods such as co-culture assays would need to be assessed, since several non-endothelial cells can also form tubes in the presence of Matrigel [91]. Genes upregulated upon *PLAGL1* knockdown were enriched for cell adhesion terms. Previous research has found that PLAGL1 targets genes involved in cellular adhesion and extracellular matrix composition [92,93]. We also observed that several immune genes were upregulated including interferons (*IRF6*, *IRF7*) [94,95], *TGFB3* [96], and *HLA-DQB1* [97]. However, the role of PLAGL1 in immunity is unknown. Further analysis could reveal a role of PLAGL1 in regulating an immune response in the placenta.

Although we performed experiments and bioinformatics analysis in both mouse placenta and a human trophoblast cell line that represent different aspects of trophoblast-endothelial interactions in placental development, we do observe similarity of PLAGL1 targets and predicted functions, as noted above. In mouse, however, we observed *Plagl1* expression in endothelial cells, whereas in human we investigated trophoblast cells. It is possible PLAGL1 has a role in both cell types in both species, and we could not capture this with the specific timepoint we investigated in mouse, and the specific cell line we used in human. We note, however that there are also similarities in the genes identified in our study and PLAGL1 target genes identified in other studies. For example, a previous study analyzed 15 imprinted genes associated with *Plagl1* and checked if they displayed significantly different expression in the embryonic liver (e18.5) of WT mice compared to the liver in *Plagl1* knockout mice [65]. Of these 15 imprinted genes, we found that *CDKN1C*, *DCN*, *GATM*, *GRB10*, *MEG3*, and *MEST* were upregulated in the HTR-8/SVneo *PLAGL1* knockdown cells, while others, including *SLC38A4*, *IGF2*, and *IGF2R* have downregulated family members in the HTR-8/SVneo cells. We found other known imprinted targets of PLAGL1 to be disrupted in the HTR-8/SVneo *PLAGL1* knockdown cells, such as *RASGRF1* [98]. *TP73* and *WT1*, which were also misregulated upon *PLAGL1* knockdown, were previously found to be highly methylated in prostate cancer, along with *PLAGL1* [99].

Although many roles of PLAGL1 have been well studied in different tissues and cells, including its ability to regulate migration [100] in neurons and proliferation [38] in several tissues, a role in the placenta has not been thoroughly described. We provide evidence that PLAGL1 alters angiogenic gene expression and function. Proper remodeling of maternal blood vessels and formation of fetal blood vessels are critical for efficient transport of nutrients and oxygen between the maternal and fetal blood supply. Errors in this process could lead to several complications. Therefore, PLAGL1 may be an interesting target of analysis when understanding the pathogenesis of pregnancy diseases.

4. Materials and Methods

4.1. Filtering of e7.5 and e9.5 Placenta RNA-seq Data

We obtained previously analyzed and published e7.5 and e9.5 placenta RNA-seq data from the Gene Expression Omnibus (GSE65808) [13]. TFs were identified from the Animal Transcription Factor Database(v2.0) [101]. TFs were considered highly expressed and retained for analysis if they had an FPKM ≥ 10 at e9.5 and were significantly upregulated compared to e7.5 (fold ≥ 2 and q-value ≤ 0.05).

4.2. Binding Site Predictions

To identify the potential binding sites of the highly expressed TFs, we used a curated library of position weight matrices as previously described [28], only retaining those with high information content (≥10). Using the PRISM phylofootprinting method [28], we predicted binding sites for the highly expressed TFs in previously defined e9.5-specific enhancers, downloaded from the Gene Expression Omnibus (GSE65807) [13]. We filtered the predictions using the parameters as described in [102], only keeping significant (p-value ≤ 0.05) predictions with a match threshold of at least 0.8 that are conserved in the human genome (hg19).

Motifs with a similarity threshold of 0.8 were grouped, and only the motif with the highest number of occurrences was kept for further analysis. We determined the number of e7.5 and e9.5 enhancers with each motif and then calculated the fold by determining the proportion of enhancers containing a particular motif at e9.5 divided by the proportion at e7.5. Significance was determined using the hypergeometric test with Bonferroni correction for multiple comparisons.

4.3. *Plagl1* Timepoint Expression qPCR

Animal experiments for this qPCR and for RNA-Scope were approved by the Iowa State University Institutional Animal Care and Use Committee (Protocol IACUC-18–350) and conformed to the NIH Guidelines for the Care and Use of Laboratory Animals. Placenta tissue was collected from e7.5 and e9.5 timed-pregnant mice, as described previously [13], for a total of three biological replicates per timepoint. For e7.5, 12 placentas were collected per biological replicate and for e9.5, one placenta was collected per biological replicate in 500 ul PBS supplemented with 1X protease inhibitor. RNA was isolated using the Purelink RNA Micro Scale kit (Thermofisher Scientific 12183016) following the recommended protocol for purifying RNA from animal tissue. Isolated RNA concentrations were measured using the Nanodrop lite (Thermofisher Scientific ND-LITE, Waltham, MA, USA) and 200–400 ng of RNA was converted into cDNA for each biological replicate with the Applied Biosystems cDNA Reverse Transcription kit (Thermofisher Scientific 4368814). *Plagl1* expression was quantified by RT-qPCR on the CFX connect real time PCR system (BioRad 1855201) with 6ng of cDNA per biological replicate. *Polr2a* was used for normalization of *Plagl1* expression (ΔCT). Primer sequences and efficiencies can be found in Supplemental Table S5.

4.4. RNA-Scope

RNAscope was performed by the Comparative Pathology Core at Iowa State University College of Veterinary Medicine using the RNAscope 2.5 High Definition (HD)-Red Assay (Advanced Cell Diagnostics Cat #322350) according to manufacturer's instructions. Briefly, for each biological replicate, e9.5 implantation sites were dissected from timed-pregnant CD-1 mice obtained from Charles Rivers Labs. Each implantation site was embedded in paraffin and was cut at 5 μm thickness, mounted, deparaffinized and treated with hydrogen peroxide prior to target retrieval, and then treated with protease plus. Mm-Plagl1 (cat # 462941) and control probes were applied to slides for hybridization followed by six rounds of amplification. The signal was detected with Fast RED and slides were counterstained with hematoxylin. *Plagl1* signal appears as red on the slide. Control slides included positive control probe Pentidylpropyl isomerase B (Mm-PPIB cat # 313911) used to determine RNA viability in the samples, and negative control probe DapB (cat # 310043) to detect nonspecific staining.

4.5. Immunohistochemistry Staining

Paraffin embedded tissue sections, collected as described in the RNA-scope methods section, were incubated at 60 °C for 20 min. Immediately after incubation, the sections were deparaffinized with three washes in xylene for 5 min each, followed by three washes in 100% ethanol for 3 min each, one wash in 95% ethanol for 1 min and rinsing the slide under running deionized water (DI) for 3 min. Antigen retrieval was achieved by placing the slides in sodium citrate buffer (pH 6) for 40 min at

100 °C in a water bath. The slides, along with sodium citrate buffer were allowed to cool down to room temperature (RT) for 30 min and washed two times in PBST buffer (0.1% tween20 in PBS pH 7.4) for 5 min each wash. The sections were blocked with hydrogen peroxide (Fisher scientific NC0185217) for 10 min and washed two times with PBST buffer, 5 min each. The sections were incubated at RT for 1 h in blocking buffer (1% DMSO and 1% BSA in PBS buffer). After draining the blocking buffer, primary staining was performed using the CD34 antibody (Abcam ab81289, Cambridge, UK) at a 1:400 dilution in blocking buffer, and slides with primary antibody were incubated overnight at 4 °C in a humidified chamber. Following primary antibody staining, the sections were washed in six changes of PBST for 5 min each and incubated for 1 h at RT in secondary Goat anti-rabbit HRP antibody (Abcam ab6112) at a 1:750 dilution in blocking buffer. This was followed with three washes in PBST, for 5 min each. The staining was visualized using DAB for 10 min (Fisher scientific NC9276270, Hampton, NH, USA) following the vendors recommended protocol and washing away excess DAB under running DI water for 5 min. The sections were counter stained with Mayer's Hematoxylin (Fisher scientific 5031794) for 7 min at RT and incubated in bluing agent (Fisher scientific 22050114) for 2 min. The sections were washed in DI water and patted dry before mounting.

4.6. *GDM Placenta qPCR*

All animal procedures were approved by the Baylor College of Medicine institutional animal care and use committee and performed in accordance with NIH Guide for the Care and Use of Laboratory Animals. Seven week old C57BL/6J female mice were placed on either a 10% kcal/fat, 0% kcal/sucrose control diet (Ctrl) or a 45% kcal/fat, 17% kcal/sucrose diet (GDM) one week prior to and throughout pregnancy to induce GDM like symptoms as previously described [45]. All females were placed with a proven breeder male for one night and then examined for copulatory plugs in the morning. Plug positive females were considered pregnant and morning of plug positive was designated as day 0.5 of pregnancy. At embryonic day 17.5, dams were euthanized. For GDM mice, 10 placentas were dissected from 5 dams. For control mice, 11 placentas were dissected from six dams. Placenta were individually flash frozen. Flash frozen placenta were thawed overnight in RNA*later*-ICE and RNA extracted using PureLink RNA Mini Kit. Quality of RNA was checked on a 1% Agarose gel. RNA was converted to cDNA using the High-Capacity cDNA Reverse Transcription kit (ThermoFisher Scientific 4368814). Primers used for *Plagl1* and *Polr2a* (mouse) can be found in Supplemental Table S5.

Sex of the placenta was determined as previously described [103] using PCR to amplify either *Rbm31x* and *Rbm31y*. Samples were then run on a 1% agarose gel to determine sex (Supplemental Table S3).

4.7. *TissueEnrich Analysis*

PLAGL1 expression data for human tissues was obtained from the 'Tissue-specific Genes' option of the TissueEnrich [49] webtool. Expression values were based on data from the Human Protein Atlas.

4.8. *Human Protein Atlas Placenta Analysis*

To obtain the percentage of trophoblast and endothelial cells in human placenta samples used for experiments, we used the eight available samples (sample IDS: 200, 202, 203, 204, 375, 385, 398, 413) from the Human Protein Atlas [48]. Cell percentages were averaged across samples.

4.9. *siRNA knockdown and qPCR*

HTR-8/SVneo (ATCC® CRL3271™) cells were cultured in RPMI-1640 (ATCC 30-2001) with 5% FBS. Cells were seeded at 15×10^4 cells/well in a 6-well plate and grown for 48 h before transfecting. Transfections were carried out with the Lipofectamine RNAiMAX Transfection Reagent (Fisher Scientific 13778150) and with *PLAGL1* siRNAs (ThermoFisher Scientific 4392420-s10602, s10603) or a negative control (ThermoFisher Scientific 4390843). We performed a media change 24 h after transfection. RNA was extracted 48 h after transfection using the Qiagen Mini RNA kit and quality was determined using the Bioanalyzer Total RNA nano analysis kit (Agilent, Santa Clara, CA, USA) and all RNA Integrity Numbers were greater than 8. Concentrations were determined using the Nanodrop and 1000 ng of RNA from eight biological replicates was collected for RNA-seq. 400 ng of RNA was then converted to cDNA using the High-Capacity cDNA Reverse Transcription kit (ThermoFisher Scientific 4368814). *PLAGL1* knockdown was quantified by Real-Time qPCR. *GAPDH* was used for normalization and percent knockdown of *PLAGL1* was calculated using the ΔΔCT method. Primer sequences and efficiency values can be found in Supplemental Table S5.

4.10. HTR-8/SVneo RNA-seq

Libraries were prepared for all eight replicates by the Iowa State DNA facility using the NEBNext Ultra II Directional library prep kit, unique dual index plate, and poly(A) mRNA magnetic isolation module, following manufacturer's protocol. Samples were run on the HiSeq 3000, across three lanes using single-end 50 bp reads. Reads were combined across lanes, aligned to the hg19 genome using HISAT2 [104] (v2.1.0; default parameters) (Supplemental Table S6), and transcript abundance was calculated using htseq-count from the HTseq [105] package. Significantly differentially expressed genes were identified using DESeq2 [60] (1.28.1, default parameters) and defined as genes with a fold of at least 1.5 and FDR $\leq$ 0.05. All raw and processed RNA-seq data have been made available in the GEO repository, under the data accession GSE154577.

4.11. Tube Formation Assay

Tube formation assays were performed as previously described with minor modifications [106]. 150 μL of Matrigel (Fisher Scientific CB40234A) was used to coat wells of a 24-well plate and solidified at 37 °C for 1–2 h. Cells were seeded at a density of 75,000 cells/well in RPMI-1640 media, 48 h after treatment with the *PLAGL1* siRNA or a scrambled control. Images were taken after 10 h of incubation on the Matrigel using light microscopy from three random fields of each well. Cord formation ability was calculated by counting the number of meshes and the total branch length using the Angiogenesis Analyzer plugin from ImageJ.

4.12. Cloning

Primers to amplify putative PLAGL1 enhancer regions were designed using the mm9 genome and cloned into the PGL4.23 vector using the ligation independent cloning method as previously described [107] with a modification. Target enhancer regions were amplified with NEB Q5 DNA polymerase (NEB M0491S) using primers listed in Supplemental Table S7. For each enhancer target, three colonies were selected to identify positive clones by colony PCR and sequenced with Applied Biosystems 3730xl DNA Analyzer (DNA Facility, Iowa State University).

4.13. Enhancer Testing with a Dual Glow Luciferase Assay

HTR-8/SVneo cells were seeded at 50,000 cells/well and grown for 24 h on a 24-well plate before being transfected with *PLAGL1* or the negative control siRNA. 24 h after this, cells were transfected again using Jetprime reagent (VWR 89129-924) with a control plasmid or an enhancer cloned into the PGL4.23 vector. The following day the plates were read following the manufacturer's protocol for the Dual-glow Luciferase Assay kit (Fisher Scientific PRE2920). To determine the significance of the relative luciferase assay change, firefly luciferase values were normalized to renilla luciferase values and then the knockdown was compared to negative control values for the same replicate using a *t*-test. Experiments were run with four biological replicates for siRNA 1.

4.14. Ontology Analysis

Gene ontology enrichment analysis for enhancers was carried out using GREAT (v3.0.0) [25] with the GO biological process, mouse phenotype, and disease ontologies. Default parameters were used for region-gene associations, unless otherwise specified, and all analyses show top terms, ranked by FDR. Ontology enrichment for sets of genes was carried out using WebGestalt using over-representation analysis and the biological process database, unless otherwise specified. The reference set for comparison was the protein-coding genome and other settings were left as default. All analyses show top terms, ranked by FDR.

4.15. Statistical Analysis

Experiments were repeated in triplicate and results are displayed as the mean ± SE unless otherwise indicated. *p*-values were calculated using student's t-test unless otherwise specified.

Supplementary Materials: The following are available online at http://www.mdpi.com/1422-0067/21/21/8317/s1, Figure S1: Transcription factor analysis, Table S1: E9.5-specific enhancer gene ontology terms, Figure S2: RNAscope showing *Plagl1* expression in the allantois and labyrinth, Table S2: E9.5 highly expressed, upregulated transcription factors, Figure S3: Disease ontology for downregulated genes, Table S3: Samples used for *Plagl1* expression in GDM placentas, Figure S4: PLAGL1 target gene and enhancer activity, Table S4: DESeq2 output, Table S5: List of primers used for RT-qPCR, Table S6: HTR-8/SVneo HISAT2 results, Table S7: Target enhancer primers.

Author Contributions: Conceptualization, R.R.S. and G.T.; Methodology, R.R.S. and G.T.; Validation, R.R.S., R.A.A. and H.K.; Formal Analysis, R.R.S., G.T.; Investigation, R.R.S., R.A.A., H.K., K.A.P., L.C.S.; Resources, K.A.P., L.C.S.; Data Curation, R.S.S.; Writing—Original Draft Preparation, R.R.S. and G.T.; Writing—Review & Editing, R.R.S., R.A.A., H.K., K.A.P., L.C.S., and G.T.; Visualization, R.R.S., G.T.; Supervision, G.T.; Project Administration, G.T.; Funding Acquisition, G.T. All authors have read and agreed to the published version of the manuscript.

Funding: Research reported in this publication was supported by the Eunice Kennedy Shriver National Institute of Child Health & Human Development of the National Institutes of Health under Award Number R01HD096083 (to GT) and Award Number R03HD090220 (to KAP). The content is solely the responsibility of the authors and does not necessarily represent the official views of the National Institutes of Health.

Acknowledgments: We would like to acknowledge Rachel Phillips and the Comparative Pathology Core at Iowa State University College of Veterinary Medicine for help with tissue fixation and performing RNAscope experiments, the Roy J. Caver High Resolution Microscopy Facility at Iowa State University for use of the Zeiss Observer microscope, and the Iowa State University DNA Facility for providing sequencing for RNA-seq experiments. We would also like to acknowledge the Research IT group at Iowa State University (http://researchit.las.iastate.edu) for providing servers and IT support.

Conflicts of Interest: The authors declare no conflict of interest.

Abbreviations

e	Embryonic day
TF	Transcription factor
PLAGL1	pleiomorphic adenoma gene-like 1
GDM	gestational diabetes mellitus
FPKM	fragments per kilobase of transcript per million of mapped reads
TPM	transcript per million
GREAT	Genomic Regions Enrichment of Annotations Tool
Ctrl	Control diet

References

1. Guttmacher, A.E.; Maddox, Y.T.; Spong, C.Y. The Human Placenta Project: Placental structure, development, and function in real time. *Placenta* **2014**, *35*, 303–304. [CrossRef] [PubMed]
2. Gagnon, R. Placental insufficiency and its consequences. *Eur. J. Obstet. Gynecol. Reprod. Biol.* **2003**, *110*, S99–S107. [CrossRef] [PubMed]
3. PrabhuDas, M.; Bonney, E.; Caron, K.; Dey, S.; Erlebacher, A.; Fazleabas, A.; Fisher, S.; Golos, T.; Matzuk, M.; McCune, J.M.; et al. Immune mechanisms at the maternal-fetal interface: Perspectives and challenges. *Nat. Immunol.* **2015**, *16*, 328–334. [CrossRef] [PubMed]
4. Al-Enazy, S.; Ali, S.; Albekairi, N.; El-Tawil, M.; Rytting, E. Placental control of drug delivery. *Adv. Drug Deliv. Rev.* **2017**, *116*, 63–72. [CrossRef] [PubMed]

5. Hemberger, M.; Cross, J.C. Genes governing placental development. *Trends Endocrinol. Metab.* **2001**, *12*, 162–168. [CrossRef] [PubMed]
6. Winn, V.D.; Gormley, M.; Fisher, S.J. The Impact of Preeclampsia on Gene Expression at the Maternal-Fetal Interface. *Pregnancy Hypertens* **2011**, *1*, 100–108. [CrossRef] [PubMed]
7. Grigsby, P.L. Animal Models to Study Placental Development and Function throughout Normal and Dysfunctional Human Pregnancy. *Semin. Reprod. Med.* **2016**, *34*, 11–16. [CrossRef]
8. Adamson, S.L.; Lu, Y.; Whiteley, K.J.; Holmyard, D.; Hemberger, M.; Pfarrer, C.; Cross, J.C. Interactions between trophoblast cells and the maternal and fetal circulation in the mouse placenta. *Dev. Biol.* **2002**, *250*, 358–373. [CrossRef]
9. Watson, E.D.; Cross, J.C. Development of structures and transport functions in the mouse placenta. *Physiology* **2005**, *20*, 180–193. [CrossRef]
10. Kleinjan, D.A.; Van Heyningen, V. Long-range control of gene expression: Emerging mechanisms and disruption in disease. *Am. J. Hum. Genet.* **2005**, *76*, 8–32. [CrossRef]
11. Herz, H.M. Enhancer deregulation in cancer and other diseases. *Bioessays* **2016**, *38*, 1003–1015. [CrossRef]
12. Abdulghani, M.; Jain, A.; Tuteja, G. Genome-wide identification of enhancer elements in the placenta. *Placenta* **2019**, *79*, 72–77. [CrossRef]
13. Tuteja, G.; Chung, T.; Bejerano, G. Changes in the enhancer landscape during early placental development uncover a trophoblast invasion gene-enhancer network. *Placenta* **2016**, *37*, 45–55. [CrossRef]
14. Kamiya, M.; Judson, H.; Okazaki, Y.; Kusakabe, M.; Muramatsu, M.; Takada, S.; Takagi, N.; Arima, T.; Wake, N.; Kamimura, K.; et al. The cell cycle control gene ZAC/PLAGL1 is imprinted–a strong candidate gene for transient neonatal diabetes. *Hum. Mol. Genet.* **2000**, *9*, 453–460. [CrossRef]
15. Iglesias-Platas, I.; Martin-Trujillo, A.; Petazzi, P.; Guillaumet-Adkins, A.; Esteller, M.; Monk, D. Altered expression of the imprinted transcription factor PLAGL1 deregulates a network of genes in the human IUGR placenta. *Hum. Mol. Genet.* **2014**, *23*, 6275–6285. [CrossRef]
16. Ritter, M.; Buechler, C.; Boettcher, A.; Barlage, S.; Schmitz-Madry, A.; Orso, E.; Bared, S.M.; Schmiedeknecht, G.; Baehr, C.H.; Fricker, G.; et al. Cloning and characterization of a novel apolipoprotein A-I binding protein, AI-BP, secreted by cells of the kidney proximal tubules in response to HDL or ApoA-I. *Genomics* **2002**, *79*, 693–702. [CrossRef]
17. Fischer, A.; Schumacher, N.; Maier, M.; Sendtner, M.; Gessler, M. The Notch target genes Hey1 and Hey2 are required for embryonic vascular development. *Genes. Dev.* **2004**, *18*, 901–911. [CrossRef]
18. Malfait, F.; Symoens, S.; De Backer, J.; Hermanns-Le, T.; Sakalihasan, N.; Lapiere, C.M.; Coucke, P.; De Paepe, A. Three arginine to cysteine substitutions in the pro-alpha (I)-collagen chain cause Ehlers-Danlos syndrome with a propensity to arterial rupture in early adulthood. *Hum. Mutat.* **2007**, *28*, 387–395. [CrossRef]
19. Shih, Y.P.; Liao, Y.C.; Lin, Y.; Lo, S.H. DLC1 negatively regulates angiogenesis in a paracrine fashion. *Cancer Res.* **2010**, *70*, 8270–8275. [CrossRef]
20. Chiu, Y.H.; Yang, M.R.; Wang, L.J.; Chen, M.H.; Chang, G.D.; Chen, H. New insights into the regulation of placental growth factor gene expression by the transcription factors GCM1 and DLX3 in human placenta. *J. Biol. Chem.* **2018**, *293*, 9801–9811. [CrossRef]
21. Nawathe, A.R.; Christian, M.; Kim, S.H.; Johnson, M.; Savvidou, M.D.; Terzidou, V. Insulin-like growth factor axis in pregnancies affected by fetal growth disorders. *Clin. Epigenet.* **2016**, *8*, 11. [CrossRef]
22. Wang, J.; Qiu, Q.; Haider, M.; Bell, M.; Gruslin, A.; Christians, J.K. Expression of pregnancy-associated plasma protein A2 during pregnancy in human and mouse. *J. Endocrinol.* **2009**, *202*, 337–345. [CrossRef]
23. Schaubach, B.M.; Wen, H.Y.; Kellems, R.E. Regulation of murine Ada gene expression in the placenta by transcription factor RUNX1. *Placenta* **2006**, *27*, 269–277. [CrossRef] [PubMed]
24. Dupressoir, A.; Vernochet, C.; Bawa, O.; Harper, F.; Pierron, G.; Opolon, P.; Heidmann, T. Syncytin-A knockout mice demonstrate the critical role in placentation of a fusogenic, endogenous retrovirus-derived, envelope gene. *Proc. Natl. Acad. Sci. USA* **2009**, *106*, 12127–12132. [CrossRef]
25. McLean, C.Y.; Bristor, D.; Hiller, M.; Clarke, S.L.; Schaar, B.T.; Lowe, C.B.; Wenger, A.M.; Bejerano, G. GREAT improves functional interpretation of cis-regulatory regions. *Nat. Biotechnol.* **2010**, *28*, 495–501. [CrossRef]
26. Haider, S.; Pollheimer, J.; Knöfler, M. Notch signalling in placental development and gestational diseases. *Placenta* **2017**, *56*, 65–72. [CrossRef]

27. Durkin, M.E.; Avner, M.R.; Huh, C.G.; Yuan, B.Z.; Thorgeirsson, S.S.; Popescu, N.C. DLC-1, a Rho GTPase-activating protein with tumor suppressor function, is essential for embryonic development. *FEBS Lett.* **2005**, *579*, 1191–1196. [CrossRef]
28. Wenger, A.M.; Clarke, S.L.; Guturu, H.; Chen, J.; Schaar, B.T.; McLean, C.Y.; Bejerano, G. PRISM offers a comprehensive genomic approach to transcription factor function prediction. *Genome Res.* **2013**, *23*, 889–904. [CrossRef]
29. Guturu, H.; Doxey, A.C.; Wenger, A.M.; Bejerano, G. Structure-aided prediction of mammalian transcription factor complexes in conserved non-coding elements. *Philos. Trans. R. Soc. Lond. B Biol. Sci.* **2013**, *368*, 20130029. [CrossRef]
30. Basyuk, E.; Cross, J.C.; Corbin, J.; Nakayama, H.; Hunter, P.; Nait-Oumesmar, B.; Lazzarini, R.A. MurineGcm1 gene is expressed in a subset of placental trophoblast cells. *Dev. Dyn.* **1999**, *214*, 303–311. [CrossRef] [PubMed]
31. Lu, J.; Zhang, S.; Nakano, H.; Simmons, D.G.; Wang, S.; Kong, S.; Wang, Q.; Shen, L.; Tu, Z.; Wang, W.; et al. A positive feedback loop involving Gcm1 and Fzd5 directs chorionic branching morphogenesis in the placenta. *PLoS Biol.* **2013**, *11*, e1001536. [CrossRef]
32. Barak, Y.; Sadovsky, Y.; Shalom-Barak, T. PPAR Signaling in Placental Development and Function. *PPAR Res.* **2008**, *2008*, 142082. [CrossRef]
33. Nadra, K.; Quignodon, L.; Sardella, C.; Joye, E.; Mucciolo, A.; Chrast, R.; Desvergne, B. PPARgamma in placental angiogenesis. *Endocrinology* **2010**, *151*, 4969–4981. [CrossRef] [PubMed]
34. Garnier, V.; Traboulsi, W.; Salomon, A.; Brouillet, S.; Fournier, T.; Winkler, C.; Desvergne, B.; Hoffmann, P.; Zhou, Q.Y.; Congiu, C.; et al. PPARgamma controls pregnancy outcome through activation of EG-VEGF: New insights into the mechanism of placental development. *Am. J. Physiol. Endocrinol. Metab.* **2015**, *309*, E357–E369. [CrossRef]
35. Vaiman, D.; Calicchio, R.; Miralles, F. Landscape of transcriptional deregulations in the preeclamptic placenta. *PLoS ONE* **2013**, *8*, e65498. [CrossRef]
36. Czubryt, M.P.; Lamoureux, L.; Ramjiawan, A.; Abrenica, B.; Jangamreddy, J.; Swan, K. Regulation of cardiomyocyte Glut4 expression by ZAC1. *J. Biol. Chem.* **2010**, *285*, 16942–16950. [CrossRef]
37. Rodriguez-Henche, N.; Jamen, F.; Leroy, C.; Bockaert, J.; Brabet, P. Transcription of the mouse PAC1 receptor gene: Cell-specific expression and regulation by Zac1. *Biochim. Biophys. Acta* **2002**, *1576*, 157–162. [CrossRef]
38. Abdollahi, A.; Bao, R.; Hamilton, T.C. LOT1 is a growth suppressor gene down-regulated by the epidermal growth factor receptor ligands and encodes a nuclear zinc-finger protein. *Oncogene* **1999**, *18*, 6477–6487. [CrossRef]
39. Uuskula, L.; Mannik, J.; Rull, K.; Minajeva, A.; Koks, S.; Vaas, P.; Teesalu, P.; Reimand, J.; Laan, M. Mid-gestational gene expression profile in placenta and link to pregnancy complications. *PLoS ONE* **2012**, *7*, e49248. [CrossRef]
40. Arora, R.; Papaioannou, V.E. The murine allantois: A model system for the study of blood vessel formation. *Blood* **2012**, *120*, 2562–2572. [CrossRef]
41. Sasaki, T.; Mizuochi, C.; Horio, Y.; Nakao, K.; Akashi, K.; Sugiyama, D. Regulation of hematopoietic cell clusters in the placental niche through SCF/Kit signaling in embryonic mouse. *Development* **2011**, *138*, 1875. [CrossRef]
42. Rossant, J.; Cross, J.C. Placental development: Lessons from mouse mutants. *Nat. Rev. Genet.* **2001**, *2*, 538–548. [CrossRef] [PubMed]
43. Colomiere, M.; Permezel, M.; Riley, C.; Desoye, G.; Lappas, M. Defective insulin signaling in placenta from pregnancies complicated by gestational diabetes mellitus. *Eur. J. Endocrinol.* **2009**, *160*, 567–578. [CrossRef]
44. Troncoso, F.; Acurio, J.; Herlitz, K.; Aguayo, C.; Bertoglia, P.; Guzman-Gutierrez, E.; Loyola, M.; Gonzalez, M.; Rezgaoui, M.; Desoye, G.; et al. Gestational diabetes mellitus is associated with increased pro-migratory activation of vascular endothelial growth factor receptor 2 and reduced expression of vascular endothelial growth factor receptor 1. *PLoS ONE* **2017**, *12*, e0182509. [CrossRef]
45. Pennington, K.A.; van der Walt, N.; Pollock, K.E.; Talton, O.O.; Schulz, L.C. Effects of acute exposure to a high-fat, high-sucrose diet on gestational glucose tolerance and subsequent maternal health in mice. *Biol. Reprod.* **2017**, *96*, 435–445. [CrossRef]
46. Rosenfeld, C.S. Sex-Specific Placental Responses in Fetal Development. *Endocrinology* **2015**, *156*, 3422–3434. [CrossRef]

47. Alur, P. Sex Differences in Nutrition, Growth, and Metabolism in Preterm Infants. *Front. Pediatr.* **2019**, *7*, 22. [CrossRef] [PubMed]
48. Uhlen, M.; Fagerberg, L.; Hallstrom, B.M.; Lindskog, C.; Oksvold, P.; Mardinoglu, A.; Sivertsson, A.; Kampf, C.; Sjostedt, E.; Asplund, A.; et al. Proteomics. Tissue-based map of the human proteome. *Science* **2015**, *347*, 1260419. [CrossRef]
49. Jain, A.; Tuteja, G. TissueEnrich: Tissue-specific gene enrichment analysis. *Bioinformatics* **2019**, *35*, 1966–1967. [CrossRef]
50. Jain, A.; Ezashi, T.; Roberts, R.M.; Tuteja, G. Deciphering transcriptional regulation in human embryonic stem cells specified towards a trophoblast fate. *Sci. Rep.* **2017**, *7*, 17257. [CrossRef]
51. Renaud, S.J.; Chakraborty, D.; Mason, C.W.; Rumi, M.A.; Vivian, J.L.; Soares, M.J. OVO-like 1 regulates progenitor cell fate in human trophoblast development. *Proc. Natl. Acad. Sci. USA* **2015**, *112*, E6175–E6184. [CrossRef]
52. Ferreira, L.M.; Meissner, T.B.; Mikkelsen, T.S.; Mallard, W.; O'Donnell, C.W.; Tilburgs, T.; Gomes, H.A.; Camahort, R.; Sherwood, R.I.; Gifford, D.K.; et al. A distant trophoblast-specific enhancer controls HLA-G expression at the maternal-fetal interface. *Proc. Natl. Acad. Sci. USA* **2016**, *113*, 5364–5369. [CrossRef]
53. Yabe, S.; Alexenko, A.P.; Amita, M.; Yang, Y.; Schust, D.J.; Sadovsky, Y.; Ezashi, T.; Roberts, R.M. Comparison of syncytiotrophoblast generated from human embryonic stem cells and from term placentas. *Proc. Natl. Acad. Sci. USA* **2016**, *113*, E2598–E2607. [CrossRef]
54. Lee, B.; Kroener, L.L.; Xu, N.; Wang, E.T.; Banks, A.; Williams, J., 3rd; Goodarzi, M.O.; Chen, Y.I.; Tang, J.; Wang, Y.; et al. Function and Hormonal Regulation of GATA3 in Human First Trimester Placentation. *Biol. Reprod.* **2016**, *95*, 113. [CrossRef] [PubMed]
55. Johnsen, G.M.; Basak, S.; Weedon-Fekjaer, M.S.; Staff, A.C.; Duttaroy, A.K. Docosahexaenoic acid stimulates tube formation in first trimester trophoblast cells, HTR8/SVneo. *Placenta* **2011**, *32*, 626–632. [CrossRef]
56. Basak, S.; Duttaroy, A.K. Leptin induces tube formation in first-trimester extravillous trophoblast cells. *Eur. J. Obstet. Gynecol. Reprod. Biol.* **2012**, *164*, 24–29. [CrossRef]
57. Das, M.K.; Basak, S.; Ahmed, M.S.; Attramadal, H.; Duttaroy, A.K. Connective tissue growth factor induces tube formation and IL-8 production in first trimester human placental trophoblast cells. *Eur. J. Obstet. Gynecol. Reprod. Biol.* **2014**, *181*, 183–188. [CrossRef] [PubMed]
58. Li, Y.; Zhu, H.; Klausen, C.; Peng, B.; Leung, P.C. Vascular Endothelial Growth Factor-A (VEGF-A) Mediates Activin A-Induced Human Trophoblast Endothelial-Like Tube Formation. *Endocrinology* **2015**, *156*, 4257–4268. [CrossRef]
59. Highet, A.R.; Buckberry, S.; Mayne, B.T.; Khoda, S.M.; Bianco-Miotto, T.; Roberts, C.T. First trimester trophoblasts forming endothelial-like tubes in vitro emulate a 'blood vessel development' gene expression profile. *Gene Expr. Patterns* **2016**, *21*, 103–110. [CrossRef]
60. Love, M.I.; Huber, W.; Anders, S. Moderated estimation of fold change and dispersion for RNA-seq data with DESeq2. *Genome Biol.* **2014**, *15*, 550. [CrossRef]
61. Shih, Y.P.; Takada, Y.; Lo, S.H. Silencing of DLC1 upregulates PAI-1 expression and reduces migration in normal prostate cells. *Mol. Cancer Res.* **2012**, *10*, 34–39. [CrossRef]
62. D'Amato, L.; Dell'Aversana, C.; Conte, M.; Ciotta, A.; Scisciola, L.; Carissimo, A.; Nebbioso, A.; Altucci, L. ARHGEF3 controls HDACi-induced differentiation via RhoA-dependent pathways in acute myeloid leukemias. *Epigenetics* **2015**, *10*, 6–18. [CrossRef] [PubMed]
63. Ruiz-Alcaraz, A.J.; Liu, H.K.; Cuthbertson, D.J.; McManus, E.J.; Akhtar, S.; Lipina, C.; Morris, A.D.; Petrie, J.R.; Hundal, H.S.; Sutherland, C. A novel regulation of IRS1 (insulin receptor substrate-1) expression following short term insulin administration. *Biochem. J.* **2005**, *392*, 345–352. [CrossRef]
64. Hansen, T.V.; Hammer, N.A.; Nielsen, J.; Madsen, M.; Dalbaeck, C.; Wewer, U.M.; Christiansen, J.; Nielsen, F.C. Dwarfism and impaired gut development in insulin-like growth factor II mRNA-binding protein 1-deficient mice. *Mol. Cell Biol.* **2004**, *24*, 4448–4464. [CrossRef]
65. Varrault, A.; Gueydan, C.; Delalbre, A.; Bellmann, A.; Houssami, S.; Aknin, C.; Severac, D.; Chotard, L.; Kahli, M.; Le Digarcher, A.; et al. Zac1 regulates an imprinted gene network critically involved in the control of embryonic growth. *Dev. Cell* **2006**, *11*, 711–722. [CrossRef] [PubMed]
66. Hoffmann, A.; Spengler, D. Role of ZAC1 in transient neonatal diabetes mellitus and glucose metabolism. *World J. Biol. Chem.* **2015**, *6*, 95–109. [CrossRef]

67. Barak, Y.; Nelson, M.C.; Ong, E.S.; Jones, Y.Z.; Ruiz-Lozano, P.; Chien, K.R.; Koder, A.; Evans, R.M. PPARγ Is Required for Placental, Cardiac, and Adipose Tissue Development. *Mol. Cell* **1999**, *4*, 585–595. [CrossRef]
68. Roberts, A.W.; Robb, L.; Rakar, S.; Hartley, L.; Cluse, L.; Nicola, N.A.; Metcalf, D.; Hilton, D.J.; Alexander, W.S. Placental defects and embryonic lethality in mice lacking suppressor of cytokine signaling 3. *Proc. Natl. Acad. Sci. USA* **2001**, *98*, 9324–9329. [CrossRef]
69. Yue, F.; Cheng, Y.; Breschi, A.; Vierstra, J.; Wu, W.; Ryba, T.; Sandstrom, R.; Ma, Z.; Davis, C.; Pope, B.D.; et al. A comparative encyclopedia of DNA elements in the mouse genome. *Nature* **2014**, *515*, 355–364. [CrossRef]
70. Azzi, S.; Sas, T.C.; Koudou, Y.; Le Bouc, Y.; Souberbielle, J.C.; Dargent-Molina, P.; Netchine, I.; Charles, M.A. Degree of methylation of ZAC1 (PLAGL1) is associated with prenatal and post-natal growth in healthy infants of the EDEN mother child cohort. *Epigenetics* **2014**, *9*, 338–345. [CrossRef] [PubMed]
71. Hoyo, C.; Daltveit, A.K.; Iversen, E.; Benjamin-Neelon, S.E.; Fuemmeler, B.; Schildkraut, J.; Murtha, A.P.; Overcash, F.; Vidal, A.C.; Wang, F.; et al. Erythrocyte folate concentrations, CpG methylation at genomically imprinted domains, and birth weight in a multiethnic newborn cohort. *Epigenetics* **2014**, *9*, 1120–1130. [CrossRef] [PubMed]
72. House, J.S.; Mendez, M.; Maguire, R.L.; Gonzalez-Nahm, S.; Huang, Z.; Daniels, J.; Murphy, S.K.; Fuemmeler, B.F.; Wright, F.A.; Hoyo, C. Periconceptional Maternal Mediterranean Diet Is Associated With Favorable Offspring Behaviors and Altered CpG Methylation of Imprinted Genes. *Front. Cell Dev. Biol.* **2018**, *6*, 107. [CrossRef]
73. Yan, Z.; Li, Q.; Zhang, L.; Kang, B.; Fan, W.; Deng, T.; Zhu, J.; Wang, Y. The growth and development conditions in mouse offspring derived from ovarian tissue cryopreservation and orthotopic transplantation. *J. Assist. Reprod. Genet.* **2020**, *37*, 923–932. [CrossRef]
74. Clark, J.; Martin, E.; Bulka, C.M.; Smeester, L.; Santos, H.P.; O'Shea, T.M.; Fry, R.C. Associations between placental CpG methylation of metastable epialleles and childhood body mass index across ages one, two and ten in the Extremely Low Gestational Age Newborns (ELGAN) cohort. *Epigenetics* **2019**, *14*, 1102–1111. [CrossRef]
75. Mao, J.; Pennington, K.A.; Talton, O.O.; Schulz, L.C.; Sutovsky, M.; Lin, Y.; Sutovsky, P. In Utero and Postnatal Exposure to High Fat, High Sucrose Diet Suppressed Testis Apoptosis and Reduced Sperm Count. *Sci. Rep.* **2018**, *8*, 7622. [CrossRef]
76. Talton, O.O.; Bates, K.; Salazar, S.R.; Ji, T.; Schulz, L.C. Lean maternal hyperglycemia alters offspring lipid metabolism and susceptibility to diet-induced obesity in micedagger. *Biol. Reprod.* **2019**, *100*, 1356–1369. [CrossRef] [PubMed]
77. Clark, K.L.; Talton, O.O.; Ganesan, S.; Schulz, L.C.; Keating, A.F. Developmental origins of ovarian disorder: Impact of maternal lean gestational diabetes on the offspring ovarian proteome in micedagger. *Biol. Reprod.* **2019**, *101*, 771–781. [CrossRef]
78. Zhang, Y.; Zhang, T.; Chen, Y. Comprehensive Analysis of Gene Expression Profiles and DNA Methylome reveals Oas1, Ppie, Polr2g as Pathogenic Target Genes of Gestational Diabetes Mellitus. *Sci. Rep.* **2018**, *8*, 16244. [CrossRef]
79. Yang, H.; He, B.; Yallampalli, C.; Gao, H. Fetal macrosomia in a Hispanic/Latinx predominant cohort and altered expressions of genes related to placental lipid transport and metabolism. *Int. J. Obes. (Lond.)* **2020**, *44*, 1743–1752. [CrossRef]
80. Retnakaran, R.; Kramer, C.K.; Ye, C.; Kew, S.; Hanley, A.J.; Connelly, P.W.; Sermer, M.; Zinman, B. Fetal sex and maternal risk of gestational diabetes mellitus: The impact of having a boy. *Diabetes Care* **2015**, *38*, 844–851. [CrossRef]
81. Le Moullec, N.; Fianu, A.; Maillard, O.; Chazelle, E.; Naty, N.; Schneebeli, C.; Gerardin, P.; Huiart, L.; Charles, M.A.; Favier, F. Sexual dimorphism in the association between gestational diabetes mellitus and overweight in offspring at 5–7 years: The OBEGEST cohort study. *PLoS ONE* **2018**, *13*, e0195531. [CrossRef]
82. Knabl, J.; de Maiziere, L.; Huttenbrenner, R.; Hutter, S.; Juckstock, J.; Mahner, S.; Kainer, F.; Desoye, G.; Jeschke, U. Cell Type- and Sex-Specific Dysregulation of Thyroid Hormone Receptors in Placentas in Gestational Diabetes Mellitus. *Int. J. Mol. Sci.* **2020**, *21*, 4056. [CrossRef]
83. Wang, Y.; Bucher, M.; Myatt, L. Use of Glucose, Glutamine and Fatty Acids for Trophoblast Respiration in Lean, Obese and Gestational Diabetic Women. *J Clin Endocrinol Metab* **2019**. [CrossRef]

84. Eksteen, M.; Heide, G.; Tiller, H.; Zhou, Y.; Nedberg, N.H.; Martinez-Zubiaurre, I.; Husebekk, A.; Skogen, B.R.; Stuge, T.B.; Kjaer, M. Anti-human platelet antigen (HPA)-1a antibodies may affect trophoblast functions crucial for placental development: A laboratory study using an in vitro model. *Reprod. Biol. Endocrinol.* **2017**, *15*, 28. [CrossRef]
85. Hide, T.; Takezaki, T.; Nakatani, Y.; Nakamura, H.; Kuratsu, J.; Kondo, T. Sox11 prevents tumorigenesis of glioma-initiating cells by inducing neuronal differentiation. *Cancer Res.* **2009**, *69*, 7953–7959. [CrossRef] [PubMed]
86. Godlewski, J.; Krazinski, B.E.; Kowalczyk, A.E.; Kiewisz, J.; Kiezun, J.; Kwiatkowski, P.; Sliwinska-Jewsiewicka, A.; Maslowski, Z.; Kmiec, Z. PLAGL1 (ZAC1/LOT1) Expression in Clear Cell Renal Cell Carcinoma: Correlations with Disease Progression and Unfavorable Prognosis. *Anticancer Res.* **2016**, *36*, 617–624. [PubMed]
87. Cvetkovic, D.; Pisarcik, D.; Lee, C.; Hamilton, T.C.; Abdollahi, A. Altered expression and loss of heterozygosity of the LOT1 gene in ovarian cancer. *Gynecol. Oncol.* **2004**, *95*, 449–455. [CrossRef]
88. Hacker, E.; Muller, K.; Whiteman, D.C.; Pavey, S.; Hayward, N.; Walker, G. Reduced expression of IL-18 is a marker of ultraviolet radiation-induced melanomas. *Int. J. Cancer* **2008**, *123*, 227–231. [CrossRef]
89. Okae, H.; Toh, H.; Sato, T.; Hiura, H.; Takahashi, S.; Shirane, K.; Kabayama, Y.; Suyama, M.; Sasaki, H.; Arima, T. Derivation of Human Trophoblast Stem Cells. *Cell Stem. Cell* **2018**, *22*, 50–63.e56. [CrossRef]
90. Kalkunte, S.; Lai, Z.; Tewari, N.; Chichester, C.; Romero, R.; Padbury, J.; Sharma, S. In vitro and in vivo evidence for lack of endovascular remodeling by third trimester trophoblasts. *Placenta* **2008**, *29*, 871–878. [CrossRef]
91. Donovan, D.; Brown, N.J.; Bishop, E.T.; Lewis, C.E. Comparison of three in vitro human 'angiogenesis' assays with capillaries formed in vivo. *Angiogenesis* **2001**, *4*, 113–121. [CrossRef] [PubMed]
92. Al Adhami, H.; Evano, B.; Le Digarcher, A.; Gueydan, C.; Dubois, E.; Parrinello, H.; Dantec, C.; Bouschet, T.; Varrault, A.; Journot, L. A systems-level approach to parental genomic imprinting: The imprinted gene network includes extracellular matrix genes and regulates cell cycle exit and differentiation. *Genome Res.* **2015**, *25*, 353–367. [CrossRef] [PubMed]
93. Varrault, A.; Dantec, C.; Le Digarcher, A.; Chotard, L.; Bilanges, B.; Parrinello, H.; Dubois, E.; Rialle, S.; Severac, D.; Bouschet, T.; et al. Identification of Plagl1/Zac1 binding sites and target genes establishes its role in the regulation of extracellular matrix genes and the imprinted gene network. *Nucleic Acids Res.* **2017**, *45*, 10466–10480. [CrossRef] [PubMed]
94. Savitsky, D.; Tamura, T.; Yanai, H.; Taniguchi, T. Regulation of immunity and oncogenesis by the IRF transcription factor family. *Cancer Immunol. Immunother.* **2010**, *59*, 489–510. [CrossRef]
95. Joly, S.; Rhea, L.; Volk, P.; Moreland, J.G.; Dunnwald, M. Interferon Regulatory Factor 6 Has a Protective Role in the Host Response to Endotoxic Shock. *PLoS ONE* **2016**, *11*, e0152385. [CrossRef]
96. Komai, T.; Okamura, T.; Inoue, M.; Yamamoto, K.; Fujio, K. Reevaluation of Pluripotent Cytokine TGF-β3 in Immunity. *Int. J. Mol. Sci.* **2018**, *19*, 2261. [CrossRef]
97. Fitzmaurice, K.; Hurst, J.; Dring, M.; Rauch, A.; McLaren, P.J.; Gunthard, H.F.; Gardiner, C.; Klenerman, P.; Irish, H.C.V.R.C.; Swiss, H.I.V.C.S. Additive effects of HLA alleles and innate immune genes determine viral outcome in HCV infection. *Gut* **2015**, *64*, 813–819. [CrossRef]
98. Daniel, G.; Schmidt-Edelkraut, U.; Spengler, D.; Hoffmann, A. Imprinted Zac1 in neural stem cells. *World J. Stem. Cells* **2015**, *7*, 300–314. [CrossRef]
99. Jacobs, D.I.; Mao, Y.; Fu, A.; Kelly, W.K.; Zhu, Y. Dysregulated methylation at imprinted genes in prostate tumor tissue detected by methylation microarray. *BMC Urol.* **2013**, *13*, 37. [CrossRef]
100. Adnani, L.; Langevin, L.M.; Gautier, E.; Dixit, R.; Parsons, K.; Li, S.; Kaushik, G.; Wilkinson, G.; Wilson, R.; Childs, S.; et al. Zac1 Regulates the Differentiation and Migration of Neocortical Neurons via Pac1. *J. Neurosci.* **2015**, *35*, 13430–13447. [CrossRef]
101. Zhang, H.M.; Chen, H.; Liu, W.; Liu, H.; Gong, J.; Wang, H.; Guo, A.Y. AnimalTFDB: A comprehensive animal transcription factor database. *Nucleic Acids Res.* **2012**, *40*, D144–D149. [CrossRef]
102. Starks, R.R.; Biswas, A.; Jain, A.; Tuteja, G. Combined analysis of dissimilar promoter accessibility and gene expression profiles identifies tissue-specific genes and actively repressed networks. *Epigenet. Chromatin* **2019**, *12*, 16. [CrossRef]
103. Tunster, S.J. Genetic sex determination of mice by simplex PCR. *Biol. Sex Differ.* **2017**, *8*, 31. [CrossRef]

104. Kim, D.; Langmead, B.; Salzberg, S.L. HISAT: A fast spliced aligner with low memory requirements. *Nat. Methods* **2015**, *12*, 357–360. [CrossRef]
105. Anders, S.; Pyl, P.T.; Huber, W. HTSeq—A Python framework to work with high-throughput sequencing data. *Bioinformatics* **2015**, *31*, 166–169. [CrossRef]
106. Lei, D.; Deng, N.; Wang, S.; Huang, J.; Fan, C. Upregulated ARRDC3 limits trophoblast cell invasion and tube formation and is associated with preeclampsia. *Placenta* **2020**, *89*, 10–19. [CrossRef]
107. Tuteja, G.; Moreira, K.B.; Chung, T.; Chen, J.; Wenger, A.M.; Bejerano, G. Automated discovery of tissue-targeting enhancers and transcription factors from binding motif and gene function data. *PLoS Comput. Biol.* **2014**, *10*, e1003449. [CrossRef]

International Journal of
Molecular Sciences

Article

Identifying Stabilin-1 and Stabilin-2 Double Knockouts in Reproduction and Placentation: A Descriptive Study

Soon-Young Kim [1,2], Eun-Hye Lee [1], Eun Na Kim [3], Woo-Chan Son [3], Yeo Hyang Kim [4,5], Seung-Yoon Park [6], In-San Kim [7,8] and Jung-Eun Kim [1,2,*]

1 Department of Molecular Medicine, Cell and Matrix Research Institute, School of Medicine, Kyungpook National University, Daegu 41944, Korea; ksygood741@naver.com (S.-Y.K.); coramdeoeh@nate.com (E.-H.L.)
2 BK21 Plus KNU Biomedical Convergence Program, Department of Biomedical Science, Kyungpook National University, Daegu 41944, Korea
3 Department of Pathology, Asan Medical Center, University of Ulsan College of Medicine, Seoul 05505, Korea; hk1997@naver.com (E.N.K.); wcson@amc.seoul.kr (W.-C.S.)
4 Department of Pediatrics, School of Medicine, Kyungpook National University, Daegu 41944, Korea; kimyhmd@knu.ac.kr
5 Division of Pediatric Cardiology, Kyungpook National University Children's Hospital, Daegu 41404, Korea
6 Department of Biochemistry, School of Medicine, Dongguk University, Gyeongju 38066, Korea; psyoon@dongguk.ac.kr
7 Center for Theragnosis, Biomedical Research Institute, Korea Institute of Science and Technology, Seoul 02792, Korea; iskim14@kist.re.kr
8 KU-KIST Graduate School of Converging Science and Technology, Korea University, Seoul 02841, Korea
* Correspondence: kjeun@knu.ac.kr; Tel.: +82-53-420-4949

Received: 20 August 2020; Accepted: 29 September 2020; Published: 30 September 2020

Abstract: The placenta undergoes reconstruction at different times during fetal development to supply oxygen and nutrients required throughout pregnancy. To accommodate the rapid growth of the fetus, small spiral arteries undergo remodeling in the placenta. This remodeling includes apoptosis of endothelial cells that line spiral arteries, which are replaced by trophoblasts of fetal origin. Removal of dead cells is critical during this process. Stabilin-1 (Stab1) and stabilin-2 (Stab2) are important receptors expressed on scavenger cells that absorb and degrade apoptotic cells, and Stab1 is expressed in specific cells of the placenta. However, the role of Stab1 and Stab2 in placental development and maintenance remain unclear. In this study, we assessed Stab1 and Stab2 expression in the placenta and examined the reproductive capacity and placental development using a double-knockout mouse strain lacking both Stab1 and Stab2 (Stab1/2 dKO mice). Most pregnant Stab1/2 dKO female mice did not produce offspring and exhibited placental defects, including decidual hemorrhage and necrosis. Findings of this study offer the first description of the phenotypic characteristics of placentas and embryos of Stab1/2 dKO females during pregnancy, suggesting that Stab1 and Stab2 are involved in placental development and maintenance.

Keywords: Stabilin-1; Stabilin-2; double knockout; placenta; decidua; hemorrhage

1. Introduction

Stabilin-1 (Stab1), belongs to a family of scavenger receptors, is a type 2 macrophage marker and is expressed in tissue macrophages and different endothelial cell subtypes [1–3]. Stabilin-2 (Stab2) belongs to the same family of scavenger receptors and shares 55% identity with Stab1 at the protein level [1,2,4,5]. Both Stab1 and Stab2 are expressed in macrophages and endothelial cells of various organs, including the liver and spleen [6,7]. Stab1 acts as a scavenger receptor for clearing debris and apoptotic cells through macrophage phagocytosis and is involved in the transcytosis of several components, including placental lactogen (PL) [1,3,8,9]. Stab1 modulates angiogenesis, supports leukocyte adhesion and transmigration, and regulates immune responses [1,3,9]. Stab2 is involved in clearing noxious blood factors and degrading apoptotic cells [10]. Stab2 also regulates extracellular matrix turnover and hyaluronan resorption to maintain body fluids, and participates in the defense against bacterial infections and recruitment of lymphocytes [6,7].

During pregnancy, the placenta undergoes structural changes to transfer maternal nutrients, oxygen, metabolites, and hormones to the fetus at different stages of development. These alterations are essential for the development and growth of the fetus [9,11,12]. One such change that occurs in early pregnancy is that the diameter of maternal blood vessels increases to supply a massive amount of maternal blood through the placenta at a low blood pressure. This phenomenon is called spiral artery remodeling, which is an essential change in the development of the placenta. During spiral artery remodeling, fetal origin trophoblasts invade maternal spiral arteries and replace endothelial cells undergoing apoptosis [12,13]. Apoptotic cells and debris are eliminated by placental macrophages and uterine natural killer (uNK) cells [12,13]. Rapid removal of apoptotic cells and debris is important for spiral artery remodeling. Inhibition of such removal hinders spiral blood vessel formation [12,13].

Stab1 and Stab2 proteins are expressed in endothelial cells and macrophages, which play a predominant role in artery remodeling in the placenta during pregnancy [12]. These proteins are closely related to factors mediating cell signaling pathways that regulate angiogenesis [4,9,14]. Functional studies of Stab1 and Stab2 in tissue macrophages have demonstrated their importance in the phagocytosis of apoptotic cells, suggesting that Stab1 and Stab2 contribute to placental reconstruction during pregnancy [1–3,10]. Interestingly, several studies have reported that Stab1 is expressed in placental endothelial cells and macrophages during pregnancy, and that it might be involved in cell adhesion and molecular scavenging in placental macrophages [1,3,9]. Previous studies have also suggested that Stab1 regulates differences in the concentration of mouse PL in maternal and fetal blood circulation [1,9]. However, the role of Stab1 in the placenta remains poorly understood and, to the best of our knowledge, the role of Stab2 in the placenta has not been reported yet.

In this study, we analyzed the mRNA expression levels of Stab1 and Stab2 in the placenta and compared them with those in the liver. In addition, we generated a double-knockout mouse strain lacking both Stab1 and Stab2 (Stab1/2 dKO mice) and observed defective reproduction in pregnant Stab1/2 dKO female mice. Furthermore, to determine the causes underlying these reproductive defects, we conducted a descriptive study to phenotypically characterize reproduction and placentation in pregnant Stab1/2 dKO female mice. This study provides the first description of Stab1/2 dKO affecting reproductive capacity and placental development, suggesting that Stab1 and Stab2 play an important role in placental reconstruction and remodeling during pregnancy.

2. Results

2.1. Stab1 and Stab2 mRNA Expression in Mouse Placenta

To determine whether Stab1 and Stab2 are expressed in the mouse placenta, mRNA expression levels of Stab1 and Stab2 in mouse placentas were measured using quantitative polymerase chain reaction (qPCR) and reverse transcription-PCR (RT-PCR) and compared with their expression levels in the liver as a positive control. Stab1 expression was 1.4-fold higher in the placenta than in the liver, whereas Stab2 expression in the placenta was half of the expression level in the liver, as determined using qPCR (Figure 1A) and confirmed using RT-PCR (Figure 1B).

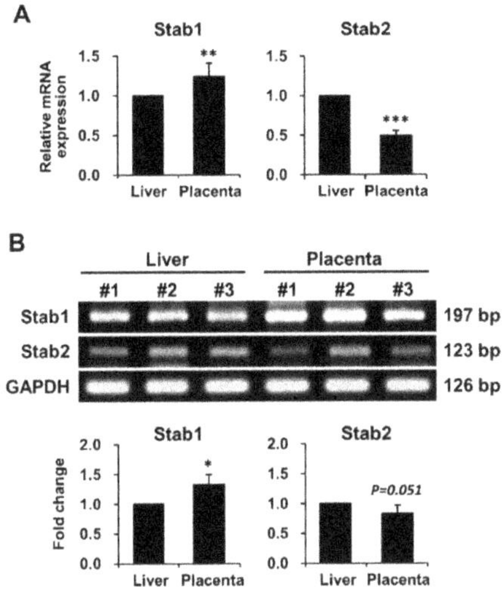

Figure 1. mRNA expression levels of Stab1 and Stab2 in WT female mouse placentas. The mRNA expression levels of Stab1 and Stab2 in mouse placentas were measured using qPCR (**A**) and RT-PCR (**B**) and were compared with those in the liver used as a positive control. The relative expression levels shown in (**A**) were plotted against those of the respective genes in the liver, which were set to 1.0. Intensities of individual bands obtained via RT-PCR shown in (**B**) were determined using ImageJ. Data were normalized to Gapdh expression and calculated as fold change relative to the expression level in the liver, which was set to 1.0. Numbers in (**B**) indicate the representative mice used in the experiment. *, $p < 0.05$; **, $p < 0.01$; ***, $p < 0.001$ versus the positive control; n = 3.

2.2. Viability of Stab1/2 dKO Female Mice during Pregnancy

After intercrossing male and female Stab1/2 dKO mice to generate Stab1/2 dKO embryos (Figure 2), some pregnant Stab1/2 dKO females died during pregnancy without any physical sign of wounds, whereas other pregnant females abruptly lost their visible pregnancies before delivery. Mating between Stab1/2 dKO mice was performed at 7–8 weeks of age. The females were separated from the males into a different cage at week 2 after mating, and their pregnancy was confirmed by visual observation (swollen belly) every day. A total of seven Stab1/2 dKO females were used in the study and each of them showed 1–3 times visually identifiable pregnancies. Of these, 78% of visually identifiable pregnancies were lost without delivery, including loss due to maternal death. During pregnancy, some females exhibited abnormal behaviors and physical features, including reduced belly size, lethargic movements and responses to external stimuli, and premature delivery. To understand why pregnant Stab1/2 dKO female mice spontaneously died or lost their progeny, mice were sacrificed and dissected immediately when they exhibited these abnormal signs. Upon dissection of the skin and peritoneum, uteri of Stab1/2 dKO female mice were collected at embryonic day 13.5 (E13.5) and E17.5. Some portions of the uterine horns at E17.5 were filled with blood (Figure 3A). This phenomenon was also observed in embryos of uteri of Stab1/2 dKO pregnant mice at E13.5 (Figure S1). Moreover, the uteri of Stab1/2 dKO female had unevenly sized embryos or undefined tissue masses (Figures S1 and S2). Some pregnant Stab1/2 dKO female mice showed embryo resorptions; however, it was difficult to determine the number of embryos that were present before embryo resorption into the uterus. Instead, the number of viable implantation sites in the uteri of WT and Stab1/2 dKO female mice at E17.5 was determined (Figure S3). Chorioamniotic membranes surrounded undefined lumps, and bleeding was also observed. After dissecting these chorioamniotic membranes in pregnant mice with embryos, brown blood spreading and embryos were visible (Figure 3B–D). The shape of each placenta was maintained, but surfaces of some placentas were surrounded by decomposed membranes. The lower abdomen of each embryo was connected to the placenta through an umbilical cord that was degraded in some embryos (Figure 3C). Three of seven Stab1/2 dKO females used in this study died during the first delivery, indicating a maternal mortality rate of 42.86% at first delivery in Stab1/2 dKO female mice (Figure 3E). Stab1/2 dKO male mice had no abnormalities before or after mating and no death until the age of 50 weeks (n = 6, data not shown).

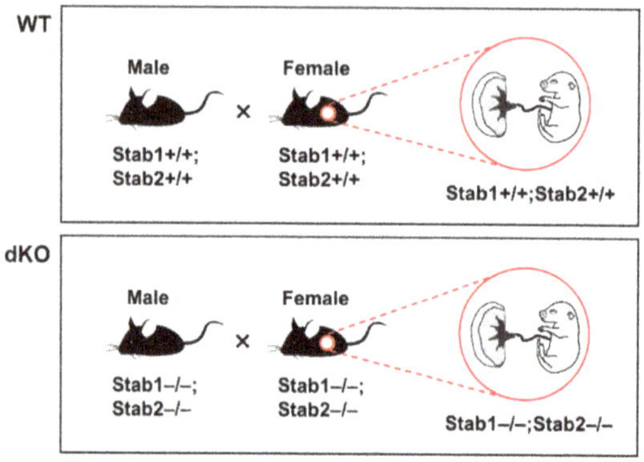

Figure 2. Pedigree and genotypes for obtaining embryos and placentas from pregnant females for the experiments.

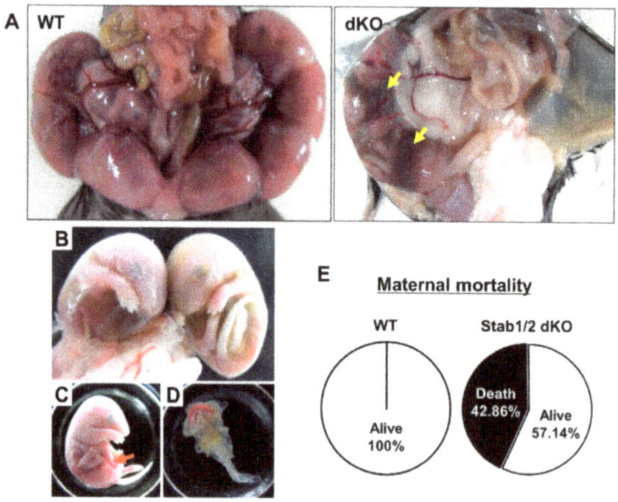

Figure 3. Embryos from pregnant Stab1/2 dKO mice at E17.5. (**A**) Uterine horns with embryos in WT and Stab1/2 dKO mice. Yellow arrows in Stab1/2 dKO indicate blood within the uterine horns. (**B**) Embryos isolated from the uterine horns of a Stab1/2 dKO female mouse. (**C**) Embryo and (**D**) placenta from a Stab1/2 dKO female mouse. The fetus only maintained the outer shape in the lower body and the fetal interior was degraded near the umbilical cord. Red arrow in (**C**) denotes abnormal abdomen in an embryo. (**E**) Maternal mortality of Stab1/2 dKO mice at the first delivery. n = 7.

It was difficult to obtain a large number of Stab1/2 dKO mice required for this study because of spontaneous deaths of pregnant Stab1/2 dKO female mice or loss of their progeny. To determine the cause of these phenomena, male mice with genotypes of Stab1 heterozygous and Stab2 knockout were intercrossed with female mice carrying the same genotypes. Female mice bred well, and the total number of offspring per litter was 4–7 (data not shown). However, the number of offsprings with the Stab1/2 dKO genotype was extremely low compared with the number of pups with other genotypes (Figure S4). These results indicate that both Stab1 and Stab2 are important in the placenta and in the embryonic development during pregnancy as well as in the pregnant female mouse itself.

2.3. Phenotypic and Histological Defects of Placentas from Stab1/2 dKO Females

To characterize the phenotypic defects observed, the placentas and embryos from pregnant WT and Stab1/2 dKO mice were harvested at E17.5 (Figure 4A). Observation of E17.5 fetuses in the WT and Stab1/2 dKO uteri revealed that some Stab1/2 dKO fetuses appeared normal, while others were difficult to observe as they were either degraded or resorbed. Interestingly, some Stab1/2 dKO fetuses were normal in shape, although their placentas and other extraembryonic structures appeared defective with brownish discoloration (Figure 4B).

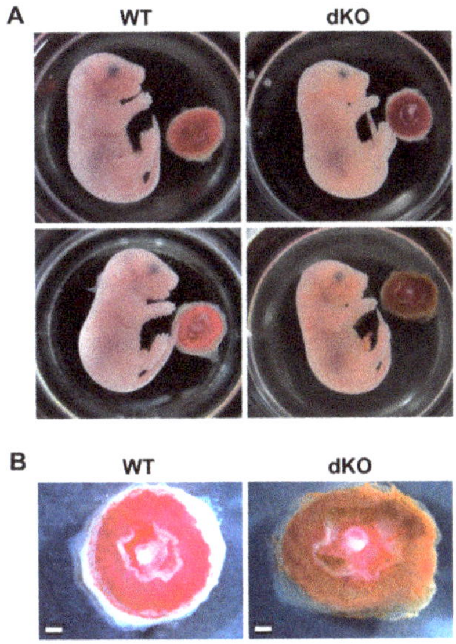

Figure 4. Embryo and placenta from pregnant WT and Stab1/2 dKO mice. (**A**) WT and Stab1/2 dKO embryos at E17.5 and associated placentas. (**B**) High-magnification images of the placentas derived from WT and Stab1/2 dKO mice. Scale bar, 1 mm; n = 3.

Histological analysis of placental tissues was performed to investigate external defects of placentas in Stab1/2 dKO mice. In both WT and Stab1/2 dKO mice, placentas had three layers: decidua, junctional zone, and labyrinth zone (Figure 5A). The length of each placental zone was quantified over the total length of the placenta expressed as a percentage. In Stab1/2 DKO placenta, the decidual length was increased relative to the total length of the placenta (Figure 5A). Glycogen trophoblast cells accumulate glycogen in the junctional zone of mouse placentas until E12.5, invade interstitially into the decidua until E18.5, and finally decrease in number [11,15]. A higher number of glycogen-containing clear cells was observed in the junctional zones of placentas obtained from Stab1/2 dKO mice than in those from WT mice based on hematoxylin and eosin (H&E) staining (Figure 5B). Periodic acid-Schiff (PAS) staining confirmed a higher number of cells containing glycogen in the junctional zones of placentas derived from Stab1/2 dKO mice than those from WT mice (Figure 6A,B). PAS-positive stained area per total tissue area was analyzed using ImageJ. Glycogen cells were significantly increased in the placenta of Stab1/2 dKO mice compared with that of WT mice (Figure 6C). Decidual hemorrhage, decidual necrosis, and fibrin deposition were also observed in the placental decidua of Stab1/2 dKO mice (Figures 5A and 7). Deciduae of placentas derived from Stab1/2 dKO mice were thicker than those of WT mice due to decidual and retroplacental hemorrhage.

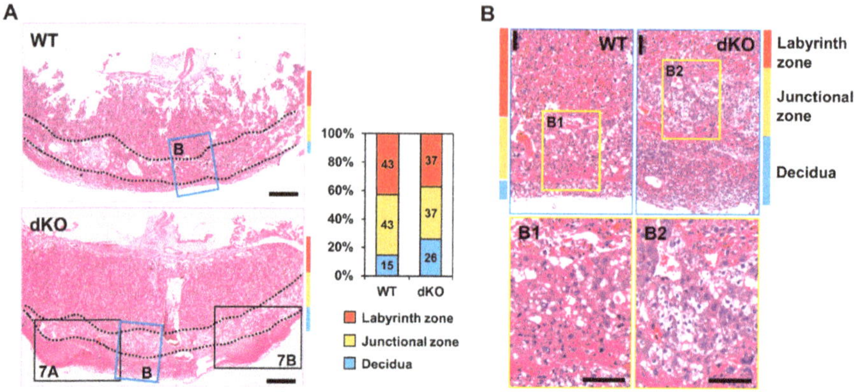

Figure 5. Histological analysis of WT and Stab1/2 dKO placentas. (**A**) H&E staining showing the overall shape and structure of placentas from WT and Stab1/2 dKO mice. Three layers of the placenta are indicated by colored lines, with red indicating the labyrinth zone, yellow indicating the junctional zone, and blue indicating the decidua. Quantitative analysis of the length of each placental zone. The percentages indicated for decidua, junctional zone, and labyrinth zone were quantified over the total length (100%) of the placenta. Scale bar, 500 μm. (**B**) High-magnification images of the blue rectangles in (**A**). (B1, B2) High-magnification images of the junctional zone in (**B**). Scale bar, 100 μm.

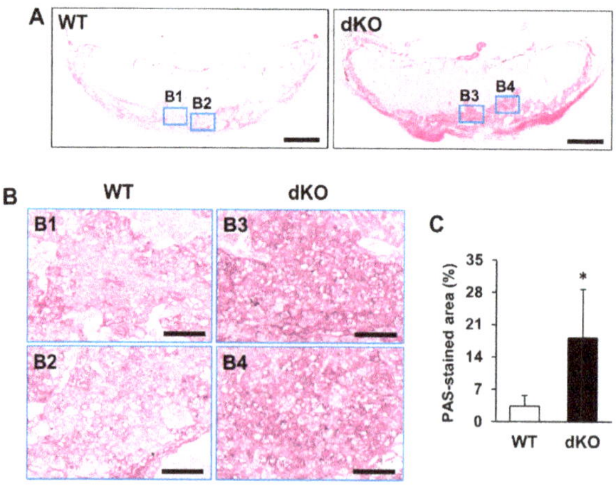

Figure 6. Histological analysis of WT and Stab1/2 dKO placentas based on PAS staining. (**A**) Overall shape of the PAS-stained placentas derived from WT and Stab1/2 dKO mice. Scale bar, 1 mm. (**B1–B4**) High-magnification images of the blue rectangles in (**A**). Scale bar, 100 μm. (**C**) Percentage of PAS-stained positive area to the total tissue area was analyzed using ImageJ. *, $p < 0.05$ versus WT; n = 3.

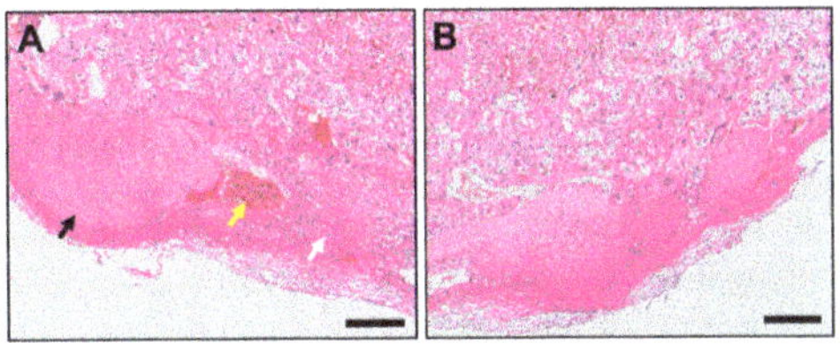

Figure 7. Histological analysis of the decidua of Stab1/2 dKO placenta. (**A**,**B**) High-magnification images of the black rectangles in the placenta from Stab1/2 dKO mouse shown in Figure 5A. In (**A**), black, yellow, and white arrows indicate decidual hemorrhage, maternal artery, and decidual necrosis with fibrin deposition, respectively. Scale bar, 200 μm.

3. Discussion

This study provides a phenotype description of reproduction, placentation, and embryo development in Stab1/2 dKO female mice. The phenotype observed in these mice was similar to human placental abruption. In pregnant mice lacking both Stab1 and Stab2, the placental decidua layer was expanded due to decidual hemorrhage and necrosis. During pregnancy, the separation of the placenta and endometrial walls results in placental abruption and the loss of nutrient and oxygen supply from the mother, leading to miscarriage [16,17]. Decidual bleeding occurs due to abnormalities in the early development of spiral arteries, which is one of the causes of placental destruction [16,17]. Spiral artery remodeling occurs in the decidua, which is comprised of specialized endometrial stromal cells to sustain the pregnancy. The decidua is made of maternal blood vessels as well as numerous leukocytes such as decidual macrophages and uNK cells known to contribute to artery remodeling [12,13,18–20]. Artery remodeling occurs through a specific mechanism in which trophoblasts replace endothelial cells of arteries and blood vessels in transit to spiral arteries [12,13,18]. Formation of spiral arteries and replacement of trophoblasts depends on the apoptosis of endothelial cells and degradation of the extracellular matrix by decidual cells [12,13,18]. Decidual macrophages and uNK cells produce various matrix metalloproteinases, angiogenic factors, and cytokines that induce endothelial cell apoptosis [9,12,19]. These cells scavenge residues due to spiral artery formation and modulate immune responses in the placenta [12,13,18–20]. Stab1 and Stab2 participate in the elimination of apoptotic cells via macrophage phagocytosis [1,2,4,5]. Stab1 and Stab2 are involved in the regulation of the extracellular environment and modulation of angiogenesis [1,4,9,14]. Stab2 also plays a role in maintaining fluidity of the blood that passes through arteries [7]. Schledzewski et al. (2011) previously investigated Stab1/2 dKO mice, which were generated by crossing Stab1 KO mice with Stab2 KO mice [10]. These mouse lines were generated using different strategies for targeting Stab1 and Stab2, and thus differ from the ones used in this study [8,21]. Their Stab1/2 dKO mice exhibited significantly lower survival rate, impaired liver blood clearance capacity, and increased kidney fibrosis compared with WT mice [10]. However, their study did not focus on elucidating the cause of the low survival rate in Stab1/2 dKO mice. Moreover, no specific physical feature was reported in each Stab1 KO and Stab2 KO mouse line [8,21] used in the present study. Here, Stab1/2 dKO progeny obtained from Stab1/2 dKO female mice were fewer than the progeny derived from WT female mice. As it was difficult to obtain a sufficient number of Stab1/2 dKO mice, no further studies were conducted to investigate the detailed molecular mechanisms underlying Stab1/2-deficient placental defects. Moreover, due to increased body waste that occurs in pregnancy, kidney function is important and metabolic imbalances such as gestational diabetes might develop in some cases [22]. Thus, it is necessary to understand the role of Stab1 and Stab2 in kidney function during pregnancy. Here, we present a possibility that Stab1 and Stab2 are associated with deciduous hemorrhage in the placenta. These findings suggest that Stab1 and Stab2 deficiency leads to a failure in clearing apoptotic cells following arterial remodeling, thereby affecting the angiogenic function of macrophages and blood fluidity. Such failure eventually results in placental defects. Additional studies are warranted to determine how Stab1 and Stab2 deficiency induces reproduction defects and to further elucidate the series of events leading to placental hemorrhage due to Stab1 and Stab2 deficiency. Our results suggest that decidual hemorrhage due to Stab1 and Stab2 deficiency can lead to serious defects in the placenta as well as reproduction failure.

The junctional zone of the placenta contains glycogen cells that accumulate glycogen during pregnancy [11]. These cells migrate into the maternal decidua after E12.5, and act as an energy reservoir for the mother and/or fetus during late-stage gestation in mice [11,13]. The number of glycogen cells declines by half by E18.5 [11]. The placentas at E17.5 carried a higher number of glycogen cells in the Stab1/2 dKO female mice than in WT mice, suggesting that the lack of Stab1 and Stab2 in the placenta hinders the accumulation and use of glycogen. Issues associated with glycogen accumulation in the placenta often occur due to inhibition of trophoblast invasion and spiral artery remodeling or due to suppression of vascular formation and expansion [11]. In particular, blood vessels and conduits are associated with abnormal formation of junctional zones [23]. Defects in the junctional zone can alter the endocrine environment in the placenta, resulting in systemic effects on both the fetus and the mother [23]. The role of Stab1 and Stab2 in the regulation of glycogen accumulation and use in the placenta remains unclear and is difficult to define. However, histological data suggest that the lack of Stab1 and Stab2 leads to defects in glycogen metabolism in mouse placenta. In addition, Stab1 and Stab2 deficiency may lead to alterations in placental morphology and function due to abnormal regulation of glycogen, at least in part. The junctional zone is an endocrine zone that secretes several pregnancy hormones [13]. Stab1 is involved in the transcytosis of placental hormones, including PL [1,9]. PL, which has a similar structure and function as the growth hormone, regulates lactogenesis, maintenance of the uterine wall, and the synthesis of progesterone during late pregnancy [1,9]. Moreover, PL induces lipolysis, amino acid mobilization, and glycogenolysis to facilitate nutrient delivery to the fetus and promotes glycogen accumulation in the junctional zone by glycogen trophoblasts. Interestingly, the concentration of mouse PL differs between maternal (10 μg/mL) and fetal (0.5 μg/mL) circulation [9]. This difference occurs due to the lack of direct connection between maternal and fetal circulatory systems. Placental macrophages are present in these circulatory systems, and PL concentration is controlled by the transcytosis of placental macrophages. Stab1 is a receptor involved in PL transcytosis of placental macrophages [9]. Stab1 deficiency in the placenta can disrupt the regulation of PL concentrations in maternal and fetal circulation, ultimately leading to placental abnormalities [9]. Moreover, Stab1 and Stab2 are involved in cell migration and extracellular matrix regulation, suggesting that they contribute to the proper migration of placental cells during pregnancy [9]. The present study also suggests that Stab1 and Stab2 are important to regulate the function of the junctional zone.

This study has a limitation that it could not conclude the role of Stab1 and Stab2 in placental development because it was difficult to obtain sufficient Stab1/2 dKO mice. Further studies are needed to ascertain that the placental changes observed in Stab1/2 dKO females are due to the lack of Stab1 and Stab 2 expression using a placental-specific Stab1/2 knockout mouse model. Furthermore, embryo transfer studies transplanting Stab1/2 dKO blastocysts into WT surrogates will be used to better determine the role of Stab1 and Stab2 in mammalian placentation. Moreover, additional work is needed to determine the detailed molecular mechanism(s) underlying this process. In conclusion, we observed that Stab1/2 dKO female mice exhibited reproduction failure owing to placental defects, including hemorrhage in the decidua and abnormalities in the junctional zone. These findings provide the first description of the reproductive capacity and phenotypic characteristics of placenta and embryos due to Stab1 and Stab2 deficiency in mice.

4. Materials and Methods

4.1. RNA Extraction, RT-PCR, and qPCR

Placentas and livers were isolated from WT pregnant mice at E17.5. The samples were immersed in TRIzol reagent (Ambion, Austin, TX, USA) and lysed to extract RNA using Qiagen tissue lyser II, as previously described [5]. RNA was then reverse transcribed into cDNA using a Reverse Transcription Master Premix (Elpis Biotech, Daejeon, Korea) for RT-PCR and qPCR. RT-PCR was performed at an annealing temperature of 64 °C for Stab1 and 60 °C for Stab2 (35 cycles), and 60 °C for Gapdh as a loading control (25 cycles). Results of RT-PCR were analyzed using ImageJ (National Institute of Health, Bethesda, MD, USA). qPCR was performed using Applied Biosystems StepOnePlus qPCR system (Life technologies Software v2.3; Life technologies, Carlsbad, CA, USA) under the following conditions: initial denaturation at 95 °C for 10 min, 40 cycles of denaturation at 95 °C for 15 s and amplification at 60 °C for 1 min, and a final cooling step. Results of qPCR were analyzed using the comparative cycle threshold (C_T) method. Primer sets used for this experiment are as follows: *Stab1*, 5′-TGC GAC ATC CAC ACC AAG TT-3′ and 5′-TGA ACC ACA TCC TTC CAG CA-3′; *Stab2*, 5′-AGC TGC TGC CTT TAA TCC TCA-3′ and 5′-ACT CCG TCT TGA TGG TTA GAG TA-3′; and *Gapdh*, 5′-GCA TCT CCC TCA CAA TTT CCA-3′ and 5′-GTG CAG CGA ACT TTA TTG ATG G-3′.

4.2. Animals and Sampling

Stab1/2 dKO mice were generated by crossing Stab1 KO mice with Stab2 KO mice [8,21]. Male and female Stab1/2 dKO mice were intercrossed to generate Stab1/2 dKO embryos and placentas (Figure 2). The mating of males and females for pregnancy was performed at the ratio of 1:1 in a single cage; and 2 weeks after mating, the females were separated into different cages and followed until birth or pregnancy loss (observation for at least 20 days after separation from the males). The date of pregnancy of the pregnant females was determined by the normal gross fetal anatomy of embryos obtained in their uteri and the comparison based on the text "The atlas of mouse development (edited by M. H. Kaufman, Elsevier academic press)". The genotypes of mice and/or embryos were analyzed using PCR and genomic DNA isolated from tails or yolk sacs. Primer sets and PCR amplification conditions for Stab1 and Stab2 were previously described [8,21]. WT females or embryos were used as controls. The mice were bred and maintained in climate-controlled (20–25 °C), specific pathogen-free conditions with a 12 h/12 h light/dark cycle and were allowed free access to standard mouse diet and water. All procedures concerning animal experiments were conducted with the approval of the Institutional Animal Care and Use Committee of Kyungpook National University (Approval number KNU-2016-0011, Daegu, Korea).

4.3. Histological Analysis

Embryos and placentas were isolated from pregnant female mice at E17.5. Placentas were fixed overnight in 4% paraformaldehyde at 4 °C, washed in distilled water, dehydrated through an ethanol series, embedded in paraffin, and sectioned to 3-μm thickness using a microtome (Leica Biosystems, Wetzlar, Germany). These sections were stained with H&E and PAS. For PAS staining, after deparaffinization and rehydration, sections were dipped in 0.5% periodic acid (Sigma-Aldrich, St. Louis, MO, USA) for 5 min and then treated with Schiff's reagent (Sigma-Aldrich) for 15 min. The stained sections were observed under a microscope slide scanner (Microscope Central, Feasterville, PA, USA), and PAS-positive stained area per total tissue area was analyzed by using ImageJ (NIH, Bethesda, MD, USA).

4.4. Statistical Analysis

Data were analyzed by Student's *t*-test and presented as means ± standard deviation (SD). A p value < 0.05 was considered statistically significant.

Supplementary Materials: Supplementary Materials can be found at http://www.mdpi.com/1422-0067/21/19/7235/s1. Figure S1: Uterine horns with E13.5 embryos in pregnant Stab1/2 dKO mice, Figure S2: Images of unevenly sized embryos or undefined tissue in pregnant Stab1/2 dKO mice compared to normal-sized embryo and placenta of pregnant WT mice at E17.5, Figure S3: Images of the uteri or E17.5 embryos in WT and Stab1/2 dKO females and the number of viable implantation sites, Figure S4: Pedigree and the number of pups generated by crossing between male and female mice with Stab1 heterozygous and Stab2 knockout genotypes.

Author Contributions: S.-Y.K., S.-Y.P., I.-S.K., and J.-E.K. designed the study; S.-Y.K., E.-H.L., E.N.K., and W.-C.S. prepared the samples and performed experiments; E.N.K., Y.H.K., S.-Y.P., I.-S.K., and J.-E.K. analyzed the data; and S.-Y.K., Y.H.K., and J.-E.K. wrote the manuscript. All authors have read and agreed to the published version of the manuscript.

Funding: This research was supported by Basic Science Research Program through the National Research Foundation of Korea (NRF) funded by the Ministry of Science, ICT & Future Planning (2016R1A2B4010043, 2019R1A2C1003398).

Conflicts of Interest: The authors declare no conflict of interest.

References

1. Kzhyshkowska, J. Multifunctional receptor stabilin-1 in homeostasis and disease. *Sci. World J.* **2010**, *10*, 2039–2053. [CrossRef] [PubMed]
2. Politz, O.; Gratchev, A.; McCourt, P.A.; Schledzewski, K.; Guillot, P.; Johansson, S.; Svineng, G.; Franke, P.; Kannicht, C.; Kzhyshkowska, J.; et al. Stabilin-1 and -2 constitute a novel family of fasciclin-like hyaluronan receptor homologues. *Biochem. J.* **2002**, *362*, 155–164.
3. Palani, S.; Maksimow, M.; Miiluniemi, M.; Auvinen, K.; Jalkanen, S.; Salmi, M. Stabilin-1/CLEVER-1, a type 2 macrophage marker, is an adhesion and scavenging molecule on human placental macrophages. *Eur. J. Immunol.* **2011**, *41*, 2052–2063. [CrossRef] [PubMed]
4. Hirose, Y.; Saijou, E.; Sugano, Y.; Takeshita, F.; Nishimura, S.; Nonaka, H.; Chen, Y.R.; Sekine, K.; Kido, T.; Nakamura, T.; et al. Inhibition of Stabilin-2 elevates circulating hyaluronic acid levels and prevents tumor metastasis. *Proc. Natl. Acad. Sci. USA* **2012**, *109*, 4263–4268. [CrossRef] [PubMed]
5. Kim, S.Y.; Lee, E.H.; Park, S.Y.; Choi, H.; Koh, J.T.; Park, E.K.; Kim, I.S.; Kim, J.E. Ablation of Stabilin-1 enhances bone-resorbing activity in osteoclast in vitro. *Calcif. Tissue Int.* **2019**, *105*, 205–214. [CrossRef] [PubMed]
6. Tamura, Y.; Adachi, H.; Osuga, J.; Yahagi, N.; Sekiya, M.; Okazaki, H.; Tomita, S.; Iizuka, Y.; Shimano, H.; Nagai, R.; et al. FEEL-1 and FEEL-2 are endocytic receptors for advanced glycation end products. *J. Biol. Chem.* **2003**, *278*, 12613–12617. [CrossRef] [PubMed]
7. Falkowski, M.; Schledzewski, K.; Hansen, B.; Goerdt, S. Expression of stabilin-2, a novel fasciclin-like hyaluronan receptor protein, in murine sinusoidal endothelia, avascular tissues, and at solid/liquid interfaces. *Histochem. Cell Biol.* **2003**, *120*, 361–369. [CrossRef]
8. Lee, W.; Park, S.Y.; Yoo, Y.; Kim, S.Y.; Kim, J.E.; Kim, S.W.; Seo, Y.K.; Park, E.K.; Kim, I.S.; Bae, J.S. Macrophagic Stabilin-1 restored disruption of vascular integrity caused by sepsis. *Thromb. Haemost.* **2018**, *118*, 1776–1789. [CrossRef] [PubMed]
9. Kzhyshkowska, J.; Gratchev, A.; Schmuttermaier, C.; Brundiers, H.; Krusell, L.; Mamidi, S.; Zhang, J.; Workman, G.; Sage, E.H.; Anderle, C.; et al. Alternatively activated macrophages regulate extracellular levels of the hormone placental lactogen via receptor-mediated uptake and transcytosis. *J. Immunol.* **2008**, *180*, 3028–3037. [CrossRef] [PubMed]
10. Schledzewski, K.; Géraud, C.; Arnold, B.; Wang, S.; Gröne, H.J.; Kempf, T.; Wollert, K.C.; Straub, B.K.; Schirmacher, P.; Demory, A.; et al. Deficiency of liver sinusoidal scavenger receptors stabilin-1 and -2 mice causes glomerulofibrotic nephropathy via impaired hepatic clearance of noxious blood factors. *J. Clin. Investig.* **2011**, *121*, 703–714. [CrossRef] [PubMed]
11. Akison, L.K.; Nitert, M.D.; Clifton, V.L.; Moritz, K.M.; Simmons, D.G. Review: Alterations in placental glycogen deposition in complicated pregnancies: Current preclinical and clinical evidence. *Placenta* **2017**, *54*, 52–58. [CrossRef] [PubMed]
12. Hazan, A.D.; Smith, S.D.; Jones, R.L.; Whittle, W.; Lye, S.J.; Dunk, C.E. Vascular-leukocyte interactions: Mechanisms of human decidual spiral artery remodeling in vitro. *Am. J. Pathol.* **2010**, *177*, 1017–1030. [CrossRef] [PubMed]
13. Rai, A.; Cross, J.C. Development of the hemochorial maternal vascular spaces in the placenta through endothelial and vasculogenic mimicry. *Dev. Biol.* **2014**, *387*, 131–141. [CrossRef] [PubMed]
14. Rost, M.S.; Sumanas, S. Hyaluronic acid receptor Stabilin-2 regulates Erk phosphorylation and arterial-venous differentiation in zebrafish. *PLoS ONE* **2014**, *9*, e88614. [CrossRef] [PubMed]
15. Coan, P.M.; Conroy, N.; Burton, G.J.; Ferguson-Smith, A.C. Origin and characteristics of glycogen cells in the developing murine placenta. *Dev. Dyn.* **2006**, *235*, 3280–3294. [CrossRef] [PubMed]

16. Oyelese, Y.; Ananth, C.V. Placental abruption. *Obstet. Gynecol.* **2006**, *108*, 1005–1016. [CrossRef] [PubMed]
17. Tikkanen, M. Etiology, clinical manifestations, and prediction of placental abruption. *Acta Obstet. Gynecol. Scand.* **2010**, *89*, 732–740. [CrossRef] [PubMed]
18. Mori, M.; Bogdan, A.; Balassa, T.; Csabai, T.; Szekeres-Bartho, J. The decidua-the maternal bed embracing the embryo-maintains the pregnancy. *Semin. Immunopathol.* **2016**, *38*, 635–649. [CrossRef]
19. Herington, J.L.; Bany, B.M. Effect of the conceptus on uterine natural killer cell numbers and function in the mouse uterus during decidualization. *Biol. Reprod.* **2007**, *76*, 579–588. [CrossRef]
20. Simmons, D.G.; Fortier, A.L.; Cross, J.C. Diverse subtypes and developmental origins of trophoblast giant cells in the mouse placenta. *Dev. Biol.* **2007**, *304*, 567–578. [CrossRef]
21. Park, S.Y.; Yun, Y.; Lim, J.S.; Kim, M.J.; Kim, S.Y.; Kim, J.E.; Kim, I.S. Stabilin-2 modulates the efficiency of myoblast fusion during myogenic differentiation and muscle regeneration. *Nat. Commun.* **2016**, *7*, 10871. [CrossRef] [PubMed]
22. Soma-Pillay, P.; Nelson-Piercy, C.; Tolppanen, H.; Mebazaa, A. Physiological changes in pregnancy. *Cardiovasc. J. Afr.* **2016**, *27*, 89–94. [CrossRef] [PubMed]
23. Woods, L.; Perez-Garcia, V.; Hemberger, M. Regulation of placental development and its impact on fetal growth—new insights from mouse models. *Front. Endocrinol.* **2018**, *9*, 570. [CrossRef] [PubMed]

International Journal of *Molecular Sciences*

MDPI

Review

Placental Dysfunction in Assisted Reproductive Pregnancies: Perinatal, Neonatal and Adult Life Outcomes

Claudio Manna [1,2], Valentina Lacconi [1,3], Giuseppe Rizzo [1,4], Antonino De Lorenzo [1] and Micol Massimiani [1,3,*]

1 Department of Biomedicine and Prevention, University of Rome Tor Vergata, Via Montpellier 1, 00133 Rome, Italy; claudiomanna55@gmail.com (C.M.); valelcc@gmail.com (V.L.); giuseppe.rizzo@uniroma2.it (G.R.); delorenzo@uniroma2.it (A.D.L.)
2 Biofertility Center of Reproductive Medicine, Viale degli Eroi di Rodi 214, 00128 Rome, Italy
3 Saint Camillus International University of Health Sciences, Via di Sant'Alessandro 8, 00131 Rome, Italy
4 Department of Obstetrics & Gynecology Moscow Russian Federation, The First I.M. Sechenov Moscow State Medical University, Trubetskaya Street, House 8, Building 2, 119991 Moscow, Russia
* Correspondence: micol.massimiani@unicamillus.org; Tel.: +39-06-7259-6151

Abstract: Obstetric and newborn outcomes of assisted reproductive technology (ART) pregnancies are associated with significative prevalence of maternal and neonatal adverse health conditions, such as cardiovascular and metabolic diseases. These data are interpreted as anomalies in placentation involving a dysregulation of several molecular factors and pathways. It is not clear which extent of the observed placental alterations are the result of ART and which originate from infertility itself. These two aspects probably act synergically for the final obstetric risk. Data show that mechanisms of inappropriate trophoblast invasion and consequent altered vascular remodeling sustain several clinical conditions, leading to obstetric and perinatal risks often found in ART pregnancies, such as preeclampsia, fetal growth restriction and placenta previa or accreta. The roles of factors such as VEGF, GATA3, PIGF, sFLT-1, sEndoglin, EGFL7, melatonin and of ART conditions, such as short or long embryo cultures, trophectoderm biopsy, embryo cryopreservation, and supraphysiologic endometrium preparation, are discussed. Inflammatory local conditions and epigenetic influence on embryos of ART procedures are important research topics since they may have important consequences on obstetric risk. Prevention and treatment of these conditions represent new frontiers for clinicians and biologists involved in ART, and synergic actions with researchers at molecular levels are advocated.

Keywords: placenta; placental dysfunction; assisted reproduction techniques; infertility; preeclampsia; intrauterine growth restriction; trophoblast invasion; PlGF; sFLT-1; EGFL7

Citation: Manna, C.; Lacconi, V.; Rizzo, G.; De Lorenzo, A.; Massimiani, M. Placental Dysfunction in Assisted Reproductive Pregnancies: Perinatal, Neonatal and Adult Life Outcomes. *Int. J. Mol. Sci.* **2022**, *23*, 659. https://doi.org/10.3390/ijms23020659

Academic Editor: Micheline Misrahi

Received: 23 November 2021
Accepted: 5 January 2022
Published: 8 January 2022

1. Introduction

There is a rapidly increasing interest in placentation in obstetric adverse outcome, neonatal and adult life health after infertility, and its related therapies. Abnormal placentation is a common finding in the infertile population, even among those couples conceiving spontaneously after a period of infertility. A higher risk of preterm birth (PTB) and low birth weight (LBW) has been found in these pregnancies [1]. Moreover, abnormal placentation and obstetric complications such as PTB, preeclampsia (PE), and fetal growth restriction (FGR) have been associated with endometriosis, a common factor of infertility [2]. Another maternal condition at risk for the development of placental anomalies, commonly associated with infertility, is polycystic ovary syndrome (PCOS), in which the prevalence of gestational diabetes mellitus (GDM) is significantly increased [3]. Of note, GDM is an independent risk factor for the onset of placental disorders with altered structure, function and hypertrophic growth of the organ [4].

An important actual focus in reproductive medicine, neonatology and population health is that up to 6% (range between 0.2% and 6.4%) of European births is conceived by assisted reproductive technology (ART) [5] and concerns existing health conditions of individuals born following ART. ART is a group of in vitro techniques used to treat moderate and severe infertility, including in vitro fertilization (IVF), intracytoplasmic sperm injection (ICSI), frozen embryo transfer (FET), oocyte donation (OD), blastocyst culture, intrauterine insemination, and preimplantation genetic testing for aneuploidy (PGT-A). Unfortunately, each of these techniques may represent a possible confounding factor to the identification of a precise relationship between ART procedures and obstetric or neonatal outcomes, including children and their adulthood health. Many studies report an evident increase of obstetrical risk and perinatal complications with ART, especially for hypertensive disorders of pregnancy (HDP), from gestational hypertension to eclampsia [6–13]. A recent meta-analysis of 50 cohort studies including 161,370 ART and 2,280,241 spontaneously conceived singleton pregnancies found increased risks for several obstetric complications. Significant worst outcomes regarded pregnancy-induced hypertension, placenta previa, abruption, antepartum hemorrhage, oligohydramnios, cesarean delivery, PTB, very low birth weight (VLBW), LBW and perinatal mortality and morbidity [14].

The role of the placenta in these conditions has been demonstrated in several studies. It is well documented that ART can be associated with changes in placental morphology and structure, growth dynamics, imprinted and non-imprinted genes, and other aspects regulating placentation [15]. Several studies demonstrate that incidence of placenta previa was significantly higher in ART than in spontaneous conception (OR 3.14, 95% CI [8]; RR 3.71, 95% CI [16]). Moreover, the placentas from ART pregnancies presented a significantly greater weight and higher placental weight-to-birth weight ratio [17,18]. Some observations indicate that placentas of pregnancies obtained by ART have an increased placental thickness and higher incidence of hematomas [19]. It has been also demonstrated that in children conceived by ART the frequency of imprinting disorders is higher than expected in the general population. ART is associated with increased risk of epigenetic alterations influencing gene expression and DNA methylation in early development and in the placenta, eventually triggering diseases and affecting long-term health [20–22]. These findings support the hypothesis of a primary ART responsibility for adverse perinatal outcomes with evident suspicions for anomalies of placentation, although fertility treatments themselves are often associated with impaired placentation and related adverse pregnancy outcomes [1]. Infertile couples who did not undergo ART showed an increased risk of adverse obstetric outcomes, such as placental abruption, fetal loss and GDM [23].

Based on these observations, the main question to address is whether placental alterations (from macroscopic to molecular level) observed in treated infertile patients are the result of ART or if they originate from infertility itself. A deeper investigation of the underlying molecular basis of reduced placental functions in infertility condition and especially after ART treatment is then needed.

In this review, we provide a summary of the most recent literature on placental dysfunction associated with obstetric, perinatal outcomes in singletons born after NON-ART and ART treatments. ART and infertility may be the cause of common dysregulated pathways, including changes in trophoblast invasion, environmental conditions, vascular defects, chronic inflammation, and oxidative stress. Each of them can be prevalent or coexist with others with different and common molecular pathways in a grading and timing that can configurate many clinical features, leading often to neonatal–adult and maternal consequences. This review is based on systematic reviews (SRs), large cohort studies, meta-analyses, and genetic, epigenetic and molecular studies.

2. Abnormal Placentation and ART: Molecular Factors and Involved Signaling

Abnormal placentation may present in a variety of phenotypes, severity, clinical conditions and consequences as the result of several types of infertility treatments and techniques used in ART. For these reasons, it becomes difficult to linearly relate placental influence to obstetrical and perinatal (or neonatal) outcomes after ART.

An altered expression of factors and molecules involved in proper placental development, leading to impaired trophoblast invasion and subsequent reduced vascular remodeling and placenta hypoperfusion, sustain several clinical conditions leading to obstetric and perinatal risks often found in ART pregnancies, such as PE, FGR and placenta previa or accrete (Figure 1) [24–28]. Syncytiotrophoblast stress has been associated with a dysregulated expression of placental growth factor (PlGF) and soluble fms-like tyrosine kinase 1 (sFLT-1) [29,30]. Circulating levels of the anti-angiogenic factor sFLT-1 are increased, and those of PlGF are decreased even before the onset of the clinical symptoms of PE and FGR [31–35]. The increased ratio sFLT-1:PlGF is thought to contribute to the systemic endothelial response and correlate with the severity of FGR and PE. A recent study indicates that the release of sFLT-1 from the placenta is regulated by the epidermal growth factor receptor (EGFR) pathway and the mitochondrial electron transport chain and its downstream pathways, both significantly increased in preeclamptic placentas [36]. The inhibition of these signaling pathways significantly reduces sFLT-1 release from primary cytotrophoblast cells [36]. Vrooman demonstrated in the rat model that an embryo culture from the 1-cell to blastocyst stage increased levels of sFLT-1 together with placental overgrowth, reduced fetal weight, and lower placental DNA methylation [37].

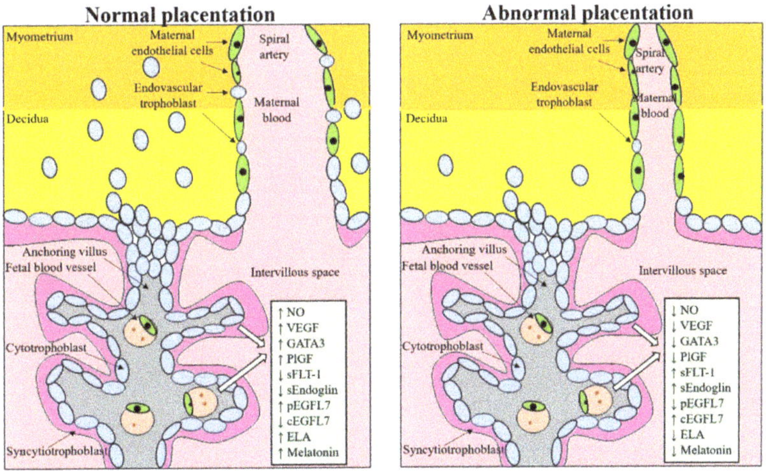

Figure 1. Molecules regulating proper placental development, whose dysregulation is involved in an abnormal placental development. NO, nitric oxide; VEGF, vascular endothelial growth factor; GATA3, GATA binding protein 3; PlGF, placental growth factor; sFLT-1, soluble fms-like tyrosine kinase 1; sEndoglin, soluble endoglin; pEGFL7, placental epidermal growth factor-like domain 7; cEGFL7, circulating epidermal growth factor-like domain 7; ELA, elabela.

Over the last decades, several other factors have been demonstrated to be altered in pregnancy disorders associated with abnormal placentation (e.g., soluble endoglin (sEndoglin), PlGF and epidermal growth factor-like domain 7 (EGFL7)), with the aim to create a panel of markers to allow an earlier and more precise diagnosis of PE. sEndoglin has been shown to be upregulated in abnormal placenta, typical of FGR and PE, and released into the maternal blood, where it acts as antiangiogenic factor by inhibiting transforming growth factor-beta (TGF-β) signaling in the vasculature. sEndoglin markedly increased, beginning 2 to 3 months before clinical manifestations of PE [38,39]. EGFL7, originally discovered as a largely endothelial restricted gene, has been recently demonstrated to be expressed in the placenta and involved in placental angiogenesis and trophoblast migration [40–42]. Altered EGFL7 expression is associated with abnormal placentation and systemic maternal endothelial dysfunction, observed in PE [40,43,44]. Maternal treatment with nitric oxide (NO) donors increases placental EGFL7 levels and improves maternal hemodynamic state and perinatal outcome [45,46]. In the maternal circulation, endothelial dysfunction and abnormal hemodynamic state are associated with increased levels of EGFL7, which return to control levels after NO treatment [44]. Moreover, EGFL7 dosage in maternal circulation allows to discriminate between PE and FGR [43]. Although there are no data correlating EGFL7 with ART, dosage of its circulating levels could help identify ART-associated abnormal placentation.

In FGR and placental insufficiency, the levels of the hormone melatonin are significantly reduced, and this decrease is correlated with proinflammatory activities of tumor necrosis factor alpha (TNF-α), interleukine-1beta (IL-1β), and IL-6 [47]. Melatonin is an antioxidant factor and an anti-inflammatory agent [48,49]. Several studies support the production of this hormone in the ovary as a whole [50,51], the granulosa cells, including those making up the cumulus oophorus [52,53], and the oocyte [54]. In women undergoing ART, melatonin significantly increased the implantation rate [55]; a similar result was obtained in women affected by PCOS undergoing intrauterine insemination [56]. Melatonin crosses the cell membrane, thus interacting with intracellular molecules via different signaling pathways and displaying scavenger functions [57]. Melatonin upregulates the primary implantation receptors, ErbB1 and ErbB4, and significantly reduces intracellular ROS in mouse blastocysts (increasing the embryo total antioxidant capacity) and promotes mitosis of the inner cell mass and trophectoderm cells [58]. Recently, it has been demonstrated that the supplementation of culture medium with melatonin (10^{-9} mol/L) improves the growth of mouse parthenogenetic embryo potentially by promoting cell cycle progression [59]. Despite all these functions, melatonin is not present in culture media used in ART. We could speculate that the addition of melatonin to the ART media may be beneficial.

3. Pathophysiology of the Placenta in Pregnancy Complications and ART Pregnancies

Altered hormonal milieu, epigenetic changes, immune activity dysregulation, dysmetabolism and inflammation are all conditions attributable to ART and associated with an abnormal placentation. This section provides a theoretical basis on placental changes caused by ART that are most strongly associated with an increased risk of fetal, neonatal and long-term diseases.

3.1. Altered Hormonal Milieu Effect on Placental Development

The interpretation of the effects of ARTs on pregnancy and the possible associated pathologies is complicated by the supra-physiological hormonal levels in the recipient, as a consequence of the ovarian stimulation treatment [37]. Bourgain and Devroey in 2003 suggested that increased hormone blood levels might alter the timing of endometrial receptivity, with possible suboptimal embryo implantation and development [60]. It was also suspected, by studies in animal models, that high estrogen levels exert a detrimental effect on spiral artery remodeling by the trophoblast [61]. Moreover, it has been observed that, in FET procedures, levels of progesterone higher than 32.5 ng/mL on the day of embryo transfer, were associated with no pregnancy [62]. This cut off level seems improbable in ovarian controlled stimulation conditions of fresh IVF cycles, where many corpora lutei are functioning and progesterone level, together with estradiol, are much higher. Conversely, the lack of hormones might also affect placental development by creating conditions leading to increased risks for HDP, as it has been suggested for FET following hormonal endometrial preparation [63,64]. Using these protocols, ovulation does not occur, and the corpus luteum, which secretes important protective vasoactive substances, is not formed [9,10,65]. Among these factors, relaxin is a potent vasodilator promoting vascular compliance [66] and facilitating the adaptation of the maternal cardiovascular system to pregnancy [67]. Moreover, in programmed FET cycles, the use of estradiol and progesterone has been recently associated with obstetric adverse outcomes, i.e., PTB, LBW, and PE, which are small for gestational age (SGA) and large for gestational age (LGA) when compared to natural FET cycle [9]. Thyroid hormone (TH) has been also demonstrated as an important player in reproduction; both hypothyroidism and hyperthyroidism have been associated with subfertility, recurrent miscarriages, and adverse pregnancy outcomes [68,69]. It has been recently demonstrated that TH regulates protease expression and activation of Notch signaling in implantation and embryo development [70]. Moreover, levothyroxine (LT4) treatment has a positive effect on conception capacity and reduces miscarriages when administered to euthyroid women with autoimmune thyroid disease affected by recurrent miscarriages [71].

Several common infertility conditions, such as endometriosis and PCOS, often treated with ART, are characterized by a hormonal dysregulation that affects proper placentation. In endometriosis, a defective deep placentation may derive from functional abnormalities of the eutopic endometrium, as well as an imbalance in endocrine and inflammatory markers [72]. PTB, placenta previa, placental abruption, gestational hypertension, PE, LBW, SGA, cesarean delivery, postpartum hemorrhage, and stillbirth have been significantly associated with endometriosis in a systematic review and meta-analyses, including in 39 studies [73]. PCOS is another condition leading to infertility, in which the defective trophoblast invasion and placentation may be caused by mother's hyperandrogenism [74,75]. Testosterone can act directly on trophoblast invasion, with modifications of placenta morphology and function [76,77]. PCOS is a chronic low-grade inflammation associated with metabolic dysfunction that is enhanced in pregnancy by the induction of an endothelial dysfunction [78]. This condition might in turn prevent normal remodeling of spiral vessels and the physiologic decrease of uterine artery impedance, thus reducing the depth of endovascular trophoblast invasion. As a consequence of these patterns, in patients with PCOS the placental weight, thickness, density and volume was found significantly reduced [76]. In women with PCOS undergoing IVF, the risk of adverse pregnancy outcomes was found to be significantly elevated. In a recent meta-analysis including 29 observational studies, Sha et al. (2019) described an increased risk for GDM, pregnancy-induced hypertension (PIH), LGA, miscarriage and PTB in women with PCOS undergoing ART [79].

3.2. Epigenetic Changes after ART Techniques Are Associated with Altered Gene Expression in the Placenta and Congenital Imprinting Disorders

Epigenetic regulatory mechanisms can originate from several sources: direct DNA methylation, non-coding RNA, imprinting, and post-translational modification of histone proteins and chromatin remodeling [80]. According to some clinical studies, it was speculated that prolonged exposure to extracorporeal environment might predispose human embryos to disorders of genomic imprinting and to epigenetic modification [81]. Each stage of embryo development, from fertilization to blastocyst formation, also in humans, can undergo to epigenetic changes and consequent differential gene expression in the placenta [82,83]. In the pre-implantation period, the embryonic epigenome is entirely reprogrammed by specific methylations [84], leading to an increase in the risk of gene expression alterations under specific circumstances. For example, there is evidence that aberrant fetal DNA methylation might cause abnormal development of the placenta and recurrent spontaneous abortion [85].

We must consider that ART procedures occur simultaneously to this extensive epigenetic reprogramming and we cannot rule out that the stress involved in ART might affect the establishment and/or maintenance of genomic imprinting (Figure 2). Actually, more than 100 imprinted genes have been found to be strictly involved in the growth and development of embryos [15]. In placentas from IVF pregnancies, a different methylation pattern was observed in comparison to non-IVF gestations and environmental conditions were considered responsible for these findings (Figure 2) [82,83]. IVF disturbs the DNA methylation of the imprinting control region of *H19/Igf2*, *Snrpn*, *Peg3*, and *Kcnq1ot1* genes, inducing morphological alterations in the placenta and an increased risk of adult metabolic syndrome [86]. Studies using animal models have demonstrated a dysregulation of several maternally and paternally expressed imprinted genes (e.g., *Kcnq1ot1* involved in placental growth, *Peg10* that is required for the differentiation of placental spongiotrophoblast and labyrinth, *Peg11/Rtl1* involved the development of placental labyrinth and nutrient passive transport, *Sfmbt2* playing a key role in the maintenance of trophoblast cell types) in the IVF group as compared to the control group [87–89]. Most of these genes play an important role in the proper placental morphology and function. Moreover, specific transcription factors, such as GATA binding protein 3 (GATA3), produced by trophectoderm at blastocyst stage, may be responsible for tissue or cell altered invasion and migration under the influence of the local environment [90].

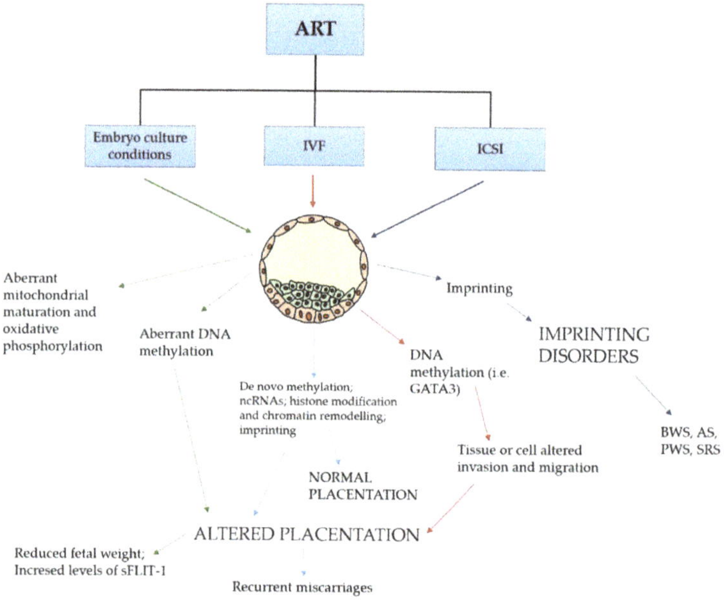

Figure 2. ART techniques affecting epigenetics in the pre-implantation embryo. Epigenetic dysregulation is involved in an abnormal placentation and genetic imprinting disorders. sFLT-1 denotes soluble fms-like tyrosine kinase 1, GATA3 GATA binding protein 3, BWS Beckwith–Wiedemann syndrome, AS Angelman syndrome, PWS Prader–Willi syndrome, SRS Silver–Russell syndrome.

Consequently, particular attention was paid to blastocyst stage transfer and its possible adverse effect on perinatal outcome. In this regard, there are data from animal studies demonstrating that in vitro fertilization and embryo culture can strongly impact the placental transcriptome [91,92]. In a multicenter double-blind randomized controlled trial, including 836 couples, a significative reduction of birthweight (158 g) and an increase of PTB (8.6% vs. 2.2%) was associated with one of the two culture media used for embryo culture [93]. Moreover, it has been demonstrated that in vitro culture conditions may obstacle mitochondrial maturation and oxidative phosphorylation, inducing cells to increase glycolysis with the aim of maintaining energetic homeostasis: this pathway is represented by the so-called "Warburg effect" [94]. This adaptation pathway may induce epigenetic changes with possible adverse consequences on fetal growth (Figure 2) [94]. There is evidence that the longer the in vitro embryo culture lasts (i.e., blastocyst transfer in comparison to cleavage stage), the more epigenetic changes that can occur [81,94], especially in the developing placenta, although this is not surprising. In this respect, we should consider that this organ develops from the outer trophoblastic layer and consequently is more sensible to epigenetic regulatory changes caused by environmental manipulation and influences during ART. Placental overgrowth was demonstrated by Vroman in the rat model, where embryo culture from 1-cell to blastocyst stage showed adverse outcomes. Reduced fetal weight and lower placental DNA methylation, together with the abovementioned increased levels of sFLT-1, were observed [37]. Additional evidence of aberrant DNA methylation following in vitro embryo culture has been obtained using the bovine model [95]. In this study, 2-, 8- and 16-cell stage embryos were flushed from the oviducts and allowed to further develop in vitro up to the blastocyst stage and their methylome investigated. It was clear that each step of the in vitro culture gave a contribution to epigenetic alterations and that the highest levels resulted around the time of early genomic activation of embryo. In this regard, a recent discovery demonstrated that human embryonic genomic activation initiates at one-cell stage [96].

However, it would be important to distinguish the potential contribution of epigenetic problems related to infertility from epigenetic alteration deriving from the culture conditions [97]. Besides the ART procedures themselves, differential methylation has also been identified in placentas of pregnancies conceived from couples with infertility utilizing different fertility treatments [82,98]. Differential DNA methylation in placentas from pregnancies conceived with different fertility treatments, even as early as the late first trimester, has been described [99].

Furthermore, ART can influence epigenetic regulation of placental formation and function by changing the embryonic environment, placental gene expression, and placental adaptive response to embryonic development [100]. In this respect, Zhang et al. [91] analyzed genes in placental samples obtained from the IVF and control groups, identifying seventeen upregulated and nine downregulated genes. Moreover, when compared to normal pregnancies, the transcriptome profile of first-trimester placental villi appeared significantly altered in IVF-ET pregnancies. A total of 3405 differentially regulated genes in the placenta following IVF (>2-fold change; $p < 0.05$) was identified by Zhao et al. [92], with 1910 upregulated and 1495 downregulated. The study revealed that these genes are involved in more than 50 biological processes and pathways: coagulation cascades, immune responses, transmembrane signaling, metabolism, cell cycle, stress control, invasion, and angiogenesis, all playing an important role in early pregnancy. From a clinical point of view, many studies have provided useful data on placental pathologies linked to the in vitro environment in which embryo develops. A Swedish population-based registry study from 2002 through 2013 showed an increased risk of placenta previa after blastocyst transfer compared to cleavage-stage transfer (OR 2.18; 95% CI 1.79–2.65) and spontaneous conception (OR 6.38; 95% CI 5.31–7.66) [101]. For FET, higher birthweight, and higher risk of LGA and VLGA were found in blastocyst vs. cleavage stage transfer [102–104]. Similar data were reported by Huang (2020), comparing results from day 3 vs. day 7 embryo in 4489 infertile women from 2006 to 2017 [105].

An example of the invasiveness of the in vitro reproductive machinery is the introduction in the clinical practice of the ICSI, which overcomes severe male infertility problems [106], injecting a single sperm into the cytoplasm of a mature oocyte. The mechanical introduction of one sperm escaping the natural control of the fertilization process by the oocyte has raised several criticisms. Reasonable concerns were expressed considering that animal studies on the safety of this practice were limited in comparison with those performed by Edwards before Louise Brown was born in 1978. The revision of some large experience and reviews did not report or demonstrated important consequences of this procedure on the newborn [107], and thus far, the cost–benefit ratio seems reasonable. However, abnormal spermatozoa prepared for ICSI showed improper methylation patterns at imprinting regions, leading to conceive embryos affected by genetic imprinting disorders. Not surprisingly, the alterations in the sperm might explain why ICSI is the fertilization procedure that induced the highest number of differentially expressed genes [108].

An increased reported incidence of autism spectrum disorders in babies born from ART might find some possible source in epigenetic alterations [82]. Numerous studies have reported an increased risk of congenital imprinting disorders in children conceived by ART (Figure 2) [109,110]. In 2019, a wide epidemiological study in Japan found 3.44-, 4.46-, and 8.91-fold increased frequencies of Prader–Willi syndrome (PWS), Beckwith–Wiedemann syndrome (BWS), and Silver–Russell syndrome (SRS) after ART, respectively [111]. In particular, BWS, which is a clinically and genetically heterogeneous overgrowth disease, is caused by epigenetic defects at the 11p15 chromosomal region and a four- to six-fold increased risk of its outcome in ART pregnancies has been reported [22].

3.3. Immune Dysregulation at the Maternal-Fetal Interface

Increased immune activity at the maternal–fetal interface and significant histological and immunohistochemical differences were observed in placentas from pregnancies obtained by IVF of heterologous oocytes, compared to those obtained using homologous oocytes [112], and may be the consequence of a host versus graft rejection-like condition. This aspect was further investigated using 3D ultrasound analysis. This technique demonstrated a more marked reduction of first-trimester placental volume in pregnancies obtained using donor oocytes [113]. A recent retrospective cohort study, including 1114 singleton pregnancies, of which 105 conceived with IVF, further supports the involvement of the immune system in pregnancy complications associated with ART [114]. A higher incidence of villitis was observed in placentas from pregnancies obtained following IVF (16.2% vs. 8.3%; $p = 0.007$).

3.4. Mechanical Stress on Embryo and Placental Development (the Case of PGT-A)

PGT-A is an ART technique that is increasingly performed, considering that it represents 40% of all IVF cycles in the USA [115]. Its aim is to discover if produced embryos are aneuploid and, in this case, they should not be transferred into the uterus, in order to prevent implantation failure and miscarriages. PGT-A preliminarily requires the in vitro development of embryos up to the blastocyst stage, in order to collect 5 to 10 trophoblast cells. Embryos are then frozen while waiting for the genetic analysis results. Embryos with no or few aneuploidies are thawed and transferred (FET) in the uterine cavity after endometrial hormonal preparation using estrogen and progesterone. In this condition, natural ovulation and formation of the corpus luteum are suppressed. Similar to PGT-A, preimplantation genetic test for monogenic diseases (PGT-M) allows to screen for embryos affected by monogenic diseases.

Some authors have expressed concerns on the possibility that trophectoderm biopsy might disturb embryo development, leading to potential adverse consequences on obstetric outcomes [65,116]. It has been hypothesized that removal of cells from the tissue that will become the placenta does not affect embryo implantation rate but may negatively influence later phases of placentation. Placenta-derived anomalies, such as placenta previa and accreta, HDP, and FGR might be consequences of this abnormal placentation [117–120].

In a comparative study investigating obstetric and neonatal outcome following PGT-A in blastocyst-stage biopsy with frozen embryo transfer and cleavage-stage biopsy with fresh embryo transfer, Jing et al. (2016) found that PGT-A-FET was associated with a higher incidence of gestational hypertension (9.0% vs. 2.3%; $p = 0.017$) [121]. These data might indicate a higher risk of placental injury as a consequence of PGT-A procedure on the trophectoderm. Similar indications came from the results of a study on 345 singleton and 76 twin deliveries after PGT-M. PGT-M showed an increased risk of obstetric complications, when compared with pregnancies conceived spontaneously or by IVF without PGT [122]. In particular, these data showed a higher rate of HDP (6.9%), compared with spontaneous conception patients (2.3%) and the IVF group (4.7%). It was reported, as well, a higher rate of SGA neonates (12.4%), in comparison with those from spontaneous conception group (3.9%) and the IVF patients (4.5%). Moreover, Makhijani et al. (2021) observed a significantly higher probability of developing HDP in patients who underwent PGT, compared with those who did not (adjusted odds ratio (aOR) 1.943, 95% CI 1.072–3.521; $p = 0.029$) [123]. In this study, the potential influence of other factors causing HDP, such as the hormonal endometrial preparation was considered. In their binary logistic regression model, they found that, with the adjusted odds of HDP, the risk associated with trophectoderm biopsy remained significantly higher. A recent large single-center study found a two-fold higher risk of pregnancy-related hypertension following 241 blastocyst biopsy and frozen transfer than in 515 non-biopsied controls [123].

Conversely, is it worthwhile to look for embryonic aneuploidies. It was shown that, among embryos with aneuploid cells, 31% were of meiotic origin and 74% related to mitotic aneuploidies, with maximum aneuploidy originating by days 4–5 after fertilization, falling to 5–6% by day 7 [124]. Consequently, we should consider the environment influence on the mitotic origin of aneuploidies in developing embryos. From a strict clinical point of view, PGT-A does not improve IVF outcomes nor does it reduce miscarriage rates [125]. Moreover, healthy babies have been born after the transfer of aneuploid mosaic embryos [126–128]. This is not surprising since it is well known the self-correction ability of aneuploid embryos. It has been demonstrated that, in mice, a *p*-53-dependent autophagy-mediated apoptotic process is able to eliminate aneuploid cells [129]. In humans, this process is promoted by the bone morphogenetic protein-4 (BMP4) [130]. These mechanisms, together with the fact that the results obtained from a trophectoderm biopsy cannot be representative of the whole embryo [131], contradict the assumption that all aneuploid embryos should be eliminated and constitutes a strong criticism for the PGT-A clinical utility [125]. The Scientific and Clinical Advances Advisory Committee of HFEA recently changed the rating of PGT-A to "red", meaning that there is no evidence to show that the treatment is effective and safe [125].

4. Metabolic and Cardiovascular Consequences of Placental Dysregulation in Mothers Following ART

Studies demonstrated that vascular dysfunctions in IVF pregnancies were similar to those identified when babies were conceived naturally from mothers affected by PE [132,133]. Sundheimer and Pisarska explain that the size of the placental bed and successful trophoblast invasion and spiral artery remodeling determine maternal blood flow. With these premises, abnormal placentation associated with infertility can represent a consequent marker of overall health for both the mother and her offspring [134]. It has been well known for a long time that inadequate trophoblast invasion and poor spiral artery remodeling may cause reduced vascular perfusion in pregnancy [135].

Women of advanced maternal age are generally at increased risk of GDM, HDP, operative and cesarean deliveries [136,137]; however, those who conceive with the use of IVF are at increased risk of retained placenta, suggesting that placentation abnormalities may contribute to maternal morbidity, and this may be more pronounced in women with infertility [138].

Fertility treatments (IVF and NON-IVF combined) have been associated with an increased risk of severe maternal morbidity (SMM), defined as unexpected outcomes of labor and delivery that result in significant short- or long-term consequences to a woman's health after [139]. In a population-based cohort study of 114,409 singleton pregnancies with conceptions dating from 11 January 2013 to 10 January 2014 in Ontario, IVF was associated with an increased risk of SMM (rate 11.3/1000; aRR 1.89, 95% CI: 1.06–3.39) [140]. In this study, the authors also noted supra-additive effects of high body mass index (BMI) and IVF on the risk of PE and GDM. The link between fertility treatment and SMM was also confirmed in another cohort study conducted from Ontario between 2006 and 2012, demonstrating that women undergoing fertility treatment are at higher risk of SMM with a relative risk of 1.39 (95% confidence interval (CI) 1.23–1.56). Indicators of SMM were considered severe postpartum hemorrhage, maternal admission to an intensive care unit, puerperal sepsis, hysterectomy and cardiac conditions [141]. The American Heart

Association considers a history of PE or HDP as a major risk factor in these women for the development of cardiovascular diseases (CVD) [142].

In summary, cardiovascular and metabolic diseases derived from impaired placentation following infertility treatments may have long-term consequences on both the mother and the newborn.

5. Health Risk in Infancy as a Consequence of Placental Dysfunction Following ART

According to the "developmental origins of health and disease theory" [143] and the Barker hypothesis of the developmental origins of chronic adult disease [144], placental dysfunction and abnormal fetal development may have long-term consequences on the neonate and his development, from childhood to adult life. A higher risk of diabetes mellitus and CVD were found in children whose mothers had PE and HDP, by long-term follow up observations [145,146]. A track of this condition is represented by the finding of reduced endothelial function in mothers and children after PE [147].

Evidence suggests, at least in animal studies, that IVF can promote endothelial dysfunction with increased vessel stiffness in offspring. This can induce arterial hypertension in vivo and an epigenetic origin of the condition has been suggested [132]. IVF causality is also suggested by the significantly elevated activity of enzymatic regulators in the cardiovascular system of IVF offspring [148]. Additionally, there is increasing evidence that offspring conceived by IVF displays a level of vascular dysfunction similar to that seen in children spontaneously conceived by mothers with PE [132,133]. Of note, in 2015 in a randomized case-control trial, antioxidant administration to IVF children was able to improve NO bioavailability and responsiveness of both systemic and pulmonary circulation [149], thus indicating that vascular dysfunctions induced by ART are reversible in young people.

Regarding the increased metabolic risk later in adult life, IVF-conceived mice showed decreased glucose tolerance, increased fasting glucose levels, and reduced insulin-stimulated protein kinase B (Akt) phosphorylation in the liver later in their adult life [150]. These findings are similar to those seen in humans. Some individuals conceived by ART have shown LBW, higher weight, height, and BMI in late infancy [151] and fasting insulin levels [152]. Adolescents conceived by ART showed significant differences in growth kinetics, glucose levels, and blood pressure in comparison with those conceived naturally by sub-fertile parents [153,154], suggesting that these alterations may be caused by ART procedures. Development delay at birth (probably caused by epigenetic alterations) and accelerated growth in late infancy, with metabolic alterations in adolescence, may predispose these individuals to chronic diseases such as obesity and type 2 diabetes [155,156]. An epigenetic programming of metabolism during prenatal and postnatal periods as a response to imprinting alterations that occurred during early embryonic development has been hypothesized [156,157].

6. Simultaneous Action of Factors Dysregulating Normal Placentation

Infertility and its treatments, including ART and its various procedures (ICSI, blastocyst culture + FET + PGT-A with ovarian stimulation or hormonal preparation), may individually or synergistically participate in the dysregulation of embryo and placenta development. Consequently, it is probably impossible to separate the relative contribution of the many factors influencing embryo, placental, and newborn health when an infertile woman is treated with the many types of ART techniques. These techniques may represent confounding factors for a full understanding of previous studies and future ones to be designed.

Each technique may lead to important epigenetic changes and differential gene expression in the placenta [82,83] or damage the developing embryos with thermal, oxidative and mechanical stressing actions [82,98,158].

Pivotal reproductive hormones, such as human chorionic gonadotropin, progesterone and estradiol, are found at high concentrations at the maternal–fetal interface during vessel remodeling. This observation allows to hypothesize that such relevant vascular transformation may be under the influence of these hormones, which may control trophoblast migration [90,159]. A potential role in placental function dysregulation has been proposed for estrogens. In this respect, elevated estrogen exposure (as in controlled ovarian hyperstimulation) has been associated with higher rates of LBW and FGR [160].

Are perinatal adverse outcomes in infertile patients the consequence of ART or are they the consequence of the underlying infertility? A possible answer to this basic question may come from infertile couples conceiving spontaneously with no ART treatments; in these couples, a higher risk of PTB and LBW has been reported [1]. However, an increasing number of observations show vascular dysfunctions similar to those observed in PE in IVF offspring [132,133], and a common mechanism of action could be postulated.

The induction of defective methylation and consequent alteration of gene expression, which may impair placentation [85], can be caused by oxidative stress [161,162], as in endometriosis and by the altered hormonal milieu, i.e., supraphysiologic estradiol levels due to ovarian hyperstimulation [97]. Particular negative effects on embryo development may occur in PGT-A procedures, in which the epigenetic risk for long in vitro culture may be added to the trophectoderm mechanical stress. If the patient is also affected by PCOS, this adverse metabolic condition for the placenta would considerably increase the overall obstetric, perinatal and postnatal final risk [76,77]. All these factors could exert additive effects, leading to the pathologic condition.

7. Final Remarks

The huge and increasing number of ART cycles in the world, together with the discovery of epigenetic, obstetric, maternal and newborn associated risk, should raise concerns in the medical and scientific communities. Observing the three main world ART registers (United States of America, Europe, Australia and New Zealand) during the period 1997–2016, we acknowledge that the numbers of recorded ART treatments increased considerably (5.3-fold in Europe, 4.6-fold in the USA, 3.0-fold in Australia and New Zealand [163]. In 2016, the ICSI technique overcome IVF and accounted for 50% to 60% of all cycles of in vitro insemination. A sharp rise in the number of freeze-all cycles has been observed both in Australia and New Zealand (reaching 26.5% of all oocyte retrieval cycles for IVF and ICSI in 2016) and in the USA (19.2% in 2016). It is relevant to highlight that the majority of the freeze-all cycles involve blastocyst stage embryos. Practically, FET has overcome fresh embryo transfer in the USA and Australia. PGT-A in the USA is performed in 40% of IVF cycles with an increasing trend [115] and, in Australia, represents the fourth most used type of ART technique [163]. This context should be considered when trying to investigate related and prospective ART risks in a growing infertile population. From a clinical point of view and according to the presented data, we can imagine an increasing risk with this pattern: ovarian stimulation < IVF < ICSI < Blastocyst culture < PGT-A. On this basis, we could speculate the higher risk in a patient with PCOS or endometriosis that performs PGT-A.

Considering the many factors and metabolic pathways that we are continuously discovering to be involved in reproductive processes, from fertilization to embryo implantation and feto-placental development, we are far from completing the entire and extremely complex picture. Our experimental findings are able to shed light on small and specific tiles of the complex mosaic regarding many reproductive patterns, often not well known or even completely obscure to our understanding and knowledge.

These observations open the question that regard our medical action, that is, the prevention of consequences linked to artificial procedure that move too far away from physiologic patterns. Not completely knowing the interrelated quite complex molecular pathways skipped or artificially modified, we cannot pharmacologically correct them in a more rational and appropriate way.

With these premises, the prevention of impaired placentation might be considered before starting an ART procedure with a metabolic improvement to women (e.g., dietary and pharmacologic tools in case of PCOS and insulin resistance); additionally, a secondary prevention might be represented by the right choice of therapeutic techniques and protocols (e.g., less invasive embryo culture/procedures and mild stimulation) and with a more personalized obstetric control before pathologic conditions of the placenta become clinically evident as hypertensive or metabolic disorders. If pathologic placentation produces detectable effect starting on the fifth or sixth gestational week, we should not wait until the second or third trimester to observe its clinical consequences, and first trimester miscarriages might not be of genetic origin. A higher prevalence of spontaneous abortion with ART should prompt us to investigate both embryo aneuploidies and placental pathology [164].

In conclusion, it is reasonable to continue research on all key points of the reproductive process, and it is logical to try to correct or prevent (using medical procedures) what is becoming pathologic on the basis of our increasing knowledge. However, maybe it is wiser not to stray too far from the physiology of reproduction. Those who are dealing with ART are requested to study with more interest the physiology and the pathology of the placenta, not only from the usual obstetric point of view. Balancing the pro and cons of our reproductive interventions and observing long-term health of the mother and offspring, remain important topics, although much has to be investigated on biological and clinical levels. With this aim, collaborative studies between clinicians and biologists should be strongly encouraged.

Author Contributions: Conceptualization, C.M. and M.M.; writing—original draft preparation, C.M. and M.M.; writing—review and editing, C.M., V.L., G.R. and M.M.; visualization, G.R. and A.D.L.; supervision, A.D.L. and M.M. All authors have read and agreed to the published version of the manuscript.

Funding: This research received no external funding.

Institutional Review Board Statement: Not applicable.

Informed Consent Statement: Not applicable.

Data Availability Statement: Not applicable.

Conflicts of Interest: The authors declare no conflict of interest.

Abbreviations

PTB	preterm birth
LBW	low birth weight
PE	preeclampsia
FGR	fetal growth restriction
PCOS	polycystic ovary syndrome
GDM	gestational diabetes mellitus
ART	assisted reproductive technology
IVF	in vitro fertilization
ICSI	intracytoplasmic sperm injection
FET	frozen embryo transfer
OD	oocyte donation
PGT-A	preimplantation genetic testing for aneuploidy
HDP	hypertensive disorders of pregnancy
VLBW	very low birth weight
PlGF	placental growth factor
sFLT-1	soluble fms-like tyrosine kinase 1
EGFR	epidermal growth factor receptor
sENDOGLIN	soluble endoglin
EGFL7	epidermal growth factor-like domain 7
TGF-β	transforming growth factor-beta
NO	nitric oxide
TNF-α	tumor necrosis factor-alpha

IL	interleukin
SGA	small for gestational age
LGA	large for gestational age
TH	thyroid hormone
LT4	levothyroxine
PIH	pregnancy induced hypertension
PGT-M	preimplantation genetic test for monogenic diseases
BMP-4	bone morphogenetic protein-4
CVD	cardiovascular disease
PWS	Prader–Willi syndrome
AS	Angelman syndrome
BWS	Beckwith–Wiedemann syndrome
SRS	Silver–Russell syndrome
SMM	severe maternal morbidity

References

1. Messerlian, C.; Maclagan, L.; Basso, O. Infertility and the risk of adverse pregnancy outcomes: A systematic review and meta-analysis. *Hum. Reprod.* **2013**, *28*, 125–137. [CrossRef]
2. Healy, D.L.; Breheny, S.; Halliday, J.; Jaques, A.; Rushford, D.; Garrett, C.; Talbot, J.M.; Baker, H.W. Prevalence and risk factors for obstetric haemorrhage in 6730 singleton births after assisted reproductive technology in Victoria Australia. *Hum. Reprod.* **2010**, *25*, 265–274. [CrossRef]
3. Palomba, S.; de Wilde, M.A.; Falbo, A.; Koster, M.P.; La Sala, G.B.; Fauser, B.C. Pregnancy complications in women with polycystic ovary syndrome. *Hum. Reprod. Update* **2015**, *21*, 575–592. [CrossRef]
4. Vambergue, A.; Fajardy, I. Consequences of gestational and pregestational diabetes on placental function and birth weight. *World J. Diabetes* **2011**, *2*, 196–203. [CrossRef]
5. De Geyter, C.; Calhaz-Jorge, C.; Kupka, M.S.; Wyns, C.; Mocanu, E.; Motrenko, T.; Scaravelli, G.; Smeenk, J.; Vidakovic, S.; Goossens, V.; et al. ART in Europe, 2014: Results generated from European registries by ESHRE: The European IVF-monitoring consortium (EIM) for the European society of human reproduction and Embryology (ESHRE). *Hum. Reprod.* **2018**, *33*, 1586–1601. [CrossRef]
6. Helmerhorst, F.M.; Perquin, D.A.M.; Donker, D.; Keirse, M.J.N.C. Perinatal outcome of singletons and twins after assisted conception: A systematic review of controlled studies. *BMJ* **2004**, *328*, 261–264. [CrossRef]
7. Pandey, S.; Shetty, A.; Hamilton, M.; Bhattacharya, S.; Maheshwari, A. Obstetric and perinatal outcomes in singleton pregnancies resulting from IVF/ICSI: A systematic review and meta-analysis. *Hum. Reprod. Update* **2012**, *18*, 485–503. [CrossRef]
8. Luke, B. Pregnancy and birth outcomes in couples with infertility with and without assisted reproductive technology: With an emphasis on US population-based studies. *Am. J. Obstet. Gynecol.* **2017**, *217*, 270–281. [CrossRef]
9. Ginström Ernstad, E.; Wennerholm, U.; Khatibi, A.; Petzold, M.; Bergh, C. Neonatal and maternal outcome after frozen embryo transfer; increased risks in programmed cycles. *Am. J. Obstet. Gynecol.* **2019**, *221*, 126.e1–126.e18. [CrossRef]
10. Saito, K.; Kuwahara, A.; Ishikawa, T.; Morisaki, N.; Miyado, M.; Miyado, K.; Fukami, M.; Miyasaka, N.; Ishihara, O.; Irahara, M.; et al. Endometrial preparation methods for frozen-thawed embryo transfer are associated with altered risks of hypertensive disorders of pregnancy, placenta accreta, and gestational diabetes mellitus. *Hum. Reprod.* **2019**, *34*, 1567–1575. [CrossRef]
11. Sunderam, S.; Kissin, D.M.; Zhang, Y.; Folger, S.G.; Boulet, S.L.; Warner, L.; Callaghan, W.M.; Barfield, W.D. Assisted reproductive technology surveillance—United States, 2016. *MMWR Surveill. Summ.* **2019**, *68*, 1–23. [CrossRef]
12. Luke, B.; Brown, M.B.; Eisenberg, M.L.; Callan, C.; Botting, B.J.; Pacey, A.; Alastair, G. In vitro fertilization and risk for hypertensive disorders of pregnancy: Associations with treatment parameters. *Am. J. Obstet. Gynecol.* **2020**, *222*, 350.e1–350.e13. [CrossRef]
13. Manna, C. *Maternal-Fetal Medicine, Practical Aspects II*; Arduini, D., Palermo, M.S.F., Eds.; AMOLCA Publishing House: Medellin, Colombia, 2021.
14. Qin, J.; Liu, X.; Sheng, X.; Wang, H.; Gao, S. Assisted reproductive technology and the risk of pregnancy-related complications and adverse pregnancy outcomes in singleton pregnancies: A meta- analysis of cohort studies. *Fertil. Steril.* **2016**, *105*, 73–85. [CrossRef]
15. Xiang, M.; Chen, S.; Zhang, X.; Ma, Y. Placental diseases associated with assisted reproductive technology. *Reprod. Biol.* **2021**, *21*, 100505. [CrossRef]
16. Yang, X.; Li, Y.; Li, C.; Zhang, W. Current overview of pregnancy complications and live-birth outcome of assisted reproductive technology in mainland China. *Fertil. Steril.* **2014**, *101*, 385–391. [CrossRef] [PubMed]
17. Haavaldsen, C.; Tanbo, T.; Eskild, A. Placental weight in singleton pregnancies with and without assisted reproductive technology: A population study of 536,567 pregnancies. *Hum. Reprod.* **2012**, *27*, 576–582. [CrossRef] [PubMed]
18. Reig, A.; Seli, E. The association between assisted reproductive technologies and low birth weight. *Curr. Opin. Obstet. Gynecol.* **2019**, *31*, 183–187. [CrossRef]
19. Joy, J.; Gannon, C.; McClure, N.; Cooke, I. Is Assisted Reproduction Associated with Abnormal Placentation? *Pediatr. Dev. Pathol.* **2012**, *15*, 306–314. [CrossRef]

20. Ochoa, E. Alteration of Genomic Imprinting after Assisted Reproductive Technologies and Long-Term Health. *Life* **2021**, *11*, 728. [CrossRef] [PubMed]
21. Kopca, T.; Tulay, P. Association of Assisted Reproductive Technology Treatments with Imprinting Disorders. *Glob. Med. Genet.* **2021**, *8*, 1–6. [CrossRef]
22. Fontana, L.; Tabano, S.; Maitz, S.; Colapietro, P.; Garzia, E.; Gerli, A.G.; Sirchia, S.M.; Miozzo, M. Clinical and Molecular Diagnosis of Beckwith-Wiedemann Syndrome with Single- or Multi-Locus Imprinting Disturbance. *Int. J. Mol. Sci.* **2021**, *22*, 3445. [CrossRef]
23. Kroener, L.; Wang, E.T.; Pisarska, M.D. Predisposing Factors to Abnormal First Trimester Placentation and the Impact on Fetal Outcomes. *Semin. Reprod. Med.* **2016**, *34*, 27–35. [CrossRef]
24. Brosens, I.; Pijnenborg, R.; Vercruysse, L.; Romero, R. The "Great Obstetrical Syndromes" are associated with disorders of deep placentation. *Am. J. Obstet. Gynecol.* **2011**, *204*, 193–201. [CrossRef]
25. Fisher, S.J. Why is placentation abnormal in preeclampsia? *Am. J. Obstet. Gynecol.* **2015**, *213* (Suppl. 4), S115–S122. [CrossRef]
26. Burton, G.J.; Yung, H.W.; Cindrova-Davies, T.; Charnock-Jones, D.S. Placental endoplasmic reticulum stress and oxidative stress in the pathophysiology of unexplained intrauterine growth restriction and early onset preeclampsia. *Placenta* **2009**, *30*, S43–S48. [CrossRef] [PubMed]
27. Jackson, M.R.; Walsh, A.J.; Morrow, R.J.; Mullen, J.B.; Lye, S.J.; Ritchie, J.W. Reduced placental villous tree elaboration in small-for-gestational-age pregnancies: Relationship with umbilical artery Doppler waveforms. *Am. J. Obstet. Gynecol.* **1995**, *172*, 518–525. [CrossRef]
28. Redman, C.W.; Sargent, I.L. Latest advances in understanding preeclampsia. *Science* **2005**, *308*, 1592–1594. [CrossRef] [PubMed]
29. Benschop, L.; Schalekamp-Timmermans, S.; Broere-Brown, Z.A.; Roeters van Lennep, J.E.; Jaddoe, V.W.V.; Roos-Hesselink, J.W.; Ikram, M.K.; Steegers, E.A.P.; Robert, J.M.; Gandley, R.E. Placental growth factor as an indicator of maternal cardiovascular risk after pregnancy. *Circulation* **2019**, *139*, 1698–1709. [CrossRef]
30. Redman, C.W.; Staff, A.C. Preeclampsia, biomarkers, syncytiotrophoblast stress, and placental capacity. *Am. J. Obstet. Gynecol.* **2015**, *213* (Suppl. 4), S9.e1–S9.e4. [CrossRef]
31. Zeisler, H.; Llurba, E.; Chantraine, F.; Vatish, M.; Staff, A.C.; Sennström, M.; Olovsson, M.; Brennecke, S.P.; Stepan, H.; Allegranza, D.; et al. Pre-dictive Value of the sFlt-1:PlGF Ratio in Women with Suspected Preeclampsia. *N. Engl. J. Med.* **2016**, *374*, 13–22. [CrossRef]
32. Stepan, H.; Unversucht, A.; Wessel, N.; Faber, R. Predictive value of maternal angiogenic factors in second trimester pregnancies with abnormal uterine perfusion. *Hypertension* **2007**, *49*, 818–824. [CrossRef]
33. Hertig, A.; Berkane, N.; Lefevre, G.; Toumi, K.; Marti, H.P.; Capeau, J.; Uzan, S.; Rondeau, E. Maternal serum sFLT-1 concentration is an early and reliable predictive marker of preeclampsia. *Clin. Chem.* **2004**, *50*, 1702–1703. [CrossRef]
34. Tidwell, S.C.; Ho, H.N.; Chiu, W.H.; Torry, R.J.; Torry, D.S. Low maternal serum levels of placenta growth factor as an antecedent of clinical preeclampsia. *Am. J. Obstet. Gynecol.* **2001**, *184*, 1267–1272. [CrossRef] [PubMed]
35. Thadhani, R.; Mutter, W.P.; Wolf, M.; Levine, R.J.; Taylor, R.N.; Sukhatme, V.P.; Ecker, J.; Karumanchi, S.A. First trimester placental growth factor and solu-ble fms-like tyrosine kinase 1 and risk for preeclampsia. *J. Clin. Endocrinol. Metab.* **2004**, *89*, 770–775. [CrossRef] [PubMed]
36. Hastie, R.; Brownfoot, F.C.; Pritchard, N.; Hannan, N.J.; Cannon, P.; Nguyen, V.; Palmer, K.; Beard, S.; Tong, S.; Kaitu'u-Lino, T.J. EGFR (Epidermal Growth Factor Receptor) Signaling and the Mitochondria Regulate sFlt-1 (Soluble FMS-Like Tyrosine Kinase-1) Secretion. *Hypertension* **2019**, *73*, 659–670. [CrossRef]
37. Vrooman, L.A.; Rhon-Calderon, E.A.; Chao, O.Y.; Nguyen, D.K.; Narapareddy, L.; Dahiya, A.K.; Putt, M.E.; Schultz, R.M.; Bartolomei, M.S. Assisted reproductive technologies induce temporally specific placental defects and the PE risk marker sFLT-1 in mouse. *Development* **2020**, *29*, 147. [CrossRef]
38. Cui, L.; Shu, C.; Liu, Z.; Tong, W.; Cui, M.; Wei, C.; Tang, J.J.; Liu, X.; Hu, J.; Jiang, J.; et al. The expression of serum sEGFR, sFlt-1, sEndoglin and PLGF in preeclampsia. *Pregnancy Hypertens.* **2018**, *13*, 127–132. [CrossRef]
39. Levine, R.J.; Lam, C.; Qian, C.; Yu, K.F.; Maynard, S.E.; Sachs, B.P.; Sibai, B.M.; Epstein, F.H.; Romero, R.; Thadhani, R.; et al. Soluble endoglin and other circulating antiangiogenic factors in preeclampsia. *N. Engl. J. Med.* **2006**, *355*, 992–1005. [CrossRef]
40. Lacko, L.A.; Massimiani, M.; Sones, J.L.; Hurtado, R.; Salvi, S.; Ferrazzani, S.; Davisson, R.L.; Campagnolo, L.; Stuhlmann, H. Novel expression of EGFL7 in placental trophoblast and endothelial cells and its implication in preeclampsia. *Mech. Dev.* **2014**, *133*, 163–176. [CrossRef] [PubMed]
41. Massimiani, M.; Vecchione, L.; Piccirilli, D.; Spitalieri, P.; Amati, F.; Salvi, S.; Ferrazzani, S.; Stuhlmann, H.; Campagnolo, L. Epidermal growth factor-like domain 7 promotes migration and invasion of human trophoblast cells through activation of MAPK, PI3K and NOTCH signaling pathways. *Mol. Hum. Reprod.* **2015**, *21*, 435–451. [CrossRef]
42. Lacko, L.A.; Hurtado, R.; Hinds, S.; Poulos, M.G.; Butler, J.M.; Stuhlmann, H. Altered feto-placental vascularization, feto-placental malperfusion and fetal growth restriction in mice with Egfl7 loss of function. *Development.* **2017**, *144*, 2469–2479. [CrossRef]
43. Massimiani, M.; Salvi, S.; Tiralongo, G.M.; Moresi, S.; Stuhlmann, H.; Valensise, H.; Lanzone, A.; Campagnolo, L. Circulating EGFL7 distinguishes between FGR and PE: An observational case-control study. *Sci. Rep.* **2021**, *11*, 17919. [CrossRef]
44. Massimiani, M.; Lacko, L.A.; Burke Swanson, C.S.; Salvi, S.; Argueta, L.B.; Moresi, S.; Ferrazzani, S.; Gelber, S.E.; Baergen, R.N.; Toschi, N.; et al. Increased circulating levels of Epidermal Growth Factor-like Domain 7 in pregnant women affected by preeclampsia. *Transl. Res.* **2019**, *207*, 19–29. [CrossRef]

45. Massimiani, M.; Tiralongo, G.M.; Salvi, S.; Fruci, S.; Lacconi, V.; La Civita, F.; Mancini, M.; Stuhlmann, H.; Valensise, H.; Campagnolo, L. Treatment of pregnancies complicated by intrauterine growth restriction with nitric oxide donors increases placental expression of Epidermal Growth Factor-Like Domain 7 and improves fetal growth: A pilot study. *Transl. Res.* **2021**, *228*, 28–41. [CrossRef] [PubMed]
46. Valensise, H.; Vasapollo, B.; Novelli, G.P.; Giorgi, G.; Verallo, P.; Galante, A.; Arduini, D. Maternal and fetal hemodynamic effects induced by nitric oxide donors and plasma volume expansion in pregnancies with gestational hypertension complicated by intrauterine growth restriction with absent end diastolic flow in the umbilical artery. *Ultrasound Obstet. Gynecol.* **2008**, *31*, 55–64. [CrossRef]
47. Berbets, A.; Koval, H.; Barbe, A.; Albota, O.; Yuzko, O. Melatonin decreases and cytokines increase in women with placental insufficiency. *J. Matern.-Fetal Neonatal Med.* **2019**, *25*, 1–6. [CrossRef] [PubMed]
48. Chuffa, L.G.; Lupi, L.A., Jr.; Seiva, F.R.; Martinez, M.; Domeniconi, R.F.; Pinheiro, P.F.; Dos Santos, L.D.; Martinez, F.E. Quantitative proteomic profiling reveals that diverse metabolic pathways are influenced by melatonin in an in vivo model of ovarian carcinoma. *J. Proteome Res.* **2016**, *15*, 3872–3882. [CrossRef]
49. De Almeida Chuffa, L.G.; Seiva, F.R.F.; Cucielo, M.S.; Silveira, H.S.; Reiter, R.J.; Lupi, L.A. Mitochondrial functions and melatonin: A tour of the reproductive cancers. *Cell. Mol. Life Sci.* **2019**, *76*, 837–863. [CrossRef] [PubMed]
50. Itoh, M.T.; Ishizuka, B.; Kudo, Y.; Fusama, S.; Amemiya, A.; Sumi, Y. Detection of melatonin and serotonin N-acetyltransferase and hydroxyindole-Omethyltransferase activities in the rat ovary. *Mol. Cell Endocrinol.* **1997**, *136*, 7–14. [CrossRef]
51. Itoh, M.T.; Ishizuka, B.; Kuribayashi, Y.; Amemiya, A.; Sumi, Y. Melatonin, its precursors, and synthesizing enzyme activities in the human ovary. *Mol. Hum. Reprod.* **1999**, *5*, 402–408. [CrossRef] [PubMed]
52. El-Raey, M.; Geshi, M.; Somfai, T.; Kaneda, M.; Hirako, M.; Abdel-Ghaffar, A.E.; Sosa, G.A.; El-Roos, M.E.A.; Nagai, T. Evidence of melatonin synthesis in the cumulus oocyte complexes and its role in enhancing oocyte maturation in vitro in cattle. *Mol. Reprod. Dev.* **2011**, *78*, 250–262. [CrossRef]
53. Amireault, P.; Dube, F. Serotonin and its antidepressant-sensitive transport in mouse cumulus-oocyte complexes and early embryos. *Biol. Reprod.* **2005**, *73*, 358–365. [CrossRef]
54. Sakaguchi, K.; Itoh, M.T.; Takahaashi, N.; Tarumi, W.; Ishizuka, B. The rat oocyte synthesises melatonin. *Reprod. Fertil. Dev.* **2013**, *25*, 674–682. [CrossRef]
55. Tamura, H.; Takasaki, A.; Miwa, I.; Taniguchi, K.; Maekawa, R.; Asada, H.; Taketani, T.; Matsuoka, A.; Yamagata, Y.; Shimamura, K.; et al. Oxidative stress impairs oocyte quality and melatonin protects oocytes from free radical damage and improves fertilization rate. *J. Pineal Res.* **2008**, *44*, 280–287. [CrossRef] [PubMed]
56. Zheng, M.; Tong, J.; Li, W.P.; Chen, Z.J.; Zhang, C. Melaton in concentration in follicular fluid is correlated with antral follicle count (AFC) and in vitro fertilization (IVF) outcomes in women undergoing assisted reproductive technology (ART) procedures. *Gynecol. Endocrinol.* **2018**, *34*, 446–450. [CrossRef] [PubMed]
57. De Almeida Chuffa, L.G.; Seiva, F.R.F.; Cucielo, M.S.; Silveira, H.S.; Reiter, R.J.; Lupi, L.A. Clock genes and the role of melatonin in cancer cells: An overview. *Melatonin Res.* **2019**, *2*, 133–157. [CrossRef]
58. Moshkdanian, G.; Moghani-Ghoroghi, F.; Pasbakhsh, P.; Nematollahi-Mahani, S.N.; Najafi, A.; Kashani, S.R. Melatonin upregulates ErbB1 and ErbB4, two primary implantation receptors, in pre-implantation mouse embryos. *Iran. J. Basic Med. Sci.* **2017**, *20*, 655–661. [CrossRef]
59. Pan, B.; Qazi, I.H.; Guo, S.; Yang, J.; Qin, J.; Lv, T.; Zang, S.; Zhang, Y.; Zeng, C.; Meng, Q.; et al. Melatonin improves the first cleavage of parthenogenetic embryos from vitrified-warmed mouse oocytes potentially by promoting cell cycle progression. *J. Anim. Sci. Biotechnol.* **2021**, *12*, 84. [CrossRef]
60. Bourgain, C.; Devroey, P. The endometrium in stimulated cycles for IVF. *Hum. Reprod. Update* **2003**, *9*, 515–522. [CrossRef]
61. Bonagura, T.W.; Pepe, G.J.; Enders, A.C.; Albrecht, E.D. Suppression of extravillous trophoblast vascular endothelial growth factor expression and uterine spiral artery invasion by estrogen during early baboon pregnancy. *Endocrinology* **2008**, *149*, 5078–5087. [CrossRef]
62. Alyasin, A.; Agha-Hosseini, M.; Kabirinasab, M.; Saeidi, H.; Nashtaei, M.S. Serum progesterone levels greater than 32,5 ng/ml on the day of embryo transfer are associated with lower live birth rate after artificial endometrial preparation: A prospective study. *Reprod. Biol. Endocrinol.* **2021**, *19*, 24. [CrossRef]
63. von Versen-Höynck, F.; Schaub, A.M.; Chi, Y.Y.; Chiu, K.H.; Liu, J.; Lingis, M.; Stan Williams, R.; Rhoton-Vlasak, A.; Nichols, W.W.; Fleischmann, R.R.; et al. Increased preeclampsia risk and reduced aortic compliance with in vitro fertilization cycles in the absence of a corpus luteum. *Hypertension* **2019**, *73*, 640–649. [CrossRef]
64. Makhijani, R.; Bartels, C.; Godiwala, P.; Bartolucci, A.; Nulsen, J.; Grow, D.; Benadiva, C.; Engmann, L. Maternal and perinatal outcomes in programmed versus natural vitrified-warmed blastocyst transfer cycles. *Reprod. Biomed. Online* **2020**, *41*, 300–308. [CrossRef] [PubMed]
65. Zhang, W.Y.; von Versen-Höynck, F.; Kapphahn, K.I.; Fleischmann, R.R.; Zhao, Q.; Baker, V.L. Maternal and neonatal outcomes associated with trophectoderm biopsy. *Fertil. Steril.* **2019**, *112*, 283–290. [CrossRef] [PubMed]
66. Marshall, S.A.; Leo, C.H.; Senadheera, S.N.; Girling, J.E.; Tare, M.; Parry, L.J. Relaxin deficiency attenuates pregnancy-induced adaptation of the mesenteric artery to angiotensin II in mice. *Am. J. Physiol. Regul. Integr. Comp. Physiol.* **2016**, *310*, R847–R857. [CrossRef] [PubMed]

67. von Versen-Höynck, F.; Strauch, N.K.; Liu, J.; Chi, Y.Y.; Keller-Woods, M.; Conrad, K.P.; Baker, V.L. Effect of mode of conception on maternal serum relaxin, creatinine, and sodium concentrations in an infertile population. *Reprod. Sci.* **2019**, *26*, 412–419. [CrossRef]
68. Obregon, M.J.; Mallol, J.; Pastor, R.; Morreale de Escobar, G.; Escobar del Rey, F. L-thyroxine and 3,5,3′-triiodo-L-thyronine in rat embryos before onset of fetal thyroid function. *Endocrinology* **1984**, *114*, 305–307. [CrossRef]
69. van den Boogaard, E.; Vissenberg, R.; Land, J.A.; van Wely, M.; van der Post, J.A.; Goddijn, M.; Bisschop, P.H. Significance of (sub)clinical thyroid dysfunction and thyroid autoimmunity before conception and in early pregnancy: A systematic review. *Hum. Reprod. Update* **2011**, *17*, 605–619. [CrossRef] [PubMed]
70. Piccirilli, D.; Baldini, E.; Massimiani, M.; Camaioni, A.; Salustri, A.; Bernardini, R.; Centanni, M.; Ulisse, S.; Moretti, C.; Campagnolo, L. Thyroid hormone regulates protease expression and activation of Notch signaling in implantation and embryo development. *J. Endocrinol.* **2018**, *236*, 1–12. [CrossRef]
71. Dal Lago, A.; Galanti, F.; Miriello, D.; Marcoccia, A.; Massimiani, M.; Campagnolo, L.; Moretti, C.; Rago, R. Positive Impact of Levothyroxine Treatment on Pregnancy Outcome in Euthyroid Women with Thyroid Autoimmunity Affected by Recurrent Miscarriage. *J. Clin. Med.* **2021**, *10*, 2105. [CrossRef]
72. Brosens, I.; Brosens, J.J.; Fusi, L.; Al-Sabbagh, M.; Kuroda, K.; Benagiano, G. Risks of adverse pregnancy outcome in endometriosis. *Fertil. Steril.* **2012**, *98*, 30–35. [CrossRef]
73. Breintoft, K.; Pinnerup, R.; Henriksen, T.; Rytter, D.; Uldbjerg, N.; Forman, A.; Arendt, L. Endometriosis and Risk of Adverse Pregnancy Outcome: A Systematic Review and Meta-Analysis. *J. Clin. Med.* **2021**, *10*, 667. [CrossRef]
74. Jindal, P.; Regan, L.; Fourkala, E.O.; Rai, R.; Moore, G.; Goldin, R.D.; Sebire, N.J. Placental pathology of recurrent spontaneous abortion: The role of histopathological examination of products of conception in routine clinical practice: A mini review. *Hum. Reprod.* **2007**, *22*, 313–316. [CrossRef] [PubMed]
75. Longtine, M.S.; Nelson, D.M. Placental dysfunction and fetal programming: The importance of placental size, shape, histopathology, and molecular composition. *Semin. Reprod. Med.* **2011**, *29*, 187–196. [CrossRef]
76. Palomba, S.; Russo, T.; Falbo, A.; Di Cello, A.; Tolino, A.; Tucci, L.; La Sala, G.B.; Zullo, F. Macroscopic and microscopic findings of the placenta in women with polycystic ovary syndrome. *Hum. Reprod.* **2013**, *28*, 2838–2847. [CrossRef] [PubMed]
77. Palomba, S.; Russo, T.; Falbo, A.; Di Cello, A.; Amendola, G.; Mazza, R.; Tolino, A.; Zullo, F.; Tucci, L.; La Sala, G.B. Decidual endovascular trophoblast invasion in women with polycystic ovary syndrome: An experimental case-control study. *J. Clin. Endocrinol. Metab.* **2012**, *97*, 2441–2449. [CrossRef]
78. Palomba, S.; Falbo, A.; Russo, T.; Battista, L.; Tolino, A.; Orio, F.; Zullo, F. Uterine blood flow in pregnant patients with polycystic ovary syndrome: Relationships with clinical outcomes. *BJOG* **2010**, *117*, 711–721. [CrossRef] [PubMed]
79. Sha, T.; Wang, X.; Cheng, W.; Yan, Y. A meta-analysis of pregnancy-related outcomes and complications in women with polycystic ovary syndrome undergoing IVF. *Reprod. Biomed. Online* **2019**, *39*, 281–293. [CrossRef]
80. Maccani, M.A.; Marsit, C.J. Epigenetics in the Placenta. *Am. J. Reprod. Immunol.* **2009**, *62*, 78–89. [CrossRef]
81. Maheshwari, A.; Hamilton, M.; Bhattacharya, S. Should we be promoting embryo transfer at blastocyst stage? *Reprod. Biomed. Online* **2016**, *32*, 142–146. [CrossRef] [PubMed]
82. Katari, S.; Turan, N.; Bibikova, M.; Erinle, O.; Chalian, R.; Foster, M.; Gaughan, J.P.; Coutifaris, C.; Sapienza, C. DNA methylation and gene expression differences in children conceived in vitro or in vivo. *Hum. Mol. Genet.* **2009**, *18*, 3769–3778. [CrossRef]
83. Nelissen, E.C.; van Montfoort, A.P.; Dumoulin, J.C.; Evers, J.L. Epigenetics and the placenta. *Hum. Reprod. Update* **2011**, *17*, 397–417. [CrossRef]
84. Robins, J.C.; Marsit, C.J.; Padbury, J.F.; Sharma, S.S. Endocrine disruptors, environmental oxygen, epigenetics and pregnancy. *Front. Biosci.* **2011**, *3*, 690–700. [CrossRef]
85. Novakovic, B.; Rakyan, V.; Ng, H.K.; Manuelpillai, U.; Dewi, C.; Wong, N.C.; Morley, R.; Down, T.; Beck, S.; Craig, J.M.; et al. Specific tumour-associated methylation in normal human term placenta and first-trimester cytotrophoblasts. *Hum. Reprod.* **2008**, *14*, 547–554. [CrossRef]
86. de Waal, E.; Mak, W.; Calhoun, S.; Stein, P.; Ord, T.; Krapp, C.; Coutifaris, C.; Schultz, R.M.; Bartolomei, M.S. In vitro culture increases the frequency of stochastic epigenetic errors at imprinted genes in placental tissues from mouse concepti produced through assisted reproductive technologies. *Biol. Reprod.* **2014**, *90*, 22. [CrossRef] [PubMed]
87. Li, B.; Chen, S.; Tang, N.; Xiao, X.; Huang, J.; Jiang, F.; Huang, X.; Sun, F.; Wang, X. Assisted reproduction causes reduced fetal growth associated with downregulation of paternally expressed imprinted genes that enhance fetal growth in mice. *Biol. Reprod.* **2016**, *94*, 45. [CrossRef]
88. Xiang, M.; Ma, Y.; Lei, H.; Wen, L.; Chen, S.; Wang, X. In vitro fertilization placenta overgrowth in mice is associated with downregulation of the paternal imprinting gene H19. *Mol. Reprod. Dev.* **2019**, *86*, 1940–1950. [CrossRef] [PubMed]
89. Chen, S.; Sun, F.Z.; Huang, X.; Wang, X.; Tang, N.; Zhu, B.; Li, B. Assisted reproduction causes placental maldevelopment and dysfunction linked to reduced fetal weight in mice. *Sci. Rep.* **2015**, *5*, 10596. [CrossRef]
90. Lee, B.; Kroener, L.L.; Xu, N.; Wang, E.T.; Banks, A.; Williams, J.; Goodarzi, M.O.; Chen, Y.I.; Tang, J.; Wang, Y.; et al. Function and Hormonal Regulation of GATA3 in Human First Trimester Placentation. *Biol. Reprod.* **2016**, *95*, 113. [CrossRef] [PubMed]
91. Zhang, Y.; Cui, Y.; Zhou, Z.; Sha, J.; Li, Y.; Liu, J. Altered global gene expressions of human placentae subjected to assisted reproductive technology treatments. *Placenta* **2010**, *31*, 251–258. [CrossRef]

92. Zhao, L.; Zheng, X.; Liu, J.; Zheng, R.; Yang, R.; Wang, Y.; Sun, L. The placental transcriptome of the first- trimester placenta is affected by in vitro fertilization and embryo transfer. *Reprod. Biol. Endocrinol.* **2019**, *17*, 50. [CrossRef]
93. Kleijkers, S.H.M.; Mantikou, E.; Slappendel, E.; Consten, D.; van Echten-Arends, J.; Wetzels, A.M.; van Wely, M.; Smits, L.J.M.; van Montfoort, A.P.A.; Repping, S.; et al. Influence of embryo culture medium (G5 and HTF) on pregnancy and perinatal outcome after IVF: A multicenter RCT. *Hum. Reprod.* **2016**, *31*, 2219–2230. [CrossRef]
94. Cagnone, G.; Sirard, M.A. The embryonic stress response to in vitro culture: Insight from genomic analysis. *Reproduction* **2016**, *152*, R247–R261. [CrossRef] [PubMed]
95. Salilew-Wondim, D.; Saeed-Zidane, M.; Hoelker, M.; Gebremedhn, S.; Poirier, M.; Pandey, H.O.; Tholen, E.; Neuhoff, C.; Held, E.; Besenfelder, U.; et al. Genome-wide DNA methylation patterns of bovine blastocysts derived from in vivo embryos subjected to in vitro culture before, during or after embryonic genome activation. *BMC Genom.* **2018**, *19*, 424. [CrossRef]
96. Asami, M.; Lam, B.; Ma, M.K.; Rainbow, K.; Braun, S.; VerMilyea, M.D.; Yeo, G.; Perry, A. Human embryonic genome activation initiates at the one-cell stage. *Cell Stem Cell* **2021**. [CrossRef]
97. Berntsen, S.; Söderström-Anttila, V.; Wennerholm, U.B.; Laivuori, H.; Loft, A.; Oldereid, N.B.; Romundstad, L.B.; Bergh, C.; Pinborg, A. The health of children conceived by ART: "The chicken or the egg?". *Hum. Reprod. Update* **2019**, *25*, 137–158. [CrossRef] [PubMed]
98. Song, S.; Ghosh, J.; Mainigi, M.; Turan, N.; Weinerman, R.; Truongcao, M.; Coutifaris, C.; Sapienza, C. DNA methylation differences between in vitro- and in vivo-conceived children are associated with ART procedures rather than infertility. *Clin. Epigenet.* **2015**, *7*, 41. [CrossRef]
99. Xu, N.; Barlow, G.M.; Cur, J.; Wang, E.T.; Lee, B.; Akhlaghpour, M.; Kroener, L.; Williams, J.; Rotter, J.I.; Chen, Y.I.; et al. Comparison of genome-wide and gene-specific DNA methylation profiling in first trimester chorionic villi from pregnancies conceived with infertility treatments. *Reprod. Sci.* **2017**, *24*, 996–1004. [CrossRef] [PubMed]
100. Choux, C.; Carmignac, V.; Bruno, C.; Sagot, P.; Vaiman, D.; Fauque, P. The placenta: Phenotypic and epigenetic modifications induced by Assisted Reproductive Technologies throughout pregnancy. *Clin. Epigenet.* **2015**, *7*, 87. [CrossRef]
101. Ginstrom Ernstad, E.; Bergh, C.; Khatibi, A.; Kallen, K.B.; Westlander, G.; Nilsson, S.; Wennerholm, U.B. Neonatal and maternal outcome after blastocyst transfer: A population-based registry study. *Am. J. Obstet. Gynecol.* **2016**, *214*, 378.e1–378.e10. [CrossRef]
102. Makinen, S.; Soderstrom-Anttila, V.; Vainio, J.; Suikkari, A.M.; Tuuri, T. Does long in vitro culture promote large for gestational age babies? *Hum. Reprod.* **2013**, *28*, 828–834. [CrossRef]
103. Ishihara, O.; Araki, R.; Kuwahara, A.; Itakura, A.; Saito, H.; Adamson, G.D. Impact of frozen-thawed single-blastocyst transfer on maternal and neonatal outcome: An analysis of 277,042 single-embryo transfer cycles from 2008 to 2010 in Japan. *Fertil. Steril.* **2014**, *101*, 128–133. [CrossRef]
104. Zhu, J.; Lin, S.; Li, M.; Chen, L.; Lian, Y.; Liu, P.; Qiao, J. Effect of in vitro culture period on birthweight of singleton newborn. *Hum. Reprod.* **2014**, *29*, 448–454. [CrossRef]
105. Huang, J.; Yang, X.; Wu, J.; Kuang, Y.; Wang, Y. Impact of Day 7 Blastocyst Transfer on Obstetric and Perinatal Outcome of Singletons Born After Vitrified-Warmed Embryo Transfer. *Front. Physiol.* **2020**, *11*, 74. [CrossRef] [PubMed]
106. Palermo, G.; Joris, H.; Devroey, P.; Van Steirteghem, A.C. Pregnancies after intracytoplasmic injection of single spermatozoon into an oocyte. *Lancet* **1992**, *340*, 17–18. [CrossRef]
107. Pereira, N.; O'Neill, C.; Lu, V.; Rosenwaks, Z.; Palermo, G.D. The safety of intracytoplasmic sperm injection and long-term outcomes. *Reproduction* **2017**, *154*, F61–F70. [CrossRef] [PubMed]
108. Feuer, S.K.; Liu, X.; Donjacour, A.; Simbulan, R.; Maltepe, E.; Rinaudo, P. Transcriptional Signatures throughout Development: The Effects of Mouse Embryo Manipulation In Vitro. *Reproduction* **2017**, *153*, 107–122. [CrossRef]
109. Hiura, H.; Okae, H.; Miyauchi, N.; Sato, F.; Sato, A.; Van De Pette, M.; John, R.M.; Kagami, M.; Nakai, K.; Soejima, H.; et al. Characterization of DNA Methylation Errors in Patients with Imprinting Disorders Conceived by Assisted Reproduction Technologies. *Hum. Reprod.* **2012**, *27*, 2541–2548. [CrossRef]
110. Hiura, H.; Okae, H.; Chiba, H.; Miyauchi, N.; Sato, F.; Sato, A.; Arima, T. Imprinting Methylation Errors in ART. *Reprod. Med. Biol.* **2014**, *13*, 193–202. [CrossRef]
111. Hattori, H.; Hiura, H.; Kitamura, A.; Miyauchi, N.; Kobayashi, N.; Takahashi, S.; Okae, H.; Kyono, K.; Kagami, M.; Ogata, T.; et al. Association of Four Imprinting Disorders and ART. *Clin. Epigenet.* **2019**, *11*, 1–12. [CrossRef]
112. Gundogan, F.; Bianchi, D.W.; Scherjon, S.A.; Roberts, D.J. Placental pathology in egg donor pregnancies. *Fertil. Steril.* **2010**, *93*, 397–404. [CrossRef] [PubMed]
113. Rizzo, G.; Aiello, E.; Pietrolucci, M.E.; Arduini, D. Placental volume and uterine artery Doppler evaluation at 11 + 0 to 13 + 6 weeks' gestation in pregnancies conceived with in-vitro fertilization: Comparison between autologous and donor oocyte recipients. *Ultrasound Obstet. Gynecol.* **2016**, *47*, 726–731. [CrossRef] [PubMed]
114. Ganer Herman, G.; Tamayev, L.; Feldstein, O.; Bustan, M.; Rachimiel, Z.; Schreiber, L.; Raziel, A.; Bar, J.; Kovo, M. Placental-related disorders of pregnancy and IVF: Does placental histological examination explain the excess risk? *Reprod. Biomed. Online* **2020**, *41*, 81–87. [CrossRef]
115. Munne, S. Status of preimplantation genetic testing and embryo selection. *Reprod. Biomed. Online* **2018**, *37*, 393–396. [CrossRef] [PubMed]
116. Paulson, R.J. Outcome of in vitro fertilization cycles with preimplantation genetic testing for aneuploidies: Let's be honest with one another. *Fertil. Steril.* **2019**, *112*, 1013–1014. [CrossRef]

117. American College of Obstetricians and Gynecologists. ACOG Practice Bulletin No. 134: Fetal growth restriction. *Obstet. Gynecol.* **2013**, *121*, 1122–1133. [CrossRef]
118. American College of Obstetricians and Gynecologists. ACOG Committee Opinion No. 743: Low dose aspirin use in pregnancy. *Obstet. Gynecol.* **2018**, *132*, e44–e52. [CrossRef]
119. American College of Obstetricians and Gynecologists. ACOG Practice Bulletin No. 202: Gestational hypertension and pre-eclampsia. *Obstet. Gynecol.* **2019**, *133*, 211–218. [CrossRef]
120. American College of Obstetricians and Gynecologists and Society for the Maternal-Fetal Medicine. Obstetrics care consensus: Placenta accreta spectrum. *Obstet. Gynecol.* **2018**, *132*, 259–275. [CrossRef] [PubMed]
121. Jing, S.; Luo, K.; He, H.; Lu, C.; Zhang, S.; Tan, Y.; Gong, F.; Lu, G.; Lin, G. Obstetric and neonatal outcomes in blastocyst-stage biopsy with frozen embryo transfer and cleavage-stage biopsy with fresh embryo transfer after preimplantation genetic diagnosis/screening. *Fertil. Steril.* **2016**, *106*, 105–112. [CrossRef] [PubMed]
122. Feldman, B.; Orvieto, R.; Weisel, M.; Aizer, A.; Meyer, R.; Haas, J.; Kirshenbaum, M. Obstetric and Perinatal Outcomes in Pregnancies Conceived After Preimplantation Genetic Testing for Monogenetic Diseases. *Obstet. Gynecol.* **2020**, *136*, 782–791. [CrossRef]
123. Makhijani, R.; Bartels, C.B.; Godiwala, P.; Bartolucci, A.; DiLuigi, A.; Nulsen, J.; Grow, D.; Benadiva, C.; Engmann, L. Impact of trophectoderm biopsy on obstetric and perinatal outcomes following frozen–thawed embryo transfer cycles. *Hum. Reprod.* **2021**, *36*, 340–348. [CrossRef]
124. Starostic, M.R.; Sosina, O.A.; McCoy, R.C. Single-cell analysis of human embryos reveals diverse patterns of aneuploidy and mosaicism. *Genome Res.* **2020**, *30*, 814–835. [CrossRef] [PubMed]
125. Gleicher, N.; Pasquale, P.; Brivanlou, A. Preimplantation Genetic Testing for Aneuploidy—A Castle Built on Sand. *Trends Mol. Med.* **2021**, *27*, 731–742. [CrossRef]
126. Gleicher, N.; Vidali, A.; Braverman, J.; Kushnir, V.A.; Albertini, D.F.; Barad, D.H. Further evidence against use of PGS in poor prognosis patients: Report of normal births after transfer of embryos reported as aneuploid. *Fertil. Steril.* **2015**, *104*, E59. [CrossRef]
127. Greco, E.; Minasi, M.G.; Fiorentino, F. Healthy babies after intrauterine transfer of mosaic aneuploid blastocysts. *N. Engl. J. Med.* **2015**, *373*, 2089–2090. [CrossRef] [PubMed]
128. Patrizio, P.; Shoham, G.; Shoham, Z.; Leong, M.; Barad, D.H.; Gleicher, N. Worldwide live births following transfer of chromosomally "abnormal" embryos after PGT/A. Results of worldwide web-based survey. *J. Assist. Reprod. Genet.* **2019**, *36*, 1599–1607. [CrossRef]
129. Singla, S.; Iwamoto-Stohl, L.K.; Zhu, M.; Zernicka-Goetz, M. Autophagy-mediated apoptosis eliminates aneuploid cells in a mouse model of chromosome mosaicism. *Nat. Commun.* **2020**, *11*, 2958. [CrossRef]
130. Yang, M.; Rito, T.; Naftaly, J.; Hu, J.; Albertini, D.F.; Barad, D.H.; Brivanlou, A.H.; Gleicher, N. Self-correction of mosaicism in human embryos and gastruloids. *Nat. Cell. Biol.* **2020**, *114*, E14–E15. [CrossRef]
131. Preimplantation Genetic Testing—Good Practice Recommendations of the European Society of Human Reproduction and Embryology (ESHRE). 2020, p. 124. Available online: www.eshre.eu/guidelines (accessed on 23 November 2021).
132. Rexhaj, E.; Paoloni-Giacobino, A.; Rimoldi, S.F.; Fuster, D.G.; Anderegg, M.; Somm, E.; Bouillet, E.; Allemann, Y.; Sartori, C.; Scherrer, U. Mice generated by in vitro fertilization exhibit vascular dysfunction and shortened life span. *J. Clin. Investig.* **2013**, *123*, 5052–5060. [CrossRef]
133. Jayet, P.Y.; Rimoldi, S.F.; Stuber, T.; Salmòn, C.S.; Hutter, D.; Rexhaj, E.; Thalmann, S.; Schwab, M.; Turini, P.; Sartori-Cucchia, C.; et al. Pulmonary and systemic vascular dysfunction in young offspring of mothers with preeclampsia. *Circulation.* **2010**, *122*, 488–494. [CrossRef]
134. Sundheimer, L.; Pisarska, M. Abnormal Placentation Associated with Infertility as a Marker of Overall Health. *Semin. Reprod. Med.* **2017**, *35*, 205–216. [CrossRef] [PubMed]
135. Brosens, I.A.; Robertson, W.B.; Dixon, H.G. The role of the spiral arteries in the pathogenesis of preeclampsia. *Obstet. Gynecol. Annu.* **1972**, *1*, 177–191. [CrossRef] [PubMed]
136. Cunningham, F.G.; Leveno, K.J. Childbearing among older women—The message is cautiously optimistic. *N. Engl. J. Med.* **1995**, *333*, 1002–1004. [CrossRef]
137. Kenny, L.C.; Lavender, T.; McNamee, R.; O'Neill, S.M.; Mills, T.; Khashan, A.S. Advanced maternal age and adverse pregnancy outcome: Evidence from a large contemporary cohort. *PLoS ONE* **2013**, *8*, e56583. [CrossRef] [PubMed]
138. Jackson, S.; Hong, C.; Wang, E.T.; Alexander, C.; Gregory, K.D.; Pisarska, M.D. Pregnancy outcomes in very advanced maternal age pregnancies: The impact of assisted reproductive technology. *Fertil. Steril.* **2015**, *103*, 76–80. [CrossRef]
139. Wang, E.T.; Ozimek, J.A.; Greene, N.; Ramos, L.; Vyas, N.; Kilpatrick, S.J.; Pisarska, M.D. Impact of fertility treatment on severe maternal morbidity. *Fertil. Steril.* **2016**, *106*, 423–426. [CrossRef] [PubMed]
140. Dayan, N.; Joseph, K.S.; Fell, D.B.; Laskin, C.A.; Basso, O.; Park, A.L.; Luo, J.; Guan, J.; Ray, J.G. Infertility treatment and risk of severe maternal morbidity: A propensity score-matched cohort study. *CMAJ* **2019**, *191*, E118–E127. [CrossRef] [PubMed]
141. Dayan, N.; Fell, D.B.; Guo, Y.; Wang, H.; Velez, M.P.; Spitzer, K.; Laskin, C.A. Severe maternal morbidity in women with high BMI in IVF and unassisted singleton pregnancies. *Hum. Reprod.* **2018**, *33*, 1548–1556. [CrossRef] [PubMed]

142. Mosca, L.; Benjamin, E.J.; Berra, K.; Bezanson, J.L.; Dolor, R.J.; Lloyd-Jones, D.M.; Newby, L.K.; Piña, I.L.; Roger, V.L.; Shaw, L.J.; et al. Effectiveness-based guidelines for the prevention of cardiovascular disease in women—2011 update: A guideline from the American Heart Association. *J. Am. Coll. Cardiol.* **2011**, *57*, 1404–1423. [CrossRef] [PubMed]
143. Briana, D.D.; Germanou, K.; Boutsikou, M.; Boutsikou, T.; Athanasopoulos, N.; Marmarinos, A.; Gourgiotis, D.; Malamitsi-Puchner, A. Potential prognostic biomarkers of cardiovascular disease in fetal macrosomia: The impact of gestational diabetes. *J. Matern.-Fetal Neonatal. Med.* **2018**, *31*, 895–900. [CrossRef]
144. Barker, D.J. The developmental origins of chronic adult disease. *Acta Paediatr. Suppl.* **2004**, *93*, 26–33. [CrossRef] [PubMed]
145. Männistö, T.; Mendola, P.; Vääräsmäki, M.; Järvelin, M.R.; Hartikainen, A.L.; Pouta, A.; Suvanto, E. Elevated blood pressure in pregnancy and subsequent chronic disease risk. *Circulation* **2013**, *127*, 681–690. [CrossRef]
146. Brown, M.C.; Best, K.E.; Pearce, M.S.; Waugh, J.; Robson, S.C.; Bell, R. Cardiovascular disease risk in women with pre-eclampsia: Systematic review and meta-analysis. *Eur. J. Epidemiol.* **2013**, *28*, 1–19. [CrossRef] [PubMed]
147. Kvehaugen, A.S.; Dechend, R.; Ramstad, H.B.; Troisi, R.; Fugelseth, D.; Staff, A.C. Endothelial function and circulating biomarkers are disturbed in women and children after PE. *Hypertension* **2011**, *58*, 63–69. [CrossRef] [PubMed]
148. Watkins, A.J.; Platt, D.; Papenbrock, T.; Wilkins, A.; Eckert, J.J.; Kwong, W.Y.; Osmond, C.; Hanson, M.; Fleming, T.P. Mouse embryo culture induces changes in postnatal phenotype including raised systolic blood pressure. *Proc. Natl. Acad. Sci. USA* **2007**, *104*, 5449–5454. [CrossRef]
149. Rimoldi, S.F.; Sartori, C.; Rexhaj, E.; Bailey, D.M.; de Marchi, S.F.; McEneny, J.; von Arx, R.; Cerny, D.; Duplain, H.; Germond, M.; et al. Antioxidants improve vascular function in children conceived by assisted reproductive technologies: A randomized double-blind placebo-controlled trial. *Eur. J. Prev. Cardiol.* **2015**, *22*, 1399–1407. [CrossRef] [PubMed]
150. Chen, M.; Wu, L.; Zhao, J.; Wu, F.; Davies, M.J.; Wittert, G.A. Altered Glucose Metabolism in Mouse and Humans Conceived by IVF. *Diabetes* **2014**, *63*, 3189–3198. [CrossRef]
151. Ceelen, M.; Van Weissenbruch, M.M.; Prein, J.; Smit, J.J.; Vermeiden, J.P.W.; Spreeuwenberg, M.; Van Leeuwen, F.E.; Delemarre-Van De Waal, H.A. Growth during Infancy and Early Childhood in Relation to Blood Pressure and Body Fat Measures at Age 8–18 Years of IVF Children and Spontaneously Conceived Controls Born to Subfertile Parents. *Hum. Reprod.* **2009**, *24*, 2788–2795. [CrossRef]
152. Guo, X.Y.; Liu, X.M.; Jin, L.; Wang, T.T.; Ullah, K.; Sheng, J.Z.; Huang, H.F. Cardiovascular and Metabolic Profiles of Offspring Conceived by Assisted Reproductive Technologies: A Systematic Review and Meta-Analysis. *Fertil. Steril.* **2017**, *107*, 622–631. [CrossRef]
153. Ceelen, M.; Van Weissenbruch, M.M.; Roos, J.C.; Vermeiden, J.P.W.; Van Leeuwen, F.E.; Delemarre-van De Waal, H.A. Body Composition in Children and Adolescents Born after in Vitro Fertilization or Spontaneous Conception. *J. Clin. Endocrinol. Metab.* **2007**, *92*, 3417–3423. [CrossRef] [PubMed]
154. Ceelen, M.; Van Weissenbruch, M.M.; Vermeiden, J.P.W.; Van Leeuwen, F.E.; Delemarre-Van De Waal, H.A. Cardiometabolic Differences in Children Born after In Vitro Fertilization: Follow-up Study. *J. Clin. Endocrinol. Metab.* **2008**, *93*, 1682–1688. [CrossRef] [PubMed]
155. Forse, T.; Eriksson, J.; Tuomilehto, J.; Reunanen, A.; Osmond, C.; Barker, D. The Fetal and Childhood Growth of Persons Who Develop Type 2 Diabetes. *Ann. Intern. Med.* **2000**, *133*, 176–182. [CrossRef] [PubMed]
156. Waterland, R.A.; Garza, C. Potential Mechanisms of Metabolic Imprinting That Lead to Chronic Disease. *Am. J. Clin. Nutr.* **1999**, *69*, 179–197. [CrossRef]
157. Millership, S.J.; Van de Pette, M.; Withers, D.J. Genomic Imprinting and Its Effects on Postnatal Growth and Adult Metabolism. *Cell Mol. Life Sci.* **2019**, *76*, 4009–4021. [CrossRef]
158. Wale, P.L.; Gardner, D.K. The effects of chemical and physical factors on mammalian embryo culture and their importance for the practice of assisted human reproduction. *Hum. Reprod. Update* **2016**, *22*, 2–22. [CrossRef] [PubMed]
159. Chen, J.Z.; Sheehan, P.M.; Brennecke, S.P.; Keogh, R.J. Vessel remodelling, pregnancy hormones and extravillous trophoblast function. *Mol. Cell Endocrinol.* **2012**, *349*, 138–144. [CrossRef]
160. Imudia, A.N.; Awonuga, A.O.; Doyle, J.O.; Kaimal, A.J.; Wright, D.L.; Toth, T.L.; Styer, A.K. Peak serum estradiol level during controlled ovarian hyperstimulation is associated with increased risk of small for gestational age and PE in singleton pregnancies after in vitro fertilization. *Fertil. Steril.* **2012**, *97*, 1374–1379. [CrossRef]
161. Jones, M.L.; Mark, P.J.; Mori, T.A.; Keelan, J.A.; Waddell, B.J. Maternal dietary omega-3 fatty acid supplementation reduces placental oxidative stress and increases fetal and placental growth in the rat. *Biol. Reprod.* **2013**, *88*, 37. [CrossRef]
162. Myatt, L. Review: Reactive oxygen and nitrogen species and functional adaptation of the placenta. *Placenta* **2010**, *31*, S66–S69. [CrossRef] [PubMed]
163. De Geyter, C.; Wyns, C.; Calhaz-Jorge, C.; de Mouzon, J.; Ferraretti, A.P.; Kupka, M.; Nyboe Andersen, A.; Nygren, K.G.; Goossens, V. 20 years of the European IVF-monitoring Consortium registry: What have we learned? A comparison with registries from two other regions. *Hum. Reprod.* **2020**, *35*, 2832–2849. [CrossRef] [PubMed]
164. Wang, J.X.; Norman, R.J.; Wilcox, A.J. Incidence of spontaneous abortion among pregnancies produced by assisted reproductive technology. *Hum. Reprod.* **2004**, *19*, 272–277. [CrossRef] [PubMed]

Table 2. Overview of flow culture approaches to study trophoblast differentiation and fusion.

Shear Stress/ Flow Rate	Cells Used	Co-Cultivation	Incubation Time	Reference
30 μL/h	JEG-3	HUVECs	68 h	Lee et al. (2015) [28] **
0.001–0.12 dyn/cm^2 2–5 μL/min	BeWo	-	15 min–12 h	Miura et al. (2015) [16] *
	HVTs	-		
5.2 dyn/cm^2	JEG-3	HBMECs	10–21 days	McConkey et al. (2016) [27] ***
1.67 μL/min	BeWo b30	HPVECs	72 h	Blundell et al. (2016) [29] **
1 dyn/cm^2 5.19 mL/min	human primary term trophoblasts	-	15 min–72 h	Lecarpentier et al. (2016) [21] *
0.001–1 dyn/cm^2 ≈0.5 mL/min	BeWo	-	96 h	Sanz et al. (2019) [17] *
0.1, 0.2, 0.5 mL/min	rTSCs	-	48 h	

human villous trophoblasts (HVTs), rabbit trophoblastic stem cells (rTSCs), human umbilical vein endothelial cells (HUVECs), human brain microvascular endothelial cells (HBMECs), human primary placental villous endothelial cells (HPVECs), * circulating flow loop; ** one-time inlet-to-outlet flow system; *** rotating wall vessel (RWV) bioreactor.

Thus, study questions, such as to how fluidic flow and shear stress influences different aspects of trophoblast physiology (e.g., differentiation and fusion (Table 2) and migration (Table 1)), should be addressed with a most appropriate setting of the flow system.

3. The Influence of Fluid Shear Stress on Trophoblast Turnover and Differentiation

Human placenta development relies on a tightly controlled villous trophoblast turnover, which involves proliferation, differentiation, and fusion of mononucleated cytotrophoblasts with the overlying syncytiotrophoblast [14]. This process guarantees that the syncytiotrophoblast is continuously supplied with cytoplasm and organelles derived from the fusing cytotrophoblasts. Acquisition of newly incorporated cell components is balanced by a concomitant release of apoptotic material as syncytial knots from the syncytiotrophoblast surface into the maternal circulation [30]. Effects of flow and fluid shear stress on trophoblast turnover, in particular on cell differentiation, are manifested by changed cell morphology. Fluid shear stress is suggested to activate signaling pathways involved in trophoblast differentiation and syncytialization by increasing levels of intracellular cyclic adenosine monophosphate (cAMP) and activated cAMP response element-binding protein (CREB) (Table 2, [21]). Activation of cAMP signaling induces upregulation of transcription factor glial cell missing 1 (GCM1) and its downstream targets syncytin-1 (ERVW-1) and syncytin-2 (ERVFRD-1), both of which well-accepted fusogens involved in trophoblast syncytialization [30]. Concurrent to syncytialization is the loss of epithelial junctional- and cytoskeletal proteins, such as E-cadherin, desmoplakin, and α-fodrin [31]. While previous flow culture experiments clearly showed a network of continuous and well-defined junctional complexes in unstimulated trophoblasts after three days perfusion (Table 2, [29]), knowledge on cytoskeleton remodeling upon syncytialization under flow rates is rather limited.

Increased wall shear stress is suggested to act at the villous surface in the inflow regions of the intervillous space, where high wall shear stress could damage the villous trophoblast or at least affect its cellular turnover. This disturbance may be reflected in enhanced trophoblast shedding and elevated levels of free fetal DNA in the maternal circulation [32]. Paradoxically, increasing shear stress has been suggested to have a protective effect against induced apoptotic death in trophoblast cell lines [18]. This has been shown in mononucleated (i.e., undifferentiated) JAR and SGHPL-4 cells, which underwent less apoptosis when cultured under 3 dyn/cm^2 than those in 0.5 dyn/cm^2

cultures [18]. Moreover, trophoblasts have been shown to have a survival advantage over endothelial cells. Trophoblasts cultured on human umbilical vein endothelial cells (HUVECs) monolayers at 0.5 or 3 dyn/cm^2 significantly induced apoptosis in directly adjacent HUVECs, by Fas/Fas-ligand mediated mechanisms [18].

In addition to apoptosis, fluid shear stress is suggested to influence trophoblast fusion. Previous studies with BeWo cells and rabbit trophoblastic stem cells (rTSCs) showed increased cell fusion under fluid shear stress. This has been demonstrated for rabbit trophoblastic stem cells at flow rates of 0.1 ml/min, 0.2 ml/min, and 0.5 ml/min, respectively. The authors of the study described that fusion of rabbit trophoblasts occurred between more than two cells, while in the case of BeWo it was mainly a fusion of only two cells (Table 2, [17]). However, at this point it should be noted that BeWo cells occasionally contain two nuclei, and that intercellular fusion must be distinguished from endoreduplication [33], which represents replication of the nuclear genome in the absence of mitosis, and therefore results in an elevated nuclear gene content and polyploidy. However, besides syncytialization, other signs of trophoblast differentiation have been observed when cells were cultured under fluidic flow. According to Miura et al., BeWo cells and human villous trophoblasts react on fluid shear stress by abundant formation of microvilli, which vary in lengths depending on the flow rate (Table 2, [16]). At the center of the chamber, where the shear stress was low (0.001 dyn/cm^2), microvilli were long (>2 μm); whereas they were shortened (<2 μm) in the area at the inlet or outlet of the chamber with high shear stress (0.1 dyn/cm^2). In agreement with this observation, ezrin—a member of the ezrin-radixin-moesin (ERM) family, which plays a major role in formation and/or maintenance of actin-based cell surface structures—was predominantly detected at the apical membrane of the cells.

The observation of fluid shear stress-induced microvilli formation has recently been confirmed in rabbit trophoblast stems cells, which were cultured on a collagen gel in the presence of flow (Table 2, [17]). Subsequent transcriptome analysis of the rabbit trophoblasts, showed enrichment in pathways regulating actin cytoskeleton and sphingolipid metabolism, which has been suggested to account for the increased formation of microvilli during differentiation [17].

4. The Influence of Fluid Shear Stress on Trophoblast Metabolism

While the transfer of gases and some other solutes occurs by flow limited diffusion, nutrients, and waste products have to be actively transported across the placental barrier. The extent of nutrient transfer and, hence, of fetal supply is determined by many factors, including placental morphology as well as uteroplacental and fetoplacental blood flow [34]. Perazzolo et al. suggested that the relationship between maternal blood flow and villous structure affects the efficiency of placental uptake and transfer, and moreover, that flow rate may be the major determinant of it [35]. However, besides uteroplacental blood flow, barrier thickness, and concentration gradients, factors such as transporter expression and metabolism of the villous trophoblast influence the dynamics of placental transfer. In fact, a growing body of evidence suggests substantial differences in metabolism in cells cultured under flow, when compared to static conditions. Accordingly, fluidic flow increased the accumulation and size of lipid droplets in rabbit trophoblasts [17], suggesting that morphological differentiation was accompanied by metabolic changes. Consistent with morphological differentiation of rabbit trophoblasts and their increased lipid droplet accumulation, a number of genes of the peroxisome proliferator-activated receptor (PPAR) signaling pathway were upregulated in response to shear stress. Amongst these genes, perilipin 2 (PLIN2), cytochrome P450 1B1 (CYP1B1), and angiopoietin-like 4 (ANGPTL4) are reported to be involved in lipid metabolism, transport and storage (Table 2, [17]), suggesting enhanced metabolic turnover in trophoblasts cultured under flow. In addition to its importance for the formation of lipid droplets, PLIN2 is necessary for trophoblast viability when exposed to fatty acids. This has been shown by overexpression of PLIN2 in human primary term trophoblasts that were exposed to a mixture of linoleic acid and oleic acid [36].

Along with lipid metabolism, trophoblastic glucose uptake is affected by flow, as shown by significantly increased uptake of the fluorescent glucose analog 2-[N-(7-nitrobenz-2-oxa-1,

3-diazol-4-yl)amino]-2-deoxy-D-glucose (2-NBDG) into BeWo cells exposed to fluid shear stress. The increased glucose uptake could be explained by slightly increased mRNA expression of the glucose transporter type 1, GLUT1 (encoded by *SLC2A1*), which predominantly localized to the apical membrane of cells and cell–cell contact regions after overnight exposure to fluid shear stress [16].

5. The Influence of Fluid Shear Stress on Trophoblast Endocrine Activity

Soon after implantation, the developing placenta starts producing hormones to adapt the maternal physiology to the progressing pregnancy. The highly differentiated syncytiotrophoblast is the predominant cell type involved in placental hormone synthesis, and thus it is obvious that maternal blood flow may not only affect trophoblast differentiation and metabolism, but also its endocrine activity. Recent studies indicate that fluidic flow induces both, syncytialization and production of the classical pregnancy hormone human chorionic gonadotropin (hCG) in BeWo cells [17]. Moreover, laminar and continuous fluid shear stress of 1 dyn/cm^2 promotes placental growth factor (PGF) upregulation in primary human term trophoblasts, which underwent spontaneous differentiation and fusion during a 48 h pre-culture under static conditions (Table 2, [21]). At the same time, secretion of soluble fms-like tyrosine kinase-1 (sFlt-1)—suggested as a biomarker to predict the risk of developing preeclampsia [37]—was only slightly, but not significantly increased by fluid shear stress [21].

Moreover, fluid shear stress is suggested to affect the intracellular availability of cortisol, a glucocorticoid that is regulated by 11β-hydroxysteroid dehydrogenase enzymes (11β-HSDs). In human trophoblast cells, this has been shown for 11β-HSD2 (encoded by *HSD11B2*), an NAD$^+$-dependent enzyme that oxidizes cortisol to the inactive metabolite cortisone. In JEG-3 cells, a unidirectional flow environment with varying fluid shear stress equal to a maximum of 5 dyn/cm^2 reduced 11β-HSD2 mRNA expression and activity, which was reversed to basal levels by discontinuation of the shear stress [15]. Conversion of cortisol to cortisone by 11β-HSD2 is suggested to protect cells from the growth-inhibiting and/or pro-apoptotic effects of cortisol, particularly during embryonic development. Thus, fluid shear stress could be one of the underlying causes of enhanced cortisol levels and reduced 11β-HSD2 activity in fetal growth retardation [38].

6. The Influence of Fluid Shear Stress on Trophoblast Motility

Early after blastocyst implantation and initial development of primary placental villi, extravillous trophoblasts detach from villi and start to invade into the decidual stroma. Once these extravillous trophoblasts have eroded uterine spiral arteries (now referred to as endovascular trophoblasts [1]), they are suggested to migrate along the luminal surfaces of the vessels and remodel them by interdigitating between the endothelial cells [14]. Thereby, endothelial cells are increasingly replaced and most of the musculoelastic tissue in the vessel walls dissolve, resulting in low-resistance vessels to guarantee constant and maximal uteroplacental blood flow at the transition from the first to second trimester of pregnancy. Hence it is more than likely that fluidic flow not only influences physiology of the villous trophoblast, but also that of endovascular trophoblasts located in invaded spiral arteries. Importantly, with the erosion of the uterine vessel walls and the phenotypic switch from interstitial extravillous trophoblasts into endovascular trophoblasts, cells are exposed to a much higher fluid shear stress than the subpopulation of villous trophoblasts that are faced with rather low shear stress in the intervillous space. In the context of spiral artery remodeling, a vast majority of studies used co-cultures of extravillous trophoblast cell lines (Table 1) with endothelial cells, such as HUVEC [14,18,20,28] and human uterine microvascular endothelial cells (UtMVEC) [23,24]. On the basis of such co-cultures, as well as trophoblast monocultures, a growing body of evidence suggests that in particular motility of endovascular trophoblasts is controlled by flow. However, it should be noted here that used extravillous trophoblast cell lines (Table 1) seem to be the right choice for such studies, although it remains to be discussed to what extent they resemble the endovascular phenotype.

Initial experiments with early gestation macaque trophoblast cells exposed to flow, showed that cells exhibit clear migration in the direction of flow as well as shape changes that involve extension and retraction of filopodia at its leading edge (Table 1, [22]). In doing so, macaque trophoblast migration velocity and movements increased with the magnitude of the applied shear stress (from 7.5 and 15 up to 30 dyn/cm^2). These observations have been confirmed by others, showing that migration of the human first trimester trophoblasts occur generally in the direction of flow (up to 30 dyn/cm^2), with only a few cells migrating against the flow stream [19]. The enhanced motility under fluid shear stress seems to be accompanied by increased expression of integrin β1, which mediates adhesion of human and macaque trophoblast cells to endothelial cells in vitro [39]. This has been shown in previous migration experiments, suggesting that factors expressed on the surface of uterine endothelial cells and factors released by the endothelium regulate trophoblast migration under flow [23]. While the extent of migration against flow at higher and more physiological shear stress levels (15 and 30 dyn/cm^2) decreased significantly for macaque trophoblasts alone, migration against flow remained virtually unchanged for trophoblasts co-cultured on human uterine microvascular endothelial cells. Hence, migration behavior of trophoblasts cultured under flow clearly depends on the substrate they are seeded.

When SGHPL-4 trophoblasts were cultured on the extracellular matrix glycoprotein fibronectin or a layer of endothelial cells, they did not undergo directional migration in 0.5 and 2 dyn/cm^2 cultures. However, under conditions of 4 and 6 dyn/cm^2, trophoblasts migrated with the direction of flow [14]. In contrast, another study suggests that the average migration velocity of human first trimester trophoblasts cultured on a type I rat collagen-coated surface increased almost linearly with increasing shear stress (from 7.5 up to 30 dyn/cm^2) (Table 1, [20]), whereas migration velocity remained almost unchanged at all levels of shear stress when cells were cultured on endothelial cells. Similar behavior was observed for the displacement of trophoblasts by flow. The absolute x-direction displacement increased with increasing shear stress, when trophoblasts were cultured on a collagen-coated surface. However, when cultured on top of an endothelial cell monolayer, first trimester trophoblasts showed a strong ability to withstand displacement by flow. Subsequent flow experiments with neutralizing antibodies suggested that integrin β1 regulated adhesion of trophoblasts to endothelial cells under shear stress [20]. However, another study using SGHPL-4 cells suggested that fluid shear stress did not affect expression of the adhesion molecules E-selectin and integrins α4, β1, and αVβ3 [14]. Different cell types and particularly varying fluid shear stress values exposed to cells may explain discrepancies in motility and expression of adhesion molecules. Accordingly, James et al. recently suggested a scenario that extravillous trophoblasts in early first trimester do not undergo directional migration under very low shear stress, whereas later on, as shear stress increases with complete dissolution of trophoblast plugs, the number of trophoblasts are stimulated to migrate in the direction of flow [14].

7. Outlook—Future Directions

In recent years a number of excellent studies have yielded in an increasing body of valuable knowledge, substantiating an important influence of maternal blood flow on the (patho-)physiology of different trophoblast subtypes. The application of different flow culture systems, including customized micro-scale 3D flow chambers, has revealed substantial differences of trophoblasts cultured under flow, when compared to cells under static conditions (Figure 2).

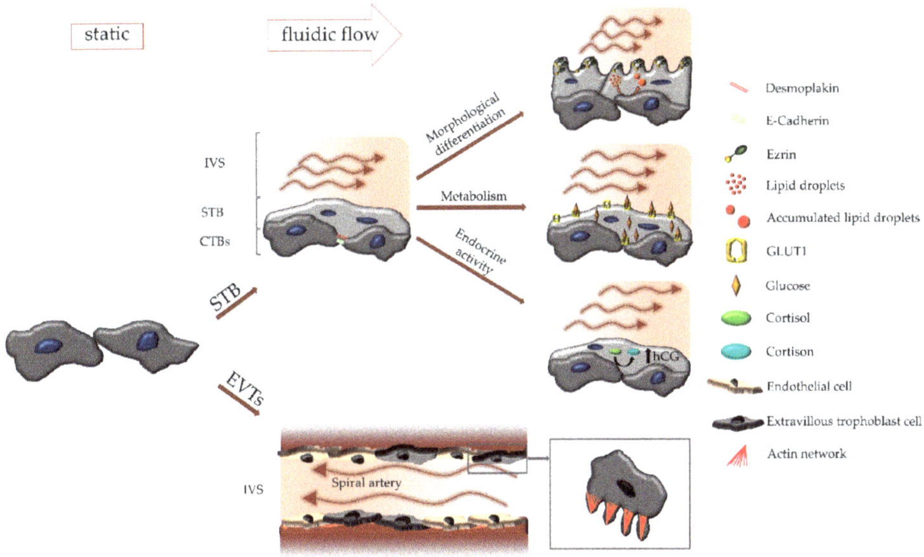

Figure 2. Influence of fluidic flow on different trophoblast subtypes. Fluidic flow regulates differentiation and physiology of both, the syncytiotrophoblast (STB) and extravillous trophoblasts (EVTs). During syncytialization, cell-cell contact proteins, such as desmoplakin and E-cadherin are downregulated, and the structural protein ezrin relocalizes from the basal to the apical side. Formation and appearance of microvilli on the apical surface as well as accumulation of lipid droplets in the STB are influenced by fluidic flow. Moreover, fluidic flow affects expression and localization of GLUT1, secretion of hCG as well as conversion of cortisol to cortisone. In the EVT subpopulation, formation of filopodia, and hence migratory behavior is regulated by fluidic flow as well.

Some important key parameters should be considered when setting up a trophoblast flow system *:

(1) Reservoir

- a single reservoir intercalated in a circulating flow system
- a single reservoir with fresh medium and a tank for consumed medium
- additional reservoirs with buffer between pump and chamber to damp flow

(2) Pump

- Peristaltic pump for circulating flow loop systems
- Syringe pump or electropneumatic pump for one-time inlet-to-outlet flow systems
- Rotating wall vessel bioreactor (rotation is responsible for distribution of medium)

(3) Chamber

- Commercial flow chamber (laminar or turbulent)
- Customized Chamber (laminar or turbulent)
- Chambers created with 3D printers (suitable for sterilization)

(4) Tubings

- Size is related to flow rate

(5) Culture medium

- Serum-free (lower viscosity)
- With serum (higher viscosity)

(6) Microscope

- continuous microscopic recording of cultivated cells

* after setup, flow systems should be calibrated prior to each experiment

Besides fluidic flow, a low oxygen environment is another crucial key regulator of trophoblast- and placenta development. In the first trimester of human gestation, prior to fully established uteroplacental blood flow, the mean pO2 in the developing placenta is suggested to be less than 20mmHg. At the beginning of the second trimester, when extravillous trophoblast plugs almost completely disintegrate, pO2 rises to values above 50 mmHg and stays at about this level until delivery [40,41]. Thus, future trends in trophoblast flow culture should incorporate appropriate oxygen conditions in the experimental setup. Moreover, a physiological pulsatile blood flow, leading to temporal and spatial variations of the wall shear stress can be considered to result in different phenotypes and functions of cells. While studies in the field of vascular tissue engineering increasingly acknowledge the importance of temporal and spatial variations of the wall shear stress on endothelial cells, this aspect has not yet been considered for trophoblast culture so far.

Since composition of the used culture medium and serum supplementation will greatly influence the trophoblastic phenotype, care should be taken in medium selection to meet specific requirements of different trophoblast subtypes. Recent studies suggest that extravillous trophoblasts penetrate and invade uterine glands (endoglandular trophoblasts), which are thereby opened towards the intervillous space from the very beginning of pregnancy onwards [2]. Hence, studying the consequences of uterine secretion products on trophoblasts cultured under flow could be another exciting field of research. Implementation of such specific culture media should be thoroughly characterized by continuous monitoring of component consumption by the cells.

Investigation of the "trophoblast mechanosome" will be another intriguing future direction, which should include gestational age-dependent variations, fetal sex, and maternal confounders such as smoking, diabetes, and/or obesity. Characterization of the trophoblastic mechanosome and pathologic aberrations thereof could be subject to therapeutic interventions. However, development of such drugs should not only include the therapeutic potential, but also the consequences on human pregnancy. Transfer of mechanosome modulators through the placental barrier may have adverse effects on the fetoplacental circulation, leading for example to disruption of the balanced pressure gradient between the maternal and fetal circulations [7].

Author Contributions: writing—original draft preparation, B.A.B., J.G., and M.G.; writing—review and editing, B.A.B., J.G., and M.G.; visualization, B.A.B., J.G.; supervision, M.G.; funding acquisition, M.G. All authors have read and agreed to the published version of the manuscript.

Funding: M. Gauster was supported by the Austrian Science Fund (FWF): P 29639, P 33554, I 3304, and through the PhD program Inflammatory Disorders in Pregnancy (DP-iDP) by the Austrian Science Fund (FWF): Doc 31-B26. Moreover, M. Gauster was supported by funds of the Oesterreichische Nationalbank (Austrian Central Bank, Anniversary Fund, project number: 18175).

Acknowledgments: Open Access Funding by the Austrian Science Fund (FWF).

Conflicts of Interest: The authors declare no conflict of interest.

Abbreviations

11β-HSDs	11β-hydroxysteroid dehydrogenase enzymes
2-NBDG	2-[N-(7-nitrobenz-2-oxa-1,3-diazol-4-yl)amino]-2-deoxy-D-glucose
ANGPTL4	Angiopoietin-like 4
cAMP	Cyclic adenosine monophosphate
CREB	Activated cAMP response element binding protein
CTBs	Cytotrophoblasts
CYP1B1	Cytochrome P450 1B1
ERM	Ezrin-Radixin-Moesin
ERVFRD-1	Syncytin-2
ERVW-1	Syncytin-1
EVTs	Extravillous trophoblasts
GCM1	Transcription factor glial cell missing 1
GD	Gestational day
GLUT1	Glucose transporter 1
HBMECs	Human brain microvascular endothelial cells
hCG	Human chorionic gonadotropin
HPVECs	Human primary placental villous endothelial cells
HUVECs	Human umbilical vein endothelial cells
HVTs	Human villous trophoblasts
IVS	Intervillous space
PGF	Placental growth factor
PLIN2	Perilipin 2
PPAR	Peroxisome proliferator-activated receptor
rTSCs	Rabbit trophoblastic stem cells
RWV	Rotating wall vessel
sFlt-1	Soluble fms-like tyrosin kinase-1
STB	Syncytiotrophoblast
UtMVECs	Uterine microvascular endothelial cells

References

1. Moser, G.; Weiss, G.; Sundl, M.; Gauster, M.; Siwetz, M.; Lang-Olip, I.; Huppertz, B. Extravillous trophoblasts invade more than uterine arteries: Evidence for the invasion of uterine veins. *Histochem. Cell Biol.* **2017**, *147*, 353–366. [CrossRef] [PubMed]
2. Moser, G.; Weiss, G.; Gauster, M.; Sundl, M.; Huppertz, B. Evidence from the very beginning: Endoglandular trophoblasts penetrate and replace uterine glands in situ and in vitro. *Hum. Reprod.* **2015**, *30*, 2747–2757. [CrossRef] [PubMed]
3. Windsperger, K.; Dekan, S.; Pils, S.; Golletz, C.; Kunihs, V.; Fiala, C.; Kristiansen, G.; Knöfler, M.; Pollheimer, J. Extravillous trophoblast invasion of venous as well as lymphatic vessels is altered in idiopathic, recurrent, spontaneous abortions. *Hum. Reprod.* **2017**, *32*, 1208–1217. [CrossRef] [PubMed]
4. Roberts, V.H.J.; Morgan, T.K.; Bednarek, P.; Morita, M.; Burton, G.J.; Lo, J.O.; Frias, A.E. Early first trimester uteroplacental flow and the progressive disintegration of spiral artery plugs: New insights from contrast-enhanced ultrasound and tissue histopathology. *Hum. Reprod.* **2017**, *32*, 2382–2393. [CrossRef] [PubMed]
5. Burton, G.J.; Woods, A.W.; Jauniaux, E.; Kingdom, J.C.P. Rheological and physiological consequences of conversion of the maternal spiral arteries for uteroplacental blood flow during human pregnancy. *Placenta* **2009**, *30*, 473–482. [CrossRef] [PubMed]
6. Wareing, M. Effects of oxygenation and luminal flow on human placenta chorionic plate blood vessel function. *J. Obstet. Gynaecol. Res.* **2012**, *38*, 185–191. [CrossRef]
7. Morley, L.C.; Beech, D.J.; Walker, J.J.; Simpson, N.A.B. Emerging concepts of shear stress in placental development and function. *Mol. Hum. Reprod.* **2019**, *25*, 329–339. [CrossRef]
8. Chatterjee, S. Endothelial Mechanotransduction, Redox Signaling and the Regulation of Vascular Inflammatory Pathways. *Front. Physiol.* **2018**, *9*, 524. [CrossRef]

9. Xie, Y.; Wang, F.; Zhong, W.; Puscheck, E.; Shen, H.; Rappolee, D.A. Shear stress induces preimplantation embryo death that is delayed by the zona pellucida and associated with stress-activated protein kinase-mediated apoptosis. *Biol. Reprod.* **2006**, *75*, 45–55. [CrossRef]
10. Pirt, S.J. Studies on cells in culture. Continuous-flow culture of mammalian cells. *Proc. R. Soc. Med.* **1963**, *56*, 1061–1062.
11. Yen, E.H.; Melcher, A.H. A continuous-flow culture system for organ culture of large explants of adult tissue: Effect of oxygen tension on mouse molar periodontium. *In Vitro* **1978**, *14*, 811–818. [CrossRef] [PubMed]
12. Huber, D.; Oskooei, A.; Casadevall I Solvas, X.; Demello, A.; Kaigala, G.V. Hydrodynamics in Cell Studies. *Chem. Rev.* **2018**, *118*, 2042–2079. [CrossRef] [PubMed]
13. Young, E.W.K.; Beebe, D.J. Fundamentals of microfluidic cell culture in controlled microenvironments. *Chem. Soc. Rev.* **2010**, *39*, 1036–1048. [CrossRef] [PubMed]
14. James, J.L.; Cartwright, J.E.; Whitley, G.S.; Greenhill, D.R.; Hoppe, A. The regulation of trophoblast migration across endothelial cells by low shear stress: Consequences for vascular remodelling in pregnancy. *Cardiovasc. Res.* **2012**, *93*, 152–161. [CrossRef]
15. Lanz, C.-B.; Causevic, M.; Heiniger, C.; Frey, F.J.; Frey, B.M.; Mohaupt, M.G. Fluid Shear Stress Reduces 11ss-Hydroxysteroid Dehydrogenase Type 2. *Hypertension* **2001**, *37*, 160–169. [CrossRef]
16. Miura, S.; Sato, K.; Kato-Negishi, M.; Teshima, T.; Takeuchi, S. Fluid shear triggers microvilli formation via mechanosensitive activation of TRPV6. *Nat. Commun.* **2015**, *6*, 8871. [CrossRef]
17. Sanz, G.; Daniel, N.; Aubrière, M.-C.; Archilla, C.; Jouneau, L.; Jaszczyszyn, Y.; Duranthon, V.; Chavatte-Palmer, P.; Jouneau, A. Differentiation of derived rabbit trophoblast stem cells under fluid shear stress to mimic the trophoblastic barrier. *Biochim. Biophys. Acta Gen. Subj.* **2019**, *1863*, 1608–1618. [CrossRef]
18. James, J.L.; Whitley, G.S.; Cartwright, J.E. Shear stress and spiral artery remodelling: The effects of low shear stress on trophoblast-induced endothelial cell apoptosis. *Cardiovasc. Res.* **2011**, *90*, 130–139. [CrossRef]
19. Liu, W.; Fan, Y.; Deng, X.; Li, N.; Guan, Z. Effect of flow-induced shear stress on migration of human trophoblast cells. *Clin. Biomech.* **2008**, *23* (Suppl. 1), S112–S117. [CrossRef]
20. Liu, W.; Fan, Y.; Deng, X.; Guan, Z.; Li, N. Adhesion behaviors of human trophoblast cells by contact with endothelial cells. *Colloids Surf. B Biointerfaces* **2009**, *71*, 208–213. [CrossRef]
21. Lecarpentier, E.; Atallah, A.; Guibourdenche, J.; Hebert-Schuster, M.; Vieillefosse, S.; Chissey, A.; Haddad, B.; Pidoux, G.; Evain-Brion, D.; Barakat, A.; et al. Fluid Shear Stress Promotes Placental Growth Factor Upregulation in Human Syncytiotrophoblast Through the cAMP-PKA Signaling Pathway. *Hypertension* **2016**, *68*, 1438–1446. [CrossRef] [PubMed]
22. Soghomonians, A.; Barakat, A.I.; Thirkill, T.L.; Blankenship, T.N.; Douglas, G.C. Effect of shear stress on migration and integrin expression in macaque trophoblast cells. *Biochim. Biophys. Acta* **2002**, *1589*, 233–246. [CrossRef]
23. Soghomonians, A.; Barakat, A.I.; Thirkill, T.L.; Douglas, G.C. Trophoblast migration under flow is regulated by endothelial cells. *Biol. Reprod.* **2005**, *73*, 14–19. [CrossRef] [PubMed]
24. Cao, T.C.; Thirkill, T.L.; Wells, M.; Barakat, A.I.; Douglas, G.C. Trophoblasts and shear stress induce an asymmetric distribution of icam-1 in uterine endothelial cells. *Am. J. Reprod. Immunol.* **2008**, *59*, 167–181. [CrossRef]
25. Lecarpentier, E.; Bhatt, M.; Bertin, G.I.; Deloison, B.; Salomon, L.J.; Deloron, P.; Fournier, T.; Barakat, A.I.; Tsatsaris, V. Computational Fluid Dynamic Simulations of Maternal Circulation: Wall Shear Stress in the Human Placenta and Its Biological Implications. *PLoS ONE* **2016**, *11*, e0147262. [CrossRef]
26. James, J.L.; Saghian, R.; Perwick, R.; Clark, A.R. Trophoblast plugs: Impact on utero-placental haemodynamics and spiral artery remodelling. *Hum. Reprod.* **2018**, *33*, 1430–1441. [CrossRef]
27. McConkey, C.A.; Delorme-Axford, E.; Nickerson, C.A.; Kim, K.S.; Sadovsky, Y.; Boyle, J.P.; Coyne, C.B. A three-dimensional culture system recapitulates placental syncytiotrophoblast development and microbial resistance. *Sci. Adv.* **2016**, *2*, e1501462. [CrossRef]
28. Lee, J.S.; Romero, R.; Han, Y.M.; Kim, H.C.; Kim, C.J.; Hong, J.-S.; Huh, D. Placenta-on-a-chip: A novel platform to study the biology of the human placenta. *J. Matern. Fetal Neonatal Med.* **2016**, *29*, 1046–1054. [CrossRef]
29. Blundell, C.; Tess, E.R.; Schanzer, A.S.R.; Coutifaris, C.; Su, E.J.; Parry, S.; Huh, D. A microphysiological model of the human placental barrier. *Lab Chip* **2016**, *16*, 3065–3073. [CrossRef]

30. Huppertz, B.; Gauster, M. Trophoblast fusion. *Adv. Exp. Med. Biol.* **2011**, *713*, 81–95. [CrossRef]
31. Gauster, M.; Siwetz, M.; Orendi, K.; Moser, G.; Desoye, G.; Huppertz, B. Caspases rather than calpains mediate remodelling of the fodrin skeleton during human placental trophoblast fusion. *Cell Death Differ.* **2010**, *17*, 336–345. [CrossRef]
32. Roth, C.J.; Haeussner, E.; Ruebelmann, T.; Koch, F.V.; Schmitz, C.; Frank, H.-G.; Wall, W.A. Dynamic modeling of uteroplacental blood flow in IUGR indicates vortices and elevated pressure in the intervillous space - a pilot study. *Sci. Rep.* **2017**, *7*, 40771. [CrossRef] [PubMed]
33. Gauster, M.; Huppertz, B. The paradox of caspase 8 in human villous trophoblast fusion. *Placenta* **2010**, *31*, 82–88. [CrossRef] [PubMed]
34. Desoye, G.; Gauster, M.; Wadsack, C. Placental transport in pregnancy pathologies. *Am. J. Clin. Nutr.* **2011**, *94*, 1896S–1902S. [CrossRef]
35. Perazzolo, S.; Lewis, R.M.; Sengers, B.G. Modelling the effect of intervillous flow on solute transfer based on 3D imaging of the human placental microstructure. *Placenta* **2017**, *60*, 21–27. [CrossRef]
36. Bildirici, I.; Schaiff, W.T.; Chen, B.; Morizane, M.; Oh, S.-Y.; O'Brien, M.; Sonnenberg-Hirche, C.; Chu, T.; Barak, Y.; Nelson, D.M.; et al. PLIN2 Is Essential for Trophoblastic Lipid Droplet Accumulation and Cell Survival During Hypoxia. *Endocrinology* **2018**, *159*, 3937–3949. [CrossRef] [PubMed]
37. Zeisler, H.; Llurba, E.; Chantraine, F.; Vatish, M.; Staff, A.C.; Sennström, M.; Olovsson, M.; Brennecke, S.P.; Stepan, H.; Allegranza, D.; et al. Predictive Value of the sFlt-1:PlGF Ratio in Women with Suspected Preeclampsia. *N. Engl. J. Med.* **2016**, *374*, 13–22. [CrossRef]
38. Shams, M.; Kilby, M.D.; Somerset, D.A.; Howie, A.J.; Gupta, A.; Wood, P.J.; Afnan, M.; Stewart, P.M. 11Beta-hydroxysteroid dehydrogenase type 2 in human pregnancy and reduced expression in intrauterine growth restriction. *Hum. Reprod.* **1998**, *13*, 799–804. [CrossRef]
39. Thirkill, T.L.; Douglas, G.C. The vitronectin receptor plays a role in the adhesion of human cytotrophoblast cells to endothelial cells. *Endothelium* **1999**, *6*, 277–290. [CrossRef]
40. Huppertz, B.; Gauster, M.; Orendi, K.; König, J.; Moser, G. Oxygen as modulator of trophoblast invasion. *J. Anat.* **2009**, *215*, 14–20. [CrossRef]
41. Jauniaux, E.; Watson, A.L.; Hempstock, J.; Bao, Y.P.; Skepper, J.N.; Burton, G.J. Onset of maternal arterial blood flow and placental oxidative stress. A possible factor in human early pregnancy failure. *Am. J. Pathol.* **2000**, *157*, 2111–2122. [CrossRef]

International Journal of
Molecular Sciences

MDPI

Review

Preimplantation Genetic Testing: Where We Are Today

Ermanno Greco [1,2], Katarzyna Litwicka [1,*], Maria Giulia Minasi [1], Elisabetta Cursio [1], Pier Francesco Greco [1] and Paolo Barillari [1]

1 Reproductive Medicine, Villa Mafalda, 00199 Rome, Italy; ergreco1@virgilio.it (E.G.); mg.minasi@gmail.com (M.G.M.); elicur@hotmail.it (E.C.); piergreco@hotmail.com (P.F.G.); paolo.barillari@villamafalda.com (P.B.)

2 UniCamillus, International Medical University, 00131 Rome, Italy

* Correspondence: litwicka@yahoo.it

Received: 18 May 2020; Accepted: 16 June 2020; Published: 19 June 2020

Abstract: Background: Preimplantation genetic testing (PGT) is widely used today in in-vitro fertilization (IVF) centers over the world for selecting euploid embryos for transfer and to improve clinical outcomes in terms of embryo implantation, clinical pregnancy, and live birth rates. Methods: We report the current knowledge concerning these procedures and the results from different clinical indications in which PGT is commonly applied. Results: This paper illustrates different molecular techniques used for this purpose and the clinical significance of the different oocyte and embryo stage (polar bodies, cleavage embryo, and blastocyst) at which it is possible to perform sampling biopsies for PGT. Finally, genetic origin and clinical significance of embryo mosaicism are illustrated. Conclusions: The preimplantation genetic testing is a valid technique to evaluated embryo euploidy and mosaicism before transfer.

Keywords: preimplantation genetic screening; in-vitro fertilization; biopsy; euploid embryo; implantation; pregnancy; endometrium; mosaicism

1. Introduction

IVF is a reproductive technique whose success rate depends on several steps: ovarian stimulation, egg retrieval, embryo culture, and transfer. Embryo implantation is one of the most critical point in every IVF program and transfer of a vital embryo in a receptive endometrium is essential for achieving a pregnancy in an assisted reproduction cycle. Despite many improvements reached today the process of embryo implantation is still very ineffective [1].

Therefore, the selection of the best embryo to transfer is the main challenge to face, mostly when a single embryo transfer (SET) program is adopted for different clinical reasons. As currently practiced, the embryo that is chosen for transfer is selected on morphologic grading criteria, which has significant inter- and intraobserver variability [2]. At the cleavage stage, the number of cells, their symmetry, and the presence of cellular fragments are evaluated. At the blastocyst stage, the evaluated parameters are blastocyst expansion and the inner cell mass and trophoectoderm appearance. Today, there is a wide consensus that the microscopic appearance of an embryo is weakly correlated with its viability [3,4]. Thus, a variety of non-invasive methods, such as time-lapse imaging for embryo morphokinetics [5], proteomic [6], and metabolomic [7] study, was proposed to assess the embryo quality. Extending embryo culture to the blastocyst stage was shown to improve outcomes from SET [8], although morphologically normal blastocysts still retain a significant risk of aneuploidy [9–12]. Therefore, the clinical outcomes from SET have been demonstrated to be lower in several randomized controlled trials performed to date and confirmed by subsequent meta-analysis [13,14]. The transfer of multiple embryos is frequently utilized in clinical practice to improve the chance of implantation, but this approach increases the risk of multiple pregnancies [15,16].

At the same time, several studies have demonstrated that embryo aneuploidy is the most important reason of IVF failure, enhancing the importance of preimplantation genetic testing for aneuploidies (PGT-A) as a method for selecting chromosomally healthy embryos [17–19]. Aneuploidies in human embryos are strictly correlated with female age [20] and are derived from chromosomal errors that can occur at different levels. Meiotic errors occur during oogenesis: the prolonged arrest of oocyte development in prophase results in a degradation of the meiotic apparatus. Mitotic errors happen after fertilization, usually during the first mitotic divisions and lead to embryo mosaicism. Sperm aneuploidies, generally correlated with sperm quality and DNA fragmentation, are less common if compared to oocytes ones, but their incidence in embryo aneuploidy is reported to be high [21].

PGT-A was introduced for the first time in the 1993 to select euploid embryos to transfer and improve assisted reproductive results [22]. However, the first generation PGT was demonstrated to be less effective in improving IVF live birth (LB) rates and reducing miscarriage rates [23] mainly due to the incomplete assessment of chromosomal status and undiagnosed mosaicism deriving from post-zygotic cleavage division errors in day-3 embryo [24]. In fact, in the beginning this screening was performed using Fluorescence in Situ Hybridization (FISH), which analyzed only a reduced number of chromosomes. The need to investigate embryos ploidy status led to the development of different techniques for the analysis of the whole chromosomal panel, such as Array-Comparative Genomic Hybridization (aCGH), Next Generation Sequencing (NGS), and Real Time Quantitative Polymerase Chain Reaction (rtq-PCR). The biopsies for the analysis can be removed from the oocyte, collecting the first and/or second polar body or from the cleavage stage embryo, removing some blastomeres or from the blastocyst, collecting some trophoectoderm cells. These techniques can be applied for different indications in which the transfer of euploid embryo might improve the clinical outcomes.

2. PGT Techniques

2.1. Polar Body Biopsy

Polar bodies (PB) are by-products of the meiotic divisions of the oocyte. They have no reproductive function and can be easily removed without affecting embryo development. PB biopsy can be applied in a single step or two step method: the first consist in the simultaneous removal of the first and second PB sixteen h after insemination, whereas the two steps method entails two different biopsies, one prior to intracitoplasmatic sperm injection (ICSI), with the removal of the first PB, and the other sixteen h after insemination, with the removal of the second PB. The single step procedure would seem to be more convenient, since pronuclei detection allows for analyzing only fertilized oocytes, reducing costs and time wasting. Furthermore, combining the first and second PB biopsy could result in an improved abnormalities detection rate [25].

However, the PB biopsy only provides maternal genetic information and does not consider parental or mitotic division abnormalities [26]. European Society of Human Reproduction and Embryology started in 2012 a multicenter randomized clinical trial to establish the effectiveness of PGT-A performed with PB biopsy. The aim of the study was to evaluate whether the analysis of 23 chromosomes in the first and second polar body, and the selection of euploid embryos for transfer, increased live birth rate within one year, in women in advanced maternal age as compared to cycles without PGT-A. From June 2012 to December 2016, 205 women were assigned to cycles with PGT-A, and 191 to cycles without PGT-A (control group). However, the LB rate was not different among the two groups: 50 out of 205 (24%) in the PGT-A group and 45 out of 191 (24%) in the control group [27].

PB biopsy has the benefit of providing a long time to perform genetic testing without requiring embryo cryopreservation despite being time-consuming [25,26] and less cost-effective per LB rate [28] and, therefore, it is the only genetic testing strategy available in many countries with legal restriction on embryo genetic assessment and cryopreservation.

2.2. Blastomere Biopsy

Blastomere biopsy is usually performed when the embryo is made of about six or eight cells, which usually happens 72 h after insemination. The first step to perform the biopsy is to open the zona using tyrode acid, mechanical piercing, or laser-assisted hatching. Laser assisted zona drilling and the use of calcium-magnesium free media to weaken cell cohesion is the most widespread procedure according to the report of ESHRE PGT consortium in 2011 [29].

It is possible to remove one to two blastomeres. Two cells biopsy is more accurate, but it could affect embryo vitality, since it results in the removal of about 30% of the whole embryo. One cell biopsy, on the other side, could result in misleading or incorrect diagnosis [30]. Other studies [31,32] have suggested that removing of more blastomeres has negative effects on embryo development, which leads to reduced implantation rates, but it provides a higher diagnostic efficiency when compared with the removal of only one cell.

However, this technique is compatible with fresh embryo transfer on day 5 or 6 of embryo development, given that genetic results will usually be available one to two days after blastomere biopsy [31].

2.3. Trophoectoderm Biopsy

The blastocyst is composed of two different cell types: the inner cell mass, which will evolve in fetal tissues, and the trophoectoderm (TE) considered the precursor of future placenta. The advantages correlated to TE biopsy are mainly three: first of all, TE is not involved in fetus formation, as it will form extra-embryonic tissues. The second benefit is that blastocyst stage embryos have already activated their genome, allowing for a more accurate analysis [25]. Finally, a sample of about five-eight cells is needed for the test, determining a loss of about 10% of all of the cells forming the blastocyst (about 100–150). When compared to the cell mass loss determined by the removal of two blastomere, this procedure seems to be much less invasive [33]. Blastocyst biopsy also implicates other practical advantages: embryos vitrified at this stage show a higher survival rate if compared with cleavage stage embryos. Therefore, it allows for postponing the transfer and even adopt a single embryo transfer policy reducing multiple pregnancies [34].

Three main approaches can be followed for TE biopsy: the first consist in opening the zona pellucida at cleavage stage using a laser-assisted drilling and then waiting for the formation of expanded or herniating blastocysts on day 5 [35]. Cleavage-stage zona drilling is performed to obtain a faster biopsy on herniating blastocysts and reduce the chance of sudden collapse. Although being widely adopted, this procedure presents two main limitations; it entails two sessions of embryo manipulation outside the incubator and there is the concrete risk of having the inner cell mass herniating outside the zona. The second approach to TE biopsy is to leave the embryo in culture until blastocyst full expansion and then open the zona immediately before the biopsy, with assisted laser hatching. This strategy requires a single intervention on the embryo and the zona can be opened in a region far from the inner cell mass, reducing its involvement in the biopsy process. The last method takes advantages of both the previous approaches: it consists of opening the zona when the blastocyst is fully expanded and then waiting for the TE herniation. Figure 1 shows the blastocysts biopsy laser assisted steps.

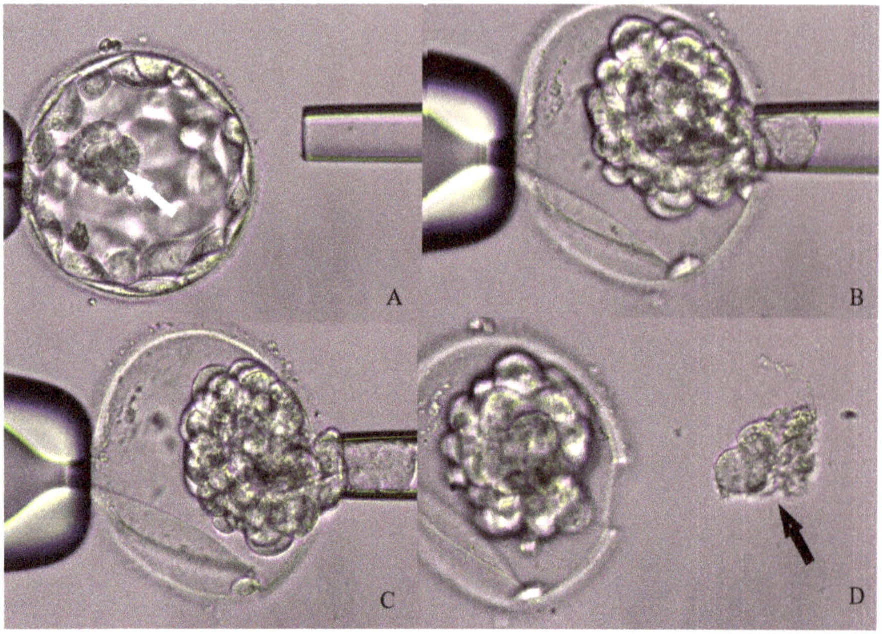

Figure 1. In the figure, the sequence of a blastocyst biopsy laser assisted is shown. The blastocyst is initially orientated by mean of the holding pipette in order to keep the inner cell mass as far as possible from the site of biopsy (**A**). Subsequently, the biopsy pipette is introduced through a hole performed with laser in the zona pellucida and a little number of trophectoderm cells are gently aspirated (**B**,**C**). Finally, the removed trophectoderm cells is transferred in a tube for the genetic analysis (**D**). The white narrow indicates the inner cell mass. The black narrow indicates the removed trophectoderm cells.

It has been demonstrated that the biopsy protocol might affect clinical outcomes [36]. The approach entailing sequential hatching and biopsy results in a significantly higher survival rate after thawing, implantation, clinical pregnancy, and LB rate if compared to the cleavage stage hatching approach. However, day 3 pre-hatching, extends the time of exposure outside the incubator and the risk of having a blastocyst herniating from the inner cell mass requiring extra manipulation during the biopsy. Furthermore this procedure allows a better synchronization with the natural expanding process of the blastocysts that could take place on day 5, 6, or 7. This technique is also cost-effective, since leaving the embryo undisturbed from fertilization to blastocyst formation allows for the employment of single-step media and time-lapse incubation protocol.

Another controversial theme regarding TE biopsy is whether day 6 and day 7 blastocysts should be analyzed or not. A study by Piccolomini and co-workers [37] investigated if slow development might reflect embryo ploidy status. This group compared blastocyst biopsy performed on day 5 versus day 6 and reported similar aneuploidy rate (61.4% on day 5 vs. 69.9% on day 6). The study by Taylor et al. [38] evidenced that day 5 blastocysts had a significantly higher chance of being euploid than day 6 blastocysts (54.6% vs. 42.8%). Both of the studies concluded that blastocysts formed on day 6 and have the same chance of resulting in a live birth rate as those formed on day 5. The study by Hernandez-Nieto et al. [39] found that the rate of embryo euploidy was significantly lower in day 7 blastocysts when compared to day 5 or day 6 cohorts (40.5% vs. 54.7% vs. 52.9%, respectively). In his study there was also a significant decrease in the odds of implantation, clinical pregnancy, and LB, but no association with pregnancy loss in patients who transferred day 7 biopsied euploid blastocysts.

Although day 5 blastocysts may have the higher euploid rates, its relationship with embryo development is still unclear [4,40,41]. On the other hand, day 7 blastocysts can be viable, of top morphology, euploid, and result in a healthy live birth. Therefore, culturing embryos an additional day increases the number of embryos useable per IVF cycle and provides further opportunity for patients who have only a few or low-quality blastocysts. These findings underlined the importance of performing biopsy of all blastocysts available independently of their morphology or growth-timing.

2.4. Non-Invasive PGT

Embryo biopsy, performed at every developmental stage, is an invasive process that might condition IVF results. There are two alternatives to invasive biopsy: blastocentesis, consisting in the analysis of the blastocyst fluid (BF), and the examination of the spent culture media. The sampling of BF is performed on the opposite side of the inner cell mass, leaving the embryo fully collapsed [42,43]. Because dynamic collapse and re-expansion of the cavity is a phenomenon routinely observed during laboratory practice, the loss of the BF should not be detrimental to the embryo [44,45]. The aspiration of the BF does not affect embryo architecture, which results in high survival rates of both good and poor morphology embryos [46].

In 2013, Palini et al. [47], using real-time PCR, reported the presence of DNA fragments in BF obtained from day 5 blastocysts. The investigation of these DNA fragments allowed for the identification, with a 95% accuracy, of male embryos, detecting the specific Y-linked protein. Another study, conducted in 2015 by Tobler et al. [48], analyzed BF from 96 embryos: embryonic DNA was recovered and analyzed, using whole genome amplification (WGA), followed by aCGH in 63% of the samples. The results were concordant with those of the matched inner cell mass karyotypes only in 48.3% of the analyzed embryos. This induced the authors to recommend not to use blastocentesis as an alternative approach for PGT. Therefore, the failure of amplification rates after blastocentesis are a lot much higher if compared with those of the traditional TE biopsy [49].

On the contrary, Gianaroli et al. [50] reported the detection of embryonic DNA in 76.5% of the samples, with a diagnosis concordance rate of 97.4%, when compared to the correspondent TE biopsy. Although the analysis of BF seems to be a promising alternative to invasive PGT, further studies must be conducted. It is important to establish whether the loss of the BF could alter cell to cell communication, or the communication of the embryo with its environment. Furthermore, it is still unknown if the DNA material obtained from the blastocentesis is representative of the embryo DNA.

The analysis of spent culture media is another non-invasive alternative to traditional PGT. It consists in the analysis of cell free DNA that can be found in the media, due to its release from the embryo after apoptosis, necrosis or active release as macrovescicle. The DNA detected might have different origins: embryonal, paternal, maternal or extra-DNA. This technique does not require any experienced embryologist, since the embryo remains untouched, but it may be affected by several variables: culture drops volume, group or single embryo culture, the use of sequential or single-step media and fresh or thawed embryos culture [51,52].

A study conducted by Ho et al. [51] in 2018, compared PGT-A on spent culture media before and after blastulation, to determine the best timing for the collection of the media. The media was collected both on day 3 and day 5: DNA was detected in about 97% of samples that were collected on day 3, but only 39% of the samples provided enough material to yield a PGT-A result. On day 5, an amplification was obtained in 97% of the samples, of which 80% produced enough material for a reliable PGT-A. This study suggests the existence of an optimal collection date, to obtain more DNA quantity and a more solid analysis. The main risks that are associated with this kind of PGT-A are gentic contamination (maternal, paternal or lab personnel), the higher rate of apoptosis in aneuploid embryos, and its dependence on laboratory routine.

Many studies have been conducted in the last years to evaluate this technique and many of them reported moderate success rates [51–53]. Although very promising, minimally or non-invasive techniques for analyzing embryo ploidy are still at a premature stage, needing further study and standardization protocols.

3. PGT Molecular Techniques

The aCGH technique allows for detecting variations in the number of copies and rearrangements of each of the 24 chromosomes when comparing the biopsied genetic material with a reference sample. After amplification by WGA the sample is labelled with fluorescent probes and hybridized to a DNA microarray. The color adopted by each spot after hybridization allows for identifying chromosomal loss or gain. A laser scanner and a data processing software are used to detect fluorescence and analyse aneuploidy and chromosomal rearrangements [54].

Single Nucleotide Polymorphism Array (SNP) is performed using an array setup consisting in DNA hybridization, fluorescence microscopy, and solid surface DNA capture. SNP found in the analyzed sample are compared with SNP of maternal and paternal derivation to assess the ploidy status [55].

Real Time Quantitative Polymerase Chain Reaction (rtq-PCR) is an assay based on polymerase chain reaction; it can identify the whole chromosome asset detecting the copy number of each chromosome. To determine the copy number, it compares three or four locus-specific amplicons along each chromosome to a reference gene from the same chromosome. It is unable to identify chromosome aberrations and uniparental disomy, but can identify triploidy [56].

Next Generation Sequencing (NGS) is the latest approach for pre-implantation genetic testing. The first step of this protocol consists in WGA as for aCGH. After genome amplification, a bar-coding procedure takes place, in which different samples are labelled with specific sequences. This process allows for combining from 24 to 96 biopsies (depending on the sequencing platform adopted) in a sequencing run and this optimized cost per sequenced embryo. Each sequence is then compared to a reference human genome and a specific software is used to identify copy number variations and large deletions or duplications [57–59].

The availability of different molecular techniques opened a debate about their sensitivity and reliability for PGT. The study by Friedenthal et al. documented that the implantation and ongoing pregnancy/live birth rate were significantly higher in women undergoing NGS (71,6% and 62,0%, respectively) as compared to those evaluated by aCGH (64.6% and 54.4%, respectively). Moreover, the comparison between NGS and aCGH group documented significantly less biochemical pregnancies (8.7% vs. 15.1%) and a similar rate of spontaneous abortion (12.4% vs. 12.7%). Based on these findings, the authors concluded that PGT using NGS significantly improves clinical outcomes with respect to the aCGH technique and suggested that NGS could be more effective in identifying mosaic embryos and those with partial aneuploidies or triploidy [60].

Friedenthal et al. [61] investigated clinical error rates in frozen-thawed embryo transfer (FTET) cycles of single euploid embryo diagnosed by NGS (1151 cases) and aCGH (846 cycles). The clinical error rates in the NGS cohort resulted in being lower with respect aCGH: 0.7% vs. 1.3% per embryo, 1% vs. 2% per pregnancy with gestational sac, 0.1% vs. 0.4% per ongoing pregnancy/live birth rate, and 13.3% vs. 23.3% per spontaneous abortion. Therefore, although NGS and aCGH are highly sensitive methods for PGT, clinicians must still consider the chance of errors occurring.

4. Current Insications for PGT

4.1. Advanced Maternal Age

Chromosomal abnormalities in oocytes and embryos constantly increase with maternal age and reduced ovarian reserve [20,62]. However, a significant proportion of aneuploid embryos presents a good morphology [9,16,63–66] and many recent studies report that good quality blastocysts can result in aneuploid embryos in up to 44.9% of the cases [20,67,68].

A study by Franasiak et al. [20], based on more than 15,000 TE biopsies, demonstrated that the aneuploidy rate was about 20–27% between the ages of 26 and 30. Chromosomal abnormalities increased steadily from age 31 through age 43, and aneuploidy rate was then stabilized at approximately 85%. The no-euploid rate was lowest and equivalent in the embryos that were obtained from women aged 26–38 year, but a statistically significant increase of this rate was noted for successive age group. Jiang et al. [69] divided patients undergoing PGT-A into three groups according to Anti-Müllerian hormone (AMH) level: <1.50 ng/mL, 1.50–5.60 ng/mL, and ≥5.60 ng/mL. There was a significant difference in aneuploidy rate between the three AMH groups (66.7% vs. 42.9% vs. 50.0%, respectively). After age stratification, the aneuploidy rate was still significantly different among AMH groups with a similar trend in women ≥35 years old, suggesting that low AMH level was associated with increased risk of embryo aneuploidy only in women of advanced age.

In women with advanced maternal age (AMA), therefore, PGT-A assessment should be considered [69–71]. The most recent studies showed that embryo selection through aneuploidy screening significantly increases the chance for implantation and decreases miscarriage rate [67,72]. Being able to select the euploid embryo with higher implantation potential allows for limiting the number of embryos transferred per cycle, decreasing the chance of twins and high order multiple gestation [73]. Moreover, some studies have shown that when an euploid embryos is transferred, the implantation rate remain similar, regardless of increasing maternal age [74,75].

Rubio et al. [76] published the results of a multi center randomized two-arms trial: a PGT-A group undergoing 24-chromosome screening on day-3 embryos, followed by blastocyst transfer, and a control group without PGT-A, in which blastocyst transfer were developed in women with advanced maternal age (38–41 years). In PGT-A group, 78.6% of embryos were aneuploid, a total of 37 pregnancies were achieved with only one clinical miscarriage, resulting in a delivery rate per transfer of 52.9%. In contrast, a total of 41 pregnancies were obtained in the control group, with 16 miscarriages, and a delivery rate per transfer of 24.2%. The authors observed a significantly lower miscarriage rate in the PGT-A group (2.7% vs. 39.0%) and a reduced time to achieve pregnancy (4.5 weeks vs. 5.8 weeks) when compared to the control group.

Sacchi et al. [77], in an observational-cohort study with two years follow-up, including a total of 2538 couples, confirmed that PGT improves clinical outcomes in patients with AMA when compared to controls, as documented by higher LBR (40.3% vs. 11.0%), reduced pregnancy loss (3.6% vs. 22.6%), and multiple pregnancy rate (0% vs. 11.1%). Multivariate analysis showed no negative impact of PGT-A related interventions on the cumulative delivery rate and on neonatal outcomes.

Extensive counseling based on biological and clinical data should be provided to women older than 43 years, because of their very low odds of success and high risk for embryo aneuploidies. Ubaldi et al. [78] evaluated the efficacy of PGT-A in women older than 44 years with a good ovarian reserve. Blastocyst development was obtained in 102 (68.0%) out of 150 cycles, but only 21 (14.0%) of them resulted in being euploid. Specifically, no euploid blastocyst was found in patients older than 46, whereas the euploidy rate was 14.4% and 4.5% in the group aged 44.0–44.9 and 45.0–45.9, respectively. The delivery rate was 57.1% per transfer. However, the delivery rate per cycle was 10.6% in patients aged 44.0–44.9 years and 2.6% in patients aged 45.0–45.9 years.

Lee et al. [79] evaluated the efficacy of PGT in women with an age included between 40 and 43; they compared the pregnancy outcomes from traditional fresh IVF cycles with day 5 embryo transfers, FTET cycles without PGT, and PGT-FTET cycles (only euploid embryos). The implantation and LB rate in PGT-FTET cycles (50.9% and 45.5%) was significantly higher than for unscreened embryos transferred in fresh (23.8% and 15.8%) or no PGT-FTET (25.4% and 19%) cycles. There were significant differences in live birth rate per embryo transferred for the three groups: 45.5% for PGT-FTET, 15.8% for fresh transfers, and 19% for No-PGT FTET. The incidence of pregnancy loss was 38.1% for fresh cycles and only 10.7% for PGT-FTET.

Outcome data coming from a total of 8175 SETs after PGT-A and embryo cryopreservation [80] evidenced that age-related decline in reproductive efficiency can be reduced selecting euploid embryos suitable for the transfer. However, the implantation rates are negatively correlated with maternal age and, also after adjusting for confounders, women 38 years or older had a significantly lower implantation rate than those under 35. These differences were also highlighted in clinical pregnancy and LB rates. Therefore, the observed impact of aging is not correlated with embryos ploidy.

Taking into considerations all of these factors, it can be affirmed that, in the AMA patient, the influence of embryo aneuploidies on infertility is relevant, and both the safety of the procedure and the reduction of the time to pregnancy achievement are of foremost importance. Patients undergoing PGT-A must be counseled that the proportion of embryos likely to be aneuploid grows in age-specific ranges, evidencing the possibility of obtaining only aneuploid embryos. It should also be noted that obtaining one euploid blastocyst becomes more challenging with increasing maternal age, since there are many adverse factors that are associated with cycle cancellation, such as the lack of follicular development, unsuccessful oocyte retrieval, and lower blastocyst formation rate.

4.2. Recurrent Pregnancy Loss

Approximately 5% of all couples undergoing IVF treatment are affected by recurrent pregnancy loss (RPL), defined as two or more consecutive miscarriages at a gestational age up to 20 weeks. It can be explained by genetic, anatomic, endocrinologic, and immunologic abnormalities [81]. However, in more than 50% of cases, the current diagnostic procedures are not able to identify etiologic factors [82–84]. Unexplained RPL is a distressing condition for the affected couple, and a frustrating problem for the clinician since no effective therapy exists.

A retrospective cohort study including 46,939 women who underwent fetal karyotype analysis through amniocentesis or chorionic villus sampling, documented that women with no prior spontaneous abortions had a risk for aneuploidy of 1.3%. In women with one, two, or three previous spontaneous abortion this risk increased significantly to 1.6%, 1.8% and 2.1%, respectively [85]. A higher aneuploidy rate in RPL patients has also been confirmed by other authors [83–87].

PGT-A could be helpful in the treatment of couples with unexplained RPL, when considering that embryo aneuploidies could be the cause of miscarriages. Several studies using genetic testing in couples with this indication have shown a decrease in miscarriage rate [88]. Earlier studies were typically performed with the use of FISH on cleavage-stage embryos and typically tested only 7–12 chromosomes. In one meta-analysis [89], four observational studies [70,90,91] evaluated patients with RPL undergoing cleavage-stage biopsy and compared them to patients looking for a natural-conception. The spontaneous abortion rate of 9% in the first group was significantly lower than in the controls (28%).

The incidence of aneuploidies in blastocysts from patients with idiopathic RPL undergoing PGT-A and PGT for monogenic diseases (PGT-M) resulted to be increased in women aged ≤35 years (48.9% vs. 36.9%), whereas no significant increase was found in group aged >35 years (66.9% vs. 61.4%). However, despite euploid embryo transfer, young patients had a higher miscarriage rate (26.1% vs. 3.1%) suggesting that in these group of patients RPL may derived not only from genetic reasons [92].

A literature review searching for available evidences on LB and miscarriage rates after PGT-A, compared to natural conception in couples with unexplained RPL suggested that PGT-A application, might reduce the miscarriage rate when compared to natural conception (9% vs. 28%) [89]. Hodez-Wertz et al. [93] documented that, in a total of 2282 embryos analyzed, 60% were aneuploid. Euploid embryo transfers performed were 181, with an implantation rate of 45% and ongoing pregnancy rate of 92%. The miscarriage rate was found to be only 6.9%, as compared with the expected rate of 33.5% in a RPL control and 23.7% in infertile control population. Therefore, the authors concluded that PGT-A by aCGH decreases miscarriage rate in idiopathic RPL, providing, at the same time, diagnosis and treatment for these patients.

No significant differences in LB and miscarriage rates per patient given or not given PGT-A were evidenced by Sato et al. [94]: 26.8 versus 21.1% and 14.3% versus 20.0%, respectively. Live birth rate per embryo transfer was improved by the application of PGT-A (52.4 vs. 21.6%), which also had the advantage of reducing the number of embryo transfers that are required to achieve a similar number of live births as compared with cycles in which PGT was not performed.

In the study by Murugappan et al. [95], PGT-A in patients with RPL resulted in a LB rate of 53% and a clinical miscarriage rate of 7%. Expectant management had a LB rate of 67% and clinical miscarriage rate of 24%. However, the IVF/PGT-A strategy was 100-fold more expensive, costing $45,300 per live birth when compared with $418 per live birth with expectant management. According to current literature, IVF/PGT-A is a very expensive way to reduce miscarriage, without increasing the chance of achieving a live birth; further study should be conducted before recommending it as standard treatment for RPL.

Katz-Jaffe et al. [96] showed a higher incidence of aneuploid blastocysts in patients with diminished ovarian reserve (DOR), whereas Trout et al. [97] evidenced a higher incidence of DOR in RPL patients. Shahine et al. [98] examined the role PGT-A in patients with RPL and DOR, being defined as a cycle day 3 FSH > 10 IU/mL and/or antimüllerian hormone <1 ng/mL. Patients with DOR had a higher percentage of aneuploid blastocysts (57% vs. 49%) when compared to controls. In all patients aged <38 years, a higher aneuploidy rate of was observed in embryos from DOR patients, when compared to patients with normal ovarian reserve (67% vs. 53%, $p = 0.04$). In the two groups, the implantation and the live birth rates after transfer of euploid blastocysts were similar (61% vs. 59% and 54% vs. 47%, respectively), despite the difference in ovarian reserve testing. The miscarriage rate was slightly higher in the DOR group, but it was not statistically significant (14% vs. 10%). It can be affirmed that patients with RPL and DOR, transferring euploid embryos, may benefit from PGT-A, as demonstrated by outcomes that are comparable with those of patients with RPL and normal ovarian reserve testing.

4.3. Repeated Implantation Failure

Repeated implantation failure (RIF) is defined by ESHRE PGD Consortium, as the absence of a gestational sac on ultrasound at five or more weeks after three embryo transfers with high quality embryos or after the transfer of ≥10 embryos in multiple transfers [99]. RIF is not only distressing for patients who require multiple cycles of treatment, but it also increase the costs of IVF procedure. This condition is still a big challenge to the clinician, since it could have multiple causes, depending on maternal or embryonic factors [100–102]. Although not all RIF cases are due to embryonic defects, many patients have different chromosomal abnormalities [103] and the extension on these anomalies, increases with the number of previous failed IVF cycles [102]. A high level of chromosome abnormalities is present in arrested embryos, but also in embryos with good morphology [11,15,20,67]. Several approaches have been used in RIF patients to select a good embryo for the transfer, such as blastocyst culture [104] or assisted hatching, which seems to improve embryo implantation capability [105].

Many centers started to perform PGT-A on single cells biopsied from cleavage stage embryos using FISH with chromosome-specific probes [106,107]. A study by Rubio et al. [71] aimed to evaluate the role of PGT by FISH for two different indications: women <40 years with RIF and women with AMA aged 41–44 years. The patients were allocated in two arms: blastocyst transfer without PGT or PGT followed by day 5 embryo transfer. In the AMA group, a significant increase in live-birth rate per patient was found in PGT group when compared with no-PGT group (32.3% vs. 15.5%). However, in RIF patients, a trend toward higher LB rate was noted (47.9% vs. 27.9%), but without statistically significant difference.

Comparative genomic hybridization is a molecular cytogenetic technique that can be applied to single cells in inter-phase to allow simultaneous analysis of every chromosome, in contrast with FISH. However, the study by Voullaire et al. [108] including women aged 26–41 years with RIF showed that 40% of the embryos could be considered suitable for transfer, a value that is similar to that found using FISH for PGT-A (ESHRE PGD Consortium Steering Committee, 2002) [109]. Infertility in this group of patients is likely to be multi-factorial and chromosomal abnormalities could not be involved in it [108].

The aim of the study performed by Greco et al. [110] was to assess the clinical pregnancy and implantation rates, after transferring a single euploid blastocyst tested with aCGH, in a group of patients younger than 36 years with a history of RIF. These results were compared with a similar group of RIF patients in whom PGT was not performed and with a group of good prognosis patients after PGT. The euploidy rate in RIF PGT and in NO RIF PGT were 46.2% and 51.8%, respectively. Clinical pregnancy and implantation rates, respectively, were 68.3% and 68.3% in RIF PGT, 22.0% and 21.2% in RIF NO PGT, and 70.5% and 70.5% NO RIF PGT. There were no spontaneous abortions in any group. The results from RIF NO PGT were significantly lower when compared to other two groups.

The similar clinical results after single euploid blastocysts transfer in good prognosis and RIF patients at their first IVF attempt demonstrated that embryo aneuploidies may be the most relevant cause of implantation failure [110].

4.4. Male Factor Infertility

Although embryonic aneuploidies mainly arise from maternal genome, some aneuploidies may derive from the spermatozoa [21]. Males with abnormal karyotype and Y chromosome deletions tend to produce spermatozoa with an unbalanced chromosome complement. Other several factors, such as varicocele, chemotherapy, age, and lifestyle, may also negatively affect meiotic divisions during spermatogenesis [21]. The rate of abnormal sperm after FISH examination was significantly higher in patients with male infertility (55.8% vs. 15.0%) and teratozoospermia was highly correlated with aneuploidy rate for chromosome 17, as documented by Petousis et al. [111]. Therefore, PGT-A should be considered in ICSI cycles with severe male factor (SMF), including azoospermia (obstructive and non-obstructive), severe oligoastenoteratozoospermia, Klinefelter syndrome (KS), Y-chromosome microdeletion, and even in men whose semen analysis does not fulfill the current World Health Organization (WHO) criteria.

Magli et al. [112] showed that SMF might contribute to a higher rate of aneuploid blastocysts (55% aneuploidy rate with normozoospermia, 62% with oligozoospermia, and 69% with nonobstructive azoospermia (NOA)), as evidenced by the 9-chromosome FISH technique performed on day-3 embryos. In a study by Silber et al. [113], patients with oligozoospermia and azoospermia had an euploidy rate of 41% and 22%, respectively, but the analysis was still conducted with the use of 9-chromosome FISH. Coates et al. [114] performed aCGH on TE biopsies; a significant increase in sexual chromosome abnormalities was observed in oligozoospermic males both with own (6.1% vs. 5.9%) and donor (1.7% vs. 2.0%) eggs.

A cohort study based on 1219 consecutive ICSI cycles performing PGT with trophoectoderm biopsy using rtq-PCR [115] evidenced that the rate of euploid blastocyst per inseminated mature oocyte was only significantly reduced in NOA (11.1%) when compared with normozoospermic patients (16.3%), whereas the euploidy rate per biopsied blastocyst was similar among the two groups. Furthermore, no differences in gestational age and birth weight were reported among the groups. The overall prevalence of congenital malformations (2.1%) was similar to that previously reported after either IVF with and without PGT-A [116,117] or spontaneous conceptions [118,119]. Following all of these considerations, SMF was not considered to be a critical parameter for embryo ploidy status. This could be due to the intrinsic potential of the oocyte to prevent further development of aneuploid embryos before the activation of the embryonic genome [120] or to the fact that sperm-derived aneuploidies may result in an early interruption of embryo development [121].

Mazzilli et al. [115] investigated in their study only 35 cycles in women under 35 years. Kahraman et al. [107] investigated 326 cycles with 741 blastocysts biopsied and analyzed using aCGH and NGS techniques; the couples enrolled were composed of young female (≤35) with SMF partners, ranging from five million/mL to NOA.

The couples undergoing IVF cycles with testicular spermatozoa had higher aneuploidy rate with respect to the patients with SMF and controls. In addition, reduced spermatozoa count (1–5 mil/mL) was associated with aneuploidy rate of 15.6%, suggesting that SMF increases the risk of chromosomal abnormalities, regardless of the female age. The highest mosaicism rate (22.0%) was observed in cycles using testicular spermatozoa and it was significantly higher than in cycles with normal sperm parameters (9.9%). No significant difference was observed in the clinical outcomes between the control group and the SMF groups after transfer of a single euploid embryo. There was a trend toward a higher miscarriage rate in testicular sperm group.

A retrospective analysis by Tarozzi et al. [122], in which blastocysts from 340 cycles were assessed by aCGH, no differences between male (MF) and non-male factor (NMF) groups were found in terms of euploid blastocysts rate. The MF group showed a significantly higher rate of mosaic embryos (3.6% vs. 0.5%, respectively). A similar pattern of results was observed in all SMF groups considered in their complex, when compared to those without SMF (7.7% vs. 1.8%), suggesting that a compromised semen quality is associated with an increase in the mosaic blastocysts rate. The reason of these differences could be found in the fact that, during spermiogenesis, the transit of the spermatozoa in the epididymal tract favors DNA packaging, through the stabilization of the chromatine structure with protamine dephosphorylation and the formation of intra and intermolecular disulphide bridges between protamines.

The study by Scarselli et al. [123] evaluated whether the duration of ejaculatory abstinence might influence the euploid blastocysts rate. The blastocysts euploidy rate resulted in being 27.5% using spermatozoa from semen sample obtained after 2–5 days of abstinence and of 43.5% using spermatozoa from the second sample obtained 1 h after the first one. The use of potentially higher quality spermatozoa from samples with a shorter abstinence might reduce the aneuploidy rate in blastocysts.

Non-mosaic Klinefelter syndrome is associated with a male genetic disorder that often leads to NOA. The literature reports about 101 children born from non-mosaic KS fathers after successful ICSI, in most cases, healthy and without the use of PGT to select normal embryos [124]. Fetuses or embryos were diagnosed to have 47,XXY karyotype in some cases [125,126]. Due to these contradictory findings, it remains an open question as to whether embryo biopsy should be offered to couples with KS male partner. In the study by Greco et al. [127], a total of 26 ICSI cycles were performed in couples with KS and a total of 11 pregnancies and deliveries were achieved (pregnancy rate 42.3%). All of the fetal karyotypes were normal. PGT-A was not done, and the embryo chromosomal assessment was unknown at the time of transfer. It could be thought that, when the SMF is due to altered karyotype, it does not affect embryo ploidy. However, Staessen et al. [128] published an extremely convincing study on the need for PGT in KS patients. They reported data on PGT-M offered to 20 non-mosaic KS, showing that 54% of the embryos were normal but KS patients had a higher incidence of abnormalities involving chromosomes 18 and 21.

Several studies have also explored the effect of advanced paternal age on sperm aneuploidy rates. Two recent studies [129,130] noted that males above the age of 50 present more sperms with damaged DNA, low blastocyst development rate, higher global aneuploidy rate, and a higher rate of trisomies. Advanced paternal age could lead to several alterations in endocrinal and reproductive phenotypes, leading to an increase in DNA damages over years and a decrease in germ cells capability to repair these damages, inducing the production of aneuploid sperms and, consequently, abnormal embryos [131].

The study by Gat et al. [132] evaluated 177 IVF-ICSI cycles, in which PGT-A was performed on 405 blastocysts and concluded that DNA fragmentation index (DFI) was not associated with impaired embryo quality in both genetic and morphological aspects. Similar DFI-related findings were published by Bronet et al. [133], who reported comparable embryo euploidy rates in different DFI levels.

Overall, approximately 25–55% of males with extreme testicular pathologies, such as hypospermatogenesis, sperm maturation arrest, and Sertoli cell only syndrome (SCOS), and about 5–25% males with severe oligozoospermia or azoospermia present Y chromosome microdeletions. In general, deletions of the entire AZFa region invariably result in SCOS and azoospermia. Patients with deletions of the AZFb region presents maturation arrest, mostly at spermatocyte stage. Men with AZFc deletions have the most variable phenotype, which ranges from complete azoospermia to mild oligozoospermia [134]. Comparisons between the different types of microdeletions and treatments revealed that, except AZFc, cases with AZFa and AZFb microdeletions presented a poor clinical prognosis. On 125 patients, 25 of them (20.0%) had abnormal karyotypes. Ninety-four cycles with testicular sperm leaded to 19 deliveries. Twenty-nine cycles with ejaculated sperm leaded to eight births (all in couples with AZFc). Patients with such alterations should be informed about the possibility of using preimplantation genetic testing due to the reported heredity of microdeletions and a possible association with men abnormal karyotype [134].

4.5. PGT in a Good-Prognosis Patients Undergoing SET

Choosing the best embryo to transfer is crucial, especially when a single embryo transfer program is adopted for different clinical reasons [135]. The first study to prospect a successful elective SET after a rapid on-site aCGH application was performed by Yang et al. [67] in good prognosis women <35 years of age. Fifty-six patients were randomized in two groups: in the first one a morphological evaluation of the embryos was used to select the one for the transfer in combination with aCGH, in the second one, morphology was used as the only discerning parameter. The aneuploidy rate in 425 blastocysts analyzed with aCGH was 44.9%, whereas 389 blastocysts were microscopically examined in the control group. The clinical and ongoing pregnancy rates were significantly higher in the morphology plus aCGH group as compared to controls (70.9 vs. 45.8%, and 69.1 vs. 41.7%, respectively). No twin pregnancies occurred in both groups. A low miscarriage rate was noted for all of the study patients, although this was slightly lower in the PGT-A group (2.6 vs. 9.1%).

Despite an increasing acceptance of elective SET treatment, many IVF cycles continue to involve the transfer of two or more embryos. Scott et al. [32] evaluated whether blastocyst biopsy with rtq-PCR comprehensive chromosomal screening might improve IVF outcome in women under 42 years with normal ovarian reserve. The aneuploidy rate was 28% among patients who were included in the genetic testing group. Clinical implantation rate and the proportion of screened embryos that progressed to delivery (79.8% and 66.4%, respectively) were significantly higher when compared to the control group (63.2% and 47.9%, respectively).

A recently published randomized clinical trial by Ozgur et al. [136] divided 220 patients aged ≤35 years in an arm, in which a single euploid blastocyst was transferred and an arm in which single unknown-ploidy blastocysts were transferred. In the PGT-A group, 73.4% of all blastocysts were diagnosed as euploid, suggesting that the best-scoring blastocysts of the selected infertile patient (≤35 years) group had the same probability to be healthy. The live birth rate in euploid subgroup was found not to be statistically different as compared to morphology group (56.3% vs. 58.6%), which suggests that PGT-A is not able to enhance LB rate in young patients [136].

Munne et al. [137] draw the same conclusions in a study in which a total of 661 women (average age 33.7 ± 3.6 years) were randomized to PGT-A or morphology only group. The ongoing pregnancy rate was equivalent between the two arms, with no significant difference per embryo transfer (50% vs. 46%). Post hoc analysis of women aged 35–40 years showed a significant increase in the ongoing pregnancy rate per embryo transfer (51% vs. 37%), but not per intention to treat.

In a study conducted by Forman et al. [138], 43.4% of the 175 randomized patients were <35 years old, 30.9% were 35–37 years old, 19.4% were 38–40 years old, and 6.3% were 41–42 years old. The different groups showed the same clinical performance. The ongoing pregnancy rate after each patient's first transfer, whether fresh or frozen, was 60.7% after single euploid blastocyst transfer and 65.1% after two untested blastocyst transfer. It must be underlined that patients who received single euploid blastocyst transfer were nearly twice as likely to have an ongoing singleton pregnancy when compared with those with two blastocysts transfer (60.7% vs. 33.7%). This trial demonstrates that the singleton delivery rate can be improved while using a validated method to assess embryo ploidy, without compromising the overall success rate.

The transfer of single euploid blastocyst might prevent twins and higher order multiple pregnancies reducing the costs and increasing the efficacy and safety of IVF procedure. However, a higher risk of monozygotic twinning following assisted reproduction procedures (1.4%) as compared with natural conception (0.4%) has been evidenced [139]. Some studies showed a consistent association between the extended embryo culture and embryo splitting [140]. However, it is still unclear as to whether embryo biopsy represents a risk factor for monozygotic twinning [141,142].

In a retrospective cohort study from a large referral center in Belgium, the incidence of monozygotic twin births following PGT was 1.5%, and 2.1% following ICSI cycles without PGT [143]. In addition, also a systematic review, pooling results from four studies, reported no increased risk of monozygotic twinning after PGT as compared with IVF without PGT [144]. However, Kamath et al. [145] evaluated data from 207,697 SET cycles being performed mainly in women aged <35 years with no infertility diagnosis; many of the cycles were performed as part of a PGT-M or PGT-SR program. There was a significantly higher risk of zygotic splitting with PGT versus non PGT IVF cycles (2.4% vs. 1.5%), even after adjustment for potential confounders. Such contradictory findings could be due to different embryo biopsy techniques adopted in these studies (for example, cleavage stage biopsy versus trophoectoderm biopsy).

4.6. PGT-A in Donor Egg Cycles

Donor eggs are usually collected from healthy, young, and fertile women, hoping that most eggs will have normal chromosomal integrity. When considering literature data presented in the previous chapters of this review, it can be affirmed that even young women produce a significant proportion of aneuploid embryos and screening out such abnormalities could potentially increase the efficacy of donor egg (DE) cycles [146].

According to the Society of Assisted Reproductive Technology, live birth rate in patients who undergo egg donation programs is only 5–10% higher than in patients under 34 years who use their own eggs. Embryos aneuploidy is an important factor affecting clinical success in young patients undergoing IVF [67] and, therefore, it might also affect embryo implantation in ovodonation cycles. In fact, it has been reported that the aneuploidy rate of embryos obtained from donor eggs might reach 53.2% and 88.1% of the embryo aneuploidies are of a maternal source, suggesting the necessity of PGT-A in DE cycles [147].

On the other hand, in the study by Hoyos et al. [148], oocyte donors aged ≤25 had similar blastocyst formation and ploidy rates. No correlation was found between euploid blastocyst rate and donor age. There is no need to favor a specific age subgroup of young oocyte donors, given the lack of significant age-related change in blastocyst euploid rates.

Haddad et al. [149] observed that, in DE cycles with PGT-A, 39.1% of blastocysts were abnormal. The transfer of normal euploid blastocysts lead to a clinical pregnancy rate of 72.4%, an ongoing/delivery rate of 65.5%, and an implantation rate of 54.9%; these rates were slightly higher than those in the control group (66.7, 54.0, and 47.8%, respectively), but no significant difference was found. Miscarriage rate was higher in the control group (19.2%) than in the PGT group (9.5%), but, once again, the difference did not reach statistical significance. These findings suggest that PGT-A might only be necessary in some specific situations, such as the need of a single embryo transfer. In 2019, Masbou et al. [146] compared, in a retrospective cohort study, clinical outcomes in patients undergoing ED cycle with SET with and without pre-implantation genetic testing. This study revealed that PGT-A and subsequent euploid SET did not improve pregnancy outcome in DE cycles, although there was a trend toward decreasing of miscarriage rate. Overall, these results suggest that the benefits of performing PGT-A on embryos derived from young DE may be limited.

It is important to evaluate the potential effect of oocyte cryopreservation on embryo aneuploidies, especially for centers using vitrified oocytes in egg donation programs. Forman et al. [150] did not find significant difference in the rate of embryonic aneuploidy between fresh and vitrified donor oocytes. In this study, thirty-three patients who thawed 475 oocytes that had been cryopreserved for a median of 3.5 years, were compared with 66 age-matched controls who underwent IVF with PGT-A based on fresh oocytes. No differences were found in the percentage of euploid blastocysts (44.5% vs. 47.6%), implantation (65% vs. 65%), and live birth rate (62.5% vs. 55%) after 24-chromosome PGS with cryopreserved or fresh oocytes.

4.7. PGT for Monogenic Diseases

Pre-implantation genetic testing for monogenic diseases (PGT-M) is an advisable approach for couples with the risk of transmitting genetic diseases to their offspring. However, chromosomal aneuploidies can involve chromosomes that different from those that were investigated with PGT-M. The first successful attempts to perform a double factor analysis (PGT-A and PGT-M) were reported by Obradors and collaborators [151]; the aim was to improve the implantation rate selecting potentially euploid embryos free of mutations responsible for cystic fibrosis [151] or Von Hippel–Lindau syndrome [152]. However, in both case reports, first genetic screening was performed by aCGH on oocyte polar bodies for PGT-A and the second using PCR on day 3 blastomeres for PGT-M. A similar procedure was applied by Rechitsky et al. [153] in 96 cycles resulting in the transfer of 153 unaffected aneuploidy-free embryos and 32 healthy live births.

To our knowledge, the largest reported series of double genetic tests for different monogenic diseases or translocations and aneuploidy screening, performed with a single biopsy on 1122 blastocysts, was presented by Minasi et al. [12]. All of the biopsies were performed at blastocyst stage and analyzed by WGA, followed by PCR for monogenic diseases and aCGH for all cycles. The study demonstrated that, while 55.7% of the biopsied blastocysts were healthy after PGT-M analysis, only 27.5% of them were also euploid. Clinical pregnancy, implantation, and miscarriage rates that were obtained after the application of both techniques were 49.0, 47.7, and 9.9%, respectively. Without performing PGT-A in association with PGT-M, 316 blastocysts resulted in being normal or carrier for the genetic disease could have been transferred, leading to implantation failures, miscarriages, or, even, to unhealthy live births.

The value of this double screening was also explored by Goldman et al. [154] in a retrospective cohort study, including patients who underwent PGT-M with or without 24-chromosome aneuploidy screening. There were no differences between the PGT-M and aneuploidy screened group and PGT-M only group, when comparing the percentage of blastocysts affected by the single gene disorder of interest (37.0% vs. 32.8%). Despite a young mean aged population, only 25.6% of the blastocysts that resulted in negative or carriers of the single gene disorder were also euploid. In the PGT-M only group, 54.7% of embryos resulted in being suitable for transfer according to their unaffected or carrier status (p = 0.001). In patients undergoing both genetic screenings, the implantation rate was higher (75% vs. 53.3%) and miscarriage rate lower (20% vs. 40%) as compared to controls. The Authors concluded that PGT-M performed concurrently with 24-chromosome aneuploidy screening provides valuable information for embryo selection, and significantly improves SET rate.

5. Mosaicism

Mosaicism is defined as the presence of different cell lines in the same embryo. Two different kinds of mosaicism can occur: diploid/aneuploid mosaic with a mix of aneuploid and euploid cell lines and aneuploid mosaic with a mix of cell lines with different chromosomal abnormalities. There can be various types of aneuploidies in mosaic embryos: single chromosome loss or gain, complex or structural aneuploidies [155]. The origin of mosaicism is related to mitotic errors happening after fertilization at the third division stage. These mitotic errors, taking place before DNA duplication, are basically: anaphase delay, mitotic non-disjunction, accidental chromosome demolition, or premature cell division. The aneuploid cells rate depends on the time at which mitotic error happens; in embryos in which errors take place at the second cleavage stage, there will be a higher percentage of aneuploid cells [156,157]. Occasionally, mosaicism may derive from a meiotic non-disjunction event, causing a trisomic conceptus, followed by a post-zygotic event (trisomy rescue) [158,159].

FISH was first technique used to investigate the frequency of embryonic mosaicism: studies involving this method reported mosaicism rates that varied from 30% to 90% at cleavage stage, and from 18% to 46% at blastocyst stage [160]. More clear information regarding the mosaicism rate was obtained with the introduction of recent molecular techniques for genetic testing, such as SNP array, aCGH, and NGS. The mosaicism rates detected using these analyses are lower than those reported using FISH.

At cleavage stage, the incidence of mosaic diploid-aneuploid varies from 15% to 71%; at blastocyst stage, the rates are lower, with an incidence ranging from 3.9% to about 33% [65,161]. Moreover, mosaic diploid/aneuploid embryos rate, decreases with maternal age: in women, less than 35 years old, the rate is 26.6%, and in >42 years old women it is 10.5% [162].

Studies regarding the distribution of mosaic cells in the blastocyst have been conducted to establish whether there is a preferential distribution between TE and inner cell mass and a recent one demonstrated that, in low-level mosaicism, there is a preferential allocation in TE, but, in high-level mosaicism, the aneuploidies seems to be uniformly distributed in the whole embryo [163]. An important question, still open, regarding mosaic embryos, is whether data obtained from a small portion of TE cells may predict the whole embryo asset. It has been hypothesized that the mosaicism rate deduced from a single TE biopsy might not be representative of the remaining TE cells or of the inner cell mass [164].

Mosaic embryos have not been considered to be suitable for transfer and they were discarded, while considering them as aneuploid embryos. Mosaicism was supposed to be responsible for altered embryo development, thus leading to implantation failure, or resulting in congenital malformation, mental retardation, and uniparental disomy [164].

Greco et al. [165] were the first, in 2015, to demonstrate that mosaic diploid/aneuploid embryos were suitable for implantation, giving birth to healthy babies. In his study, 18 diploid/aneuploid mosaic embryos, diagnosticated by aCGH, were transferred in women who did not have any euploid blastocyst for frozen embryo transfer. The implantation rate of these embryos was 44%, with a live birth rate of 33%. Every fetus was confirmed through chorionic villi to have a normal karyotype. Since then, a great number of studies involving the transfer of mosaic embryos were developed and they all confirmed the capability of these embryos to lead to healthy live births [166,167]. Table 1 reassume the clinical outcomes after mosaic blastocysts transfer presented in the most relevant papers [155,157,165–167].

Table 1. Clinical outcomes after transfer of mosaic embryos.

References	No. of Mosaic Blastocysts	Pregnancy Rate	Implantation Rate	Miscarriage Rate
[165]	18	33%	44%	11%
[157]	44	15.4%	30.1%	55.6%
[166]	143	41%	53%	24%
[155]	78	30%	38.5%	7.8%
[167]	100	30%	38%	7%

Several studies have demonstrated that chromosomal mosaicism extent significantly affect clinical results. Spinella et al. [155] proved that blastocysts with a mosaicism rate lower than 50% had significantly higher ongoing pregnancies (40.9% vs. 15.2%) and live births (38% vs. 19%) when compared to those with a mosaicism rate over 50%. These findings were confirmed by the paper of Viotti et al. from 2019 [168]. This study assessed that complex and single segmental mosaics had better outcomes than those with multiple gain or losses of whole chromosomes [168]. Victor et al. [167] assessed that the mosaicism rate is not correlated with clinical outcomes, reporting any significant difference in embryos with mosaicism rate lower than 40% or included between 40% and 80%. Clinical outcomes were only affected by maternal age and the type of mosaicism involved, with better results in young women and segmental aneuploidies.

A still unsolved question is whether the transfer of mosaic embryos might result in congenital abnormalities or in the birth of affected babies. Over 200 pregnancies have been reported, obtained after the transfer of mosaic embryos; 42 gave birth to healthy children [155,165,167] and the other were, at the moment of the studies publication, still ongoing and apparently normal [166–168]. However, this number is still not high enough to draw any conclusion and future studies are needed. Due to the lack of these data, a transfer involving mosaic embryos should only be performed when unavoidable and after appropriate counseling [162]. Couples considering to transfer mosaic embryos should be informed about potential pregnancy risks and outcomes; it should be explained them that these transfers may be characterized by lower implantation and pregnancy rates, beyond a higher risk of genetic abnormalities and adverse pregnancy outcome.

It must be underlined a recently published case report: a blastocyst showing a 35% mosaicsim of monosomy 2 was transferred in a 39-year old woman with diminished ovarian reserve resulting in a clinical pregnancy. Amniocentesis revealed a mosaic trisomic karyotype: 46,XX(98)/47,XX + 2(2). This case demonstrates the need for a strict prenatal monitoring and diagnosis by early amniocentesis [169].

Preimplantation Genetic Diagnosis International Society (PGDIS) and Controversies in Preconception and Perinatal Genetic Diagnosis (CoGEN) have been suggested to always prefer the transfer of embryos with low level of mosaicism (20–40%), avoiding the transfer of those with potentially viable aneuplodies or intrauterine growth restriction [170,171].

Mosaicism rate is less than 1–2% in viable pregnancies suggesting the existence of a self-correction process that removes aneuploid cells from embryos after implantation [172]. Some models have been proposed to explain this phenomenon: one of them suggests cell death; the other a reduced proliferation of aneuploid cells when compared to euploid ones [173]. In 2019, a study by Popovic et al. [174], confirmed that, in human embryos, cell proliferation and death have different dynamics among euploid, aneuploid, or mosaic blastocysts. He used an extended in vitro embryo culture protocol to study the effect of mosaicsm on early preimplantation, up to 12 days post-fertilization. Blastocysts with high mosacism levels were more likely to be non-viable at this stage of development. This could confirm the self-correction model, since aneuploid cells could proliferate more slowly or undergo apoptosis, whilst euploid ones could proliferate faster to compensate. There is still no evidence, anyway, that could support this model.

6. Strategies for Euploid/Mosaic Blastocysts Transfer

Implantation is considered to be an essential step for the success of assisted reproduction techniques and mainly depends on endometrial receptivity, embryo quality, and synchrony between them. However, the process of ovarian stimulation with elevated estrogen level, together with a possible progesterone premature growth, might reduce the expression of genes involved in the implantation process and negatively modify embryo-endometrium communication [175].

These negative effects can be responsible of decreased clinical results and adverse obstetrics and perinatal outcomes [175,176]. It has been suggested, indeed, that, after a fresh embryo transfer in a stimulated IVF cycle with E_2 levels >2724 pg/mL at the time of hCG administration, the risk of abnormal placentation and low birth weight [177] as well as the risk of obstetric hemorrhage [178] is definitely higher.

Euploid embryos can be transferred in a fresh stimulated cycle, performing biopsy of expanded blastocysts on day 5 and waiting for PGT result for a fresh transfer on day 6. Unfortunately, slower growing embryos, which cannot be biopsied on day 5, will not be included in PGT and considered for the fresh transfer. On the other hand, the freeze-all strategy involves cryopreservation of all biopsied embryos, including blastocysts developed on day 5, 6, or also 7, and then waiting for the PGT results in preparation for a further FTET. Undoubtedly, the uncertainty of having euploid embryos available for fresh transfer might enhance stress perception, strongly present in all patients with infertility problems. At the same time, the awareness that a fresh transfer is not planned, and that the whole cohort of embryos will be carefully evaluated, might relieve patient concerns [179].

Embryo cryopreservation is often only considered as a strategy to enhance the overall pregnancy rate per oocyte retrieval, when surplus embryos are available. However, a randomized controlled trial by Coates et al. [179] compared the clinical outcomes from frozen and fresh embryo transfers, finding that where the implantation rate was similar (75% vs. 67%), the ongoing PR (80% vs. 61%) and LB rates (77% vs. 59%) were significantly higher in the frozen group with respect to the fresh one. Based on these observations, it is possible to suggest the freeze-all strategy as a routine IVF approach when PGT is performed.

Frozen-thawed embryos can be transferred in different protocols: natural cycle (NC), modified natural cycle (modified-NC), and hormone replacement cycle (HR), but the best strategy should be still identified [180]. The authors of a retrospective cohort study including 389 cycles with 24-chromosome day 5/6 PGT-A found that the ongoing pregnancy rate obtained from a single euploid FTET in NC was significantly higher when compared to that in HR regimen. No difference was documented in the miscarriage rates [181].

Several studies investigated whether the duration of estrogen therapy or progesterone values before euploid embryo transfer might influence clinical outcomes in HR programs. A retrospective cohort study suggested that the duration of estradiol treatment before progesterone initiation does not affect the immediate FTET results. However, every additional day of estrogen administration is associated with a reduction of gestational age at delivery, not exceeding the criteria of preterm birth [182].

Gaggiotti-Marre et al. [183] examined the role of progesterone levels on the day before euploid embryo transfer, still in HR protocol, and documented that patients with its values over 13.1 ng/mL have a significantly higher miscarriage rate and lower live birth rate. On the other hand, a minimum threshold of progesterone concentration must be reached on the day of ET. It was documented that the progesterone level on the day of ET was predictive of clinical success and resulted in being significantly higher (28.0 ng/mL) in pregnant women, when compared to those with negative pregnancy test (16.4 ng/mL). The optimal cut-off value suggested was at least 20.6 ng/mL [184].

At the same time, a prospective controlled randomized trial documented that HR and modified-NC protocol for FTET lead to comparable clinical results, as suggested by overall pregnancy (61.9% vs. 62.3%), clinical pregnancy (50.4% vs. 54.1%), and implantation rates (50.4% vs. 54.1%). The live-birth rate of 45.8% in modified-NC was comparable to that observed in HR (41.5%). These overlapping results suggest that the choice of endometrial preparation should be based on women menstrual characteristics or, otherwise, the need for FTET planning [180].

Nevertheless, a study conducted by Litwicka et al. observed that, in modified-NC, the timing of HCG administration for ovulation induction might be relevant. The HCG injection when LH reaches values higher than 13 mIU/mL is associated with significantly lower overall and clinical pregnancy rates (45.4% vs. 73.3%) with respect to those in cycles with LH < 13 mIU/mL on the day of ovulation induction (36.4% vs. 65.9%). The authors suggested that, if natural ovulation process is already started (LH level ≥ 13 mIU/mL), hCG administration should be avoided, postponing the embryo transfer after spontaneous follicular rupture [17].

7. Maternal and Neonatal Outcomes

The wide use of PGT to implement assisted reproduction results, raises a concern about the potential risks of embryo biopsy, and extended embryo culture [10,180,181]. TE biopsy removes cells that are destined to form the placenta, increasing the risk of pathological placentation, potentially responsible for pre-eclampsia and reduced fetal growth. Therefore, the comparison of adverse obstetric and neonatal outcomes in pregnancies obtained from PGT IVF cycles and traditional IVF is mandatory.

In the study by Scott et al. [185], only 30% of biopsied embryos had sustained implantation and developed into live-born infants, versus 50% of non-biopsied controls. In contrast, sustained implantation rates were equivalent (51% vs. 54%) for biopsied and control blastocysts.

Zhang et al. [176] aimed to evaluate the existence of a correlation between TE biopsy and blastocysts quality. The authors found that, in high quality blastocysts, there were no differences in the survival and implantation rates, independently of the number of cells removed. However, blastocysts with grades B and C had significantly lower implantation rate, with an increasing number of TE cells removed. Implantation potential is negatively affected by the number of TE cell removed for the analysis in blastocysts with poor morphological score.

Additionally, a reduced cumulative HCG secretion after the biopsy procedure was suggested by Dokras et al. [186]. Moreover, the authors evidenced that the removal of less than 10 cells reduce the

HCG secretion (87.6 +/− 24.8 mIU/mL), but the difference was not significant. Oppositely, when a large biopsy was performed (greater than 10 cells), the HCG levels fell to 19.9 +/− 9.1 mIU/mL. This study indicates that blastocyst biopsy might impair the blastocysts development.

Forman et al. [10] reported neonatal and obstetric outcomes, when comparing SET after blastocyst biopsy and untested double embryo transfer in a randomized controlled trial. The delivery rates were similar (69% vs. 72%) through the fresh cycle and up to one frozen transfer, with a dramatic difference in multiple births (1.6% vs. 47%). The risk of preterm delivery, low birth weight, and admission to neonatal intensive care were significantly higher after untested two-embryos transfer. The improved obstetrical and neonatal outcomes suggest that PGT of a single euploid embryo might be a valid approach to patients requiring IVF.

On the other hand, a large study examining maternal and neonatal outcomes after TE biopsy documented a statistically significant increase in the risk of preeclampsia (10.5% vs. 5.8%) and placenta previa (4.1% vs. 1.4%) among pregnancies after IVF with PGT compared to those from IVF without PGT. The incidence of gestational diabetes, preterm premature rupture of membranes, and postpartum hemorrhage were similar. In addition, no differences were noted between the groups regarding neonatal outcomes, such as gestational age at delivery, preterm birth, low birth weight, neonatal intensive care unit admission, neonatal morbidities, or birth defects. [187].

The evaluation of 1721 children born after IVF cycles, including PGT at blastocyst stage combined with FTET, only showed an increased rate of cesarean section. In singletons, the cesarean section rate was as high as 80%, reached more than 90% in twins and it was significantly increased in PGT group when compared with controls [188].

Jing et al. [189] evaluated obstetric and neonatal outcomes from pregnancies that were obtained from FTET after TE and cleavage-stage biopsy for PGT. The results demonstrated that the incidence of gestational hypertension was significantly higher after blastocyst biopsy, as compared with cleavage-stage embryo (9.0% vs. 2.3%). Birth-weight and gestational age were higher after blastocyst-stage embryo transfer when compared to cleavage-stage in twins, but no significant differences were detected in the incidence of perinatal deaths, birth defects, neonates gender, and birth-weight for gestational age in both singletons and twins.

8. Conclusions

The preimplantation genetic testing is a valid technique to evaluated embryo euploidy and mosaicism before transfer. Next generation sequencing is considered by several studies as the best molecular test and trophoectodermal biopsy at the blastocyst stage is today the most used method for embryo biopsy. Preimplantation genetic testing is currently under study for assessing its usefulness, safety, and clinical validity. The clinical application of PGT-A are mainly those conditions in which the risk of embryo aneuploidies might increase, such as advanced maternal age, recurrent pregnancy lost, repeated implantation failure, severe male infertility factor, or when a single embryo transfer is necessary. The clinical benefit of this strategy in good prognosis patients and egg donation programs should be assessed by properly designed randomized control trials, especially if single embryo transfer is requested. Maternal and neonatal outcomes seem to be reassuring but more studies are needed. Mosaic embryo should be considered for transfer after an appropriate genetic counseling for the transfer for patients without euploid embryos.

Author Contributions: Study conception, writting and final approval—E.G.; study conception, design and writing—K.L.; study conceptualization—M.G.M.; study conception, design and writing—E.C.; study conception—P.F.G.; study conception—P.B. All authors have read and agreed to the published version of the manuscript.

Funding: This research received no external funding.

Conflicts of Interest: The authors declare no conflict of interest.

Abbreviations

PGT	Preimplantation genetic testing
IVF	in-vitro fertilization
SET	Single embryo transfer
PGT-A	Preimplantation gentic testing for aneuploidies
LB	Live birth
FISH	Fluorescence in Situ Hybridization
aCGH	Array-Comparative Genomic Hybridization
NGS	Next Generation Sequencing
rtq-PCR	Real Time Quantitative Polymerase Chain Reaction
PB	Polar body
ICSI	Intracitoplasmatic sperm injection
TE	Trophoectoderm
BF	Blastocyst fluid
WGA	Whole Genome Amplification
AMH	Anti-Müllerian hormone
AMA	Advanced maternal age
FTET	Frozen.thawed embryo transfer
RPL	Recurrent pregnancy loss
PGT-M	Preimplantation gentic testing for monogenic disease
DOR	Diminished ovarian reserve
RIF	Repeated implantation failure
SMF	Severe male factor
KS	Klinefelter syndrome
WHO	World Health Organization
NOA	Nonobstructive azoospermia
DFI	DNA fragmentation index
SCOS	Sertoli cell-only syndrome
DE	Donor egg

References

1. Kupka, M.S.; Ferraretti, A.P.; de Mouzon, J.; Erb, K.; D'Hooghe, T.; Castilla, J.A.; Calhaz-Jorge, C.; De Geyter, C. Assisted reproductive technology in Europe, 2010: Results generated from European registers by ESHRE. *Hum. Reprod.* **2014**, *29*, 2099–2113. [CrossRef]
2. Paternot, G.; Devroe, J.; Debrock, S.; D'Hooghe, T.M.; Spiessens, C. Intra- and inter-observer analysis in the morphological assessment of early-stage embryos. *Reprod. Biol. Endocrinol.* **2009**, *7*, 105. [CrossRef]
3. Abeyta, M.; Behr, B. Morphological assessment of embryo viability. *Semin. Reprod. Med.* **2014**, *32*, 114–126. [CrossRef]
4. Minasi, M.G.; Colasante, A.; Riccio, T.; Ruberti, A.; Casciani, V.; Scarselli, F.; Spinella, F.; Fiorentino, F.; Varricchio, M.T.; Greco, E. Correlation between aneuploidy, standard morphology evaluation and morphokinetic development in 1730 biopsied blastocysts: A consecutive case series study. *Hum. Reprod.* **2016**, *31*, 2245–2254. [CrossRef]
5. Desai, N.; Ploskonk, S.; Goodman, L.R.; Austin, C.; Goldberg, J.; Falcone, T. Analysis of embryo morphokinetics, multinucleation and cleavage anomalies using continuous time-lapse monitoring in blastocyst transfer cycles. *Reprod. Biol. Endocrinol.* **2014**, *20*, 12–54. [CrossRef]
6. Katz-Jaffe, M.G.; McReynolds, S. Embryology in the era of proteomics. *Fertil. Steril.* **2013**, *15*, 1073–1077. [CrossRef]
7. Uyar, A.; Seli, E. Metabolomic assessment of embryo viability. *Semin. Reprod. Med.* **2014**, *32*, 141–152. [CrossRef]
8. Papanikolaou, E.G.; Camus, M.; Kolibianakis, E.M.; Van Landuyt, L.; Van Steirteghem, A.; Devroey, P. In vitro fertilization with single blastocyst-stage versus single cleavage-stage embryos. *N. Engl. J. Med.* **2006**, *354*, 1139–1146. [CrossRef]

9. Munne, S.; Chen, S. Cromosome abnormalities in over 6000 cleavage-stage embryos. *Reprod. Biomed. Online* **2007**, *14*, 628–634. [CrossRef]
10. Forman, E.J.; Hong, K.H.; Franasiak, J.M.; Scott, R.J. Obstetrical and neonatal out-comes from the BEST Trial: Single embryo transfer with aneuploidy screening improves outcomes after in vitro fertilization without compromising delivery rates. *Am. J. Obstet. Gynecol.* **2014**, *210*, 157.e1–157.e6. [CrossRef]
11. Fragouli, E.; Alfarawati, S.; Spath, K.; Jaroudi, S.; Sarasa, J.; Enciso, M. The origin and impact of embryonic aneuploidy. *Hum. Genet.* **2013**, *132*, 1001–1013. [CrossRef]
12. Minasi, M.G.; Fiorentino, F.; Ruberti, A.; Biricik, A.; Cursio, E.; Cotroneo, E.; Varricchio, M.T.; Surdo, M.; Spinella, F.; Greco, E. Genetic diseases and aneuploidies can be detected with a single blastocyst biopsy: A successful clinical approach. *Hum. Reprod.* **2017**, *32*, 1770–1777. [CrossRef]
13. Pandian, Z.; Bhattacharya, S.; Ozturk, O.; Serour, G.; Templeton, A. Number of embryos for transfer following in-vitro fertilisation or intra-cytoplasmic sperm injection. *Cochrane Database Syst. Rev.* **2009**, *2*, CD003416.
14. Gelbaya, T.A.; Tsoumpou, I.; Nardo, L.G. The likelihood of live birth and multiple birth after single versus double embryo transfer at the cleavage stage: Asystematic review and meta-analysis. *Fertil. Steril.* **2010**, *94*, 936–994. [CrossRef]
15. Fragouli, E.; Lenzi, M.; Ross, R.; Katz-Jaffe, M.; Schoolcraft, W.B.; Wells, D. Comprehensive molecular cytogenetic analysis of the human blastocyst stage. *Hum. Reprod.* **2008**, *23*, 2596–2608. [CrossRef]
16. Alfarawati, S.; Fragouli, E.; Colls, P.; Stevens, J.; Gutiérrez-Mateo, C.; Schoolcraft, W.B.; Katz-Jaffe, M.G.; Wells, D. The relationship between blastocyst morphology, chromosomal abnormality, and embryo gender. *Fertil. Steril.* **2011**, *95*, 520–524. [CrossRef]
17. Litwicka, K.; Mencacci, C.; Arrivi, C.; Varricchio, M.T.; Caragia, A.; Minasi, M.G.; Greco, E. HCG administration after endogenous LH rise negatively influences pregnancy rate in modified natural cycle for frozen-thawed-euploid-blastocyst-transfer: A pilot study. *J. Assist. Reprod. Genet.* **2018**, *35*, 449–455. [CrossRef] [PubMed]
18. Dahdouh, E.M.; Balayla, J.; García-Velasco, J.A. Impact of blastocyst biopsy and comprehensive chromosome screening technology on preimplantation genetic screening: A systematic review of randomized controlled trials. *Reprod. Biomed. Online* **2015**, *30*, 281–289. [CrossRef] [PubMed]
19. Sahin, L.; Bozkurt, M.; Sahin, H.; Gürel, A.; Yumru, A.E. Is preimplantation genetic diagnosis the ideal embryo selection method in aneuploidy screening? *Kaohsiung J. Med. Sci.* **2014**, *30*, 491–498. [CrossRef]
20. Franasiak, J.M.; Forman, E.J.; Hong, K.H.; Werner, M.D.; Upham, K.M.; Treff, N.R. The nature of aneuploidy with increasing age of the female partner: A review of 15,169 consecutive trophectoderm biopsies evaluated with comprehensive chromosomal screening. *Fertil. Steril.* **2014**, *101*, 656–663. [CrossRef]
21. Colaco, S.; Sakkas, D. Paternal factors contributing toembryoquality. *J. Assist. Reprod. Genet.* **2018**, *35*, 1953–1968. [CrossRef] [PubMed]
22. Munné, S.; Lee, A.; Rosenwaks, Z.; Grifo, J.; Cohen, J. Diagnosis of major chromosome aneuploidies in human preimplantation embryos. *Hum. Reprod.* **1993**, *8*, 2185–2191. [CrossRef] [PubMed]
23. Mastenbroek, S.; Twisk, M.; van der Veen, F.; Repping, S. Preimplantation genetic screening: A systematic review and meta-analysis of RCTs. *Hum. Reprod. Update* **2011**, *17*, 454–466. [CrossRef]
24. Geraedts, J.; Collins, J.; Gianaroli, L.; Goossens, V.; Handyside, A.; Harper, J.; Montag, M.; Repping, S.; Schmutzler, A.G. What next for preimplantation genetic screening? A polar body approach! *Hum. Reprod.* **2010**, *25*, 575–577. [CrossRef] [PubMed]
25. Schmutzler, A.G. Theory and practice of preimplantation genetic screening (PGS). *Eur. J. Med. Genet.* **2019**, *62*, 103670. [CrossRef] [PubMed]
26. Delhanty, J.D. Is the polar body approach best for pre-implantation genetic screening? *Placenta* **2011**, *32* (Suppl. 3), 68–70. [CrossRef]
27. Verpoest, W.; Staessen, C.; Bossuyt, P.M.; Goossens, V.; Altarescu, G.; Bonduelle, M.; Devesa, M.; Eldar-Geva, T.; Gianaroli, L.; Griesinger, G.; et al. Preimplantation genetic testing for aneuploidy by microarray analysis of polar bodies in advanced maternal age: A randomized clinical trial. *Hum. Reprod.* **2018**, *33*, 1767–1776. [CrossRef]
28. Neumann, K.; Sermon, K.; Bossuyt, P.; Goossens, V.; Geraedts, J.; Traeger-Synodinos, J.; Parriego, M.; Schmutzler, A.G.; van der Ven, K.; Rudolph-Rothfeld, W.; et al. An economic analysis of preimplantation genetic testing for aneuploidy by polar body biopsy in advanced maternal age. *BJOG* **2020**, *127*, 710–718. [CrossRef]

29. Harton, G.L.; Magli, M.C.; Lundin, K.; Montag, M.; Lemmen, J.; Harper, J.C.; European Society for Human Reproduction and Embryology (ESHRE); PGD Consortium/Embryology Special Interest Group. ESHRE PGD Consortium/Embryology Special Interest Group-best practice guidelines for polar body and embryo biopsy for preimplantation genetic diagnosis/screening (PGD/PGS). *Hum. Reprod.* **2011**, *26*, 41–46.
30. Capalbo, A.; Bono, S.; Spizzichino, L.; Biricik, A.; Baldi, M.; Colamaria, S.; Ubaldi, F.M.; Rienzi, L.; Fiorentino, F. Sequential comprehensive chromosome analysis onpolarbodies, blastomeres and trophoblast: Insights into female meiotic errors and chromosomal segregation in the preimplantation window of embryo development. *Hum. Reprod.* **2013**, *28*, 509–518. [CrossRef]
31. Scott, K.L.; Hong, K.H.; Scott, R.T., Jr. Selecting the optimal time to perform biopsy for preimplantation genetic testing. *Fertil. Steril.* **2013**, *100*, 608–661. [CrossRef] [PubMed]
32. Scott, R.T., Jr.; Upham, K.M.; Forman, E.J.; Hong, K.H.; Scott, K.L.; Taylor, D.; Tao, X.; Tref, N.R. Blastocyst biopsy with comprehensive chromosome screening and fresh embryo transfer significantly increases in vitro fertilization implantation and delivery rates: A randomized controlled trial. *Fertil. Steril.* **2013**, *100*, 697–703. [CrossRef] [PubMed]
33. De Vos, A.; Staessen, C.; De Rycke, M.; Verpoest, W.; Haentjens, P.; Devroey, P.; Liebaers, I.; Van de Velde, H. Impact of cleavage-stage embryo biopsy in view of PGD on human blastocyst implantation: A prospective cohort of single embryo transfers. *Hum. Reprod.* **2009**, *24*, 2988–2996. [CrossRef]
34. Zeng, M.; Su, S.; Li, L. Comparison of pregnancy outcomes after vitrification at the cleavage and blastocyst stage: A meta-analysis. *J. Assist. Reprod. Genet.* **2018**, *35*, 127–134. [CrossRef] [PubMed]
35. McArthur, S.J.; Leigh, D.; Marshall, J.T.; de Boer, K.A.; Jansen, R.P. Pregnancies and live births after trophectoderm biopsy and preimplantation genetic testing of human blastocysts. *Fertil. Steril.* **2005**, *84*, 1628–1636. [CrossRef] [PubMed]
36. Rubino, P.; Tapia, L.; Ruiz de Assin Alonso, R.; Mazmanian, K.; Guan, L.; Dearden, L.; Thiel, A.; Moon, C.; Kolb, B.; Norian, J.M.; et al. Trophectoderm biopsy protocols can affect clinical outcomes: Time to focus on the blastocyst biopsy technique. *Fertil. Steril.* **2020**, *20*, S0015–S0282. [CrossRef] [PubMed]
37. Piccolomini, M.M.; Nicolielo, M.; Bonetti, T.C.; Motta, E.L.; Serafini, P.C.; Alegretti, J.R. Does slow embryo development predict a high aneuploidy rate on trophectoderm biopsy? *Reprod. Biomed. Online* **2016**, *33*, 398–403. [CrossRef]
38. Taylor, T.H.; Patrick, J.L.; Gitlin, S.A.; Wilson, J.M.; Crain, J.L.; Griffin, D.K. Comparison of aneuploidy, pregnancy and live birth rates between day 5 and day 6 blastocysts. *Reprod. Biomed. Online* **2014**, *29*, 305–310. [CrossRef]
39. Hernandez-Nieto, C.; Lee, J.A.; Slifkin, R.; Sandler, B.; Copperman, A.B.; Flisser, E. What is the reproductive potential of day 7 euploidembryos? *Hum. Reprod.* **2019**, *34*, 1697–1706. [CrossRef]
40. Rienzi, L.; Capalbo, A.; Stoppa, M.; Romano, S.; Maggiulli, R.; Albricci, L.; Scarica, C.; Farcomeni, A.; Vajta, G.; Ubaldi, F.M. No evidence of association between blastocyst aneuploidy and morphokinetic assessment in a selected population of poor-prognosis patients: A longitudinal cohort study. *Reprod. Biomed. Online* **2015**, *30*, 57–66. [CrossRef]
41. Hammond, E.R.; Cree, L.M.; Morbeck, D.E. Should extended blastocyst culture include Day 7? *Hum. Reprod.* **2018**, *33*, 991–997. [CrossRef] [PubMed]
42. Poli, M.; Ori, A.; Child, T.; Jaroudi, S.; Spath, K.; Beck, M.; Wells, D. Characterization and quantification of proteins secreted by single human embryos prior to implantation. *EMBO Mol. Med.* **2015**, *7*, 1465–1479. [CrossRef] [PubMed]
43. Magli, M.C.; Pomante, A.; Cafueri, G.; Valerio, M.; Crippa, A.; Ferraretti, A.P.; Gianaroli, L. Preimplantation genetic testing: Polar bodies, blastomeres, trophectoderm cells, or blastocoelic fluid? *Fertil. Steril.* **2016**, *105*, 676–683. [CrossRef]
44. Marcos, J.; Pérez-Albalá, S.; Mifsud, A.; Molla, M.; Landeras, J.; Meseguer, M. Collapse of blastocysts is strongly related to lower implantation success: A time-lapse study. *Hum. Reprod.* **2015**, *30*, 2501–2508. [CrossRef]
45. Bodri, D.; Sugimoto, T.; Yao Serna, J.; Kawachiya, S.; Kato, R.; Matsumoto, T. Blastocyst collapse is not an independent predictor of reduced live birth: A time-lapse study. *Fertil. Steril.* **2016**, *105*, 1476–1483. [CrossRef]

46. Chen, S.U.; Lee, T.H.; Lien, Y.R.; Tsai, Y.Y.; Chang, L.J.; Yang, Y.S. Microsuction of blastocoelic fluid before vitrification increased survival and pregnancy of mouse expanded blastocysts, but pretreatment with the cytoskeletal stabilizer did not increase blastocyst survival. *Fertil. Steril.* **2005**, *84* (Suppl. 2), 1156–1162. [CrossRef]
47. Palini, S.; Galluzzi, L.; De Stefani, S.; Bianchi, M.; Wells, D.; Magnani, M.; Bulletti, C. Genomic DNA in human blastocoele fluid. *Reprod. Biomed. Online* **2013**, *26*, 603–610. [CrossRef]
48. Tobler, K.J.; Zhao, Y.; Ross, R.; Benner, A.T.; Xu, X.; Du, L.; Broman, K.; Thrift, K.; Brezina, P.R.; Kearns, W.G. Blastocoel fluid from differentiated blastocysts harbors embryonic genomic material capable of a whole-genome deoxyribonucleic acid amplification and comprehensive chromosome microarray analysis. *Fertil. Steril.* **2015**, *104*, 418–425. [CrossRef]
49. Handyside, A.H. Noninvasive preimplantation genetic testing: Dream or reality? *Fertil. Steril.* **2016**, *106*, 1324–1325. [CrossRef]
50. Gianaroli, L.; Magli, M.C.; Pomante, A.; Crivello, A.M.; Cafueri, G.; Valerio, M.; Ferraretti, A.P. Blastocentesis: A source of DNA for preimplantation genetic testing. Results from a pilot study. *Fertil. Steril.* **2014**, *102*, 1692–1699. [CrossRef]
51. Ho, J.R.; Arrach, N.; Rhodes-Long, K.; Ahmady, A.; Ingles, S.; Chung, K.; Bendikson, K.A.; Paulson, R.J.; McGinnis, L.K. Pushing the limits of detection: Investigation of cell-free DNA for aneuploidy screening in embryos. *Fertil. Steril.* **2018**, *110*, 467–475. [CrossRef] [PubMed]
52. Hammond, E.R.; McGillivray, B.C.; Wicker, S.M.; Peek, J.C.; Shelling, A.N.; Stone, P.; Chamley, L.W.; Cree, L.M. Characterizing nuclear and mitochondrial DNA in spent embryo culture media: Genetic contamination identified. *Fertil. Steril.* **2017**, *107*, 220–228. [CrossRef] [PubMed]
53. Capalbo, A.; Romanelli, V.; Patassini, C.; Poli, M.; Girardi, L.; Giancani, A.; Stoppa, M.; Cimadomo, D.; Ubaldi, F.M.; Rienzi, L. Diagnostic efficacy of blastocoel fluid and spent media as sources of DNA for preimplantation genetic testing in standard clinical conditions. *Fertil. Steril.* **2018**, *110*, 870–879. [CrossRef]
54. Rodrigo, L.; Mateu, E.; Mercader, A.; Cobo, A.C.; Peinado, V.; Milán, M.; Al-Asmar, N.; Campos-Galindo, I.; García-Herrero, S.; Mir, P.; et al. New tools for embryo selection: Comprehensive chromosome screening by array comparative genomic hybridization. *Biomed. Res. Int.* **2014**, *2014*, 517125. [CrossRef]
55. Northrop, L.E.; Treff, N.R.; Levy, B.; Scott, R.T., Jr. SNP microarray-based 24 chromosome aneuploidy screening demonstrates that cleavage-stage FISH poorly predicts aneuploidy in embryos that develop to morphologically normal blastocysts. *Mol. Hum. Reprod.* **2010**, *16*, 590–600. [CrossRef]
56. Treff, N.R.; Tao, X.; Ferry, K.M.; Su, J.; Taylor, D.; Scott, R.T., Jr. Development and validation of an accurate quantitative real-time polymerase chain reaction-based assay for human blastocyst comprehensive chromosomal aneuploidy screening. *Fertil. Steril.* **2012**, *97*, 819–824. [CrossRef] [PubMed]
57. Huang, J.; Yan, L.; Lu, S.; Zhao, N.; Xie, X.S.; Qiao, J. Validation of a next-generation sequencing-based protocol for 24-chromosome aneuploidy screening of blastocysts. *Fertil. Steril.* **2016**, *105*, 1532–1536. [CrossRef]
58. Vera-Rodríguez, M.; Michel, C.E.; Mercader, A.; Bladon, A.J.; Rodrigo, L.; Kokocinski, F.; Mateu, E.; Al-Asmar, N.; Blesa, D.; Simón, C.; et al. Distribution patterns of segmental aneuploidies in human blastocysts identified by next-generation sequencing. *Fertil. Steril.* **2016**, *105*, 1047–1055. [CrossRef]
59. Fiorentino, F.; Biricik, A.; Bono, S.; Spizzichino, L.; Cotroneo, E.; Cottone, G.; Cottone, G.; Kokocinski, F.; Michel, C.M. Development and validation of a next-generation sequencing-based protocol for 24- chromosome aneuploidy screening of embryos. *Fertil. Steril.* **2014**, *101*, 1375–1382. [CrossRef]
60. Friedenthal, J.; Maxwell, S.M.; Munné, S.; Kramer, Y.; McCulloh, D.H.; McCaffrey, C.; Grifo, J.A. Next generation sequencing for preimplantation genetic screening improves pregnancy outcomes compared with array comparative genomic hybridization in single thawed euploid embryo transfer cycles. *Fertil. Steril.* **2018**, *109*, 627–632. [CrossRef]
61. Friedenthal, J.; Maxwell, S.M.; Tiegs, A.W.; Besser, A.G.; McCaffrey, C.; Munné, S.; Noyes, N.; Grifo, J.A. Clinical error rates of next generation sequencing and array comparative genomic hybridization with single thawed euploid embryo transfer. *Eur. J. Med. Genet.* **2020**, *20*, 103852. [CrossRef] [PubMed]
62. Hassold, T.; Hunt, P. Maternal age and chromosomally abnormal pregnancies: What we know and what we wish we knew. *Curr. Opin. Pediatr.* **2009**, *21*, 703–708. [CrossRef]
63. Rubio, C.; Rodrigo, L.; Mercader, A.; Mateu, E.; Buendia, P.; Pehlivan, T.; Viloria, T.; Santos, D.L.; Simon, C.; Remohi, J.; et al. Impact of chromosomal abnormalities on preimplantation embryo development. *Prenat. Diagn.* **2007**, *27*, 748–756. [CrossRef] [PubMed]

64. Sandalinas, M.; Sadowy, S.; Alikani, M.; Calderon, G.; Cohen, J.; Munne, S. Developmental ability of chromosomally abnormal human embryos to develop to the blastocyst stage. *Hum. Reprod.* **2001**, *16*, 1954–1958. [CrossRef] [PubMed]
65. Magli, M.C.; Jones, G.M.; Gras, L.; Gianaroli, L.; Korman, I.; Trounson, A.O. Chromosome mosaicism in day 3 aneuploid embryos that develop to morphologically normal blastocysts in vitro. *Hum. Reprod.* **2000**, *15*, 1781–1786. [CrossRef] [PubMed]
66. Marquez, C.; Sandalinas, M.; Bahce, M.; Alikani, M.; Munne, S. Chromosome abnormalities in 1255 cleavage-stage human embryos. *Reprod. Biomed. Online* **2000**, *1*, 17–26. [CrossRef]
67. Yang, Z.; Liu, J.; Collins, G.S.; Salem, S.A.; Liu, X.; Lyle, S.S.; Peck, A.C.; Scott, E.S.; Salem, R.D. Selection of single blastocysts for fresh transfer via standard morphology assessment alone and with array CGH for good prognosis IVF patients: Results from a randomized pilot study. *Mol. Cytogenet.* **2012**, *5*, 24. [CrossRef]
68. Rabinowitz, M.; Ryan, A.; Gemelos, G.; Hill, M.; Baner, J.; Cinnioglu, C.; Banjevic, M.; Potter, D.; Petrov, D.A.; Demko, Z. Origins and rates of aneuploidy in human blastomeres. *Fertil. Steril.* **2012**, *97*, 395–400. [CrossRef]
69. Jiang, X.; Yan, J.; Sheng, Y.; Sun, M.; Cui, L.; Chen, Z.J. Low anti-Müllerian hormone concentration is associated with increased risk of embryonic aneuploidy in women of advanced age. *Reprod. Biomed. Online* **2018**, *37*, 178–183. [CrossRef]
70. Munne, S.; Chen, S.; Fischer, J.; Colls, P.; Zheng, X.; Stevens, J.; Escudero, T.; Oter, M.; Schoolcraft, B.; Simpson, J.L.; et al. Preimplantation genetic diagnosis reduces pregnancy loss in women aged 35 years and older with a history of recurrent miscarriages. *Fertil. Steril.* **2005**, *4*, 331–335. [CrossRef]
71. Rubio, C.; Bellver, J.; Rodrigo, L.; Bosch, E.; Mercader, A.; Vidal, C.; De Los Santos, M.J.; Giles, J.; Labarta, E.; Domingo, J.; et al. Preimplantation genetic screening using fluorescence in situ hybridization in patients with repetitive implantation failure and advanced maternal age: Two randomized trials. *Fertil. Steril.* **2013**, *99*, 1400–1407. [CrossRef]
72. Scott, R.T., Jr.; Ferry, K.; Su, J.; Tao, X.; Scott, K.; Treff, N.R. Comprehensive chromosome screening is highly predictive of the reproductive potential of human embryos: A prospective, blinded, nonselection study. *Fertil. Steril.* **2012**, *97*, 870–875. [CrossRef] [PubMed]
73. Grifo, J.A.; Hodes-Wertz, B.; Lee, H.L.; Amperloquio, E.; Clarke-Williams, M.; Adler, A. Single thawed euploid embryo transfer improves IVF pregnancy, miscarriage, and multiple gestation outcomes and has similar implantation rates as egg donation. *J. Assist. Reprod. Genet.* **2013**, *30*, 259–264. [CrossRef] [PubMed]
74. Harton, G.L.; Munné, S.; Surrey, M.; Grifo, J.; Kaplan, B.; McCulloh, D.H.; Griffin, D.K.; Wells, D. Diminished effect of maternal age on implantation after preimplantation genetic diagnosis with array comparative genomic hybridization. *Fertil. Steril.* **2013**, *100*, 1695–1703. [CrossRef]
75. Platteau, P.; Staessen, C.; Michiels, A.; Van Steirteghem, A.; Liebaers, I.; Devroey, P. Preimplantation genetic diagnosis for aneuploidy screening in women older than 37 years. *Fertil. Steril.* **2005**, *84*, 319–324. [CrossRef]
76. Rubio, C.; Bellver, J.; Rodrigo, L.; Castillón, G.; Guillén, A.; Vidal, C.; Giles, J.; Ferrando, M.; Cabanillas, S.; Remohí, J.; et al. In vitro fertilization with preimplantation genetic diagnosis for aneuploidies in advanced maternal age: A randomized, controlled study. *Fertil. Steril.* **2017**, *107*, 1122–1129. [CrossRef]
77. Sacchi, L.; Albani, E.; Cesana, A.; Smeraldi, A.; Parini, V.; Fabiani, M.; Poli, M.; Capalbo, A.; Levi-Setti, P.E. Preimplantation Genetic Testing for Aneuploidy Improves Clinical, Gestational, and Neonatal Outcomes in Advanced Maternal Age Patients Without Compromising Cumulative Live-Birth Rate. *J. Assist. Reprod. Genet.* **2019**, *36*, 2493–2504. [CrossRef]
78. Ubaldi, F.M.; Cimadomo, D.; Capalbo, A.; Vaiarelli, A.; Buffo, L.; Trabucco, E.; Ferrero, S.; Albani, E.; Rienzi, L.; Levi-Setti, P.E. Preimplantation genetic diagnosis for aneuploidy testing in women older than 44 years: A multicenter experience. *Fertil. Steril.* **2017**, *107*, 1173–1180. [CrossRef]
79. Lee, H.L.; McCulloh, D.H.; Hodes-Wertz, B.; Adler, A.; McCaffrey, C.; Grifo, J.A. In vitro fertilization with preimplantation genetic screening improves implantation and live birth in women age 40 through 43. *J. Assist. Reprod. Genet.* **2015**, *32*, 435–444. [CrossRef]
80. Reig, A.; Franasiak, J.; Scott, R.T., Jr.; Seli, E. The impact of age beyond ploidy: Outcome data from 8175 euploid single embryo transfers. *J. Assist. Reprod. Genet.* **2020**, *37*, 595–602. [CrossRef]
81. Jauniaux, E.; Farquharson, R.G.; Christiansen, O.B.; Exalto, N. Evidence-based guidelines for the investigation and medical treatment of recurrent miscarriage. *Hum. Reprod.* **2006**, *21*, 2216–2222. [CrossRef] [PubMed]
82. Rai, R.; Regan, L. Recurrent miscarriage. *Lancet* **2006**, *368*, 601–611. [CrossRef]

83. Stephenson, M.; Kutteh, W. Evaluation and management of recurrent early pregnancy loss. *Clin. Obstet. Gynecol.* **2007**, *50*, 132–145. [CrossRef]
84. Stephenson, M.D.; Awartani, K.A.; Robinson, W.P. Cytogenetic analysis of miscarriages from couples with recurrent miscarriage: A case-control study. *Hum. Reprod.* **2002**, *17*, 446–451. [CrossRef]
85. Bianco, K.; Caughey, A.B.; Shaffer, B.L.; Davis, R.; Norton, M.E. History of miscarriage and increased incidence of fetal aneuploidy in subsequent pregnancy. *Obstet. Gynecol.* **2006**, *107*, 1098–1102. [CrossRef]
86. Fritz, B.; Hallermann, C.; Olert, J.; Fuchs, B.; Bruns, M.; Aslan, M.; Schmidt, S.; Coerdt, W.; Müntefering, H.; Rehder, H. Cytogenetic analyses of culture failures by comparative genomic hybridisation (CGH)-Re-evaluation of chromosome aberration rates in early spontaneous abortions. *Eur. J. Hum. Genet.* **2001**, *9*, 539–547. [CrossRef]
87. Sullivan, A.E.; Silver, R.M.; LaCoursiere, D.Y.; Porter, T.F.; Branch, D.W. Recurrent fetal aneuploidy and recurrent miscarriage. *Obstet. Gynecol.* **2004**, *104*, 784–788. [CrossRef]
88. Rubio, C.; Simon, C.; Vidal, F.; Rodrigo, L.; Pehlivan, T.; Remohi, J.; Pellicer, A. Chromosomal abnormalities and embryo development in recurrent miscarriage couples. *Hum. Reprod.* **2003**, *18*, 182–188. [CrossRef]
89. Musters, A.M.; Repping, S.; Korevaar, J.C.; Mastenbroek, S.; Limpens, J.; van der Veen, F.; Goddijn, M. Pregnancy outcome after preimplantation genetic screening or natural conception in couples with unexplained recurrent miscarriage: A systematic review of the best available evidence. *Fertil. Steril.* **2011**, *95*, 2153–2157. [CrossRef]
90. Platteau, P.; Staessen, C.; Michiels, A.; Van Steirteghem, A.; Liebaers, I.; Devroey, P. Preimplantation genetic diagnosis for aneuploidy screening in patients with unexplained recurrent miscarriages. *Fertil. Steril.* **2005**, *83*, 393–397. [CrossRef] [PubMed]
91. Wilding, M.; Forman, R.; Hogewind, G.; di Matteo, L.; Zullo, F.; Cappiello, F.; Dale, B. Preimplantation genetic diagnosis for the treatment of failed in vitro fertilization–embryo transfer and habitual abortion. *Fertil. Steril.* **2004**, *81*, 1302–1307. [CrossRef] [PubMed]
92. Liu, X.Y.; Fan, Q.; Wang, J.; Li, R.; Xu, Y.; Guo, J.; Wang, Y.Z.; Zeng, Y.H.; Ding, C.H.; Cai, B.; et al. Higher chromosomal abnormality rate in blastocysts from young patients with idiopathic recurrent pregnancy loss. *Fertil. Steril.* **2020**, *113*, 853–864. [CrossRef] [PubMed]
93. Hodes-Wertz, B.; Grifo, J.; Ghadir, S.; Kaplan, B.; Laskin, C.A.; Glassner, M.; Munné, S. Idiopathic recurrent miscarriage is caused mostly by aneuploid embryos. *Fertil. Steril.* **2012**, *98*, 675–680. [CrossRef] [PubMed]
94. Sato, T.; Sugiura-Ogasawara, M.; Ozawa, F.; Yamamoto, T.; Kato, T.; Kurahashi, H.; Kuroda, T.; Aoyama, N.; Kato, K.; Kobayashi, R.; et al. Preimplantation genetic testing for aneuploidy: A comparison of live birth rates in patients with recurrent pregnancy loss due to embryonic aneuploidy or recurrent implantation failure. *Hum. Reprod.* **2019**, *34*, 2340–2348. [CrossRef] [PubMed]
95. Murugappan, G.; Ohno, M.S.; Lathi, R.B. Cost-effectiveness analysis of preimplantation genetic screening and in vitro fertilization versus expectant management in patients with unexplained recurrent pregnancy loss. *Fertil. Steril.* **2015**, *103*, 1215–1220. [CrossRef] [PubMed]
96. Katz-Jaffe, M.G.; Surrey, E.S.; Minjarez, D.A.; Gustofson, R.L.; Stevens, J.M.; Schoolcraft, W.B. Association of abnormal ovarian reserve parameters with a higher incidence of aneuploidy blastocysts. *Obstet. Gynecol.* **2013**, *121*, 71–77. [CrossRef]
97. Trout, S.W.; Seifer, D.B. Do women with unexplained recurrent pregnancy loss have higher day 3 serum FSH and estradiol values? *Fertil. Steril.* **2000**, *74*, 335–337. [CrossRef]
98. Shahine, L.K.; Marshall, L.; Lamb, J.D.; Hickok, L.R. Higher rates of aneuploidy in blastocysts and higher risk of no embryo transfer in recurrent pregnancy loss patients with diminished ovarian reserve undergoing in vitro fertilization. *Fertil. Steril.* **2016**, *106*, 1124–1128. [CrossRef]
99. Harton, G.; Braude, P.; Lashwood, A.; Schmutzler, A.; Traeger-Synodinos, J.; Wilton, L.; Harper, J.C. ESHRE PGD consortium best practice guidelines for organization of a PGD centre for PGD/preimplantation genetic screening. European Society for Human Reproduction and Embryology (ESHRE) PGD Consortium. *Hum. Reprod.* **2011**, *26*, 14–24. [CrossRef]
100. Somigliana, E.; Vigano, P.; Busnelli, A.; Paffoni, A.; Vegetti, W.; Vercellini, P. Repeated implantation failure at the crossroad between statistics, clinics and over-diagnosis. *Reprod. Biomed. Online* **2018**, *36*, 32–38. [CrossRef]

101. Coughlan, C.; Ledger, W.; Wang, Q.; Liu, F.; Demirol, A.; Gurgan, T.; Cutting, R.; Ong, K.; Sallam, H.; Li, T.C. Recurrent implantation failure: Definition and management. *Reprod. BioMed. Online* **2014**, *28*, 14–38. [CrossRef]
102. Rubio, C.; Pehlivan, T.; Rodrigo, L.; Simón, C.; Remohí, J.; Pellicer, A. Embryo aneuploidy screening for unexplained recurrent miscarriage: A minireview. *Am. J. Reprod. Immunol.* **2005**, *53*, 159–165. [CrossRef]
103. Pehlivan, T.; Rubio, C.; Rodrigo, L.; Romero, J.; Remohi, J.; Simón, C.; Pellicer, A. Impact of preimplantation genetic diagnosis on IVF outcome in implantation failure patients. *Reprod. Biomed. Online* **2003**, *6*, 232–237. [CrossRef]
104. Simón, C.; Mercader, A.; Garcia-Velasco, J.; Nikas, G.; Moreno, C.; Remohí, J.; Pellicer, A. Coculture of human embryos with autologous human endometrial epithelial cells in patients with implantation failure. *J. Clin. Endocrinol. Metab.* **1999**, *84*, 2638–2646. [CrossRef]
105. Gianaroli, L.; Magli, M.C.; Ferraretti, A.P.; Tabanelli, C.; Trombetta, C.; Boudjema, E. The role of preimplantation diagnosis for aneuploidies. *Reprod. Biomed. Online* **2002**, *4*, 31–33. [CrossRef]
106. Gianaroli, L.; Magli, M.C.; Ferraretti, A.P.; Munne, S. Preimplantation diagnosis for aneuploidies in patients undergoing in vitro fertilization with a poor prognosis: Identification of the categories for which it should be proposed. *Fertil. Steril.* **1999**, *72*, 837–844. [CrossRef]
107. Kahraman, S.; Sahin, Y.; Yelke, H.; Kumtepe, Y.; Tufekci, M.A.; Yapan, C.C.; Yesil, M.; Cetinkaya, M. High rates ofaneuploidy, mosaicism and abnormal morphokinetic development in cases with low sperm concentration. *J. Assist. Reprod. Genet.* **2020**, *37*, 629–664. [CrossRef]
108. Voullaire, L.; Wilton, L.; McBain, J.; Callaghan, T.; Williamson, R. Chromosome abnormalities identified by comparative genomic hybridization in embryos from women with repeated implantation failure. *Mol. Hum. Reprod.* **2002**, *8*, 1035–1041. [CrossRef]
109. ESHRE PGD Consortium Steering Committee. ESHRE Preimplantation Genetic Diagnosis Consortium: Data collection III (May 2001). *Hum. Reprod.* **2002**, *17*, 233–246. [CrossRef]
110. Greco, E.; Bono, S.; Ruberti, A.; Lobascio, A.M.; Greco, P.; Biricik, A.; Spizzichino, L.; Greco, A.; Tesarik, J.; Minasi, M.G.; et al. Comparative Genomic Hybridization Selection of Blastocysts for Repeated Implantation Failure Treatment: A Pilot Study. *Biomed. Res. Int.* **2014**, 457913. [CrossRef]
111. Petousis, S.; Prapas, Y.; Papatheodorou, A.; Margioula-Siarkou, C.; Papatzikas, G.; Panagiotidis, Y.; Karkanaki, A.; Ravanos, K.; Prapas, N. Fluorescence in situ hybridization sperm examination is significantly impaired in all categories of male infertility. *Andrologia* **2018**, *50*, 12847. [CrossRef]
112. Magli, M.C.; Gianaroli, L.; Ferraretti, A.P.; Gordts, S.; Fredericks, V.; Crippa, A. Paternal contribution to aneuploidy in preimplantation embryos. *Reprod. Biomed. Online* **2009**, *18*, 536–542. [CrossRef]
113. Silber, S.; Escudero, T.; Lenahan, K.; Abdelhadi, I.; Kilani, Z.; Munne, S. Chromosomal abnormalities in embryos derived from testicular sperm extraction. *Fertil. Steril.* **2003**, *79*, 30–38. [CrossRef]
114. Coates, A.; Hesla, J.S.; Hurliman, A.; Coate, B.; Holmes, E.; Matthews, R.; Mounts, E.L.; Turner, K.J.; Thornhill, A.R.; Griffin, D.K. Use of suboptimal sperm increases the risk of aneuploidy of the sex chromosomes in preimplantation blastocyst embryos. *Fertil. Steril.* **2015**, *104*, 866–872. [CrossRef]
115. Mazzilli, R.; Cimadomo, D.; Vaiarelli, A.; Capalbo, A.; Dovere, L.; Alviggi, E.; Dusi, L.; Foresta, C.; Lombardo, F.; Lenzi, A.; et al. Effect of themalefactor on the clinical outcome of intracytoplasmic sperm injection combined with preimplantation aneuploidy testing: Observational longitudinal cohort study of 1,219 consecutive cycles. *Fertil. Steril.* **2017**, *108*, 961–972. [CrossRef]
116. Liebaers, I.; Desmyttere, S.; Verpoest, W.; de Rycke, M.; Staessen, C.; Sermon, K.; Devroey, P.; Haentjens, P.; Bonduelle, M. Report on a consecutive series of 581 children born after blastomere biopsy for preimplantation genetic diagnosis. *Hum. Reprod.* **2010**, *25*, 275–282. [CrossRef]
117. Levi-Setti, P.E.; Moioli, M.; Smeraldi, A.; Cesaratto, E.; Menduni, F.; Livio, S.; Morenghi, M.; Patrizio, P. Obstetric outcome and incidence of congenital anomalies in 2351 IVF/ICSI babies. *J. Assist. Reprod. Genet.* **2016**, *33*, 711–717. [CrossRef]
118. Hobbs, C.A.; Cleves, M.A.; Simmons, C.J. Genetic epidemiology and congenital malformations: From the chromosome to the crib. *Arch. Pediatr. Adolesc. Med.* **2002**, *156*, 315–320. [CrossRef] [PubMed]
119. Matthews, T.J.; MacDorman, M.F.; Thoma, M.E. Infant mortality statistics from the 2013 period linked birth/infant death data set. *Natl. Vital. Stat. Rep.* **2015**, *64*, 1–30.
120. Ahmadi, A.; Ng, S.C. Fertilizing ability of DNA-damaged spermatozoa. *J. Exp. Zool.* **1999**, *284*, 696–704. [CrossRef]

121. Loutradi, K.E.; Tarlatzis, B.C.; Goulis, D.G.; Zepiridis, L.; Pagou, T.; Chatziioannou, E.; Grimbizis, G.F.; Papadimas, I.; Bontis, I. The effects of sperm quality on embryo development after intracytoplasmic sperm injection. *J. Assist. Reprod. Genet.* **2006**, *23*, 69–74. [CrossRef] [PubMed]
122. Tarozzi, N.; Nadalini, M.; Lagalla, C.; Coticchio, G.; Zacà, C.; Borini, A. Male factor in fertility impacts the rate of mosaic blastocysts in cycles of preimplantation genetic testing for aneuploidy. *J. Assist. Reprod. Genet.* **2019**, *36*, 2047–2055. [CrossRef]
123. Scarselli, F.; Cursio, E.; Muzzì, S.; Casciani, V.; Ruberti, A.; Gatti, S.; Greco, P.; Varricchio, M.T.; Minasi, M.G.; Greco, E. How 1h of abstinence improves sperm quality and increases embryo euploidy rate after PGT-A: A study on 106 sibling biopsied blastocysts. *J. Assist. Reprod. Genet.* **2019**, *36*, 1591–1597. [CrossRef]
124. Fullerton, G.; Hamilton, M.; Maheshwari, A. Should non-mosaic Klinefelter syndrome men be labelled as infertile in 2009? *Hum. Reprod.* **2010**, *25*, 588–597. [CrossRef]
125. Ron-El, R.; Strassburger, D.; Gelman-Kohan, S.; Friedler, S.; Raziel, A.; Appelman, Z. A 47,XXY fetus conceived after ICSI of spermatozoa from a patient with non-mosaic Klinefelter's syndrome: Case report. *Hum. Reprod.* **2000**, *15*, 1804–1806. [CrossRef]
126. Friedler, S.; Raziel, A.; Strassburger, D.; Schachter, M.; Bern, O.; Ron-El, R. Outcome of ICSI using fresh and cryopreserved-thawed testicular spermatozoa in patients with non-mosaic Klinefelter's syndrome. *Hum. Reprod.* **2001**, *16*, 2616–2620. [CrossRef]
127. Greco, E.; Scarselli, F.; Minasi, M.G.; Casciani, V.; Zavaglia, D.; Dente, D.; Tesarik, J.; Franco, G. Birth of 16 healthy children after ICSI in cases of non-mosaic Klinefelter syndrome. *Hum. Reprod.* **2013**, *28*, 1155–1160. [CrossRef]
128. Staessen, C.; Tournaye, H.; Van Assche, E.; Michiels, A.; Van Landuyt, L.; Devroey, P.; Liebaers, I.; Van Steirteghem, A. PGD in 47,XXY Klinefelter'syndrome patients. *Hum. Reprod. Update* **2003**, *9*, 319–330. [CrossRef]
129. García-Ferreyra, J.; Luna, D.; Villegas, L.; Romero, R.; Zavala, P.; Hilario, R.; Dueñas-Chacón, J. High aneuploidy rates observed in embryos derived from donated oocytes are related to male aging and high percentages of sperm DNA fragmentation. *Clin. Med. Insights Reprod. Health* **2015**, *9*, 21–27. [CrossRef]
130. García-Ferreyra, J.; Hilario, R.; Dueñas, J. High percentages of embryos with 21, 18 or 13 trisomy are related to advanced paternal age in donor egg cycles. *JBRA Assist. Reprod.* **2018**, *22*, 26–34. [CrossRef]
131. El-Domyati, M.M.; Al-Din, A.B.; Barakat, M.T.; El-Fakahany, H.M.; Xu, J.; Sakkas, D. Deoxyribonucleic acid repair and apoptosis in testicular germ cells of aging fertile men: The role of the poly [adenosine diphosphate-ribosyl] ation pathway. *Fertil. Steril.* **2009**, *1*, 2221–2229. [CrossRef]
132. Gat, I.; Tang, K.; Quach, K.; Kuznyetsov, V.; Antes, R.; Filice, M.; Zohni, K.; Librach, C. Sperm DNA fragmentation index does not correlate with blastocyst aneuploidy or morphological grading. *PLoS ONE* **2017**, *12*, e0179002. [CrossRef] [PubMed]
133. Bronet, F.; Martinez, E.; Gaytan, M.; Linan, A.; Cernuda, D.; Ariza, M.; Nogales, M.; Pacheco, A.; San Celestino, M.; Garcia-Velasco, J.A. Sperm DNA fragmentation index does not correlate with the sperm or embryo aneuploidy rate in recurrent miscarriage or implantation failure patients. *Hum. Reprod.* **2012**, *27*, 1922–1929. [CrossRef] [PubMed]
134. Stacy, C.; Deepak, M. Y Genetics of the human Y chromosomeand its association with male infertility. *Reprod. Biol. Endocrinol.* **2018**, *16*, 14.
135. Capalbo, A.; Rienzi, L.; Cimadomo, D.; Maggiulli, R.; Elliott, T.; Wright, G.; Nagy, Z.P.; Ubaldi, F.M. Correlation between standard blastocyst morphology, euploidy and implantation: An observational study in two centers involving 956 screened blastocysts. *Hum. Reprod.* **2014**, *29*, 1173–1181. [CrossRef]
136. Ozgur, K.; Berkkanoglu, M.; Bulut, H.; Yoruk, G.D.A.; Candurmaz, N.N.; Coetzee, K. Single best euploid versus single best unknown-ploidy blastocyst frozen embryo transfers: A randomized controlled trial. *J. Assist. Reprod. Genet.* **2019**, *36*, 629–636. [CrossRef]
137. Munné, S.; Kaplan, B.; Frattarelli, J.L.; Child, T.; Nakhuda, G.; Shamma, F.N.; Silverberg, K.; Kalista, T.; Handyside, A.H.; Katz-Jaffe, M.; et al. Preimplantation genetic testing for aneuploidy versus morphology as selection criteria for single frozen-thawed embryo transfer in good-prognosis patients: A multicenter randomized clinical trial. *Fertil. Steril.* **2019**, *112*, 1071–1079. [CrossRef] [PubMed]
138. Forman, E.J.; Hong, K.H.; Ferry, K.M.; Tao, X.; Taylor, D.; Levy, B.; Treff, N.R.; Scott, R.T., Jr. In vitro fertilization with single euploid blastocyst transfer: A randomized controlled trial. *Fertil. Steril.* **2013**, *100*, 100–107. [CrossRef]

139. Nakasuji, T.; Saito, H.; Araki, R.; Nakaza, A.; Nakashima, A.; Kuwahara, A.; Ishihara, O.; Irahara, M.; Kubota, T.; Yoshimura, Y.; et al. The incidence of monozygotic twinning in assisted reproductive technology: Analysis based on results from the 2010 Japanese ART national registry. *J. Assist. Reprod. Genet.* **2014**, *31*, 803–807. [CrossRef]
140. Da Costa, A.L.A.L.; Abdelmassih, S.; de Oliveira, F.G.; Abdelmassih, V.; Abdelmassih, R.; Nagy, Z.P.; Balmaceda, J.P. Monozygotic twins and transfer at the blastocyst stage after ICSI. *Hum. Reprod.* **2001**, *16*, 333–336. [CrossRef]
141. Vaughan, D.A.; Ruthazer, R.; Penzias, A.S.; Norwitz, E.R.; Sakkas, D. Clustering of monozygotic twinning in IVF. *J. Assist. Reprod. Genet.* **2016**, *33*, 19–26. [CrossRef] [PubMed]
142. Ikemoto, Y.; Kuroda, K.; Ochiai, A.; Yamashita, S.; Ikuma, S.; Nojiri, S.; Itakura, A.; Takeda, S. Prevalence and risk factors of zygotic splitting after 937,848 single embryo transfer cycles. *Hum. Reprod.* **2018**, *33*, 1984–1991. [CrossRef] [PubMed]
143. Verpoest, W.; Van Landuyt, L.; Desmyttere, S.; Cremers, A.; Devroey, P.; Liebaers, I. The incidence of monozygotic twinning following PGD is not increased. *Hum. Reprod.* **2009**, *24*, 2945–2950. [CrossRef]
144. Busnelli, A.; Dallagiovanna, C.; Reschini, M.; Paffoni, A.; Fedele, L.; Somigliana, E. Risk factors for monozygotic twinning after in vitro fertilization: A systematic review and meta-analysis. *Fertil. Steril.* **2019**, *111*, 302–317. [CrossRef] [PubMed]
145. Kamath, M.S.; Antonisamy, B.; Sunkara, S.K. Zygotic splitting following embryo biopsy: A cohort study of 207697 single-embryo transfers following IVF treatment. *BJOG* **2020**, *127*, 562–569. [CrossRef]
146. Masbou, A.K.; Friedenthal, J.B.; McCulloh, D.H.; McCaffrey, C.; Fino, M.E.; Grifo, J.A.; Licciardi, F. A Comparison of Pregnancy Outcomes in Patients Undergoing Donor Egg Single Embryo Transfers with and Without Preimplantation Genetic Testing. *Reprod. Sci.* **2019**, *26*, 1661–1665. [CrossRef]
147. Sills, E.; Li, X.; Frederick, J.L.; Khoury, C.D.; Potter, D.A. Determining parental origin of embryo aneuploidy: Analysis of genetic error observed in 305 embryos derived from anonymous donor oocyte IVF cycles. *Mol. Cytogenet.* **2014**, *7*, 68. [CrossRef]
148. Hoyos, L.R.; Cheng, C.Y.; Brennan, K.; Hubert, G.; Wang, B.; Buyalos, R.P.; Quinn, M.; Shamonki, M.J. Euploid rates amongoocytedonors: Is there an optimal age fordonation? *Assist. Reprod. Genet.* **2020**, *37*, 589–594. [CrossRef]
149. Haddad, G.; Deng, M.; Wang, C.T.; Witz, C.; Williams, D.; Griffith, J.; Skorupski, J.; Gill, J.; Wang, W.H. Assessment of aneuploidy formation in human blastocysts resulting from donated eggs and the necessity of the embryos for aneuploidy screening. *J. Assist. Reprod. Genet.* **2015**, *32*, 999–1006. [CrossRef]
150. Forman, E.J.; Li, X.; Ferry, K.M.; Scott, K.; Treff, N.R.; Scott, R.T., Jr. Oocyte vitrification does not increase the risk of embryonic aneuploidyor diminish the implantation potential of blastocysts created after intracytoplasmic sperm injection: A novel, paired randomized controlled trial using DNA fingerprinting. *Fertil. Steril.* **2012**, *98*, 644–649. [CrossRef]
151. Obradors, A.; Fernández, E.; Oliver-Bonet, M.; Rius, M.; de la Fuente, A.; Wells, D.; Benet, J.; Navarro, J. Birth of a healthy boy after a double factor PGD in a couple carrying a genetic disease and at risk for aneuploidy: Case report. *Hum. Reprod.* **2008**, *23*, 1949–1956. [CrossRef] [PubMed]
152. Obradors, A.; Fernández, E.; Rius, M.; Oliver-Bonet, M.; Martínez-Fresno, M.; Benet, J.; Navarro, J. Outcome of twin babies free of Von Hippel-Lindau disease after a double-factor preimplantation genetic diagnosis: Monogenetic mutation analysis and comprehensive aneuploidy screening. *Fertil. Steril.* **2009**, *91*, 933.e1–933.e7. [CrossRef]
153. Rechitsky, S.; Verlinsky, O.; Kuliev, A. PGD for cystic fibrosis patients and couples at risk of an additional genetic disorder combined with 24-chromosome aneuploidy testing. *Reprod. Biomed. Online* **2013**, *26*, 420–430. [CrossRef] [PubMed]
154. Goldman, K.N.; Nazem, T.; Berkeley, A.; Palter, S.; Grifo, J.A. Preimplantation Genetic Diagnosis (PGD) for Monogenic Disorders: The Value of Concurrent Aneuploidy Screening. *J. Genet. Couns.* **2016**, *25*, 1327–1337. [CrossRef] [PubMed]
155. Spinella, F.; Fiorentino, F.; Biricik, A.; Bono, S.; Ruberti, A.; Cotroneo, E.; Baldi, M.; Cursio, E.; Minasi, M.G.; Greco, E. Extent of chromosomal mosaicism influences the clinical outcome of in vitro fertilization treatments. *Fertil. Steril.* **2018**, *109*, 77–83. [CrossRef]

156. Delhanty, J.D.; Griffin, D.K.; Handyside, A.H.; Harper, J.; Atkinson, G.H.; Pieters, M.H.; Winston, R.M. Detection of aneuploidy and chromosomal mosaicism in human embryos during preimplantation sex determination by fluorescent in situ hybridisation, (FISH). *Hum. Mol. Genet.* **1993**, *2*, 1183–1185. [CrossRef] [PubMed]
157. Fragouli, E.; Alfarawati, S.; Spath, K.; Babariya, D.; Tarozzi, N.; Borini, A.; Wells, D. Analysis of implantation and ongoing pregnancy rates following the transfer of mosaic diploid-aneuploid blastocysts. *Hum. Genet.* **2017**, *136*, 805–819. [CrossRef]
158. Daphnis, D.D.; Delhanty, J.D.; Jerkovic, S.; Geyer, J.; Craft, I.; Harper, J.C. Detailed FISH analysis of day 5 human embryos reveals the mechanisms leading to mosaic aneuploidy. *Hum. Reprod.* **2005**, *20*, 129–137. [CrossRef]
159. Mantikou, E.; Wong, K.M.; Repping, S.; Mastenbroek, S. Molecular origin of mitotic aneuploidies in preimplantation embryos. *Biochim. Biophys. Acta* **2012**, *1822*, 1921–1930. [CrossRef]
160. Fragouli, E.; Alfarawati, S.; Daphnis, D.D.; Goodall, N.N.; Mania, A.; Griffiths, T.; Gordon, A.; Wells, D. Cytogenetic analysis of human blastocysts with the use of FISH, CGH and aCGH: Scientific data and technical evaluation. *Hum. Reprod.* **2011**, *26*, 480–490. [CrossRef]
161. Bielanska, M.; Tan, S.L.; Ao, A. Chromosomal mosaicism throughout human preimplantation development in vitro: Incidence, type, and relevance to embryo outcome. *Hum. Reprod.* **2002**, *17*, 413–419. [CrossRef]
162. Munné, S.; Grifo, J.; Wells, D. Mosaicism: "survival of the fittest" versus "no embryo left behind". *Fertil. Steril.* **2016**, *105*, 1146–1149. [CrossRef]
163. Capalbo, A.; Rienzi, L. Mosaicism between trophectoderm and inner cell mass. *Fertil. Steril.* **2017**, *107*, 1098–1106. [CrossRef]
164. Munné, S.; Spinella, F.; Grifo, J.; Zhang, J.; Beltran, M.P.; Fragouli, E.; Fiorentino, F. Clinical outcomes after the transfer of blastocysts characterized as mosaic by high resolution Next Generation Sequencing- further insights. *Eur. J. Med. Genet.* **2020**, *63*, 103741. [CrossRef]
165. Greco, E.; Minasi, M.G.; Fiorentino, F. Healthy babies born after intrauterine transfer of mosaic aneuploid blastocyst. *N. Engl. J. Med.* **2015**, *373*, 2089–2090. [CrossRef]
166. Munné, S.; Blazek, J.; Large, M.; Martinez-Ortiz, P.A.; Nisson, H.; Liu, E.; Tarozzi, N.; Borini, A.; Becker, A.; Zhang, J.; et al. Detailed investigation into the cytogenetic constitution and pregnancy outcome of replacing mosaic blastocysts detected with the use of high-resolution next-generation sequencing. *Fertil. Steril.* **2017**, *108*, 62–71. [CrossRef]
167. Victor, A.R.; Tyndall, J.C.; Brake, A.J.; Lepkowsky, L.T.; Murphy, A.E.; Griffin, D.K.; McCoy, R.C.; Barnes, F.L.; Zouves, C.G.; Viotti, M. One hundred mosaic embryos transferred prospectively in a single clinic: Exploring when and why they result in healthy pregnancies. *Fertil. Steril.* **2019**, *111*, 280–293. [CrossRef]
168. Viotti, M. Mosaic embryos. A comprehensive and powered analysis of clinical outcomes. *Fertil. Steril.* **2019**, *112*, e33. [CrossRef]
169. Kahraman, S.; Cetinkaya, M.; Yuksel, B.; Yesil, M.; Cetinkaya, C.P. The birth of a baby with mosaicism resulting from a known mosaic embryo transfer: A case report. *Hum. Reprod.* **2020**, *35*, 727–733. [CrossRef]
170. Cram, D.S.; Leigh, D.; Handyside, A.; Rechitsky, L.; Xu, K.; Harton, G.; Grifo, J.; Rubio, C.; Fragouli, E.; Kahraman, S.; et al. PGDIS Position Statement on the Transfer of Mosaic Embryos 2019. *Reprod. Biomed. Online* **2019**, *39* (Suppl. 1), e1. [CrossRef]
171. CoGEN 2017. COGEN Position Statement on Chromosomal Mosaicism Detected in Preimplantation Blastocyst Biopsies. Available online: https://www.ivfworldwide.com (accessed on 20 April 2018).
172. Ledbetter, D.H.; Zachary, J.M.; Simpson, J.L.; Golbus, M.S.; Pergament, E.; Jackson, L.; Mahoney, M.J.; Desnick, R.J.; Schulman, J.; Copeland, K.L. Cytogenetic results from the U.S. Collaborative Study on CVS. *Prenat. Diagn.* **1992**, *12*, 317–345. [CrossRef]
173. Santos, M.A.; Teklenburg, G.; Macklon, N.S.; Van Opstal, D.; Schuring-Blom, G.H.; Krijtenburg, P.J.; de Vreeden-Elbertse, J.; Fauser, B.C.; Baart, E.B. The fate of the mosaic embryo: Chromosomal constitution and development of Day 4, 5 and 8 human embryos. *Hum. Reprod.* **2010**, *25*, 1916–1926. [CrossRef]
174. Popovic, M.; Dhaenens, L.; Taelman, J.; Dheedene, A.; Bialecka, M.; De Sutter, P.; Chuva de Sousa Lopes, S.M.; Menten, B.; Heindryckx, B. Extended in vitro culture of human embryos demonstrates the complex nature of diagnosing chromosomal mosaicism from a single trophectoderm biopsy. *Hum. Reprod.* **2019**, *34*, 758–769. [CrossRef]

175. Shapiro, B.S.; Daneshmand, S.T.; Garner, F.C.; Aguirre, M.; Hudson, C. Clinical rationale for cryopreservation of entire embryo cohorts in lieu of fresh transfer. *Fertil. Steril.* **2014**, *102*, 3–9. [CrossRef]
176. Zhang, S.; Luo, K.; Cheng, D.; Tan, Y.; Lu, C.; He, H.; Gu, Y.; Lu, G.; Gong, F.; Lin, G. Number of biopsied trophectoderm cells is likely to affect the implantation potential of blasto- cysts with poor trophectoderm quality. *Fertil. Steril.* **2016**, *105*, 1222–1227. [CrossRef]
177. Farhi, J.; Ben-Haroush, A.; Andrawus, N.; Pinkas, H.; Sapir, O.; Fisch, B.; Ashkenazi, J. High serum oestradiol concentrations in IVF cycles increase the risk of pregnancy complications related to abnormal placentation. *Reprod. Biomed. Online* **2010**, *21*, 331–337. [CrossRef]
178. Healy, D.L.; Breheny, S.; Halliday, J.; Jaques, A.; Rushford, D.; Garrett, C.; Talbot, J.M.; Baker, H.W.B. Prevalence and risk factors for obstetric haemorrhage in 6730 singleton births after assisted reproductive technology in Victoria Australia. *Hum. Reprod.* **2010**, *25*, 265–274. [CrossRef]
179. Coates, A.; Kung, A.; Mounts, E.; Hesla, J.; Bankowski, B.; Barbieri, E.; Ata, B.; Cohen, J.; Munné, S. Optimal euploidembryo transfer strategy, fresh versus frozen, after preimplantation genetic screening with next generation sequencing: A randomized controlled trial. *Fertil. Steril.* **2017**, *107*, 723–730. [CrossRef]
180. Greco, E.; Litwicka, K.; Arrivi, C.; Varricchio, M.T.; Caragia, A.; Greco, A.; Minasi, M.G.; Fiorentino, F. The endometrial preparation for frozen-thawed euploid blastocyst transfer: A prospective randomized trial comparing clinical results from natural modified cycle and exogenous hormone stimulation with GnRH agonist. *J. Assist. Reprod. Genet.* **2016**, *33*, 873–884. [CrossRef]
181. Wang, A.; Murugappan, G.; Kort, J.; Westphal, L. Hormone replacement versus natural frozen embryo transfer for euploid embryos. *Arch. Gynecol. Obstet.* **2019**, *300*, 1053–1060. [CrossRef]
182. Sekhon, L.; Feuerstein, J.; Pan, S.; Overbey, J.; Lee, J.A.; Briton-Jones, C.; Flisser, E.; Stein, D.E.; Mukherjee, T.; Grunfeld, L.; et al. Endometrial Preparation Before the Transfer of Single, Vitrified-Warmed, Euploid Blastocysts: Does the Duration of Estradiol Treatment Influence Clinical Outcome? *Fertil. Steril.* **2019**, *111*, 1177–1185. [CrossRef] [PubMed]
183. Gaggiotti-Marre, S.; Martinez, F.; Coll, L.; Garcia, S.; Álvarez, M.; Parriego, M.; Barri, P.N.; Polyzos, N.; Coroleu, B. Low serum progesterone the day prior to frozen embryo transfer of euploid embryos is associated with significant reduction in live birth rates. *Gynecol. Endocrinol.* **2019**, *35*, 439–442. [CrossRef] [PubMed]
184. Boynukalin, F.K.; Gultomruk, M.; Turgut, E.; Demir, B.; Findikli, N.; Serdarogullari, M.; Coban, O.; Yarkiner, Z.; Bahceci, M. Measuring the serum progesterone level on the day of transfercan be an additional tool to maximize ongoing pregnancies in single euploid frozen blastocyst transfers. *Reprod. Biol. Endocrinol.* **2019**, *17*, 102. [CrossRef]
185. Scott, R.J.; Upham, K.M.; Forman, E.J.; Zhao, T.; Treff, N.R. Cleavage-stage biopsy significantly impairs human embryonic implantation potential while blasto cyst biopsy does not: A randomized and paired clinical trial. *Fertil. Steril.* **2013**, *100*, 624–630. [CrossRef]
186. Dokras, A.; Sargent, I.L.; Gardner, R.L.; Barlow, D.H. Human trophectoderm biopsy and secretion of chorionic gonadotrophin. *Hum. Reprod.* **1991**, *6*, 1453–1459. [CrossRef]
187. Zhang, W.Y.; von Versen-Höynck, F.; Kapphahn, K.I.; Fleischmann, R.R.; Zhao, Q.; Baker, V.L. Maternal and neonatal outcomes associated with trophectoderm biopsy. *Fertil. Steril.* **2019**, *112*, 283–290. [CrossRef]
188. He, H.; Jing, S.; Lu, C.F.; Tan, Y.Q.; Luo, K.L.; Zhang, S.P.; Gong, F.; Lu, G.X.; Lin, G. Neonatal outcomes of live births after blastocyst biopsy in preimplantation genetic testing cycles: A follow-up of 1721 children. *Fertil. Steril.* **2019**, *112*, 82–88. [CrossRef]
189. Jing, S.; Luo, K.; He, H.; Lu, C.; Zhang, S.; Tan, Y.; Gong, F.; Lu, G.; Lin, G. Obstetric and neonatal outcomes in blastocyst-stage biopsy with frozen embryo transfer and cleavage-stage biopsy with fresh embryo transfer after preimplantation genetic diagnosis/screening. *Fertil. Steril.* **2016**, *106*, 105–112. [CrossRef] [PubMed]

International Journal of
Molecular Sciences

MDPI

Review

The Role of LIN28-*let*-7-ARID3B Pathway in Placental Development

Asghar Ali *, Gerrit J. Bouma, Russell V. Anthony and Quinton A. Winger

Department of Biomedical Sciences, Animal Reproduction and Biotechnology Laboratory, 1683 Campus Delivery, Colorado State University, Fort Collins, CO 80523, USA; gerrit.bouma@colostate.edu (G.J.B.); russ.anthony@colostate.edu (R.V.A.); quinton.winger@colostate.edu (Q.A.W.)

* Correspondence: asghar.ali@colostate.edu

Received: 17 April 2020; Accepted: 18 May 2020; Published: 21 May 2020

Abstract: Placental disorders are a major cause of pregnancy loss in humans, and 40–60% of embryos are lost between fertilization and birth. Successful embryo implantation and placental development requires rapid proliferation, invasion, and migration of trophoblast cells. In recent years, microRNAs (miRNAs) have emerged as key regulators of molecular pathways involved in trophoblast function. A miRNA binds its target mRNA in the 3′-untranslated region (3′-UTR), causing its degradation or translational repression. Lethal-7 (*let-7*) miRNAs induce cell differentiation and reduce cell proliferation by targeting proliferation-associated genes. The oncoprotein LIN28 represses the biogenesis of mature *let-7* miRNAs. Proliferating cells have high LIN28 and low *let-7* miRNAs, whereas differentiating cells have low LIN28 and high *let-7* miRNAs. In placenta, low LIN28 and high *let-7* miRNAs can lead to reduced proliferation of trophoblast cells, resulting in abnormal placental development. In trophoblast cells, *let-7* miRNAs reduce the expression of proliferation factors either directly by binding their mRNA in 3′-UTR or indirectly by targeting the AT-rich interaction domain (ARID)3B complex, a transcription-activating complex comprised of ARID3A, ARID3B, and histone demethylase 4C (KDM4C). In this review, we discuss regulation of trophoblast function by miRNAs, focusing on the role of LIN28-*let*-7-ARID3B pathway in placental development.

Keywords: miRNA; trophoblast cells; cell proliferation; ARID3B complex

1. Introduction

Every year, more than 15 million babies are born preterm in the world. A healthy placenta is required for successful establishment of pregnancy and optimal pregnancy outcome [1,2]. Most structural and functional development of the placenta occurs during the first trimester of pregnancy which requires rapid proliferation, invasion, and migration of trophoblast cells [3]. Improper trophoblast function can result in miscarriage, pre-term labor, stillbirth, pre-eclampsia (PE), intrauterine growth restriction (IUGR), and long-term postnatal complication in the mother and fetus [4–8]. The process of rapid trophoblast proliferation and dynamic transformation in placental structure is poorly understood. Previous studies have shown the role of non-coding miRNAs in regulation of trophoblast function. Lethal-7 (*Let-7*) miRNAs are one of the most studied families of miRNAs and have a well-established role in cell proliferation, invasion, migration, differentiation, and metabolism [9,10]. *Let-7* miRNAs reduce cell proliferation by downregulating the proliferation-associated genes [11]. In highly proliferative cells, the RNA binding protein LIN28 represses the production of mature *let-7* miRNAs [12,13]. Low LIN28 and increased *let-7* miRNAs are thought to be associated with pathogenesis of PE and IUGR [11,14]. During early placental development, dysregulation of miRNAs can lead to reduced proliferation, invasion and migration of trophoblast cells, and contribute to the etiology of placental abnormalities. This review focuses on the role of different miRNAs in trophoblast function, with *let-7* miRNAs being the center of discussion.

2. Early Placental Development and Trophoblast Cells

The human blastocyst is formed at day 4–5 after fertilization and contains an outer most layer of zona pellucida, a single layer of mononuclear trophectoderm or trophoblast (TE), a blastocoel cavity, and an inner cell mass (ICM) or embryoblast. The blastocyst sheds the zona pellucida at day 7, exposing the TE [15]. The hatched blastocyst adheres to the endometrial epithelium, subsequently activating the adherent trophoblast cells to proliferate and give rise to different trophoblast lineages [16] (Figure 1). There are two prominent pathways for trophoblast lineages: syncytial pathway and invasive pathway.

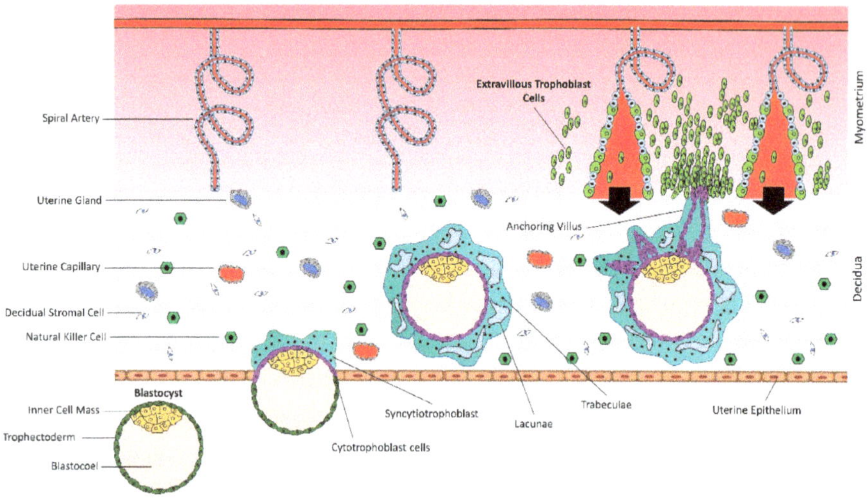

Figure 1. Early placental development and spiral artery remodeling. Human placental development starts with interaction between hatched blastocyst and uterine epithelium. The trophoblast cells that contact with the uterine epithelium transform into highly proliferative cytotrophoblasts (CTBs). Cytotrophoblasts undergo rapid proliferation and some of them fuse to form a multinucleated syncytiotrophoblast (STB). Within a few hours, STB expands and covers whole blastocyst and helps in blastocyst invasion into the uterine decidua. Continuous proliferation of CTBs results in formation of villi. Some CTBs from the tip of anchoring villi break the STB cover, invade the uterine stroma and myometrium, and transform into extravillous trophoblast cells (EVTs). EVTs remodel the spiral arteries to ensure sufficient flow of blood to the placenta.

2.1. *Syncytial Pathway*

At the site of attachment, the trophoblast cells transform into rapidly proliferating cytotrophoblasts (CTBs). CTBs undergo rapid proliferation and the newly formed CTBs fuse with each other to form a multinucleated syncytiotrophoblast (STB) [17]. Within a few hours, the STB expands and surrounds the whole blastocyst and mediates the invasion of the blastocyst in the decidualized uterine stroma [18]. The expansion of STB depends upon rapid proliferation and fusion of CTBs [19,20]. By day 21 after conception, the tertiary villi have formed as functional units of the placenta. By week five after conception, the fetoplacental circulation is fully established [21].

2.2. *Invasive Pathway*

Numerous daughter villi arise from tertiary villi, some of which extend to the maternal tissue and are called anchoring or stem villi [22–24]. The anchoring sites can be established as early as the second week of gestation [22]. At proximal ends of anchoring villi, some highly proliferating CTBs break free of the overlying STB layer and invade into maternal endometrium and myometrium [24]. As soon as the detached CTBs make contact with decidual extracellular matrix, they differentiate into interstitial extravillous trophoblast cells (iEVTs) [25]. The iEVTs reach the vascular lumen and differentiate into endovascular extravillous trophoblast cells (enEVTs) [26,27]. The enEVTs remodel the spiral arteries which includes loss of endothelial and smooth muscle cells from arterial walls and their replacement by invasive enEVTs, loss of elasticity, dilation of the arterial lumen, and loss of maternal vasomotor control on the remodeled blood vessels [19,27–29].

Spiral artery remodeling is crucial for normal placental development and supplying enough nutrients to the fetus (Figure 2). Inadequate remodeling of the spiral arteries is associated with conditions such as preeclampsia (PE), intrauterine growth restriction (IUGR)/fetal growth restriction (FGR), and recurrent miscarriage, that are harmful for both the mother and the fetus. The proliferation and differentiation of trophoblast cells continues throughout gestation. However, unlike cancerous tissues, the proliferation of trophoblast cells is strictly regulated by complex molecular pathways [30,31]. In recent studies, microRNAs (miRNAs) have been shown to play vital roles in trophoblast proliferation and early placental development.

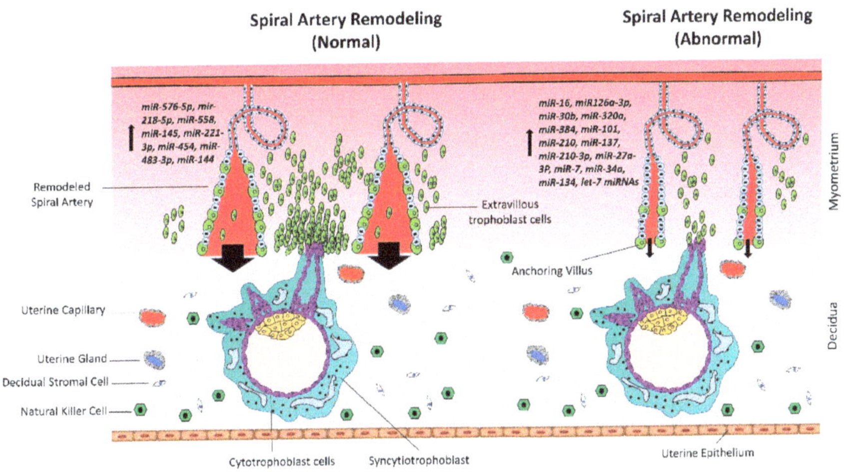

Figure 2. Normal vs. abnormal spiral artery remodeling. CTBs from anchoring villi break out of the SCT layer and enter the uterine stroma where they differentiate into extravillous trophoblasts (EVTs). Spiral artery remodeling is accomplished by invasion and migration of EVTs. EVTs replace the vascular endothelial cells, remodel the spiral arteries, and ensure sufficient flow of blood to the placenta. In placenta-associated disorders like preeclampsia, reduced proliferation of CTBs results in less availability of EVTs. This leads to insufficient remodeling of spiral arteries and reduced blood flow to the placenta. Based on different studies listed in Table 1, a different set of miRNAs is upregulated in trophoblast cells during normal vs. preeclamptic pregnancies.

3. Functional Analysis of microRNAs in Trophoblast Cells

MicroRNAs are 20–25 nucleotide-long single stranded RNAs which bind 3′-untranslated region (3′-UTR) of the target mRNA causing its degradation or translational repression [32–36]. MicroRNAs have been shown to play important roles in deciding the fate of trophoblast cells [37]. Proliferation, invasion, and migration of trophoblast cells are critical steps during early human placental development. With increasing evidence for the role of miRNAs in regulation of genes associated with cell proliferation, invasion, and migration, several studies have been conducted to investigate the role of miRNAs in placental development and pathogenesis of placenta-associated disorders.

Table 1. Gene regulation by miRNAs in trophoblast cells.

miRNA	Target Genes	Reference	Effect of Higher miRNA Expression
let-7	*ARID3A, ARID3B, HMGA1, cMYC*	[11]	Reduces proliferation and invasiveness of trophoblast cells
miR-384	*STAT3*	[38]	
miR-106b	*MMP-2*	[39]	
miR-203	*VEGFA*	[40]	
miR-520g	*MMP-2*	[41]	
miR-210	*Notch1*	[42]	
miR-16	*Notch2*	[43]	
miR-320a	*IL-4*	[44]	
miR-320a	*ERRγ*	[45]	
miR-210-3p	*FGF1*	[46]	
miR-7	*EMT-related TFs*	[47]	Reduces migration and invasion of trophoblast cells
miR-218	*SOX4*	[48]	
miR-34a-5p	*Smad4*	[49]	
miR-193b-3p	*TGF-β2*	[50]	
miR-34a	*MYC*	[51]	
miR-519d	*MMP-2*	[52]	
miR-101	*CXCL6*	[53]	
miR-34a	*Notch*	[54]	
miR-431	*ZEB1*	[55]	
miR-145-5p	*Cyr61*	[56]	
miR-134	*ITGB1*	[57]	
miR-27a-3p	*USP25*	[58]	
miR-362-3p	*Pax3*	[59]	
miR-181a-5p	*IGF2BP2*	[60]	
miR-137	*FNDC5*	[61]	
miR-30b	*MXRA5*	[62]	
miR-30a-3p	*IGF1*	[63]	
miR-18b	*HIF-1α*	[64]	
miR-299	*HDAc2*	[65]	
miR-454	*ALK7*	[66]	Increases proliferation of trophoblast cells
miR-145	*MUC1*	([67]	
miR-221-3p	*THBS2*	[68]	
miR-126a-3p	*ADAM9*	[69]	
miR-483-3p	*RB1CC1*	[70]	
miR-144	*PTEN*	[71]	
miR-518b	*Rap1b*	[72]	
miR-218-5p	*TGFB2*	[73]	Promotes endovascular extravillous trophoblast cells (enEVTs) and spiral artery remodeling

Table 1. *Cont.*

miRNA	Target Genes	Reference	Effect of Higher miRNA Expression
miR-210	*CPEB2*	[74]	Inhibits trophoblast syncytialization
miR-106a	*hCYP19A1, hGCM1*	[75]	
miR-558	*TIMP4*	[76]	Enhances invasion of trophoblast cells
miR-576-5p	*TFAP2A*	[77]	
miR-184	*WIG1*	[78]	Promotes apoptosis of trophoblast cells
miR-133	*Rho/ROCK*	[79]	
miR-152	*Bax, Bcl-2*	[80]	
miR-155	*HIF-1α*	[81]	
miR-371a-5p	*XIAP*	[82]	
miR-520	*PARP1*	[83]	
miR-34a	*BCL-2*	[84]	
miR-18a	*ER1*	[85]	
miR-182	*BRCA1*	[86]	
miR-23a	*XIAP*	[87]	
miR-101-3p	*mTOR*	[88]	
miR-96-5p	*mTOR, Bcl-2*	[88]	
miR-200c	*Wnt/β-catenin*	[89]	
miR-125a	*MCL1*	[90]	

miRNAs impose their effect by regulating the expression of different genes and the effect of a specific miRNA on the phenotype of a cell or tissue depends upon the role of genes targeted by that miRNA. Hence, depending upon the function of their target genes in trophoblast cells, some miRNAs support successful placental development by promoting trophoblast cell proliferation, invasion and migration, and inhibiting the apoptosis of trophoblast cells, whereas some miRNAs can lead to abnormal placental development by reducing cell proliferation, invasion and migration, and increasing apoptosis of trophoblast cells. Table 1 describes the genes regulated by miRNAs and their effect on functionality of trophoblast cells as described in some recent studies. All gene symbols used in Table 1 are according to the Human Genome Organization (HUGO) Gene Nomenclature.

4. Let-7 miRNAs

The lethal-7 (let-7) family of miRNAs was first discovered in 2002 as a development regulator in *Caenorhabditis elegans* [91]. The expression of let-7 miRNAs is low in undifferentiated cells and increases gradually as the cells differentiate during development [92]. Therefore, let-7 miRNAs are also referred to as differentiation-inducing miRNAs. The let-7 mutated *C. elegans* larvae do not mature to the adult stage but keep proliferating and eventually die, earning the name "lethal-7 (let-7)" for this family of miRNAs [91]. Let-7 miRNAs are highly conserved in various animal species [93], suggesting that let-7 miRNAs regulate the same molecular pathways and biological processes in different organisms. In humans, let-7 miRNAs family comprises 12 members including let-7a, let-7b, let-7c, let-7d, let-7e, let-7f, let-7g, let-7i, and miR-98 [94], which originate from eight different genomic loci [95]. Some let-7 miRNAs produced from different genomic loci at different chromosomes have the same sequence. For examples, in humans, let-7a-1, let-7a-2, and let-7a-3 have the same sequence but are encoded by loci on chromosomes 9, 11, and 12, respectively. Similarly, let-7f-1 and let-7f-2 are encoded by different genomic loci but have the same sequence [96]. Let-7 miRNAs have a common seed sequence of seven nucleotides "GAGGUAG" from nucleotide two to eight in all species, which plays an important role in recognizing miRNA response element (MRE) in 3′-UTR of their target mRNA [96]. However, differences in non-seed flanking sequence of let-7 miRNAs affect target specificity [97,98]. Presence of similar seed sequence in all let-7 miRNAs across different species suggests that let-7 miRNAs have the same mechanism for target recognition and might have overlapping targets.

The microarray analysis data from *C. elegans* show that *let-7* miRNAs regulate the expression of thousands of genes, directly and indirectly, indicating their widespread role in biological processes [12]. *Let-7* miRNAs play profound roles in embryo development, glucose metabolism, cell pluripotency and differentiation, tumorigenesis, tissue regeneration, age of onset of puberty and menopause in humans, and organ growth [99,100]. Various studies have shown that *let-7* miRNAs induce cell differentiation and act as fundamental tumor suppressors by downregulating oncogenes [13,101–103]. At early stages of cancer development, *let-7* miRNAs are downregulated and *let-7* targeted oncofetal genes (LOG) are re-expressed [104]. Comparative bioinformatics analysis shows that *let-7* miRNAs target several oncofetal genes including high mobility group AT-hook 2 (HMGA2), insulin like growth factor 2 mRNA binding protein 1 (IMP1), IMP2, IMP3, and malignancy marker nucleosome assembly protein 1 like 1 (NAP1L1) [104]. In hematopoietic stem cells, *let-7* miRNAs inhibit transforming growth factor β (TGFβ) pathway and high mobility group AT-hook 2 (HMGA2), decide the fate of these cells, and regulate cell proliferation, self-renewal and differentiation [105,106].

Let-7 miRNAs are synthesized following the same general mechanism for miRNA synthesis. The *let-7* loci are transcribed as pri-*let-7* miRNA, then processed into 67–80 nucleotide long pre-*let-7* miRNA by microprocessor complex [95]. Based on the mechanism of further processing, pre-*let-7* miRNAs are divided in two groups: Group I pre-*let-7* miRNAs (pre-*let-7a-2, 7c*, and *7e*) are processed in cytoplasm by direct action of Dicer, whereas group II pre-*let-7* miRNAs (all remaining *let-7s*) are mono-uridylated prior to processing by Dicer [107]. Action of Dicer produces 22 nucleotide long mature *let-7* miRNAs, called *let-7-5p*. As a part of miRNA induced silencing complex (miRISC), *let-7* miRNAs suppress a wide range of genes involved in development, cell proliferation, metabolism, and other important physiological processes [108]. There is no significant difference in expression of pri-*let-7* and pre-*let-7* miRNAs between undifferentiated and differentiated cells, however mature *let-7* miRNAs are high in differentiated cells compared to undifferentiated cells [109].

Mature *let-7* miRNA is a part of hairpin structure in pri- and pre-*let-7* miRNA. This hairpin structure contains mature *let-7* miRNA (*let-7-5p)* in the stem and a partially complimentary strand of nucleotides called *let-7-3p* miRNA, connected by a terminal loop region of different lengths called pre-element (preE) [110]. The process of generation of mature *let-7* miRNAs is more precisely regulated compared to the synthesis of other miRNAs. Different proteins regulate posttranscriptional biogenesis of mature *let-7* miRNA by binding the preE region of pri- and pre-*let-7* miRNAs [111]. One of the most prominent mechanism for regulation of *let-7* miRNAs biogenesis is through LIN28 [112].

5. Suppression of *let-7* miRNAs by LIN28

LIN28 is a highly conserved RNA binding protein with two paralogues, LIN28A and LIN28B. Both LIN28A and LIN28B have a cold-shock domain (CSD) at the N-terminal and two zinc knuckle domains (ZKDs) at the C-terminal [113]. LIN28 promotes cell proliferation and inhibits cell differentiation [114]. LIN28 is also involved in reprogramming of differentiated somatic cells into tumor or stem cells, hence known as oncoprotein [115,116]. Reduced expression of LIN28 in embryos results in reduced prenatal growth and development and long-term metabolic abnormalities [117]. Knockout of LIN28A in mice leads to perinatal lethality while LIN28B knockout results in postnatal growth abnormalities in males. Knockout of both LIN28A and LIN28B in mice is embryonically lethal at around E13. Conditional knockout of LIN28A and LIN28B in mice at six weeks of age does not produce any evident phenotype [118]. Collectively, these findings show that LIN28 has a more profound role during prenatal development and organogenesis.

LIN28 regulates expression of several genes either directly binding to the mRNA of target genes or by repressing the production of mature *let-7* miRNAs, later being a more prevalent mechanism [118,119]. There are conflicting theories about the localization of LIN28A and LIN28B in cells. LIN28A is predominantly localized in the cytoplasm but can be found in the nucleus as well [120,121]; however, according to another study, LIN28A is exclusively localized in the cytoplasm [122]. LIN28B has a nucleolar and nuclear localization signal, while others found it predominantly in cytoplasm with a possibility to shuttle to the nucleus [113,119,123]. LIN28A and LIN28B selectively repress the expression and maturation of *let-7* miRNAs by distinct mechanisms, without directly affecting the expression of other miRNAs [124]. LIN28A CSD binds GNGAY motif while ZKDs bind GGAG motif in the stem loop pre-*let-7* miRNA in the cytoplasm. After binding to pre-*let-7* miRNA, LIN28A recruits terminal uridylyl transferase (TUTase) Zcchc11 (also referred as TUT4). TUT4 causes polyuridylation of pre-*let-7* miRNA which blocks the cleavage of pre-*let-7* miRNA by Dicer and hence inhibits the production of mature *let-7* miRNAs [124–126]. Polyuridylated pre-*let-7* miRNA is recognized and degraded by exonuclease Dis3L2 [127]. The mechanism of *let-7* miRNA suppression by LIN28B remains controversial and there are four different theories. First, LIN28B inhibits maturation of *let-7* miRNAs by TUT4 independent mechanism. In the nucleus, LIN28B binds the pri-*let-7* miRNA by its CSD and ZKDs and inhibits its processing by a microprocessor [122]. Second, in the cytoplasm, LIN28B binds to pre-*let-7* miRNA and inhibits its processing by Dicer [128]. Third, in the cytoplasm, LIN28B binds to pre-*let-7* miRNA and leads to its polyuridylation by recruiting an unknown TUTase, leading to its degradation [121]. Fourth, in the nucleolus, LIN28B has the ability to sequester pri-*let-7* miRNAs and hence inhibits further processing to mature pre-*let-7* miRNAs [122].

In 2018, Ustianenko et al. demonstrated that LIN28 selectively regulates a subclass of *let-7* miRNAs [129]. Using single nucleotide resolution, they identified –(U)GAU- as the new binding motif of the CSD. Some pre-*let-7* miRNAs with both (U)GAU and GGAG motifs in the stem loop make a stronger and stable interaction with LIN28 and are referred to as CSD^+, while the others which do not contain (U)GAU motif are called CSD^-. The CSD^+ subclass includes pre-*let-7b*, pre-*let-7d*, pre-*let-7f-1*, pre-*let-7g*, pre-*let-7i*, and *miR-98*. The CSD^- subclass includes pre-*let-7a-1*, pre-*let-7a-2*, pre-*let-7-3*, pre-*let-7c*, pre-*let-7e*, and pre-*let-7f-2* [110,129]. Although all *let-7* miRNAs express ZKD binding GGAG motif, both LIN28A and LIN28B have shown greater binding affinity for CSD^+ pre-*let-7* miRNAs, hence leading to their polyuridylation and repression [129].

6. Gene Regulation by LIN28-*let*-7 miRNA Axis in Trophoblast Cells

Due to the profound role of *let-7* miRNAs as differentiation-inducing miRNAs, the focus of our lab is to investigate the role of LIN28-*let-7* miRNA axis in trophoblast cells. Both LIN28A and LIN28B are highly expressed in human placenta and are localized to trophoblast cells [11,130–132]. High throughput genotyping array reveals that LIN28B is paternally imprinted in human placenta [133,134]. Using single cell transcriptome profiling, Liu et al. identified 14 different cell types in human placenta and showed that paternally imprinted LIN28B has high expression in CTBs, EVTs and STB, whereas it has no to low expression in mesenchymal cells, macrophages, and blood cells in placenta [135]. They further showed that LIN28B expression in week 24 EVTs was lower compared to week 8 EVTs [135], suggesting that expression of LIN28B in trophoblast cells reduces as the pregnancy progresses. LIN28B is the main paralogue of LIN28 in human placenta and *LIN28B* mRNA is 1300-fold higher compared to *LIN28A* mRNA in term human placental tissue [14]. Immunohistochemical analysis of term human placenta shows that LIN28B expression in CTBs and STB is higher compared to placental decidual cells [14]. In 2013, Gu el al. compared the expression of miRNAs between first and third trimester human placentas [136]. They reported that along with many other miRNAs, *let-7a*, *let-7c*, *let-7d*, *let-7f*, *let-7g*, and *let-7i* are upregulated in third trimester compared to first trimester human placenta [136]. We measured *LIN28A* and *LIN28B* mRNA in first trimester (11 week) vs. term human placentas and found that *LIN28A* mRNA was nearly 700-fold higher and *LIN28B* mRNA was nearly 300-fold higher in first trimester compared to term human placenta (Figure 3). Based on these results, we suggest that increased expression of *let-7* miRNAs in term human placentas, reported by Gu et al., is due to reduced expression of *LIN28A* and *LIN28B*. Low LIN28 and higher level of *let-7* miRNAs in term placenta compared to first trimester placenta suggest that the proliferation rate of trophoblast cells is higher during the first trimester and decreases with advancement in gestational age. As LIN28-*let-7* miRNA axis regulates expression of several genes, it would not be surprising to see a difference in gene expression in first trimester vs. third trimester human placenta.

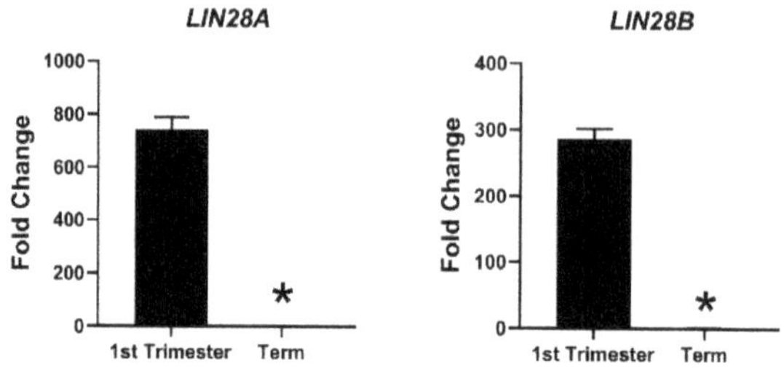

Figure 3. mRNA was extracted from first trimester (week 11) and term human placentas and *LIN28A* and *LIN28B* mRNA levels were measured using real-time RT-PCR, where * $p < 0.05$.

In IUGR pregnancies, the size of placenta is significantly smaller compared to normal pregnancies [137], which suggests the role of reduced trophoblast proliferation in etiology of IUGR. In a recently published study, we showed that term human placentas from IUGR pregnancies have low LIN28A and LIN28B, and high *let-7* miRNAs compared to term human placentas from normal pregnancies [11]. Canfield et al. reported that term human placentas from preeclamptic pregnancies have reduced LIN28B but no change in LIN28A compared to normal term placentas [14]. They further demonstrated that in first trimester human placenta, LIN28B is higher in extravillous cytotrophoblasts compared to villous trophoblast cells, indicating their role in trophoblast cell invasion [14]. Low LIN28 and high *let-7* miRNAs during the first trimester of pregnancy can lead to reduced trophoblast proliferation and invasion leading to pregnancy-related disorders.

Due to the limitation that humans cannot be used as experimental models, most studies investigating molecular mechanisms involved in human placental development are conducted using placental cell lines. Commonly used human trophoblast-derived cell lines include BeWo, ACH-3P, JEG3, JAR, Sw.71, and HTR8/SVneo. LIN28A knockdown in immortalized first trimester human trophoblast (ACH-3P) cells drives these cells towards syncytial differentiation and increases the expression of syncytiotrophoblast markers including *hCG*, *LGALS13*, and *ERVW-1* [132]. Moreover, knockdown of LIN28A increases the expression of *let-7* miRNAs including *let-7a*, *let-7c*, *let-7d*, *let-7e*, *let-7g*, and *let-7i* [132], suggesting that differentiation of cells might be due to increased levels of *let-7* miRNAs. Overexpression of LIN28B in HTR8 cells increases cell proliferation, invasion, and migration, whereas knockdown of LIN28B in JEG3 cells reduces cell proliferation [14]. In a recently published study, we further investigated the correlation between LIN28 and *let-7* miRNAs in trophoblast cells using first trimester human trophoblast-derived ACH-3P and Sw.71 cells. ACH-3P cells were generated by fusing first trimester human trophoblast cells with human choriocarcinoma cells, whereas Sw.71 cells were generated by overexpressing human telomerase reverse transcriptase (h-TERT) in first trimester human trophoblast cells [138,139]. These cell lines have contrasting levels of LIN28 and *let-7* miRNAs [11]. ACH-3P cells have high expression of LIN28A and LIN28B whereas these proteins are not detectable in Sw.71 cells [11]. The expression of all *let-7* miRNAs is 50–500-fold higher in Sw.71 cells compared to ACH-3P cells, potentially due to depleted LIN28A and LIN28B in Sw.71 cells which are major suppressors of *let-7* miRNAs [11]. The contrasting levels of LIN28 and *let-7* miRNAs between ACH-3P and Sw.71 cells are potentially due to the difference of methodology used to generate these cell lines. LIN28A knockout in ACH-3P cells increases *let-7a*, *let-7b*, *let-7c*, *let-7d*, and *let-7e*, whereas LIN28B knockout in ACH-3P cells increases *let-7a*, *let-7b*, *let-7c*, *let-7d*, *let-7e*, and *let-7i* [11]. According to another study, knockdown of LIN28B in ACH-3P cells increases *let-7c*, *let-7d*, *let-7e*, *let-7f*, and *let-7i* [140]. Double knockout of LIN28A and LIN28B in ACH-3P cells results in increased expression of all *let-7* miRNAs compared to knockout of either LIN28A or LIN28B [11]. Similarly, LIN28A overexpression in Sw.71 cells decreases *let-7d* and *let-7i*, whereas LIN28B overexpression causes reduction in all *let-7* miRNAs. However, overexpression of both LIN28A and LIN28B in Sw.71 cells results in decreased expression of all *let-7* miRNAs compared to overexpression of either LIN28A or LIN28B [11]. These results suggest that LIN28A and LIN28B work in coordination to suppress *let-7* miRNAs and manipulating one paralogue of LIN28 in human trophoblast cells might not induce a similar phenotype compared to if both paralogues are changed.

The majority of *let-7*-regulated genes are associated with cell proliferation, migration, and invasion—processes which are crucial during early human placental development. We recently demonstrated that double knockout of LIN28A and LIN28B in ACH-3P cells increases in *let-7* miRNAs and leads to reduction in expression of proliferation-associated genes including high-mobility group AT-hook 1 (*HMGA1*), MYC protooncogene (*c-MYC*), vascular endothelial growth factor A (*VEGF-A*), and Wnt family member 1 (*WNT1*). LIN28A/B knockout reduces trophoblast cell proliferation and drives them towards differentiating to a syncytiotrophoblast [11,130]. Similarly, double knockin of LIN28A/B in Sw.71 cells leads to reduction in *let-7* miRNAs and increases the expression of *HMGA1*, *c-MYC*, *VEGF-A*, and *WNT1* [11]. Other than its role in cell proliferation, VEGF-A is required at all steps of angiogenesis during placental development [21]. Reduced expression of VEGF-A due to high *let-7* miRNAs can lead to serious pregnancy complications due to impaired angiogenesis in placenta.

Several studies have demonstrated that *let-7* miRNAs bind the 3′-UTR of HMGA2 and reduce its expression in cancer cells [102,141,142]. However, in a recent study, we found a different mechanism of HMGA2 regulation in human trophoblast cells. Double knockout of LIN28A/B in ACH-3P cells increases *let-7* miRNAs but does not change HMGA2 expression [130]. Along with increased *let-7* miRNAs, LIN28A/B double knockout also increases miR-182. The exact mechanism behind increases in miR-182 in LIN28A/B double knockout ACH-3P cells is not clear. We further showed that HMGA2 expression in trophoblast cells is regulated by a transcription-repressing complex comprised of breast cancer susceptibility gene 1 (BRCA1), CtBP-interacting protein (CtIP), and zinc finger protein 350 (ZNF350). This complex, also called BRCA1 repressor complex, binds the promoter region of HMGA2 and inhibits its transcription [130]. In LIN28A/B double knockout ACH-3P cells, high miR-182 targets BRCA1 leading to inhibition of BRCA1 repressor complex and hence increases HMGA2 expression [130]. Therefore, the expected decrease in HMGA2 due to high *let-7* miRNAs is rescued by inhibition of BRCA1 repressor complex. These findings indicate that all genetic pathways demonstrated in the cancer cells might not be applicable in trophoblast cells. It further suggests that rapid proliferation of trophoblast cells during early placental development is more precisely regulated compared to cancer cells.

Although in vitro studies demonstrate the vital role of LIN28-*let-7* miRNA axis in trophoblast function, its role in placental development in vivo is not well understood. Using sheep as an experimental model, we investigated the role of LIN28-*let-7* miRNA axis in trophoblast proliferation in vivo. In sheep, the hatched blastocyst undergoes a phase of trophectoderm elongation before attachment to the uterine epithelium. The conceptus elongation is accomplished by rapid proliferation of trophoblast cells. Trophoblast proliferation is a critical process in early placental development both in humans and sheep. Trophoblast specific knockdown of LIN28A or LIN28B in sheep leads to reduced conceptus elongation due to reduced proliferation of trophoblast cells [143], suggesting that both LIN28A and LIN28B are equally important in early placental development. Knockdown of LIN28A and LIN28B leads to an increase in *let-7* miRNAs and decrease in expression of proliferation-associated genes including insulin like growth factor 2 mRNA binding proteins (IGF2BP1-3), high mobility group AT-hook 1 (*HMGA1*), AT-rich interaction domain 3B (ARID3B), and MYC protooncogene (*c-MYC*) [143]. Additionally, overexpression of LIN28A or LIN28B in immortalized ovine trophoblast cells (iOTR) reduces *let-7* miRNAs, increases the expression of proliferation associated genes, and increases cell proliferation [143]. These findings further strengthen the data from in vitro studies about the role of LIN28-*let-7* miRNA axis in proliferation of human trophoblast cells.

7. LIN28-*let-7*-ARID3B Pathway in Trophoblast Cells

AT-rich interactive domain (ARID) proteins, first recognized in 1997, are a family of 15 proteins which binds to AT-rich regions of DNA [144,145]. ARID proteins play an important role in cell proliferation, differentiation, and development, and are upregulated in tumorous tissues [146]. The subfamilies of ARID proteins include ARID1, ARID2, ARID3, ARID4, ARID5, JARID1, and JARID2. The ARID3 subfamily has three members including ARID3A, ARID3B, and ARID3C. Although most of the ARID proteins act as tumor suppressors, ARID3A and ARID3B promote tumorigenesis [144]. ARID3A inhibits cell differentiation, promotes cell proliferation, and increases survival potential of cells [146,147], whereas ARID3B promotes proliferation, invasion, and migration of cancer cells [148–150]. However, ARID3B is more widely expressed in different tissues compared to ARID3A, suggesting more involvement of ARID3B in biological functions [145]. ARID3A and ARID3B are structurally similar and bind a similar region of DNA. Both ARID3A and ARID3B have an extended central ARID domain and two conserved amino acid domains at the C-terminal, termed REKLES α and REKLES β. Only members of the ARID3 subfamily have REKLES domains [151].

In contrast to ARID3B which is exclusively localized in the nucleus, ARID3A shuttles between the nucleus and cytoplasm. Once in the nucleus, ARID3A interacts with ARID3B through REKLES β domain. Therefore localization of ARID3A in the nucleus is dependent on its interaction with ARID3B in the nucleus, suggesting the dominant role of ARID3B [151,152]. In cancer cells, ARID3A and ARID3B recruit histone demethylase 4C (KDM4C) to make a tri-protein complex, called the ARID3B complex [153]. The ARID3B complex binds in the promoter areas of stemness genes and *let-7* target genes, leading to histone demethylation by KDM4C and increased gene expression by initiation of transcription [153]. Therefore, genes regulated by the ARID3B complex also include *let-7* miRNA target genes. Liao et al. further demonstrated that both ARID3A and ARID3B are targeted by *let-7* miRNAs. Hence, other than directly targeting the mRNAs of target genes, *let-7* miRNA can indirectly regulate their target genes by targeting and reducing the expression of *ARID3A* and *ARID3B* (Figure 4) [11,153].

Both ARID3A and ARID3B have high expression in human trophoblast cells [154,155]. ARID3A knockout mice have severe structural defects in placenta [156]. In ACH-3P cells, ARID3A, ARID3B, and KDM4C make the tri-protein ARID3B complex [11]. In term human placentas from IUGR pregnancies, LIN28A and LIN28B are low, *let-7* miRNAs are high, and ARID3A and ARID3B are low, which suggest a correlation between LIN28-*let-7* miRNA axis and the ARID3B complex [11]. Due to the well-established pathway of regulation of *let-7* target genes through ARID3B complex and their role in cell proliferation, it is important to understand this phenomenon in early placental development. We recently showed the correlation between LIN28-*let-7* miRNA axis and the ARID3B complex using ACH-3P and Sw.71 cells. Double knockout of LIN28A and LIN28B in ACH-3P cells increases *let-7* miRNAs and decreases the expression ARID3A, ARID3B, and KDM4C. Similarly, double knockin of LIN28A and LIN28B in Sw.71 cells decreases *let-7* miRNAs and increases expression of ARID3A, ARID3B, and KDM4C [11]. In trophoblast cells, the ARID3B complex binds to the promoter areas of proliferation-associated *let-7* target genes including *HMGA1, c-MYC, VEGF-A,* and *WNT1,* facilitating their transcription via KDM4C mediated histone demethylation. ARID3B knockout ACH-3P cells and KDM4C cannot be recruited in the promoter regions of *HMGA1, c-MYC, VEGF-A* and *WNT1,* and expression of these genes is also significantly reduced [73]. Moreover, ARID3B knockout ACH-3P cells have a reduced proliferation rate compared to control cells [11]. Knockdown of LIN28A or LIN28B in sheep trophectoderm increases *let-7* miRNAs and reduces the expression of ARID3A and ARID3B [143], showing regulation of the ARID3B complex by LIN28-*let-7* miRNA axis in vivo. Collectively these findings show that *let-7* miRNAs target the ARID3B complex in trophoblast cells, and the ARID3B complex regulates genes with known importance in placental development.

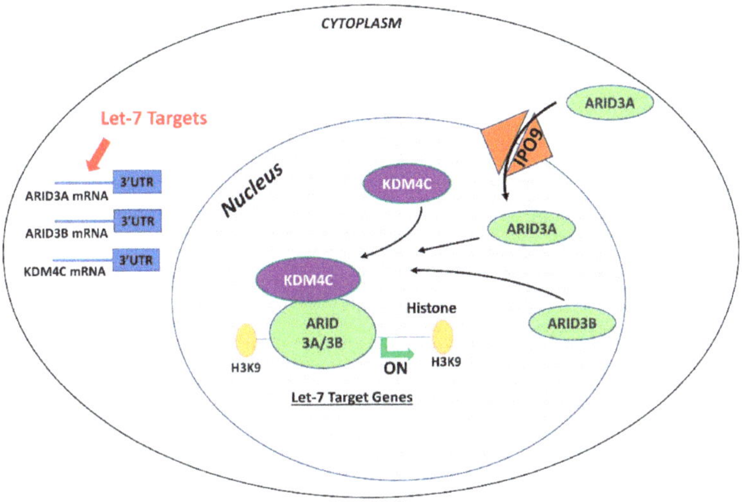

Figure 4. Gene regulation by the AT-rich interactive domain (ARID)3B complex. ARID3A is imported in the nucleus by importin 9 (IPO9), where it binds ARID3B and histone demethylase 4C (KDM4C) to form the ARID3B complex. The ARID3B complex binds in the promoter regions and activates transcription of *let-7* target genes. Other than directly binding the mRNA of their target genes, *let-7* miRNAs also target the ARID3B complex and reduce its expression, ultimately leading to reduced expression of *let-7* target genes.

8. Conclusions

For a long time, transcription activating proteins were thought to be the main regulators of gene expression in cells. However, in recent years microRNAs have emerged as "regulators of the regulators". miRNAs regulate important processes in trophoblast cells including cell proliferation, differentiation, invasion, and migration. Although identification of widespread *let-7* miRNA target genes makes them an important player in placental development, the role of other miRNAs and molecular pathways in placental development cannot be ignored. Although different studies have reported reduced LIN28 and high *let-7* miRNAs in term human placentas from preeclamptic and IUGR pregnancies, the exact cause of this dysregulation is not clear. Epigenetic modifications due to adverse uterine environment, random genetic mutations, prenatal insults like hypoxia, oxidative stress, maternal malnutrition, and gestational stress are some of the possible factors which can dysregulate LIN28-*let-7* axis in trophoblast cells. LIN28-*let-7*-ARID3B pathway regulates trophoblast cell proliferation by modulating the expression of proliferation associated genes in vitro and in vivo. The trophoblast cells with low LIN28 will have high *let-7* miRNAs and low ARID3B. There are two possible pathways of regulation of proliferation-associated genes by *let-7* miRNAs in trophoblast cells (Figure 5). One pathway involves binding of *let-7* miRNAs in 3′-UTR of their target mRNA leading to mRNA degradation or translational repression. Secondly, *let-7* miRNAs target ARID3A, ARID3B, and KDM4C to inhibit or reduce transcriptional activation of proliferation-associated genes by the ARID3B complex. Therefore, trophoblast cells with high *let-7* miRNAs will have reduced expression of proliferation factors and more of a tendency to differentiate. High *let-7* miRNAs during early placental development reduce cell proliferation, invasion, and migration of trophoblast cells, leading to placental abnormalities. *Let-7* miRNAs might be the major players in pathogenesis of placenta-associated disorders. miRNAs can be readily measured in peripheral blood, tissue biopsies, saliva, cerebrospinal fluid, urine, and other biological samples. High *let-7* miRNAs are upregulated in IUGR and preeclamptic placentas, suggesting that *let-7* miRNAs can be potential biomarkers for early diagnosis of PE and IUGR. It remains to be explored if early stage placentas from compromised pregnancies will have high *let-7* miRNAs and will the increase in *let-7* miRNAs in placenta be reflected in maternal blood. Animal models of IUGR can be used to answer these questions.

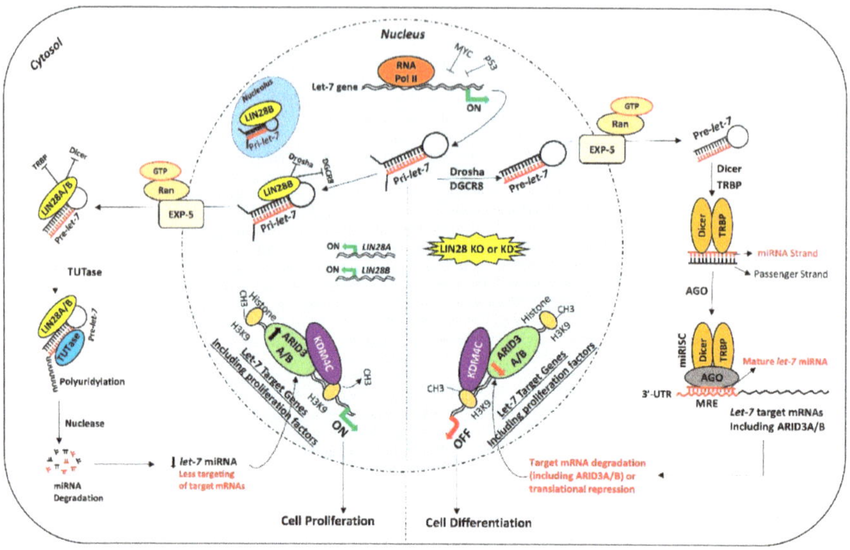

Figure 5. Proposed mechanism for gene regulation in trophoblast cells. Left panel of figure: LIN28 represses the biogenesis of mature *let-7* miRNAs by binding pri-*let-7* and pre-*let-7* miRNAs and inhibiting their processing. Due to the low level of mature *let-7* miRNAs in the cells, there will be less targeting of proliferation-associated genes. Moreover, the ARID3B complex will initiate the transcription of proliferation-associated genes and increase their expression. Increased expression of proliferation-associated genes will lead to increased cell proliferation. Right panel of figure: If LIN28 is knocked-out or knocked-down, there will be no suppression of *let-7* miRNA biogenesis. High *let-7* miRNAs will target and reduce the expression of proliferation associated genes and the ARDI3B complex, driving the cells towards differentiation.

Author Contributions: A.A. performed research, analyzed the data, and wrote the manuscript; Q.A.W., G.J.B., and R.V.A. edited the manuscript. All authors have read and agree to the published version of the manuscript.

Funding: This project was supported by Agriculture and Food Research Initiative Competitive Grant No. 2017-67015-26460 from the United States Department of Agriculture (USDA) National Institute of Food and Agriculture. This work was supported by the USDA National Institute of Food and Agriculture, Hatch Project COL00293D, accession number 1021217.

Conflicts of Interest: The authors declare no conflicts of interest.

References

1. Carter, A.M. Evolution of Placental Function in Mammals: The Molecular Basis of Gas and Nutrient Transfer, Hormone Secretion, and Immune Responses. *Physiol. Rev.* **2012**, *92*, 1543–1576. [CrossRef] [PubMed]
2. Gude, N.; Roberts, C.T.; Kalionis, B.; King, R.G. Growth and function of the normal human placenta. *Thromb. Res.* **2004**, *114*, 397–407. [CrossRef] [PubMed]
3. Hamilton, W.J.; Boyd, J.D. Development of the human placenta in the first three months of gestation. *J. Anat.* **1960**, *94*, 297–328.
4. Crocker, I.P.; Cooper, S.; Ong, S.C.; Baker, P.N. Differences in Apoptotic Susceptibility of Cytotrophoblasts and Syncytiotrophoblasts in Normal Pregnancy to Those Complicated with Preeclampsia and Intrauterine Growth Restriction. *Am. J. Pathol.* **2003**, *162*, 637–643. [CrossRef]

5. Longtine, M.S.; Chen, B.; Odibo, A.; Zhong, Y.; Nelson, D. Villous trophoblast apoptosis is elevated and restricted to cytotrophoblasts in pregnancies complicated by preeclampsia, IUGR, or preeclampsia with IUGR. *Placenta* **2012**, *33*, 352–359. [CrossRef]
6. Burton, G.J.; Fowden, A.L.; Thornburg, K.L. Placental Origins of Chronic Disease. *Physiol. Rev.* **2016**, *96*, 1509–1565. [CrossRef]
7. Barker, D.J.P. The developmental origins of well-being. *Philos. Trans. R. Soc. B Biol. Sci.* **2004**, *359*, 1359–1366. [CrossRef] [PubMed]
8. Barker, D.J.P. The developmental origins of chronic adult disease. *Acta Paediatr.* **2004**, *93*, 26–33. [CrossRef]
9. Jiang, S. A Regulator of Metabolic Reprogramming: MicroRNA Let-7. *Transl. Oncol.* **2019**, *12*, 1005–1013. [CrossRef]
10. Boyerinas, B.; Park, S.-M.; Hau, A.; Murmann, A.E.; Peter, M.E. The role of let-7 in cell differentiation and cancer. *Endocr.-Relat. Cancer* **2010**, *17*, F19–F36. [CrossRef]
11. Ali, A.; Anthony, R.V.; Bouma, G.J.; Winger, Q.A. LIN28-let-7 axis regulates genes in immortalized human trophoblast cells by targeting the ARID3B-complex. *FASEB J.* **2019**, *33*, 12348–12363. [CrossRef] [PubMed]
12. Hunter, S.E.; Finnegan, E.F.; Zisoulis, D.G.; Lovci, M.T.; Melnik-Martinez, K.V.; Yeo, G.W.; Pasquinelli, A.E. Functional Genomic Analysis of the let-7 Regulatory Network in Caenorhabditis elegans. *PLoS Genet.* **2013**, *9*, e1003353. [CrossRef] [PubMed]
13. Wang, X.; Cao, L.; Wang, Y.; Wang, X.; Liu, N.; You, Y. Regulation of let-7 and its target oncogenes (Review). *Oncol. Lett.* **2012**, *3*, 955–960. [CrossRef] [PubMed]
14. Canfield, J.; Arlıer, S.; Mong, E.F.; Lockhart, J.; Van Wye, J.; Guzeloglu-Kayisli, O.; Schatz, F.; Magness, R.R.; Lockwood, C.J.; Tsibris, J.C.M.; et al. Decreased LIN28B in preeclampsia impairs human trophoblast differentiation and migration. *FASEB J.* **2018**, *33*, 2759–2769. [CrossRef]
15. Bazer, F.W.; Spencer, T.E.; Johnson, G.A.; Burghardt, R.C.; Wu, G. Comparative aspects of implantation. *Reproduction* **2009**, *138*, 195–209. [CrossRef]
16. Herzog, M. A contribution to our knowledge of the earliest known stages of placentation and embryonic development in man. *Am. J. Anat.* **2005**, *9*, 361–400. [CrossRef]
17. Pötgens, A.; Schmitz, U.; Bose, P.; Versmold, A.; Kaufmann, P.; Frank, H.-G. Mechanisms of Syncytial Fusion: A Review. *Placenta* **2002**, *23*, S107–S113. [CrossRef]
18. Kim, S.-M.; Kim, J.-S. A Review of Mechanisms of Implantation. *Dev. Reprod.* **2017**, *21*, 351–359. [CrossRef]
19. Benirschke, K.; Burton, G.J.; Baergen, R.N. *The Pathology of the Human Placenta*; Springer Science and Business Media LLC: Berlin/Heidelberg, Germany, 2012; pp. 97–571.
20. Frank, H.-G. 10-Placental Development. In *Fetal and Neonatal Physiology*, 5th ed.; Polin, R.A., Abman, S.H., Rowitch, D.H., Benitz, W.E., Fox, W.W., Eds.; Elsevier: Amsterdam, The Netherlands, 2017; pp. 101–113.
21. Chen, D.-B.; Zheng, J. Regulation of placental angiogenesis. *Microcirculation* **2014**, *21*, 15–25. [CrossRef]
22. Vicovac, L. Trophoblast differentiation during formation of anchoring villi in a model of the early human placenta in vitro. *Placenta* **1995**, *16*, 41–56. [CrossRef]
23. Damsky, C.H.; Fitzgerald, M.L.; Fisher, S.J. Distribution patterns of extracellular matrix components and adhesion receptors are intricately modulated during first trimester cytotrophoblast differentiation along the invasive pathway, in vivo. *J. Clin. Investig.* **1992**, *89*, 210–222. [CrossRef]
24. Prakobphol, A.; Genbacev, O.; Gormley, M.; Kapidzic, M.; Fisher, S.J. A role for the L-selectin adhesion system in mediating cytotrophoblast emigration from the placenta. *Dev. Biol.* **2006**, *298*, 107–117. [CrossRef] [PubMed]
25. Kemp, B.; Kertschanska, S.; Kadyrov, M.; Rath, W.; Kaufmann, P.; Huppertz, B. Invasive depth of extravillous trophoblast correlates with cellular phenotype:A comparison of intra- and extrauterine implantation sites. *Histochem. Cell Biol.* **2002**, *117*, 401–414. [CrossRef]
26. Espinoza, J.; Romero, R.; Kim, Y.M.; Kusanovic, J.P.; Hassan, S.; Erez, O.; Gotsch, F.; Than, N.G.; Papp, Z.; Kim, C.J. Normal and abnormal transformation of the spiral arteries during pregnancy. *J. Périnat. Med.* **2006**, *34*, 447–458. [CrossRef] [PubMed]
27. Anin, S.; Vince, G.; Quenby, S. Trophoblast invasion. *Hum. Fertil.* **2004**, *7*, 169–174. [CrossRef] [PubMed]
28. Pijnenborg, R.; Vercruysse, L.; Hanssens, M. The Uterine Spiral Arteries in Human Pregnancy: Facts and Controversies. *Placenta* **2006**, *27*, 939–958. [CrossRef]
29. Lyall, F.; Bulmer, J.N.; Duffie, E.; Cousins, F.; Thériault, A.; Robson, S.C. Human Trophoblast Invasion and Spiral Artery Transformation. *Am. J. Pathol.* **2001**, *158*, 1713–1721. [CrossRef]

30. Red-Horse, K.; Zhou, Y.; Genbacev, O.; Prakobphol, A.; Foulk, R.; McMaster, M.; Fisher, S.J. Trophoblast differentiation during embryo implantation and formation of the maternal-fetal interface. *J. Clin. Investig.* **2004**, *114*, 744–754. [CrossRef]
31. Reynolds, L.; Redmer, D.A. Angiogenesis in the Placenta1. *Biol. Reprod.* **2001**, *64*, 1033–1040. [CrossRef]
32. Kobayashi, H.; Tomari, Y. RISC assembly: Coordination between small RNAs and Argonaute proteins. *Biochim. Biophys. Acta (BBA)-Gene Regul. Mech.* **2016**, *1859*, 71–81. [CrossRef]
33. Braun, J.E.; Truffault, V.; Boland, A.; Huntzinger, E.; Chang, C.-T.; Haas, G.; Weichenrieder, O.; Coles, M.; Izaurralde, E. A direct interaction between DCP1 and XRN1 couples mRNA decapping to 5′ exonucleolytic degradation. *Nat. Struct. Mol. Biol.* **2012**, *19*, 1324–1331. [CrossRef]
34. Christie, M.; Boland, A.; Huntzinger, E.; Weichenrieder, O.; Izaurralde, E. Structure of the PAN3 Pseudokinase Reveals the Basis for Interactions with the PAN2 Deadenylase and the GW182 Proteins. *Mol. Cell* **2013**, *51*, 360–373. [CrossRef] [PubMed]
35. Behm-Ansmant, I.; Rehwinkel, J.; Doerks, T.; Stark, A.; Bork, P.; Izaurralde, E. mRNA degradation by miRNAs and GW182 requires both CCR4:NOT deadenylase and DCP1:DCP2 decapping complexes. *Genome Res.* **2006**, *20*, 1885–1898. [CrossRef] [PubMed]
36. Bartel, D.P. MicroRNAs:Genomics, biogenesis, mechanism, and function. *Cell* **2004**, *116*, 281–297. [CrossRef]
37. Doridot, L.; Miralles, F.; Barbaux, S.; Vaiman, D. Trophoblasts, invasion, and microRNA. *Front. Genet.* **2013**, *4*, 4. [CrossRef] [PubMed]
38. Zhou, W.; Wang, H.; Yang, J.; Long, W.; Zhang, B.; Liu, J.; Yu, B. Down-regulated circPAPPA suppresses the proliferation and invasion of trophoblast cells via the miR-384/STAT3 pathway. *Biosci. Rep.* **2019**, *39*. [CrossRef]
39. Li, J.; Wang, J.M.; Liu, Y.H.; Zhang, Z.; Han, N.; Xue, S.H.; Wang, P. Effect of microRNA-106b on the invasion and proliferation of trophoblasts through targeting MMP-2. *Zhonghua Fu Chan Ke Za Zhi* **2017**, *52*, 327–332.
40. Liu, F.; Wu, K.; Wu, W.; Chen, Y.; Wu, H.; Wang, H.; Zhang, W. miR-203 contributes to pre-eclampsia via inhibition of VEGFA expression. *Mol. Med. Rep.* **2018**, *17*, 5627–5634. [CrossRef]
41. Jiang, L.; Long, A.; Tan, L.; Hong, M.; Wu, J.; Cai, L.; Li, Q. Elevated microRNA-520g in pre-eclampsia inhibits migration and invasion of trophoblasts. *Placenta* **2017**, *51*, 70–75. [CrossRef]
42. Wang, R.; Liu, W.; Liu, X.; Liu, X.; Tao, H.; Wu, D.; Zhao, Y.; Zou, L. MicroRNA-210 regulates human trophoblast cell line HTR-8/SVneo function by attenuating Notch1 expression: Implications for the role of microRNA-210 in pre-eclampsia. *Mol. Reprod. Dev.* **2019**, *86*, 896–907. [CrossRef]
43. Yuan, Y.; Wang, X.; Sun, Q.; Dai, X.; Cai, Y. MicroRNA-16 is involved in the pathogenesis of pre-eclampsia via regulation of Notch2. *J. Cell. Physiol.* **2019**, *235*, 4530–4544. [CrossRef] [PubMed]
44. Xie, N.; Jia, Z.; Li, L. miR-320a upregulation contributes to the development of preeclampsia by inhibiting the growth and invasion of trophoblast cells by targeting interleukin 4. *Mol. Med. Rep.* **2019**, *20*, 3256–3264. [CrossRef] [PubMed]
45. Liu, R.; Meng, Q.; Shi, Y.-P.; Xu, H.-S. Regulatory role of microRNA-320a in the proliferation, migration, invasion, and apoptosis of trophoblasts and endothelial cells by targeting estrogen-related receptor γ. *J. Cell. Physiol.* **2018**, *234*, 682–691. [CrossRef] [PubMed]
46. Li, L.; Huang, X.; He, Z.; Xiong, Y.; Fang, Q. miRNA-210-3p regulates trophoblast proliferation and invasiveness through fibroblast growth factor 1 in selective intrauterine growth restriction. *J. Cell. Mol. Med.* **2019**, *23*, 4422–4433. [CrossRef]
47. Shih, J.-C.; Lin, H.-H.; Hsiao, A.-C.; Su, Y.-T.; Tsai, S.; Chien, C.-L.; Kung, H.-N. Unveiling the role of microRNA-7 in linking TGF-β-Smad-mediated epithelial-mesenchymal transition with negative regulation of trophoblast invasion. *FASEB J.* **2019**, *33*, 6281–6295. [CrossRef] [PubMed]
48. Chen, Y.J.; Wu, P.Y.; Gao, R.Q. MiR-218 inhibits HTR-8 cells migration and invasion by targeting SOX4. *Zhongguo Ying Yong Sheng Li Xue Za Zhi* **2017**, *33*, 169–173. [PubMed]
49. Xue, F.; Yang, J.; Li, Q.; Zhou, H. Down-regulation of microRNA-34a-5p promotes trophoblast cell migration and invasion via targetting Smad4. *Biosci. Rep.* **2019**, *39*, 39. [CrossRef]
50. Zhou, X.; Li, Q.; Xu, J.; Zhang, X.; Zhang, H.; Xiang, Y.; Fang, C.; Wang, T.; Xia, S.; Zhang, Q.; et al. The aberrantly expressed miR-193b-3p contributes to preeclampsia through regulating transforming growth factor-β signaling. *Sci. Rep.* **2016**, *6*. [CrossRef]
51. Sun, M.; Chen, H.; Liu, J.; Tong, C.; Meng, T. MicroRNA-34a inhibits human trophoblast cell invasion by targeting MYC. *BMC Cell Biol.* **2015**, *16*. [CrossRef]

52. Ding, J.; Huang, F.; Wu, G.; Han, T.; Xu, F.; Weng, D.; Wu, C.; Zhang, X.; Yao, Y.; Zhu, X. MiR-519d-3p Suppresses Invasion and Migration of Trophoblast Cells via Targeting MMP-2. *PLoS ONE* **2015**, *10*, e0120321. [CrossRef]
53. Zhang, S.; Wang, Y.; Li, J.; Zhong, Q.; Li, Y. MiR-101 inhibits migration and invasion of trophoblast HTR-8/SVneo cells by targeting CXCL6 in preeclampsia. *Minerva Med.* **2019**. [CrossRef]
54. Liu, J.-J.; Zhang, L.; Zhang, F.-F.; Luan, T.; Yin, Z.-M.; Rui, C.; Ding, H.-J. Influence of miR-34a on preeclampsia through the Notch signaling pathway. *Eur. Rev. Med. Pharmacol. Sci.* **2019**, *23*, 923–931. [PubMed]
55. Yang, X.; Meng, T. MicroRNA-431 affects trophoblast migration and invasion by targeting ZEB1 in preeclampsia. *Gene* **2018**, *683*, 225–232. [CrossRef] [PubMed]
56. Wen, Z.; Chen, Y.; Long, Y.; Yu, J.; Li, M. Tumor necrosis factor-alpha suppresses the invasion of HTR-8/SVneo trophoblast cells through microRNA-145-5p-mediated downregulation of Cyr61. *Life Sci.* **2018**, *209*, 132–139. [CrossRef]
57. Zou, A.-X.; Chen, B.; Li, Q.-X.; Liang, Y.-C. MiR-134 inhibits infiltration of trophoblast cells in placenta of patients with preeclampsia by decreasing ITGB1 expression. *Eur. Rev. Med. Pharmacol. Sci.* **2018**, *22*, 2199–2206.
58. Ding, J.; Cheng, Y.; Zhang, Y.; Liao, S.; Yin, T.; Yang, J. The miR-27a-3p/USP25 axis participates in the pathogenesis of recurrent miscarriage by inhibiting trophoblast migration and invasion. *J. Cell. Physiol.* **2019**, *234*, 19951–19963. [CrossRef]
59. Wang, N.; Feng, Y.; Xu, J.; Zou, J.; Chen, M.; He, Y.; Liu, H.; Xue, M.; Feng, Y.-L. miR-362-3p regulates cell proliferation, migration and invasion of trophoblastic cells under hypoxia through targeting Pax3. *Biomed. Pharmacother.* **2018**, *99*, 462–468. [CrossRef]
60. Wu, L.; Song, W.-Y.; Xie, Y.; Hu, L.-L.; Hou, X.-M.; Wang, R.; Gao, Y.; Zhang, J.-N.; Zhang, L.; Li, W.-W.; et al. miR-181a-5p suppresses invasion and migration of HTR-8/SVneo cells by directly targeting IGF2BP2. *Cell Death Dis.* **2018**, *9*, 16. [CrossRef]
61. Peng, H.-Y.; Li, M.-Q.; Li, H.-P. MiR-137 Restricts the Viability and Migration of HTR-8/SVneo Cells by Downregulating FNDC5 in Gestational Diabetes Mellitus. *Curr. Mol. Med.* **2019**, *19*, 494–505. [CrossRef]
62. Qian, S.; Liu, R. miR-30b facilitates preeclampsia through targeting MXRA5 to inhibit the viability, invasion and apoptosis of placental trophoblast cells. *Int. J. Clin. Exp. Pathol.* **2019**, *12*, 4057–4065.
63. Niu, Z.-R.; Han, T.; Sun, X.-L.; Luan, L.-X.; Gou, W.; Zhu, X. MicroRNA-30a-3p is overexpressed in the placentas of patients with preeclampsia and affects trophoblast invasion and apoptosis by its effects on IGF-1. *Am. J. Obstet. Gynecol.* **2018**, *218*, 249. [CrossRef] [PubMed]
64. Wang, S.; Wang, X.; Weng, Z.; Zhang, S.; Ning, H.; Li, B. Expression and role of microRNA 18b and hypoxia inducible factor-1α in placental tissues of preeclampsia patients. *Exp. Ther. Med.* **2017**, *14*, 4554–4560. [CrossRef] [PubMed]
65. Gao, Y.; She, R.; Wang, Q.; Li, Y.; Zhang, H. Up-regulation of miR-299 suppressed the invasion and migration of HTR-8/SVneo trophoblast cells partly via targeting HDAC2 in pre-eclampsia. *Biomed. Pharmacother.* **2018**, *97*, 1222–1228. [CrossRef] [PubMed]
66. Shi, Z.; She, K.; Li, H.; Yuan, X.; Han, X.; Wang, Y. MicroRNA-454 contributes to sustaining the proliferation and invasion of trophoblast cells through inhibiting Nodal/ALK7 signaling in pre-eclampsia. *Chem. Interact.* **2019**, *298*, 8–14. [CrossRef] [PubMed]
67. Chi, Z.; Zhang, M. Exploration of the regulation and control mechanisms of miR-145 in trophoblast cell proliferation and invasion. *Exp. Ther. Med.* **2018**, *16*, 5298–5304. [CrossRef] [PubMed]
68. Yang, Y.; Li, H.; Ma, Y.; Zhu, X.; Zhang, S.; Li, J. MiR-221-3p is down-regulated in preeclampsia and affects trophoblast growth, invasion and migration partly via targeting thrombospondin 2. *Biomed. Pharmacother.* **2019**, *109*, 127–134. [CrossRef] [PubMed]
69. Zhao, S.; Wang, J.; Cao, Z.; Gao, L.; Zheng, Y.; Wang, J.; Liu, X. miR-126a-3p induces proliferation, migration and invasion of trophoblast cells in pre-eclampsia-like rats by inhibiting A Disintegrin and Metalloprotease 9. *Biosci. Rep.* **2019**, *39*. [CrossRef]
70. Li, J.; Fu, Z.; Jiang, H.; Chen, L.; Wu, X.; Ding, H.; Xia, Y.; Wang, X.; Tang, Q.; Wu, W. IGF2-derived miR-483-3p contributes to macrosomia through regulating trophoblast proliferation by targeting RB1CC1. *Mol. Hum. Reprod.* **2018**, *24*, 444–452. [CrossRef]

71. Xiao, J.; Tao, T.; Yin, Y.; Zhao, L.; Yang, L.; Hu, L. miR-144 may regulate the proliferation, migration and invasion of trophoblastic cells through targeting PTEN in preeclampsia. *Biomed. Pharmacother.* **2017**, *94*, 341–353. [CrossRef]
72. Liu, M.; Wang, Y.; Lu, H.; Wang, H.; Shi, X.; Shao, X.; Li, Y.-X.; Zhao, Y.; Wang, Y.-L. miR-518b Enhances Human Trophoblast Cell Proliferation Through Targeting Rap1b and Activating Ras-MAPK Signal. *Front. Endocrinol.* **2018**, *9*, 100. [CrossRef]
73. Brkić, J.; Dunk, C.; O'Brien, J.; Fu, G.; Nadeem, L.; Wang, Y.-L.; Rosman, D.; Salem, M.; Shynlova, O.; Yougbaré, I.; et al. MicroRNA-218-5p Promotes Endovascular Trophoblast Differentiation and Spiral Artery Remodeling. *Mol. Ther.* **2018**, *26*, 2189–2205. [CrossRef] [PubMed]
74. Wang, H.; Zhao, Y.; Luo, R.; Bian, X.; Wang, Y.; Shao, X.; Li, Y.-X.; Liu, M.; Wang, Y.-L. A positive feedback self-regulatory loop between miR-210 and HIF-1α mediated by CPEB2 is involved in trophoblast syncytiolization:Implication of trophoblast malfunction in preeclampsia. *Biol. Reprod.* **2019**. [CrossRef]
75. Kumar, P.; Luo, Y.; Tudela, C.; Alexander, J.M.; Mendelson, C.R. The c-Myc-Regulated MicroRNA-17~92 (miR-17~92) and miR-106a~363 Clusters Target hCYP19A1 and hGCM1 To Inhibit Human Trophoblast Differentiation. *Mol. Cell. Biol.* **2013**, *33*, 1782–1796. [CrossRef] [PubMed]
76. Cheng, D.; Jiang, S.; Chen, J.; Li, J.; Ao, L.; Zhang, Y. Upregulated long noncoding RNA Linc00261 in pre-eclampsia and its effect on trophoblast invasion and migration via regulating miR-558/TIMP4 signaling pathway. *J. Cell. Biochem.* **2019**, *120*, 13243–13253. [CrossRef]
77. Wang, X.; Peng, S.; Cui, K.; Hou, F.; Ding, J.; Li, A.; Wang, M.; Geng, L. MicroRNA-576-5p enhances the invasion ability of trophoblast cells in preeclampsia by targeting TFAP2A. *Mol. Genet. Genom. Med.* **2019**, *8*, e1025. [CrossRef] [PubMed]
78. Zhang, Y.; Zhou, J.; Li, M.-Q.; Xu, J.; Zhang, J.-P.; Jin, L.-P. MicroRNA-184 promotes apoptosis of trophoblast cells via targeting WIG1 and induces early spontaneous abortion. *Cell Death Dis.* **2019**, *10*, 223. [CrossRef] [PubMed]
79. Zhang, W.-M.; Cao, P.; Xin, L.; Zhang, Y.; Liu, Z.; Yao, N.; Ma, Y.-Y. Effect of miR-133 on apoptosis of trophoblasts in human placenta tissues via Rho/ROCK signaling pathway. *Eur. Rev. Med. Pharmacol. Sci.* **2019**, *23*, 10600–10608. [PubMed]
80. Zhang, L.; Yuan, J.-M.; Zhao, R.-H.; Wang, L.-M.; Tu, Z.-B. Correlation of MiR-152 expression with VEGF expression in placental tissue of preeclampsia rat and its influence on apoptosis of trophoblast cells. *Eur. Rev. Med. Pharmacol. Sci.* **2019**, *23*, 3553–3560. [PubMed]
81. Li, X.; Lu, J.; Dong, L.; Lv, F.; Liu, W.; Liu, G.; Zhu, W.; Diao, X. Effects of MiR-155 on trophoblast apoptosis in placental tissues of preeclampsia rats through HIF-1α signaling pathway. *Panminerva Med.* **2019**. [CrossRef]
82. Du, E.; Cao, Y.; Feng, C.; Lu, J.; Yang, H.; Zhang, Y. The Possible Involvement of miR-371a-5p Regulating XIAP in the Pathogenesis of Recurrent Pregnancy Loss. *Reprod. Sci.* **2019**, *26*, 1468–1475. [CrossRef]
83. Dong, X.; Yang, L.; Wang, H. miR-520 promotes DNA-damage-induced trophoblast cell apoptosis by targeting PARP1 in recurrent spontaneous abortion (RSA). *Gynecol. Endocrinol.* **2016**, *33*, 274–278. [CrossRef] [PubMed]
84. Guo, M.; Zhao, X.; Yuan, X.; Li, P. Elevated microRNA-34a contributes to trophoblast cell apoptosis in preeclampsia by targeting BCL-2. *J. Hum. Hypertens.* **2017**, *31*, 815–820. [CrossRef] [PubMed]
85. Yang, Y.; Zhang, S.; Li, Y.; Han, B.; Ma, Y. Inhibition of miR-18a increases expression of estrogen receptor 1 and promotes apoptosis in human HTR8 trophoblasts. *Xi Bao Yu Fen Zi Mian Yi Xue Za Zhi* **2017**, *33*, 1102–1107. [PubMed]
86. West, R.; Russ, J.E.; Bouma, G.J.; Winger, Q.A. BRCA1 regulates HMGA2 levels in the Swan71 trophoblast cell line. *Mol. Reprod. Dev.* **2019**, *86*, 1663–1670. [CrossRef]
87. Li, L.; Hou, A.; Gao, X.; Zhang, J.; Zhang, L.; Wang, J.; Li, H.; Song, Y. Lentivirus-mediated miR-23a overexpression induces trophoblast cell apoptosis through inhibiting X-linked inhibitor of apoptosis. *Biomed. Pharmacother.* **2017**, *94*, 412–417. [CrossRef]
88. Mao, Z.; Yao, M.; Li, Y.; Fu, Z.; Li, S.; Zhang, L.; Zhou, Z.; Tang, Q.; Han, X.; Xia, Y. miR-96-5p and miR-101-3p as potential intervention targets to rescue TiO2 NP-induced autophagy and migration impairment of human trophoblastic cells. *Biomater. Sci.* **2018**, *6*, 3273–3283. [CrossRef]
89. Zhang, X.; Ge, Y.-W.; Wang, Z.-X.; Xu, Q.-L.; Guo, R.; Xu, H.-Y. MiR-200c regulates apoptosis of placental trophoblasts in preeclampsia rats through Wnt/β-catenin signaling pathway. *Eur. Rev. Med. Pharmacol. Sci.* **2019**, *23*, 7209–7216.

90. Gu, Y.; Meng, J.; Zuo, C.; Wang, S.; Li, H.; Zhao, S.; Huang, T.; Wang, X.-T.; Yan, J. Downregulation of MicroRNA-125a in Placenta Accreta Spectrum Disorders Contributes Antiapoptosis of Implantation Site Intermediate Trophoblasts by Targeting MCL1. *Reprod. Sci.* **2019**, *26*, 1582–1589. [CrossRef]
91. Reinhart, B.J.; Slack, F.J.; Basson, M.; Pasquinelli, A.E.; Bettinger, J.C.; Rougvie, A.E.; Horvitz, H.R.; Ruvkun, G. The 21-nucleotide let-7 RNA regulates developmental timing in Caenorhabditis elegans. *Nature* **2000**, *403*, 901–906. [CrossRef]
92. Liu, S.; Xia, Q.; Zhao, P.; Cheng, T.; Hong, K.; Xiang, Z. Characterization and expression patterns of let-7 microRNA in the silkworm (Bombyx mori). *BMC Dev. Biol.* **2007**, *7*, 88. [CrossRef]
93. Pasquinelli, A.E.; Reinhart, B.J.; Slack, F.J.; Martindale, M.Q.; Kuroda, M.I.; Maller, B.; Hayward, D.C.; Ball, E.; Degnan, B.M.; Müller, P.; et al. Conservation of the sequence and temporal expression of let-7 heterochronic regulatory RNA. *Nature* **2000**, *408*, 86–89. [CrossRef] [PubMed]
94. Landgraf, P.; Rusu, M.; Sheridan, R.; Sewer, A.; Iovino, N.; Aravin, A.; Pfeffer, S.; Rice, A.; Kamphorst, A.O.; Landthaler, M.; et al. A Mammalian microRNA Expression Atlas Based on Small RNA Library Sequencing. *Cell* **2007**, *129*, 1401–1414. [CrossRef] [PubMed]
95. Roush, S.; Slack, F.J. The let-7 family of microRNAs. *Trends Cell Biol.* **2008**, *18*, 505–516. [CrossRef] [PubMed]
96. Lee, H.; Han, S.; Kwon, C.S.; Lee, D. Biogenesis and regulation of the let-7 miRNAs and their functional implications. *Protein Cell* **2015**, *7*, 100–113. [CrossRef]
97. Zhang, H.; Artiles, K.L.; Fire, A.Z. Functional relevance of "seed" and "non-seed" sequences in microRNA-mediated promotion of C. elegans developmental progression. *RNA* **2015**, *21*, 1980–1992. [CrossRef]
98. Kehl, T.; Backes, C.; Kern, F.; Fehlmann, T.; Ludwig, N.; Meese, E.; Lenhof, H.-P.; Keller, A. About miRNAs, miRNA seeds, target genes and target pathways. *Oncotarget* **2017**, *8*, 107167–107175. [CrossRef]
99. Shyh-Chang, N.; Zhu, H.; De Soysa, T.Y.; Shinoda, G.; Seligson, M.T.; Tsanov, K.M.; Nguyen, L.; Asara, J.M.; Cantley, L.C.; Daley, G.Q. Lin28 enhances tissue repair by reprogramming cellular metabolism. *Cell* **2013**, *155*, 778–792. [CrossRef]
100. Shyh-Chang, N.; Daley, G.Q. Lin28:Primal regulator of growth and metabolism in stem cells. *Cell Stem Cell* **2013**, *12*, 395–406. [CrossRef]
101. Caygill, E.E.; Johnston, L.A. Temporal Regulation of Metamorphic Processes in Drosophila by the let-7 and miR-125 Heterochronic MicroRNAs. *Curr. Biol.* **2008**, *18*, 943–950. [CrossRef]
102. Wagner, S.; Ngezahayo, A.; Murua Escobar, H.; Nolte, I. Role of miRNA let-7 and its major targets in prostate cancer. *Biomed Res. Int.* **2014**, *2014*, 14. [CrossRef]
103. Takamizawa, J.; Chamoto, K.; Tsuji, T.; Funamoto, H.; Kosaka, A.; Matsuzaki, J.; Sato, T.; Konishi, H.; Fujio, K.; Yamamoto, K.; et al. Reduced Expression of the let-7 MicroRNAs in Human Lung Cancers in Association with Shortened Postoperative Survival. *Cancer Res.* **2004**, *64*, 3753–3756. [CrossRef] [PubMed]
104. Boyerinas, B.; Park, S.-M.; Shomron, N.; Hedegaard, M.M.; Vinther, J.; Andersen, J.S.; Feig, C.; Xu, J.; Burge, C.B.; Peter, M.E. Identification of Let-7-Regulated Oncofetal Genes. *Cancer Res.* **2008**, *68*, 2587–2591. [CrossRef]
105. Copley, M.R.; Babovic, S.; Benz, C.; Knapp, D.J.; Beer, P.A.; Kent, D.; Wohrer, S.; Treloar, D.Q.; Day, C.; Rowe, K.; et al. The Lin28b–let-7–Hmga2 axis determines the higher self-renewal potential of fetal haematopoietic stem cells. *Nat. Cell Biol.* **2013**, *15*, 916–925. [CrossRef] [PubMed]
106. Emmrich, S.; Rasche, M.; Schöning, J.; Reimer, C.; Keihani, S.; Maroz, A.; Xie, Y.; Li, Z.; Schambach, A.; Reinhardt, D.; et al. miR-99a/100~125b tricistrons regulate hematopoietic stem and progenitor cell homeostasis by shifting the balance between TGFβ and Wnt signaling. *Genes Dev.* **2014**, *28*, 858–874. [CrossRef] [PubMed]
107. Heo, I.; Ha, M.; Lim, J.; Yoon, M.-J.; Park, J.-E.; Kwon, S.C.; Chang, H.; Kim, V.N. Mono-Uridylation of Pre-MicroRNA as a Key Step in the Biogenesis of Group II let-7 MicroRNAs. *Cell* **2012**, *151*, 521–532. [CrossRef] [PubMed]
108. Büssing, I.; Slack, F.J.; Großhans, H. let-7 microRNAs in development, stem cells and cancer. *Trends Mol. Med.* **2008**, *14*, 400–409. [CrossRef]
109. Thomson, J.M.; Newman, M.; Parker, J.S.; Morin-Kensicki, E.M.; Wright, T.; Hammond, S.M. Extensive post-transcriptional regulation of microRNAs and its implications for cancer. *Genome Res.* **2006**, *20*, 2202–2207. [CrossRef]
110. Nam, Y.; Chen, C.; Gregory, R.I.; Chou, J.J.; Sliz, P. Molecular Basis for Interaction of let-7 MicroRNAs with Lin28. *Cell* **2011**, *147*, 1080–1091. [CrossRef]

111. Zhang, P.; Elabd, S.; Hammer, S.; Solozobova, V.; Yan, H.; Bartel, F.; Inoue, S.; Henrich, T.; Wittbrodt, J.; Loosli, F.; et al. TRIM25 has a dual function in the p53/Mdm2 circuit. *Oncogene* **2015**, *34*, 5729–5738. [CrossRef]
112. Treiber, T.; Treiber, N.; Meister, G. Publisher Correction: Regulation of microRNA biogenesis and its crosstalk with other cellular pathways. *Nat. Rev. Mol. Cell Biol.* **2019**, *20*, 321. [CrossRef]
113. Guo, Y.; Chen, Y.; Ito, H.; Watanabe, A.; Ge, S.X.; Kodama, T.; Aburatani, H. Identification and characterization of lin-28 homolog B (LIN28B) in human hepatocellular carcinoma. *Gene* **2006**, *384*, 51–61. [CrossRef] [PubMed]
114. Tsialikas, J.; Romer-Seibert, J. LIN28: Roles and regulation in development and beyond. *Development* **2015**, *142*, 2397–2404. [CrossRef] [PubMed]
115. Yu, J.; Vodyanik, M.A.; Smuga-Otto, K.; Antosiewicz-Bourget, J.; Frane, J.L.; Tian, S.; Nie, J.; Jonsdottir, G.A.; Ruotti, V.; Stewart, R.; et al. Induced Pluripotent Stem Cell Lines Derived from Human Somatic Cells. *Science* **2007**, *318*, 1917–1920. [CrossRef] [PubMed]
116. Jiang, S.; Baltimore, D. RNA-binding protein Lin28 in cancer and immunity. *Cancer Lett.* **2016**, *375*, 108–113. [CrossRef] [PubMed]
117. Shinoda, G.; Shyh-Chang, N.; De Soysa, T.Y.; Zhu, H.; Seligson, M.T.; Shah, S.P.; Abo-Sido, N.; Yabuuchi, A.; Hagan, J.P.; Gregory, R.I.; et al. Fetal deficiency of lin28 programs life-long aberrations in growth and glucose metabolism. *Stem Cells* **2013**, *31*, 1563–1573. [CrossRef]
118. Mayr, F.; Heinemann, U. Mechanisms of Lin28-Mediated miRNA and mRNA Regulation—A Structural and Functional Perspective. *Int. J. Mol. Sci.* **2013**, *14*, 16532–16553. [CrossRef]
119. Zhang, J.; Ratanasirintrawoot, S.; Chandrasekaran, S.; Wu, Z.; Ficarro, S.B.; Yu, C.; Ross, C.A.; Cacchiarelli, D.; Xia, Q.; Seligson, M.; et al. LIN28 Regulates Stem Cell Metabolism and Conversion to Primed Pluripotency. *Cell Stem Cell* **2016**, *19*, 66–80. [CrossRef]
120. Viswanathan, S.; Daley, G.Q.; Gregory, R. Selective Blockade of MicroRNA Processing by Lin28. *Science* **2008**, *320*, 97–100. [CrossRef]
121. Heo, I.; Joo, C.; Cho, J.; Ha, M.; Han, J.; Kim, V.N. Lin28 Mediates the Terminal Uridylation of let-7 Precursor MicroRNA. *Mol. Cell* **2008**, *32*, 276–284. [CrossRef]
122. Piskounova, E.; Polytarchou, C.; Thornton, J.E.; Lapierre, R.J.; Pothoulakis, C.; Hagan, J.P.; Iliopoulos, D.; Gregory, R. Lin28A and Lin28B Inhibit let-7 MicroRNA Biogenesis by Distinct Mechanisms. *Cell* **2011**, *147*, 1066–1079. [CrossRef]
123. Molenaar, J.J.; Domingo-Fernández, R.; Ebus, M.E.; Lindner, S.; Koster, J.; Drabek, K.; Mestdagh, P.; van Sluis, P.; Valentijn, L.J.; van Nes, J.; et al. LIN28B induces neuroblastoma and enhances MYCN levels via let-7 suppression. *Nat. Genet.* **2012**, *44*, 1199–1206. [CrossRef] [PubMed]
124. Thornton, J.E.; Chang, H.-M.; Piskounova, E.; Gregory, R. Lin28-mediated control of let-7 microRNA expression by alternative TUTases Zcchc11 (TUT4) and Zcchc6 (TUT7). *RNA* **2012**, *18*, 1875–1885. [CrossRef] [PubMed]
125. Heo, I.; Joo, C.; Kim, Y.-K.; Ha, M.; Yoon, M.-J.; Cho, J.; Yeom, K.-H.; Han, J.; Kim, V.N. TUT4 in Concert with Lin28 Suppresses MicroRNA Biogenesis through Pre-MicroRNA Uridylation. *Cell* **2009**, *138*, 696–708. [CrossRef] [PubMed]
126. Hagan, J.P.; Piskounova, E.; Gregory, R. Lin28 recruits the TUTase Zcchc11 to inhibit let-7 maturation in mouse embryonic stem cells. *Nat. Struct. Mol. Biol.* **2009**, *16*, 1021–1025. [CrossRef]
127. Faehnle, C.R.; Walleshauser, J.; Joshua-Tor, L. Mechanism of Dis3l2 substrate recognition in the Lin28–let-7 pathway. *Nature* **2014**, *514*, 252–256. [CrossRef]
128. Rybak, A.; Fuchs, H.; Smirnova, L.; Brandt, C.; Pohl, E.E.; Nitsch, R.; Wulczyn, F.G. A feedback loop comprising lin-28 and let-7 controls pre-let-7 maturation during neural stem-cell commitment. *Nature* **2008**, *10*, 987–993. [CrossRef]
129. Ustianenko, D.; Chiu, H.-S.; Treiber, T.; Weyn-Vanhentenryck, S.M.; Treiber, N.; Meister, G.; Sumazin, P.; Zhang, C. LIN28 Selectively Modulates a Subclass of Let-7 MicroRNAs. *Mol. Cell* **2018**, *71*, 271–283.e5. [CrossRef]
130. West, R.; McWhorter, E.S.; Ali, A.; Goetzman, L.N.; Russ, J.; Gonzalez-Berrios, C.L.; Anthony, R.V.; Bouma, G.J.; A Winger, Q. HMGA2 is regulated by LIN28 and BRCA1 in human placental cells. *Biol. Reprod.* **2018**, *100*, 227–238. [CrossRef]
131. Seabrook, J.L.; Cantlon, J.D.; Cooney, A.J.; McWhorter, E.E.; Fromme, B.A.; Bouma, G.J.; Anthony, R.V.; Winger, Q.A. Role of LIN28A in mouse and human trophoblast cell differentiation. *Biol. Reprod.* **2013**, *89*, 95. [CrossRef]

132. Chan, H.-W.; Lappas, M.; Yee, S.Y.; Vaswani, K.; Mitchell, M.; Rice, G.E. The expression of the let-7 miRNAs and Lin28 signalling pathway in human term gestational tissues. *Placenta* **2013**, *34*, 443–448. [CrossRef]
133. Barbaux, S.; Gascoin-Lachambre, G.; Buffat, C.; Monnier, P.; Mondon, F.; Tonanny, M.-B.; Pinard, A.; Auer, J.; Bessières, B.; Barlier, A.; et al. A genome-wide approach reveals novel imprinted genes expressed in the human placenta. *Epigenetics* **2012**, *7*, 1079–1090. [CrossRef] [PubMed]
134. Monk, D. Genomic imprinting in the human placenta. *Am. J. Obstet. Gynecol.* **2015**, *213*, S152–S162. [CrossRef]
135. Liu, Y.; Fan, X.; Wang, R.; Lu, X.; Dang, Y.-L.; Wang, H.; Lin, H.-Y.; Zhu, C.; Ge, H.; Cross, J.C.; et al. Single-cell RNA-seq reveals the diversity of trophoblast subtypes and patterns of differentiation in the human placenta. *Cell Res.* **2018**, *28*, 819–832. [CrossRef] [PubMed]
136. Gu, Y.; Sun, J.; Groome, L.J.; Wang, Y. Differential miRNA expression profiles between the first and third trimester human placentas. *Am. J. Physiol. Metab.* **2013**, *304*, E836–E843. [CrossRef] [PubMed]
137. Souza, M.A.; Brizot, M.L.; Biancolin, S.E.; Schultz, R.; De Carvalho, M.H.B.; Francisco, R.P.V.; Zugaib, M. Placental weight and birth weight to placental weight ratio in monochorionic and dichorionic growth-restricted and non-growth-restricted twins. *Clinics* **2017**, *72*, 265–271. [CrossRef]
138. Hiden, U.; Wadsack, C.; Prutsch, N.; Gauster, M.; Weiss, U.; Frank, H.-G.; Schmitz, U.; Fast-Hirsch, C.; Hengstschläger, M.; Pötgens, A.; et al. The first trimester human trophoblast cell line ACH-3P: A novel tool to study autocrine/paracrine regulatory loops of human trophoblast subpopulations – TNF-α stimulates MMP15 expression. *BMC Dev. Biol.* **2007**, *7*, 137. [CrossRef] [PubMed]
139. Chavez, S.L.; Abrahams, V.M.; Alvero, A.B.; Aldo, P.B.; Ma, Y.; Guller, S.; Romero, R.; Mor, G. The Isolation and Characterization of a Novel Telomerase Immortalized First Trimester Trophoblast Cell Line, Swan 71. *Placenta* **2009**, *30*, 939–948. [CrossRef]
140. McWhorter, E.S.; West, R.; Russ, J.E.; Ali, A.; Winger, Q.A.; Bouma, G.J. LIN28B regulates androgen receptor in human trophoblast cells through Let-7c. *Mol. Reprod. Dev.* **2019**, *86*, 1086–1093. [CrossRef]
141. Madison, B.B.; Jeganathan, A.N.; Mizuno, R.; Winslow, M.M.; Castells, A.; Cuatrecasas, M.; Rustgi, A.K. Let-7 Represses Carcinogenesis and a Stem Cell Phenotype in the Intestine via Regulation of Hmga2. *PLoS Genet.* **2015**, *11*. [CrossRef]
142. Lee, Y.S.; Dutta, A. The tumor suppressor microRNA let-7 represses the HMGA2 oncogene. *Genome Res.* **2007**, *21*, 1025–1030. [CrossRef] [PubMed]
143. Ali, A.; Stenglein, M.D.; Spencer, T.E.; Bouma, G.J.; Anthony, R.V.; Winger, Q.A. Trophectoderm-Specific Knockdown of LIN28 Decreases Expression of Genes Necessary for Cell Proliferation and Reduces Elongation of Sheep Conceptus. *Int. J. Mol. Sci.* **2020**, *21*, 2549. [CrossRef] [PubMed]
144. Wilsker, D.; Patsialou, A.; Dallas, P.; Moran, E. ARID proteins:A diverse family of DNA binding proteins implicated in the control of cell growth, differentiation, and development. *Cell Growth Differ. Mol. Biol. J. Am. Assoc. Cancer Res.* **2002**, *13*, 95–106.
145. Hong, Z.; Wei, S.; Bi, X.; Zhao, J.; Huang, Z.; Li, Z.; Zhou, J.; Cai, J.; Chen, L.; Lin, C.; et al. Recent advances in the ARID family:Focusing on roles in human cancer. *OncoTargets Ther.* **2014**, *7*, 315–324. [CrossRef] [PubMed]
146. Fukuyo, Y.; Takahashi, A.; Hara, E.; Horikoshi, N.; Pandita, T.K.; Nakajima, T. E2FBP1 antagonizes the p16INK4A-Rb tumor suppressor machinery for growth suppression and cellular senescence by regulating promyelocytic leukemia protein stability. *Int. J. Oral Sci.* **2011**, *3*, 200–208. [CrossRef] [PubMed]
147. Fukuyo, Y.; Mogi, K.; Tsunematsu, Y.; Nakajima, T. E2FBP1/hDril1 modulates cell growth through downregulation of promyelocytic leukemia bodies. *Cell Death Differ.* **2004**, *11*, 747–759. [CrossRef]
148. Kobayashi, K.; Jakt, L.M.; Nishikawa, S.-I. Epigenetic regulation of the neuroblastoma genes, Arid3b and Mycn. *Oncogene* **2012**, *32*, 2640–2648. [CrossRef]
149. Bobbs, A.; Gellerman, K.; Hallas, W.M.; Joseph, S.; Yang, C.; Kurkewich, J.; Dahl, K.C. ARID3B Directly Regulates Ovarian Cancer Promoting Genes. *PLoS ONE* **2015**, *10*, e0131961. [CrossRef]
150. Nakahara, S.; Fukushima, S.; Yamashita, J.; Kubo, Y.; Tokuzumi, A.; Miyashita, A.; Harada, M.; Nakamura, K.; Jinnin, M.; Ihn, H. AT-rich Interaction Domain-containing Protein 3B is a New Tumour Marker for Melanoma. *Acta Derm. Venereol.* **2017**, *97*, 112–114. [CrossRef]
151. Kim, D.; Probst, L.; Das, C.; Tucker, P.W. REKLES Is an ARID3-restricted Multifunctional Domain. *J. Biol. Chem.* **2007**, *282*, 15768–15777. [CrossRef]

152. Webb, C.F.; Bryant, J.; Popowski, M.; Allred, L.; Kim, D.; Harriss, J.; Schmidt, C.; Miner, C.A.; Rose, K.; Cheng, H.-L.; et al. The ARID Family Transcription Factor Bright Is Required for both Hematopoietic Stem Cell and B Lineage Development. *Mol. Cell. Biol.* **2011**, *31*, 1041–1053. [CrossRef]
153. Liao, T.-T.; Hsu, W.-H.; Ho, C.-H.; Hwang, W.-L.; Lan, H.-Y.; Lo, T.; Chang, C.-C.; Tai, S.; Yang, M.-H. Let-7 Modulates Chromatin Configuration and Target Gene Repression through Regulation of the ARID3B Complex. *Cell Rep.* **2016**, *14*, 520–533. [CrossRef] [PubMed]
154. Uhlén, M.; Oksvold, P.; Fagerberg, L.; Lundberg, E.; Jonasson, K.; Forsberg, M.; Zwahlen, M.; Kampf, C.; Wester, K.; Hober, S.; et al. Towards a knowledge-based Human Protein Atlas. *Nat. Biotechnol.* **2010**, *28*, 1248–1250. [CrossRef] [PubMed]
155. Rhee, C.; Lee, B.-K.; Beck, S.; Anjum, A.; Cook, K.R.; Popowski, M.; Tucker, H.O.; Kim, J. Arid3a is essential to execution of the first cell fate decision via direct embryonic and extraembryonic transcriptional regulation. *Genome Res.* **2014**, *28*, 2219–2232. [CrossRef] [PubMed]
156. Rhee, C.; Edwards, M.; Dang, C.; Harris, J.; Brown, M.; Kim, J.; Tucker, H.O. ARID3A is required for mammalian placenta development. *Dev. Biol.* **2016**, *422*, 83–91. [CrossRef]

International Journal of
Molecular Sciences

MDPI

Article

Fractalkine Regulates HEC-1A/JEG-3 Interaction by Influencing the Expression of Implantation-Related Genes in an In Vitro Co-Culture Model

Ramóna Pap [1], Gergely Montskó [2,3], Gergely Jánosa [1], Katalin Sipos [1], Gábor L. Kovács [2,3,4] and Edina Pandur [1,*]

1 Department of Pharmaceutical Biology, Faculty of Pharmacy, University of Pécs, H-7624 Pécs, Hungary; pap.ramona@pte.hu (R.P.); janosa.gergely@gytk.pte.hu (G.J.); katalin.sipos@aok.pte.hu (K.S.)
2 Szentágothai Research Centre, University of Pécs, H-7624 Pécs, Hungary; montsko.gergely@pte.hu (G.M.); kovacs.l.gabor@pte.hu (G.L.K.)
3 MTA-PTE Human Reproduction Research Group, University of Pécs, H-7624 Pécs, Hungary
4 Department of Laboratory Medicine, Medical School, University of Pécs, H-7624 Pécs, Hungary
* Correspondence: edina.pandur@aok.pte.hu

Received: 9 March 2020; Accepted: 29 April 2020; Published: 30 April 2020

Abstract: Embryo implantation is a complex process regulated by a network of biological molecules. Recently, it has been described that fractalkine (CX3CL1, FKN) might have an important role in the feto–maternal interaction during gestation since the trophoblast cells express fractalkine receptor (CX3CR1) and the endometrium cells secrete fractalkine. CX3CR1 controls three major signalling pathways, PLC-PKC pathway, PI3K/AKT/NFκB pathway and Ras-mitogen-activated protein kinases (MAPK) pathways regulating proliferation, growth, migration and apoptosis. In this study, we focused on the molecular mechanisms of FKN treatment influencing the expression of implantation-related genes in trophoblast cells (JEG-3) both in mono-and in co-culture models. Our results reveal that FKN acted in a concentration and time dependent manner on JEG-3 cells. FKN seemed to operate as a positive regulator of implantation via changing the action of progesterone receptor (PR), activin receptor and bone morphogenetic protein receptor (BMPR). FKN modified also the expression of matrix metalloproteinase 2 and 9 controlling invasion. The presence of HEC-1A endometrial cells in the co-culture contributed to the effect of fractalkine on JEG-3 cells regulating implantation. The results suggest that FKN may contribute to the successful attachment and implantation of embryo.

Keywords: fractalkine; implantation; endometrium; trophoblast; bilaminar co-culture

1. Introduction

Embryo implantation, the process of attachment and invasion of the uterus endometrium by the conceptus, is a complex physiological process tightly regulated by multiple biological molecules. Implantation requires a well orchestrated interaction between maternal and foetal tissues and consists of a fine balanced cross talk of cytokines, hormones and chemokines [1,2].

Chemokines play important key roles in several physiologic and pathologic aspects of human reproductive system, among them, the menstrual cycle, ovulation, implantation, cervical ripening, preterm labour and endometriosis [3].

The chemokine fractalkine (FKN) is synthesized as a 373 amino-acid transmembrane molecule. It is the only CX3C-chemokine which has been described [4,5]. FKN exists as both membrane-anchored and soluble forms. As an individual chemokine, its function is not merely chemoattraction, but it also acts as an adhesion molecule and is capable of regulating the immune response via CX3CR1 corresponding receptor interaction [5–7].

CX3CR1 belongs to a family of G protein-coupled receptors [7], expressed on cytotoxic effector lymphocytes, including NK cells and cytotoxic T lymphocytes [8]. The FKN-CX3CR1 axis is a crucial regulator of the microglia, the immune cells of the central nervous system (CNS) [9,10].

FKN and its receptor are expressed in numerous reproductive tissues, comprising testis, uterus, ovaries and Fallopian tubes [11,12]. According to previous studies, fractalkine might have an important involvement in the feto–maternal communication during gestation since the trophoblast cells express CX3CR1 and the endometrium cells produce FKN [13,14].

CX3CR1 controls three major signalling pathways, the PLC-PKC pathway, the PI3K/AKT/NFκB pathway and the Ras-mitogen-activated protein kinases (MAPK) pathways (p38, ERK1/2 and JNK) [15]. MAPKs are one of the oldest known signal transduction pathways [16,17] regulating proliferation, growth, migration and apoptosis [18]. PLC-PKC pathway can trigger the MAPK pathway and the secretion of various hormones [15]. The PI3K/AKT signalling pathway is activated via growth factors and plays a critical role in regulating diverse cellular functions including cell growth, metabolism, proliferation, survival, transcription, protein synthesis and mitogenesis [19,20].

Cell migration and invasion are pivotal processes for endometrial cells during implantation, cell trafficking and embryonal morphogenesis [21]. The endometrium undergoes different highly regulated physiological changes during implantation [22,23]. The main supporters of migration are protein kinases, including extracellular signal-regulated protein kinase (ERK)1/2, the phosphatidylinositol 3 kinase (PI3K), the focal adhesion kinase (FAK) and other kinases that can be activated by different cytokines, growth factors and the extracellular matrix [24]. The JNK and p38 pathways are activated in mouse preimplantation development [25].

Progesterone receptor (PR) is involved in the proliferation and differentiation processes by regulating the transcription of specific genes (c-myc, p21, EGFR etc.). The human progesterone receptor has two classical isoforms—A and B—but non-classical intracellular progesterone receptor variants have also been detected (PRC, PRM, PRS and PRT) [26]. PRA and PRB are transcribed from the same gene by using two distinct promoters. The PR can be phosphorylated at basal phosphorylation sites, as well as in response to progesterone. Different protein kinases, including mitogen-activated protein kinases (MAPK), are known to phosphorylate PR at specific serine amino acid residues and modify PR's activity [27].

Trophoblast invasion into the uterus is an essential step at implantation of the human blastocyst. The action is facilitated by the degradation of the extracellular matrix of the endometrium/decidua by various proteinases, including the matrix metalloproteinases (MMPs) [28]. The tissue inhibitors of matrix metalloproteinases (TIMPs) are key regulators of the metalloproteinases. They inhibit the activity of the MMPs by binding to the highly conserved zinc-binding site of active MMP [29]. It is suggested that MMPs and their regulators control many aspects of reproductive function. The expression of MMP2 and MMP9 depends on the activation of the aforementioned MAPK pathways [30,31] that are regulated by CX3CR1, as well. MMPs have been localized most strongly to the placental bed in early pregnancy suggesting that these proteins are involved in the invasion of trophoblasts [32,33]. MMP expression is also regulated by other factors, e.g., activin signalling and BMP2 signalling pathways [34–36].

Activins secreted by endometrial cells belong to the pleiotropic family of the transforming growth factor beta (TGFβ) superfamily of cytokines [37] and are potential factors for maternal–embryo interactions [38]. Activin/Nodal signalling through activin receptors plays an important role at implantation [34,38]. Activin is antagonized by another secreted molecule, called follistatin, which is a key regulator of the biological actions of activin [37]. Follistatins are able to bind directly and irreversibly to activins, neutralizing the ligand [39]. In early development, many processes are contingent on BMPs signalling for cell growth and differentiation [40]. Some follistatin isoforms can bind bone morphogenetic proteins (BMPs) and hinder their biological activities [41].

A bilaminar co-culture system is used to study cell–cell interactions and the technique can be modified for co-culturing any variety of adherent cell types. In this system, the two cell types can connect to each other physically. Therefore, they are able to communicate with each other not only via their released signalling molecules, but also through different cell surface molecules. This way, we can get a better view of action/reaction both of the examined cells [42–45].

In this study, we focused on the molecular mechanisms governing FKN supporting HEC-1A/JEG-3 interaction and influencing the expression of implantation-related genes in an in vitro co-culture model. Understanding how these mechanisms contribute to implantation might open new targeted medical therapies to give reassurance to women suffering from its failure.

2. Results

2.1. *Effect of Fractalkine on the Viability of JEG-3 Cells in Mono- and Co-Cultures*

FKN has been demonstrated to regulate cell proliferation and inhibit apoptosis through different signalling pathways (MAPK, PI3K/AKT) in several cell types (e.g., monocytes, T cells, smooth muscle cells, glia cells, neurons) [10,46,47].

Endometrial HEC-1A cells produce FKN and trophoblast-like JEG-3 cells express fractalkine receptor, CX3CR1; therefore, we supposed that FKN might have an effect on the cell proliferation of JEG-3 cells in both mono- and co-cultures. It was revealed that viability was elevated in the monocultures of JEG-3 cells at 24 h and 48 h, but FKN did not affect the viability compared to the untreated control cells (Figure 1A), suggesting a normal cell proliferation. In case of the co-cultures, FKN influenced cell viability: at 24 h, JEG-3 cells showed significantly increased viability (Figure 1B). Comparing the viabilities of 24 h controls in mono- and co-cultures, it seems that the presence of HEC-1A cells in the co-culture affected JEG-3 proliferation, suggesting that the cell–cell interactions influenced cell cycle and viability (Figure 1A,B). These results indicate that FKN exerts different effects on mono- and co-cultures, and for the co-culture, the effect is time dependent.

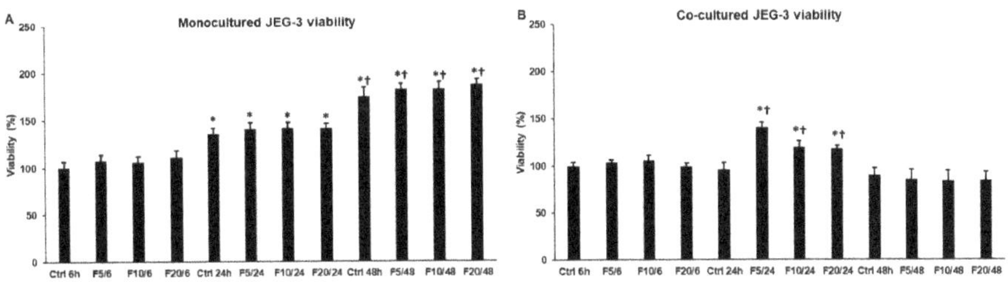

Figure 1. Cell viability determinations of the mono- and co-cultured JEG-3 cells. Viability was measured using CCK-8 cell viability assay after fractalkine treatments. Viability is expressed as percentile of the total cell number of the untreated control cells. (**A**) Viability of the monocultured JEG-3 cells and (**B**) Cell viability measurements of co-cultured JEG-3 cells. Cell viability assays were made in quintuplicate in three independent experiments. The bars represent mean values and error bars represent standard deviation (SD) for three independent experiments ($n = 3$). The * indicate $p < 0.05$ compared to the 6 h untreated control. The † indicate $p < 0.05$ compared to the 24 h untreated control. Abbreviations of fractalkine treatments: 5 ng/mL-F5; 10 ng/mL-F10; 20 ng/mL-F20.

2.2. *Fractalkine Changes the Activation of ERK1/2, p38, JNK and AKT Signalling Pathways in Mono- and Co-Cultured JEG-3 Cells*

The increased viability of the cells suggests an enhanced proliferation that is regulated by several signalling pathways. FKN is involved in the regulation of MAPK and PI3K/AKT pathways, regulators of proliferation, differentiation and apoptosis [48,49]. We examined the phosphorylation of ERK1/2, p38, JNK (MAPKs) and AKT to reveal which pathway was affected by FKN and if there were any differences between the activation of signalling pathways with time and by increasing FKN concentrations (5 ng/mL-F5; 10 ng/mL-F10; 20 ng/mL-F20).

In case of JEG-3 monoculture, F10 decreased ERK1/2 phosphorylation at 24 h while at 48 h, it increased it compared to the control cells (Figure 2A,C). Meanwhile, F20 treatment increased ERK1/2 phosphorylation at each time points (Figure 2A,C). F10 reduced p-p38 level significantly at 6 h and 24 h but increased again at 48 h (Figure 2A,D), while treatment with F20 caused elevation in p-p38 level at each time points (Figure 2A,D). In contrast with the aforementioned changes of MAPKs, F10 raised p-JNK level at 24 h. Treatment with F20 increased continuously the p-JNK level (Figure 2A,E). AKT showed different alterations compared to MAPKs due to FKN treatment. Phospho-AKT level was elevated by F10 at 24 h while F20 increased it only at 6 h (Figure 2A,F). The results show that the effect of fractalkine is concentration- and time-dependent; the higher FKN concentration (20 ng/mL) has a stronger and longer effect on the protein phosphorylation.

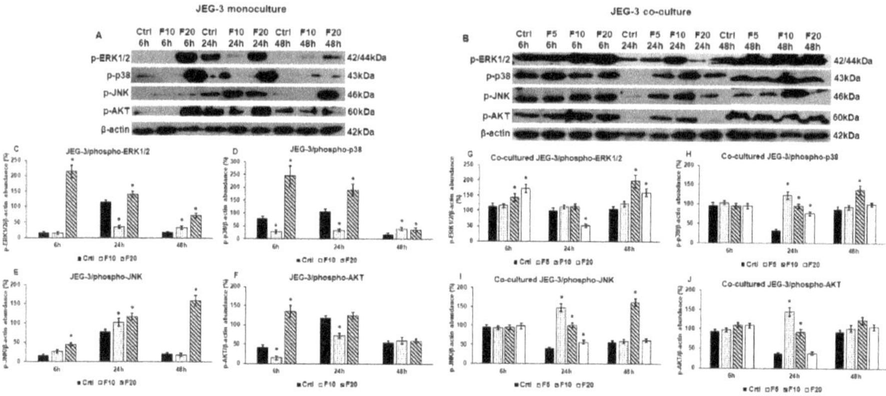

Figure 2. Western blot analyses of signalling pathways regulated by fractalkine in mono- (**A**) and co-cultures (**B**) JEG-3 cells. Cells were collected and pelleted after fractalkine treatments then cells were lysed and their protein contents were measured. The same amount of protein (10 μg) from each sample was separated by SDS-PAGE using 12% polyacrylamide gel, transferred by electroblotting to nitrocellulose membranes. The membranes were probed with anti-phospho-ERK1/2, anti-phospho-p38, anti-phospho-JNK and anti-phospho-AKT according to the manufacturer's instruction. The experiments were repeated three times. β-actin was used as loading control. (**C–F**) Optical density analyses of the examined proteins in JEG-3 monocultures. (**G–J**) Optical density analyses of the examined proteins in co-cultured JEG-3 cells. The analyses were carried out using ImageJ software; the optical densities of the examined proteins were expressed as percentage of target protein/β-actin abundance. The bars represent mean values and error bars represent standard deviation (SD) for three independent experiments ($n = 3$). The * mark $p < 0.05$ compared to the appropriate controls (6 h, 24 h and 48 h). Abbreviations of fractalkine treatments: 5 ng/mL-F5; 10 ng/mL-F10; 20 ng/mL-F20.

Regarding the co-cultures, in which the two cell types can get in contact with each other, we revealed different alterations in the protein phosphorylation patterns. Although F5 had no effect on the examined signalling pathways in the monoculture, we examined the effect of F5 on the co-cultured JEG-3 cells, too. F5 did not act on the cells at 6 h and 48 h, but it elevated p-p38, p-JNK and p-AKT levels at 24 h that correlated with the increased cell viability (Figure 2B,G–J). F10 raised p-ERK1/2 level at 6 h and 48 h (Figure 2B,G), while the other three pathways were triggered at 24 h and 48 h (Figure 2B,H–J). Interestingly, F20 treatment was less effective on the co-cultured JEG-3 cells. Phospho-ERK1/2 level increased at 6 h and 48 h that was similar to the F10 treatment (Figure 2B,G), although p-ERK1/2 level was reduced at 24 h. Phospho-p38 and p-JNK only elevated at 24 h (Figure 2B,H,I), and p-AKT level did not change during the whole experiment using F20 (Figure 2B,J).

These result reveal that in the case of co-cultured JEG-3 cells, the strongest effect was found at 24 h and at this time point, F5 treatment was the most effective. Comparing the phosphorylation of MAPKs and AKT in mono- and co-cultures it seems that FKN exerts distinct effects on the cells suggesting that the presence of HEC-1A cells in the co-culture modifies the action of FKN.

2.3. Fractalkine Exerts Different Effects on the mRNA Expression of Proliferation, Differentiation and Invasion Regulating Genes in Mono- and Co-Cultured JEG-3 Cells

Next, we investigated the mRNA expression levels of PR, which is responsible for differentiation and proliferation of trophoblast cells, and CX3CR1 to reveal if fractalkine provides an autoregulatory effect on its receptor expression on JEG-3 cells. The mRNA expression of PR significantly increased both at 6 h and at 24 h in the monoculture, and was also elevated at the same time points in the co-cultured JEG-3 cells, although there was a significant difference between mono- and co-cultures (Figure 3A). At 48 h, the mRNA expression of PR returned to the control level both in mono- and co-cultured JEG-3 cells (Figure 3A).

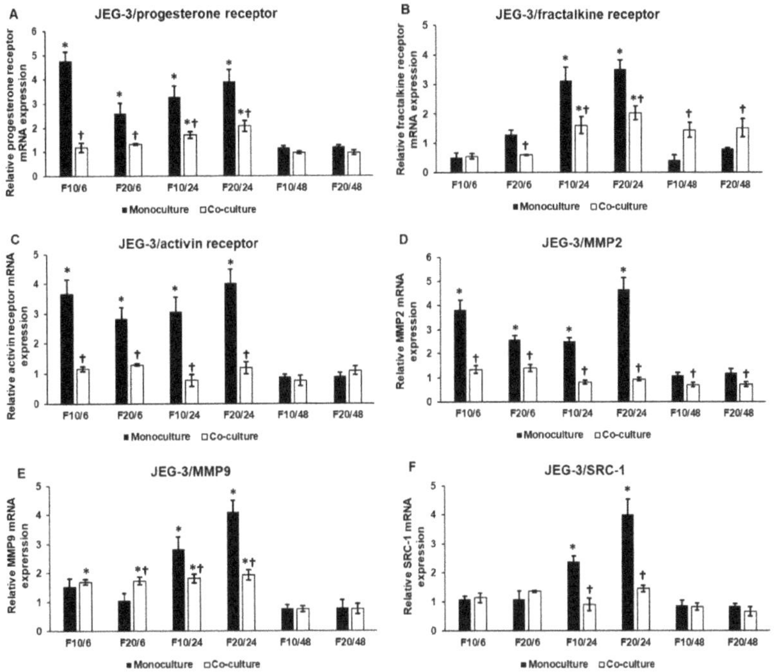

Figure 3. Effects of fractalkine treatments on the mRNA expressions of the implantation-related genes in mono- and co-cultured JEG-3 cells. Real-time PCR was performed with SYBR green protocol using gene specific primers. β-actin was used as house-keeping gene for the normalization. The relative expression of untreated controls was regarded as 1. The mRNA expressions of the treated cells were compared to their appropriate controls (6 h, 24 h or 48 h). (**A**) mRNA expression levels of progesterone receptor. (**B**) mRNA expression levels of fractalkine receptor. (**C**) mRNA expression levels of activin receptor 1B. (**D**) mRNA expression levels of MMP2. (**E**) mRNA expression levels of MMP9. (**F**) mRNA expression levels of SRC-1. The columns represent mean values and error bars represent standard deviation (SD) of three independent determinations ($n = 3$). The * indicates $p < 0.05$ compared to the untreated controls. The † shows $p < 0.05$ between mono- and co-cultures. Abbreviations of fractalkine treatments: 5 ng/mL-F5; 10 ng/mL-F10; 20 ng/mL-F20.

CX3CR1 significantly raised at 24 h in the monoculture (Figure 3B). The same phenomenon was found in the co-cultured cells, but in this case, fractalkine maintained the augmented CX3CR1 mRNA expression at 48 h, suggesting a longer effect of FKN on the co-cultured JEG-3 cells through CX3CR1 (Figure 3B).

The activin receptor is supposed to have a role in implantation [50], while matrix metalloproteinases MMP2 and MMP9 are the major contributors to normal implantation by increasing invasiveness [51].

We examined also whether FKN affected the expression of these genes. Interestingly, activin receptor was upregulated in the JEG-3 cells at 6 h and 24 h but its mRNA expression level did not change significantly in the co-cultured JEG-3 cells (Figure 3C), suggesting that the interaction between JEG-3 and HEC-1A cells may influence the effect of fractalkine treatment. The same result was found in case of MMP2 in both mono- and co-cultured JEG-3 cells (Figure 3D). MMP9 mRNA expression was triggered by FKN at 24 h in the monoculture; meanwhile in the co-cultured JEG-3, MMP9 expression began to elevate earlier, at 6 h and remained elevated at 24 h although this level was significantly lower than those in the monoculture (Figure 3E). These results suggest that MMP9 facilitates trophoblast invasion after binding to the endometrial cells while MMP2 expression is downregulated.

We also examined SRC-1 mRNA expression, the co-factor of PR, which is regulated by MAPK pathways and can promote cell differentiation. The expression analysis revealed that SRC-1 mRNA expression only increased at 24 h in the monocultured JEG-3 cells, and there was no significant alteration of SRC-1 mRNA level in the co-culture (Figure 3F).

It seems that in the monocultured JEG-3 cells, F10 exerted a higher effect at 6 h, while at 24 h, F20 was more effective on the gene expression. In case of the co-cultures, the action of FKN did not show concentration dependence.

2.4. Western Blot Analysis of the Implantation-Related Genes Reveals Alterations Between Fractalkine Treated Mono- and Co-Cultured JEG-3 Cells

After the mRNA expression analysis, it was unravelled that FKN influenced the expression of genes that are implicated in the implantation process. We also examined if these alterations appeared at protein level, suggesting that FKN regulates implantation by controlling proliferation, differentiation and invasion related proteins of trophoblast cells.

In the monocultures, PR and CX3CR1 protein levels showed delays between the elevation of mRNA and protein expressions at F10 treatment, suggesting that the lower FKN concentration has a slower effect on the JEG-3 cells. The effect of F20 lasted longer as the mRNA levels decreased at 48 h but the protein levels were still higher compared to the control cells (Figure 3A,B and Figure 4A).

Activin receptor, MMP2, MMP9 and SRC-1 protein levels correlated with the changes of the mRNA expression levels (Figure 3C–F and Figure 4A), but their expressions varied by time and by FKN concentration. Activin receptor was upregulated by both FKN concentrations at 24 h, and then, at 48 h, the protein level was significantly reduced (Figure 4A,E).

MMP2 protein level raised at 6 h and 24 h and then, at 48 h, the protein level significantly decreased (Figure 4A,F). In case of MMP9, significant elevation was observed at only 24 h (Figure 4A,G). The same alteration was found in case of SRC-1 protein at 24 h, and then F10 decreased SRC-1 level at 48 h (Figure 4A,H). These results show that in the monocultured JEG-3 cells, F20 was more effective in upregulating the protein levels.

In the case of the co-cultured JEG-3 cells, the correlation between mRNA expression and protein level is not as straightforward as for monoculture suggesting that the interaction between the two cell types influences the effect of FKN and the translational rate of the examined proteins.

FKN treatments increased the protein level of PR at 6 h and 48 h, while at 24 h, F5 still increased the expression. On the other hand F10 and F20 had an opposite effect, significantly decreasing the level of PR, showing that the effect of FKN was concentration dependent (Figure 4B,I). SRC-1, a progesterone co-receptor, was elevated by F20 at 6 h, while F10 and F20 treatments reduced SRC-1 protein level at 24 h, and elevated it at 48 h (Figure 4B,N). These changes were parallel to the alterations found at PR. In the case of CX3CR1, the strongest effect was mediated by F10 which significantly reduced CX3CR1 expression at 24 h and 48 h (Figure 4B,J). The activin receptor was downregulated by F10 and F20 at 6 h and 24 h, but it was significantly elevated at 48 h by all three FKN concentrations (Figure 4B,K).

F5 reduced MMP2 protein level at 6 h and increased it at 48 h (Figure 4B,L). F10 and F20 decreased MMP2 level at 24 h and had a reverse effect at 48 h (Figure 4B,L). Interestingly, MMP9 protein level was elevated during the whole experiment by each FKN concentration (Figure 4B,M). This suggests that MMP9 had a specific role in the invasion and may work together with MMP2 at 48 h.

Based on the results, it is proven that F10 and F20 treatments triggered parallel alterations at protein level. The only exception is CX3CR1, which was downregulated by F10 but was upregulated by F20 treatments (Figure 4B,J). The latter result suggests that the regulation of CX3CR1 by FKN is concentration dependent. The fluctuations of PR, CX3CR1 and activin receptor expression may provide protection against the overactivation of signalling pathways regulating the same signal transduction proteins, MAPKs, SMAD transcription factors, the same genes (e.g., MMPs) or cellular process (e.g., invasion). These changes in protein levels can contribute to maintaining the proper implantation mechanism of trophoblast cells.

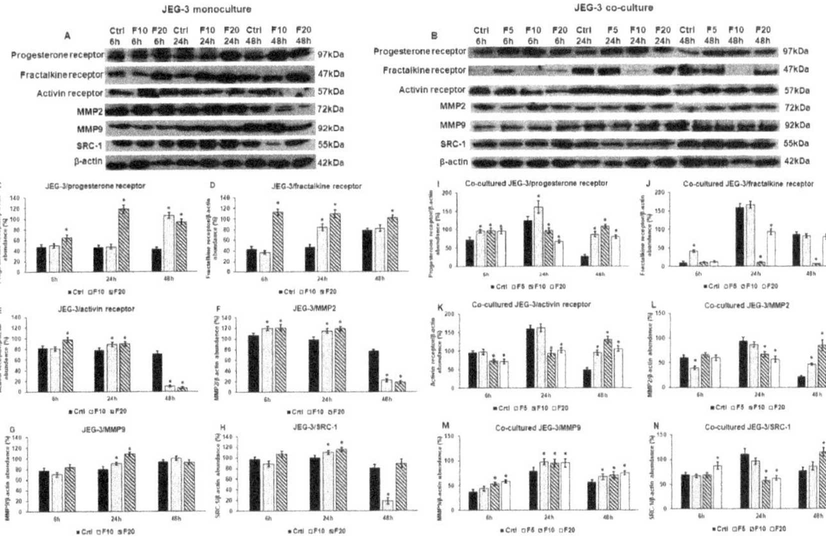

Figure 4. Western blot analyses of implantation-related proteins in fractalkine treated mono- (**A**) and co-cultured (**B**) JEG-3 cells. JEG-3 cells were collected and pelleted after fractalkine treatments. Following lysis, protein contents were measured. The same amount of protein (15 μg) from each sample was separated by SDS-PAGE using 10% or 12% polyacrylamide gel, transferred by electroblotting to nitrocellulose membranes. The membranes were probed with anti-progesterone receptor, anti-fractalkine receptor, anti-activin receptor, anti-MMP2, anti-MMP9 and anti-SRC-1 polyclonal rabbit antibodies according to the manufacturer's protocols. The experiments were repeated three times. (**C–H**) Optical density analyses of the examined proteins in JEG-3 monocultures. (**I–N**) Optical density analyses of the examined proteins in co-cultured JEG-3 cells. The analyses were carried out using ImageJ software, the optical densities of the examined proteins were expressed as percentage of target protein/β-actin abundance. The bars represent mean values and error bars represent standard deviation (SD) for three independent experiments (n = 3). The * mark $p < 0.05$ compared to the controls. Abbreviations of fractalkine treatments: 5 ng/mL-F5; 10 ng/mL-F10; 20 ng/mL-F20.

2.5. The Presence of HEC-1A Cells Contributes to the Action of Fractalkine on JEG-3 Cells by Changing the Expressions of Activin, Follistatin and BMP2

Based on our results, it seems that the presence of HEC-1A cells in the co-cultures contributed to the action of fractalkine on JEG-3 cells. Therefore, we examined whether HEC-1A-expressed implantation-related genes showed alterations suggesting a cooperation in the implantation process at fractalkine treatment. We determined the mRNA expression levels of activin and follistatin regulating activin receptor, and BMP2 regulating cell growth and differentiation. Interestingly, follistatin and activin levels changed inversely in the co-culture: increased follistatin mRNA level and decreased activin level were found at 6 h and 48 h, and decreased follistatin level and increased activin mRNA level were detected at 24 h (Figure 5A,B). The expression of BMP2 that is also regulated by follistatin binding protein was significantly higher during the whole experiment compared to the control (Figure 5C). These results suggest that the interaction between JEG-3 and HEC-1A cells influences the expressional changes occurring in JEG-3 cells in fractalkine treatment.

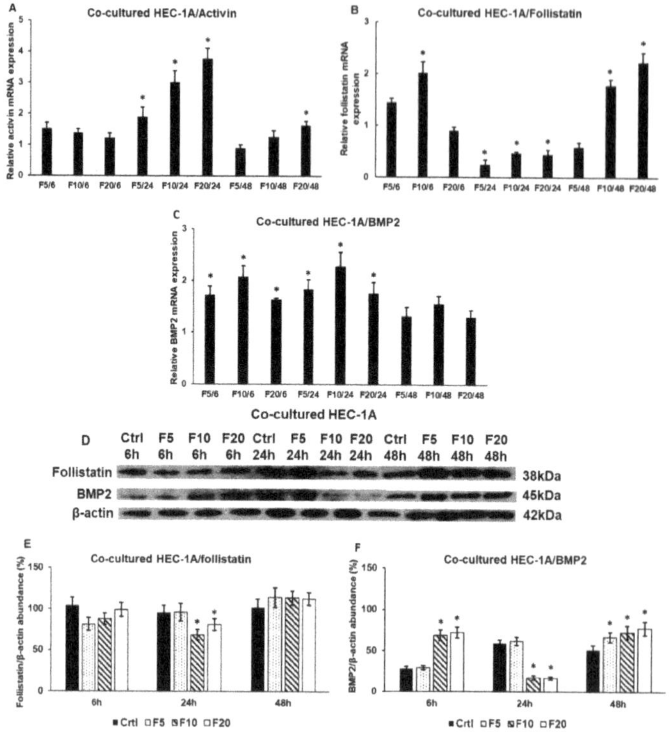

Figure 5. Expression analysis of activin, follistatin and BMP2 in co-cultured HEC-1A cells. Real-time PCR was performed with SYBR green protocol using gene specific primers. β-actin was used as house-keeping gene for the normalization. The relative expression of untreated controls was regarded as 1. The mRNA expressions of the treated cells were compared to their appropriate controls (6 h, 24 h or 48 h). HEC-1A cells were collected by trypsinization directly from the coverslips. The cells were lysed and their protein contents were measured. The same amount of protein (15 µg) from each sample was separated by SDS-PAGE using 10% or 12% polyacrylamide gel, transferred by electroblotting to nitrocellulose membranes. The membranes were incubated with anti-follistatin and anti-BMP2 polyclonal rabbit antibodies according to the manufacturer's protocols. The experiments were repeated three times. (**A–C**) mRNA expression levels of activin, follistatin and BMP2. (**D**) Western blot analyses of follistatin and BMP2. (**E**,**F**) Optical density analyses of the examined proteins in HEC-1A cells. The analyses were carried out using ImageJ software; the optical densities of the examined proteins were expressed as percentage of target protein/β-actin abundance. The columns represent mean values and error bars represent standard deviation (SD) for three independent experiments ($n = 3$). The * mark $p < 0.05$ compared to the controls. Abbreviations of fractalkine treatments: 5 ng/mL-F5; 10 ng/mL-F10; 20 ng/mL-F20.

At protein level, we found that follistatin level decreased at 24 h that may influence the regulation of activin receptor activity in JEG-3 cells. BMP2 protein level was elevated at 6 h and 48 h. The latter shows that there is a delay between mRNA expression and protein synthesis (Figure 5C,D,F). The decreasing level of BMP2 protein at 24 h is supposed to occur due to the secretion of protein into the culture medium (Figure 5D,F). Our results suggest that HEC-1A cells contribute to the implantation by altering the expression of implantation-related genes and proteins, which could contribute to the action of fractalkine in trophoblast cells.

3. Discussion

The early stages of pregnancy comprise the attachment of embryo to the uterine epithelium, invasion of embryo into the uterine stroma and the decidualization of the stroma. Multiple molecules forming a complex network regulate implantation. The interplay between endometrial tissue and conceptus is critical to induce uterine receptivity to implantation [52]. The pre- and early implantation periods require alterations in the gene expression both in the epithelium and in the trophoblast cells. These changes lead to the activation of signalling pathways regulating proliferation, growth, migration and invasion [53].

FKN expression and release were detected in the HEC-1A endometrial cell line [54] while the trophoblast cell line JEG-3 expressed CX3CR1 [14]. Recent publications revealed that FKN might regulate adhesion and migration of trophoblast cells at different stages of pregnancy [14,55] and the FKN-CX3CR1 axis was suggested to be implicated in the maternal-fetal communication [13]. In our study, we established JEG-3 monocultures modelling the pre-implantation period, when the trophoblast cells of conceptus interact with soluble fractalkine secreted by endometrial cells. Bilaminar co-cultures, in which JEG-3 and HEC-1A cells can get in physical contact with each other [10,42,56,57], were used to model the attachment of conceptus with uterine epithelium and the early implantation period. The aim of our study was to unravel the role of fractalkine in the regulation of implantation by the examination of a set of implantation-related genes in trophoblast cells (Figure S1).

Progesterone receptor (PR) is a crucial nuclear receptor regulating implantation and decidualization by the activation of its target genes [58]. The mediators (kinases and coactivators) of PR signalling are also important for successful embryo implantation [53]. During the pre-implantation period, PR initiates a complex signalling network to prepare the endometrium for embryo attachment and implantation that is called the window of receptivity. PR expression in the uterus is dynamically regulated by many factors [58].

We hypothesized that soluble FKN may influence PR expression in trophoblast cells; therefore, the mRNA and protein levels of PR were examined both in mono- and co-cultures. The mRNA expression of PR increased both in the mono- and co-cultures after fractalkine treatments. At protein level, PR showed a significant increase only using the highest FKN concentration at each time point of the JEG-3 monoculture. In the co-culture, PR protein expression revealed fluctuation. These results suggest a time dependent negative feedback in PR expression, and it is supposed that FKN acted in a concentration-dependent manner on PR.

We also examined SRC-1, the member of p160/steroid receptor coactivator family, the coregulatory molecule of PR at FKN treatment [59]. SRC-1 binds to the activated PR in the nucleus and triggers transcriptional response of PR [59]. SRC-1 significantly increased only in monocultured JEG-3 cells at 24 h both at mRNA and protein levels. In case of co-cultured JEG-3 cells, the increment of SRC-1 protein level at 24 h and 48 h showed parallel changes with the PR protein level suggesting the SRC-1 contributed to the PR action.

Activin receptor 1B is expressed by JEG-3 cells showed different expression levels after FKN treatments. Activin receptors are co-localized with activins and follistatins that are the regulators of activin receptor's activity [60]. Activin receptor regulates SMAD2/3 transcription factors that can be modulated by MAPKs as well [61]. In JEG-3 cells, mRNA expression of activin receptor was significantly elevated after 6 h and 24 h. The protein level of activin receptor followed the changes of the mRNA levels showing that FKN increased the activin receptor expression on the plasma membrane of JEG-3 cells. In the co-culture, the activin receptor B1 protein level showed fluctuation that may be caused by the paracrine effect of activin and follistatin produced by the endometrial cells [60].

MMPs play a critical role in the invasion of trophoblast cells. Among MMPs, MMP2 and 9 seem to be essential in the regulation of invasion and the behaviours of trophoblasts [62–64]. During the implantation, trophoblast cells release large amounts of MMP2 and 9 [65]. MMP expression is regulated by many factors e.g., activin signalling, BMP2 signalling and MAPK pathways [33–36]. FKN treatment increased MMP2 mRNA level at 6 h, MMP2 and MMP9 levels at 24 h only in the monocultured JEG-3 cells. MMP9 was elevated in the co-culture, proposing that MMP9 is more affective in the regulation of invasion than MMP2. At protein level, both MMP2 and MMP9 followed the alterations of mRNA expression in the JEG-3 cells. In the case of the co-cultured JEG-3 cells, we revealed that MMP2 protein level significantly elevated only at 48 h. The increased level of MMP9 protein supports the hypothesis that FKN-CX3CR1 axis acted as a regulator of MMP2 and MMP9 expressions.

These results raise the possibility that during the pre-implantation period (JEG-3 monoculture, at 24 h), fractalkine increases both MMP2 and MMP9 levels to preparing the embryo for implantation. Then at attachment and at early implantation period (co-culture model), MMP9 level increases continuously as the dominant enzyme regulating invasion, while MMP2 participates in the later steps of invasion.

CX3CR1 can be activated both by soluble and membrane-bound FKN. It seems that soluble FKN is important in JEG-3 monocultures acting through CX3CR1 regulating signalling pathways, PLC-PKC, PI3K/AKT/NFκB and MAPK pathways (p38, ERK1/2 and JNK) [15]. These pathways control the proliferation, growth, differentiation and apoptosis [18]. In the case of the co-cultures, both soluble and membrane bound FKN are present, which may contribute to the alterations of fractalkine receptor expression between mono- and co-cultured JEG-3 cells: elevated level in monocultured cells using F10 and 20 at 24 h, but reduced level using the same FKN concentrations at the same time point in co-cultured cells. Fluctuation of the CX3CR1 level in JEG-3 cells may be caused by MAPKs activated by the receptor. MAPKs may activate transcription factors (e.g., SP-1, AP-1, STATs) altering the transcriptional rate of CX3CR1 [66].

To determine the activity of downstream signalling pathways mediated by FKN-CX3CR1, we examined the phosphorylated forms of MAPKs (p38, ERK1/2, JNK) and AKT. The phosphorylation pattern was different in the two examined cell types. In monocultured JEG-3 cells p-ERK1/2 and p-p38 protein levels significantly increased at each time point using F20 although the rate of elevation showed decreasing phenomenon with time. On the contrary, p-JNK showed an increasing level with time, suggesting that p-JNK may replace the aforementioned enzymes when their amounts decrease. F10 exerted lower effect on the JEG-3 cells proposing a concentration dependence. In case of p-AKT, only F20 was able to increase its level. Co-cultured JEG-3 showed increased p-ERK1/2 levels at 6 h and 48 h while p-p38 and p-JNK elevated at 24 h proposing a cooperation between the kinases. In the case of p-AKT, we found similar changes to the p-p38 and p-JNK. The co-work of these three kinases may be responsible for the increased viability of co-cultured JEG-3 cells. Our results prove that FKN provides diverse effects on mono- and co-cultured trophoblasts and these alterations may be caused by the distinct impressions of soluble and membrane-bound FKN and may be influenced by other interactions between trophoblasts and endometrial cells.

It is proven that the presence of HEC-1A in the co-cultures contributes to the action of FKN on JEG-3 cells. This effect is probably mediated by membrane-bound fractalkine expressed by endometrial cells. We examined the expression of those genes and proteins that have been describe as the participants of implantation. We examined activin, follistatin and BMP2 expressions, which are the regulators of activin receptor and BMPR signalling. The activin and follistatin expression changed inversely in HEC-1A cells. The activin mRNA expression increased when the follistatin expression decreased and activin expression decreased when the follistatin mRNA levels were elevated. These results suggest that the interactions between the two cell types may influence the expressions of the activin receptor ligand activin and the binding protein of the ligand follistatin affecting receptor activity. The increasing activin level may contribute to the invasion capacity of trophoblast by acting on matrix metalloproteinase (MMP) expression [34].

Follistatins act as binding proteins for bone morphogenetic protein 2 (BMP2) as well, regulating BMP2 signalling through BMP receptors (BMPR) [67]. The BMP2 signalling inhibits the proliferation of trophoblasts but promotes trophoblast invasion by increasing metalloproteinase secretion [68]. BMP2 is a regulator of SMAD2/3 signalling, influencing the effect of activin receptors, and contributing to increased invasiveness [35,62]. BMP2 induces activin production as well, thus increases the activity of activin receptors [69]. In our experiments, BMP2 mRNA expression was significantly higher during the whole experiment compared to the control. At protein level, BMP2 showed fluctuation suggesting that the changing level of BMP2 in HEC-1A cells contribute to the signalling processes occurring in JEG-3 cells and may increase the action of fractalkine on trophoblast cells. Maybe the increased secretion of BMP2 affects the activity of BMPR elevating MMP expression and invasiveness of trophoblast cells.

Our experiments reveal that increasing fractalkine concentrations act differently on mono- and co-cultured JEG-3 cells. In the case of the monoculture the highest FKN concentration (20 ng/mL) was the most effective in increasing the expression of implantation-related genes. In the case of the co-culture, the intercellular interactions between HEC-1A and JEG-3 cells influenced the action of fractalkine treatment and it seems that both F10 and F20 affected the expression of the examined genes. The activated signalling pathways contributed to cell proliferation and then to the increased invasiveness. The results obtained from our in vitro co-culture experiments should be further investigated in vivo.

In this study, we examined the effect of FKN on the implantation-related genes expressed by trophoblast cells. FKN and CX3CR1 interaction is proven to be involved in the regulation of implantation influencing the expression of several genes (Figure 6). It is revealed that membrane-bound FKN and CX3CR1 receptor binding changes the expression and/or activity of PR by the activation of MAPK pathways [70]. In the co-cultures, it is revealed that HEC-1A cells alter the expressions of activin, follistatin and BMP2 regulators of activin receptor and BMP receptors. Secreted activin may bind to its receptor by paracrine way on trophoblast cells activating SMAD2/3 transcription factors that are also controlled by the BMP2/BMPR signalling pathway. SMAD2/3 pathway can upregulate MMP2 and MMP9 regulating invasion process. MAPKs activated by the FKN signal transduction pathway affect positively MMP2 and MMP9 expression, increasing the invasiveness of trophoblast cells. PR influenced by fractalkine can also modify MMP2 transcription. (Figure 6).

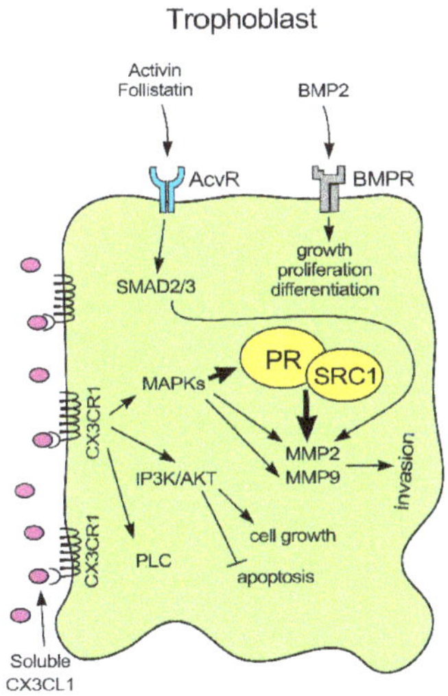

Figure 6. Mechanism of action of fractalkine on JEG-3 trophoblast cells Fractalkine binds to its cognate receptor on the surface of trophoblast cells activating MAPK and AKT pathways. MAPKs can phosphorylate PR on serine amino acid residue increasing its activity. Secreted activin may bind to activin receptor on the surface of trophoblast cells activating SMAD2/3 transcription factors that are also the targets of the BMP2/BMPR signalling pathway. SMAD2/3 transcription factors can upregulate MMP2 and MMP9 expressions. MAPKs are activated by fractalkine signal transduction pathway, act positively on MMP2 and MMP9 transcription. PR influenced by fractalkine can also affect MMP2 synthesis.

In summary, FKN acts in a concentration and time dependent manner on trophoblast cells. FKN is a positive regulator of implantation, via changing the activity of PR, activin receptor and BMPR and the expression of MMPs. The presence of endometrial cells contributes to the proper implantation as the interactions between trophoblast and endometrial cells in the co-culture changes the gene expression pattern of JEG-3 cells at fractalkine treatment. Based on our results, FKN treatment may contribute to successful attachment and implantation of the embryo.

4. Materials and Methods

4.1. Cell Cultures and Treatments

The JEG-3 human choriocarcinoma cell line is widely used to study the molecular mechanisms underlying the proliferation and invasive potential of trophoblast cells [54]. The HEC-1A adenocarcinoma derived endometrial cell line is a suitable in vitro model for non-receptive and receptive endometrium [71]. JEG-3 trophoblast-like choriocarcinoma cells (ATCC, HTB-36) were cultured in EMEM medium (Lonza Ltd. Basel, Switzerland) supplemented with 10% fetal bovine serum (FBS, EuroClone S.p.A, Pero, Italy) 1% Non-essential amino acids (NEAA, Lonza Ltd., Basel, Switzerland), 1% sodium-pyruvate (Lonza Ltd., Basel, Switzerland) and 1% Penicillin-Streptomycin (P/S, Lonza Ltd., Basel, Switzerland). HEC-1A endometrial cells (ATCC, HT-112) were maintained in McCoy's 5A medium (Corning Inc., Corning, NY, USA) supplemented with 10% FBS and 1% P/S. For the experiments, the culture medium was supplemented with charcoal/dextran treated FBS (EuroClone S.p.A, Pero, Italy). For the monocultures, JEG-3 cells were seeded onto culture dishes (60 mm, Corning Inc., Corning, NY, USA) in the appropriate culture medium and were treated after a 24 h resting period. For the co-culture experiments, JEG-3 cells were seeded on culture dishes while HEC-1A cells were placed on Thermanox coverslips (Thermo Fisher Scientific Inc., Waltham, MA, USA). After 24 h, HEC-1A cells were added to JEG-3 cells by turning the endometrial cell holding coverslips upside down facing the trophoblast-like cells [42]. This way the cells were separated only by a thin layer of 1:1 mixture of supplemented EMEM:McCoy's 5A medium. Both mono- and co-cultures were treated for 6 h, 24 h and 48 h with 5, 10 and 20 ng/mL human recombinant fractalkine (Shenandoah Biotechnology Inc., Warwick, PA, USA). Untreated mono- and co-cultured cells were used as controls.

4.2. Cell Viability Assay

The viability of JEG-3 monocultures was measured using Cell-counting Kit-8 (CCK-8) cell viability assay (Merck KGaA, Darmstadt, Germany) after the 6 h, 24 h and 48 h long fractalkine treatments. The viability of the co-culture was determined after the separation of the two cell types. Briefly, the monocultured JEG-3 cells were seeded on a 96-well culture plate. After fractalkine treatments, 10 μL of tetrazolium salt WST-8 reagent was added to each well and the plate was incubated for 1 h at 37 °C and 5% CO_2. The dehydrogenase reaction was stopped by adding 10 μL 1% sodium-dodecyl sulphate (SDS, Molar Chemicals Kft. Halásztelek, Hungary). The co-culture viability assay was performed in 24-well plate. After each treatment, JEG-3 and HEC-1A cells were separated. JEG-3 cells were incubated for 1 h at 37 °C and 5% CO_2 in the presence of 40 μL of WST-8 reagent then the reaction was halted by adding 40 μL 1% SDS to each sample. The absorbance of the mono- and co-cultured JEG-3 cells was measured at 450 nm using MultiSkan GO microplate spectrophotometer (Thermo Fisher Scientific Inc., Waltham, MA, USA). Viability was expressed as the percentile of the total cell number of the untreated control cells.

4.3. Real-Time PCR

Cells were harvested after washing three times with phosphate buffered saline (PBS, Lonza Ltd., Basel, Switzerland). Total RNA was isolated using Quick RNA mini kit (Zymo Research, Irvine, CA, USA). Complementary DNA was synthesized from 200 ng total RNA using iScript Select cDNA Synthesis Kit (Bio-Rad Inc., Hercules, CA, USA) according to the manufacturer's protocol. A gene

PLC	phospholipase C
PR	progesterone receptor
SMAD2/3	homologues of the Drosophila protein, mothers against decapentaplegic (Mad) and the *caenorhabtidis elegans* protein (Sma)
SRC-1	steroid receptor coactivator-1

References

1. Wang, H.; Zhang, S.; Lin, H.; Kong, S.; Wang, S.; Wang, H.; Armant, D.R. Physiological and molecular determinants of embryo implantation. *Mol. Asp. Med.* **2013**, *34*, 939–980.
2. Aplin, J.D.; Ruane, P.T. Embryo-epithelium interactions during implantation at a glance. *J. Cell Sci.* **2017**, *130*, 15–22. [CrossRef] [PubMed]
3. Garcia-Velasco, J.A.; Arici, A. Chemokines in human reproduction. *Immunol. Allergy Clin. N. Am.* **2002**, *22*, 567–583. [CrossRef]
4. Bazan, J.F.; Bacon, K.B.; Hardiman, G.; Wang, W.; Soo, K.; Rossi, D.; Greaves, D.R.; Zlotnik, A.; Schall, T.J. A new class of membrane-bound chemokine with a CX3C motif. *Nature* **1997**, *385*, 640–642. [CrossRef] [PubMed]
5. Pan, Y.; Lloyd, C.; Zhou, H.; Dolich, S.; Deeds, J.; Gonzalo, J.A.; Vath, J.; Gosselin, M.; Ma, J.; Dussault, B.; et al. Neurotactin, a membrane-anchored chemokine upregulated in brain inflammation. *Nature* **1997**, *387*, 611–617. [CrossRef]
6. Sheridan, G.K.; Murphy, K.J. Neuron-glia crosstalk in health and disease: Fractalkine and CX3CR1 take centre stage. *Open Biol.* **2013**, *3*, 130181. [CrossRef]
7. Wojdasiewicz, P.; Poniatowski, Ł.A.; Kotela, A.; Deszczyński, J.; Kotela, I.; Szukiewicz, D. The Chemokine CX3CL1 (Fractalkine) and its Receptor CX3CR1: Occurrence and Potential Role in Osteoarthritis. *Arch. Immunol. Ther. Exp.* **2014**, *62*, 395–403. [CrossRef]
8. Nagira, M.; Imai, T.; Yoshida, R.; Takagi, S.; Iwasaki, M.; Baba, M.; Tabira, Y.; Akagi, J.; Nomiyama, H.; Yoshie, O. A lymphocyte-specific CC chemokine, secondary lymphoid tissue chemokine (SLC), is a highly efficient chemoattractant for B cells and activated T cells. *Eur. J. Immunol.* **1998**, *28*, 1516–1523. [CrossRef]
9. Cardona, A.E.; Pioro, E.P.; Sasse, M.E.; Kostenko, V.; Cardona, S.M.; Dijkstra, I.M.; Huang, D.R.; Kidd, G.; Dombrowski, S.; Dutta, R.; et al. Control of microglial neurotoxicity by the fractalkine receptor. *Nat. Neurosci.* **2006**, *9*, 917–924. [CrossRef]
10. Pandur, E.; Tamási, K.; Pap, R.; Varga, E.; Miseta, A.; Sipos, K. Fractalkine Induces Hepcidin Expression of BV-2 Microglia and Causes Iron Accumulation in SH-SY5Y Cells. *Cell. Mol. Neurobiol.* **2019**, *39*, 985–1001. [CrossRef]
11. Hannan, N.J.; Jones, R.L.; Critchley, H.O.D.; Kovacs, G.J.; Rogers, P.A.W.; Affandi, B.; Salamonsen, L.A. Coexpression of fractalkine and its receptor in normal human endometrium and in endometrium from users of progestin-only contraception supports a role for fractalkine in leukocyte recruitment and endometrial remodeling. *J. Clin. Endocrinol. Metab.* **2004**, *89*, 6119–6129. [CrossRef] [PubMed]
12. Huang, S.; Zhao, P.; Yang, L.; Chen, Y.; Yan, J.; Duan, E.; Qiao, J. Fractalkine is expressed in the human ovary and increases progesterone biosynthesis in human luteinised granulosa cells. *Reprod. Biol. Endocrinol.* **2011**, *9*, 1–8. [CrossRef] [PubMed]
13. Kervancioglu Demirci, E.; Salamonsen, L.A.; Gauster, M. The role of CX3CL1 in fetal-maternal interaction during human gestation. *Cell Adhes. Migr.* **2016**, *10*, 189–196. [CrossRef] [PubMed]
14. Hannan, N.J.; Jones, R.L.; White, C.A.; Salamonsen, L.A. The Chemokines, CX3CL1, CCL14, and CCL4, Promote Human Trophoblast Migration at the Feto-Maternal Interface. *Biol. Reprod.* **2006**, *74*, 896–904. [CrossRef] [PubMed]
15. Bérangère Ré, D.; Przedborski, S. Fractalkine: Moving from chemotaxis to neuroprotection. *Nat. Neurosci.* **2006**, *9*, 859–861. [CrossRef] [PubMed]
16. Cargnello, M.; Roux, P.P. Activation and Function of the MAPKs and Their Substrates, the MAPK-Activated Protein Kinases. *Microbiol. Mol. Biol. Rev.* **2011**, *75*, 50–83. [CrossRef] [PubMed]
17. Widmann, C.; Gibson, S.; Jarpe, M.B.; Johnson, G.L. Mitogen-activated protein kinase: Conservation of a three-kinase module from yeast to human. *Physiol. Rev.* **1999**, *79*, 143–180. [CrossRef]

18. Sun, Y.; Liu, W.Z.; Liu, T.; Feng, X.; Yang, N.; Zhou, H.F. Signaling pathway of MAPK/ERK in cell proliferation, differentiation, migration, senescence and apoptosis. *J. Recept. Signal Transduct.* **2015**, *35*, 600–604. [CrossRef]
19. Chan, T.O.; Rittenhouse, S.E.; Tsichlis, P.N. AKT/PKB and Other D3 Phosphoinositide-Regulated Kinases: Kinase Activation by Phosphoinositide-Dependent Phosphorylation. *Annu. Rev. Biochem.* **1999**, *68*, 965–1014. [CrossRef]
20. Hers, I.; Vincent, E.E.; Tavaré, J.M. Akt signalling in health and disease. *Cell. Signal.* **2011**, *23*, 1515–1527. [CrossRef]
21. Alessandro, R.; Kohn, E.C. Signal transduction targets in invasion. *Clin. Exp. Metastasis* **2002**, *19*, 265–273. [CrossRef] [PubMed]
22. Salamonsen, L.A. Tissue injury and repair in the female human reproductive tract. *Reproduction* **2003**, *125*, 301–311. [CrossRef] [PubMed]
23. Matsumoto, H.; Nasu, K.; Nishida, M.; Ito, H.; Bing, S.; Miyakawa, I. Regulation of proliferation, motility, and contractility of human endometrial stromal cells by platelet-derived growth factor. *J. Clin. Endocrinol. Metab.* **2005**, *90*, 3560–3567. [CrossRef] [PubMed]
24. Chiu, D.; Ma, K.; Scott, A.; Duronio, V. Acute activation of Erk1/Erk2 and protein kinase B/akt proceed by independent pathways in multiple cell types. *FEBS J.* **2005**, *272*, 4372–4384. [CrossRef] [PubMed]
25. Maekawa, M.; Yamamoto, T.; Kohno, M.; Takeichi, M.; Nishida, E. Requirement for ERK MAP kinase in mouse preimplantation development. *Development* **2007**, *134*, 2751–2759. [CrossRef] [PubMed]
26. Halasz, M.; Szekeres-Bartho, J. The role of progesterone in implantation and trophoblast invasion. *J. Reprod. Immunol.* **2013**, *97*, 43–50. [CrossRef]
27. Hagan, C.R.; Daniel, A.R.; Dressing, G.E.; Lange, C.A. Role of phosphorylation in progesterone receptor signaling and specificity. *Mol. Cell. Endocrinol.* **2012**, *357*, 43–49. [CrossRef]
28. Gaide Chevronnay, H.P.; Selvais, C.; Emonard, H.; Galant, C.; Marbaix, E.; Henriet, P. Regulation of matrix metalloproteinases activity studied in human endometrium as a paradigm of cyclic tissue breakdown and regeneration. *Biochim. Biophys. Acta-Proteins Proteom.* **2012**, *1824*, 146–156. [CrossRef]
29. Goldman, S. Differential activity of the gelatinases (matrix metalloproteinases 2 and 9) in the fetal membranes and decidua, associated with labour. *Mol. Hum. Reprod.* **2003**, *9*, 367–373. [CrossRef]
30. Davidson, B.; Givant-Horwitz, V.; Lazarovici, P.; Risberg, B.; Nesland, J.M.; Trope, C.G.; Schaefer, E.; Reich, R. Matrix metalloproteinases (MMP), EMMPRIN (extracellular matrix metalloproteinase inducer) and mitogen-activated protein kinases (MAPK): Co-expression in metastatic serous ovarian carcinoma. *Clin. Exp. Metastasis* **2003**, *20*, 621–631. [CrossRef]
31. Loesch, M.; Zhi, H.Y.; Hou, S.W.; Qi, X.M.; Li, R.S.; Basir, Z.; Iftner, T.; Cuenda, A.; Chen, G. p38γ MAPK cooperates with c-Jun in trans-activating matrix metalloproteinase 9. *J. Biol. Chem.* **2010**, *285*, 15149–15158. [CrossRef] [PubMed]
32. Onogi, A.; Naruse, K.; Sado, T.; Tsunemi, T.; Shigetomi, H.; Noguchi, T.; Yamada, Y.; Akasaki, M.; Oi, H.; Kobayashi, H. Hypoxia inhibits invasion of extravillous trophoblast cells through reduction of matrix metalloproteinase (MMP)-2 activation in the early first trimester of human pregnancy. *Placenta* **2011**, *32*, 665–670. [CrossRef] [PubMed]
33. Staun-Ram, E.; Goldman, S.; Gabarin, D.; Shalev, E. Expression and importance of matrix metalloproteinase 2 and 9 (MMP-2 and -9) in human trophoblast invasion. *Reprod. Biol. Endocrinol.* **2004**, *2*, 1–13. [CrossRef] [PubMed]
34. Refaat, B. Role of activins in embryo implantation and diagnosis of ectopic pregnancy: A review. *Reprod. Biol. Endocrinol.* **2014**, *12*, 1–8. [CrossRef]
35. Zhao, H.J.; Klausen, C.; Li, Y.; Zhu, H.; Wang, Y.L.; Leung, P.C.K. Bone morphogenetic protein 2 promotes human trophoblast cell invasion by upregulating N-cadherin via non-canonical SMAD2/3 signaling. *Cell Death Dis.* **2018**, *9*, 4–15. [CrossRef] [PubMed]
36. Reddy, K.B.; Krueger, J.S.; Kondapaka, S.B.; Diglio, C.A. Mitogen-activated protein kinase (MAPK) regulates the expression of progelatinase B (MMP-9) in breast epithelial cells. *Int. J. Cancer* **1999**, *82*, 268–273. [CrossRef]
37. Wijayarathna, R.; de Kretser, D.M. Activins in reproductive biology and beyond. *Hum. Reprod. Update* **2016**, *22*, 342–357. [CrossRef]
38. Jones, R.L.; Kaitu'u-Lino, T.J.; Nie, G.; Sanchez-Partida, L.G.; Findlay, J.K.; Salamonsen, L.A. Complex expression patterns support potential roles for maternally derived activins in the establisment of pregnancy in mouse. *Reproduction* **2006**, *132*, 799–810. [CrossRef]

39. Thompson, T.B.; Lerch, T.F.; Cook, R.W.; Woodruff, T.K.; Jardetzky, T.S. The structure of the follistatin: Activin complex reveals antagonism of both type I and type II receptor binding. *Dev. Cell* **2005**, *9*, 535–543. [CrossRef]
40. Hemmati-Brivanlou, A.; Thomsen, G.H. Ventral mesodermal patterning in Xenopus embryos: Expression patterns and activities of BMP-2 and BMP-4. *Dev. Genet.* **1995**, *17*, 78–89. [CrossRef]
41. Sidis, Y.; Mukherjee, A.; Keutmann, H.; Delbaere, A.; Sadatsuki, M.; Schneyer, A. Biological activity of follistatin isoforms and follistatin-like-3 is dependent on differential cell surface binding and specificity for activin, myostatin, and bone morphogenetic proteins. *Endocrinology* **2006**, *147*, 3586–3597. [CrossRef] [PubMed]
42. Shimizu, S.; Abt, A.; Meucci, O. Bilaminar co-culture of primary rat cortical neurons and glia. *J. Vis. Exp.* **2011**, 1–5. [CrossRef] [PubMed]
43. Xu, S.Y.; Wu, Y.M.; Ji, Z.; Gao, X.Y.; Pan, S.Y. A modified technique for culturing primary fetal rat cortical neurons. *J. Biomed. Biotechnol.* **2012**, *2012*. [CrossRef] [PubMed]
44. Cook, A.; Hippensteel, R.; Shimizu, S.; Nicolai, J.; Fatatis, A.; Meucci, O. Interactions between Chemokines: Regulation of fractalkine/CX 3CL1 homeostasis by SDF/CXCL12 in cortical neurons. *J. Biol. Chem.* **2010**, *285*, 10563–10571. [CrossRef] [PubMed]
45. Richardson, S.M.; Walker, R.V.; Parker, S.; Rhodes, N.P.; Hunt, J.A.; Freemont, A.J.; Hoyland, J.A. Intervertebral Disc Cell-Mediated Mesenchymal Stem Cell Differentiation. *Stem Cells* **2006**, *24*, 707–716. [CrossRef]
46. Hatori, K.; Nagai, A.; Heisel, R.; Ryu, J.K.; Kim, S.U. Fractalkine and fractalkine receptors in human neurons and glial cells. *J. Neurosci. Res.* **2002**, *69*, 418–426. [CrossRef]
47. White, G.E.; Greaves, D.R. Fractalkine: A survivor's guide chemokines as antiapoptotic mediators. *Arterioscler. Thromb. Vasc. Biol.* **2012**, *32*, 589–594. [CrossRef]
48. Szepesi, Z.; Manouchehrian, O.; Bachiller, S.; Deierborg, T. Bidirectional Microglia–Neuron Communication in Health and Disease. *Front. Cell. Neurosci.* **2018**, *12*, 1–26. [CrossRef]
49. Zhuang, Q.; Ou, J.; Zhang, S.; Ming, Y. Crosstalk between the CX3CL1/CX3CR1 Axis and Inflammatory Signaling Pathways in Tissue Injury. *Curr. Protein Pept. Sci.* **2019**, *20*, 844–854. [CrossRef]
50. Jones, R.L.; Salamonsen, L.A.; Findlay, J.K. Potential roles for endometrial inhibins, activins and follistatin during human embryo implantation and early pregnancy. *Trends Endocrinol. Metab.* **2002**, *13*, 144–150. [CrossRef]
51. Minas, V.; Loutradis, D.; Makrigiannakis, A. Factors controlling blastocyst implantation. *Reprod. Biomed. Online* **2005**, *10*, 205–216. [CrossRef]
52. de Ruijter-Villani, M.; Stout, T. The Role of Conceptus-maternal Signalling in the Acquisition of Uterine Receptivity to Implantation in Mammals. *Reprod. Domest. Anim.* **2015**, *50*, 7–14. [CrossRef] [PubMed]
53. Wetendorf, M.; DeMayo, F.J. Progesterone receptor signaling in the initiation of pregnancy and preservation of a healthy uterus. *Int. J. Dev. Biol.* **2014**, *58*, 95–106. [CrossRef] [PubMed]
54. Hannan, N.J.; Paiva, P.; Dimitriadis, E.; Salamonsen, L.A. Models for Study of Human Embryo Implantation: Choice of Cell Lines? *Biol. Reprod.* **2010**, *82*, 235–245. [CrossRef]
55. Hannan, N.J.; Salamonsen, L.A. CX3CL1 and CCL14 Regulate Extracellular Matrix and Adhesion Molecules in the Trophoblast: Potential Roles in Human Embryo Implantation. *Biol. Reprod.* **2008**, *79*, 58–65. [CrossRef]
56. Pandur, E.; Varga, E.; Tamási, K.; Pap, R.; Nagy, J.; Sipos, K. Effect of inflammatory mediators lipopolysaccharide and lipoteichoic acid on iron metabolism of differentiated SH-SY5Y cells alters in the presence of BV-2 microglia. *Int. J. Mol. Sci.* **2019**, *20*, 17. [CrossRef]
57. Rothbauer, M.; Patel, N.; Gondola, H.; Siwetz, M.; Huppertz, B.; Ertl, P. A comparative study of five physiological key parameters between four different human trophoblast-derived cell lines. *Sci. Rep.* **2017**, *7*, 1–11. [CrossRef]
58. Franco, H.L.; Jeong, J.W.; Tsai, S.Y.; Lydon, J.P.; DeMayo, F.J. In vivo analysis of progesterone receptor action in the uterus during embryo implantation. *Semin. Cell Dev. Biol.* **2008**, *19*, 178–186. [CrossRef]
59. Szwarc, M.M.; Kommagani, R.; Lessey, B.A.; Lydon, J.P. The p160/Steroid Receptor Coactivator Family: Potent Arbiters of Uterine Physiology and Dysfunction1. *Biol. Reprod.* **2014**, *91*, 1–11. [CrossRef]
60. Jones, R.L. Expression of activin receptors, follistatin and betaglycan by human endometrial stromal cells; consistent with a role for activins during decidualization. *Mol. Hum. Reprod.* **2002**, *8*, 363–374. [CrossRef]
61. Jones, R.L.; Salamonsen, L.A.; Findlay, J.K. Activin A promotes human endometrial stromal cell decidualization in vitro. *J. Clin. Endocrinol. Metab.* **2002**, *87*, 4001–4004. [CrossRef] [PubMed]

62. Massimiani, M.; Lacconi, V.; La Civita, F.; Ticconi, C.; Rago, R.; Campagnolo, L. Molecular signaling regulating endometrium–blastocyst crosstalk. *Int. J. Mol. Sci.* **2020**, *21*, 23. [CrossRef] [PubMed]
63. Maia-Filho, V.O.A.; Rocha, A.M.; Ferreira, F.P.; Bonetti, T.C.S.; Serafini, P.; Motta, E.L.A. Matrix metalloproteinases 2 and 9 and E-cadherin expression in the endometrium during the implantation window of infertile women before in vitro fertilization treatment. *Reprod. Sci.* **2015**, *22*, 416–422. [CrossRef] [PubMed]
64. Luo, J.; Qiao, F.; Yin, X. Impact of silencing MMP9 gene on the biological behaviors of trophoblasts. *J. Huazhong Univ. Sci. Technol. Med. Sci.* **2011**, *31*, 241–245. [CrossRef] [PubMed]
65. Hess, A.P.; Hamilton, A.E.; Talbi, S.; Dosiou, C.; Nyegaard, M.; Nayak, N.; Genbecev-Krtolica, O.; Mavrogianis, P.; Ferrer, K.; Kruessel, J.; et al. Decidual Stromal Cell Response to Paracrine Signals from the Trophoblast: Amplification of Immune and Angiogenic Modulators1. *Biol. Reprod.* **2007**, *76*, 102–117. [CrossRef] [PubMed]
66. Garin, A.; Pellet, P.; Deterre, P.; Debre, P.; Combadière, C. Cloning and functional characterization of the human fractalkine receptor promoter regions. *Biochem. J.* **2002**, *368*, 753–760. [CrossRef] [PubMed]
67. Bhurke, A.S.; Bagchi, I.C.; Bagchi, M.K. Progesterone-Regulated Endometrial Factors Controlling Implantation. *Am. J. Reprod. Immunol.* **2016**, *75*, 237–245. [CrossRef] [PubMed]
68. Wang, R.N.; Green, J.; Wang, Z.; Deng, Y.; Qiao, M.; Peabody, M.; Zhang, Q.; Ye, J.; Yan, Z.; Denduluri, S.; et al. Bone Morphogenetic Protein (BMP) signaling in development and human diseases. *Genes Dis.* **2014**, *1*, 87–105. [CrossRef]
69. Zhao, H.J.; Chang, H.M.; Zhu, H.; Klausen, C.; Li, Y.; Leung, P.C.K. Bone morphogenetic protein 2 promotes human trophoblast cell invasion by inducing activin A production. *Endocrinology* **2018**, *159*, 2815–2825. [CrossRef]
70. Treviño, L.S.; Bingman, W.E.; Edwards, D.P.; NI, W. The requirement for p42/p44 MAPK activity in progesterone receptor-mediated gene regulation is target gene-specific. *Steroids* **2013**, *78*, 542–547. [CrossRef]
71. Tamm, K.; Rõõm, M.; Salumets, A.; Metsis, M. Genes targeted by the estrogen and progesterone receptors in the human endometrial cell lines HEC1A and RL95-2. *Reprod. Biol. Endocrinol.* **2009**, *7*, 1–12. [CrossRef] [PubMed]
72. Bars-Cortina, D.; Riera-Escamilla, A.; Gou, G.; Piñol-Felis, C.; Motilva, M.J. Design, optimization and validation of genes commonly used in expression studies on DMH/AOM rat colon carcinogenesis model. *PeerJ* **2019**, *2019*, 1–18. [CrossRef] [PubMed]
73. ImageJ. Available online: https://imagej.nih.gov/ij/ (accessed on 23 September 1997).

International Journal of
Molecular Sciences

MDPI

Article

Trophectoderm-Specific Knockdown of LIN28 Decreases Expression of Genes Necessary for Cell Proliferation and Reduces Elongation of Sheep Conceptus

Asghar Ali [1], Mark D. Stenglein [2], Thomas E. Spencer [3], Gerrit J. Bouma [1], Russell V. Anthony [1] and Quinton A. Winger [1,*]

1 Department of Biomedical Sciences, Animal Reproduction and Biotechnology Laboratory, 1683 Campus Delivery, Colorado State University, Fort Collins, CO 80523, USA; asghar.ali@colostate.edu (A.A.); gerrit.bouma@colostate.edu (G.J.B.); Russ.Anthony@ColoState.edu (R.V.A.)

2 Department of Microbiology, Immunology and Pathology, Colorado State University, Fort Collins, CO 80523, USA; mark.stenglein@colostate.edu

3 Animal Science Research Center, College of Agriculture, Food and Natural Resources, University of Missouri, Columbia, MO 65211, USA; spencerte@missouri.edu

* Correspondence: Quinton.winger@colostate.edu; Tel.: +970-491-7702; Fax: +970-491-3557

Received: 13 March 2020; Accepted: 3 April 2020; Published: 6 April 2020

Abstract: LIN28 inhibits *let-7* miRNA maturation which prevents cell differentiation and promotes proliferation. We hypothesized that the LIN28-*let-7* axis regulates proliferation-associated genes in sheep trophectoderm in vivo. Day 9-hatched sheep blastocysts were incubated with lentiviral particles to deliver shRNA targeting LIN28 specifically to trophectoderm cells. At day 16, conceptus elongation was significantly reduced in LIN28A and LIN28B knockdowns. *Let-7* miRNAs were significantly increased and IGF2BP1-3, HMGA1, ARID3B, and c-MYC were decreased in trophectoderm from knockdown conceptuses. Ovine trophoblast (OTR) cells derived from day 16 trophectoderm are a useful tool for in vitro experiments. Surprisingly, LIN28 was significantly reduced and *let-7* miRNAs increased after only a few passages of OTR cells, suggesting these passaged cells represent a more differentiated phenotype. To create an OTR cell line more similar to day 16 trophectoderm we overexpressed LIN28A and LIN28B, which significantly decreased *let-7* miRNAs and increased IGF2BP1-3, HMGA1, ARID3B, and c-MYC compared to control. This is the first study showing the role of the LIN28-*let-7* axis in trophoblast proliferation and conceptus elongation in vivo. These results suggest that reduced LIN28 during early placental development can lead to reduced trophoblast proliferation and sheep conceptus elongation at a critical period for successful establishment of pregnancy.

Keywords: trophectoderm; placenta; cell proliferation; *let-7* miRNAs; gene regulation

---

1. Introduction

Early placental development is one of the main factors determining perinatal fetal growth and postnatal fetal and maternal health. In humans, blastocyst implantation is an invasive process that occurs 7–9 days after fertilization [1]. Rapidly proliferating cytotrophoblast cells (CTBs) are the progenitor trophoblast cells which proliferate as well as differentiate into different trophoblast lineages throughout gestation [2]. If the balance between proliferation and differentiation of CTBs is dysregulated, it can result in severe disorders including preterm birth, intrauterine growth restriction (IUGR), and preeclampsia [3,4]. These pregnancy related disorders affect about a third of human pregnancies [5].

In sheep, the blastocyst hatches out of the zona pellucida at day 8–9 and is surrounded by a single layer of mononuclear cells called trophectoderm (TE) [6]. Instead of invading the uterus, the hatched blastocyst elongates from day 11–16 due to rapid proliferation of trophoblast cells and adopts a filamentous shape comprised of mainly extraembryonic trophoblast cells [7–9]. Conceptus elongation is critical for implantation, placentation, and successful establishment of pregnancy in sheep [10–12]. Reduced conceptus elongation and compromised placental function in domestic ruminants is one of the main causes of embryonic mortality resulting in reduced fertility [13–15]. Rapid trophoblast proliferation is an important phenomenon during early stages of pregnancy in both humans and domestic ruminants. The molecular mechanisms involved in regulating trophoblast proliferation and invasion are not well understood. Therefore, exploring the genes involved in sheep trophectoderm elongation can help to better understand the reasons for reduced fertility in domestic ruminants and to improve the diagnosis and treatment of various pregnancy-related disorders in humans.

Trophoblast proliferation and differentiation is an intensively regulated process, and the role of several genes in placental development has been studied using various in vivo and in vitro models [16–20]. The pluripotency factor LIN28 is a highly conserved RNA binding protein which is expressed in placenta and has two paralogs, LIN28A and LIN28B [21,22]. It is usually described as a protooncogene due to its ability to regulate and stabilize oncogenes at the post-transcriptional level in tumor cells [23,24]. It also inhibits the biogenesis of lethal-7 (*let-7)* miRNAs in mammalian cells by binding pri-*let-7* and pre-*let-7* [25–30]. LIN28 is low and *let-7* miRNAs are high in differentiated cells and adult tissues, hence *let-7* miRNAs are considered markers of cell differentiation [31–33]. *Let-7* miRNAs reduce the expression of different proliferation factors either by directly targeting their mRNA or through chromatin-dependent pathways by targeting the ARID3B-complex, which is comprised of AT-Rich Interaction Domain 3A (ARID3A), AT-Rich Interaction Domain 3B (ARID3B) and lysine demethylase 4C (KDM4C) [18,34]. We recently showed that term human placentas from IUGR pregnancies had reduced LIN28A and LIN28B and high *let-7* miRNAs compared to term human placentas from control pregnancies [18]. We further demonstrated that LIN28B is localized to cytotrophoblast cells in human placenta, and knockout of LIN28 in immortalized first trimester human trophoblast (ACH-3P) cells leads to an increase in *let-7* miRNAs, reduced expression of proliferation-associated genes, and reduced cell proliferation [18–20].

Insulin like growth factor 2 mRNA binding proteins (*IGF2BP1, IGF2BP2, IGF2BP3*), high mobility group AT-hook 1 (*HMGA1*), *ARID3B*, and MYC protooncogene (*c-MYC)* are all *let-7* miRNA targets with known roles in cell proliferation [18,35–41]. IGF2BPs are highly conserved RNA binding oncofetal proteins with three paralogs, IGF2BP1, IGF2BP2, and IGF2BP3 [42]. By binding different mRNAs, IGF2BPs decide the fate of those mRNAs by controlling their localization, stability, and translation [40]. Many studies have reported the role of IGF2BPs in cell proliferation, cell invasion, tumorigenesis, and embryogenesis [40–51]. IGF2BPs have also been found in sheep trophoblast cells suggesting their role in rapid proliferation of these cells [52]. HMGA1 promotes invasion of trophoblast cells and reduced levels of HMGA1 has been linked to pathogenesis of preeclampsia [53,54]. ARID3B binds with ARID3A and KDM4C to form the ARID3B-complex. The ARID3B-complex plays a vital role in cell proliferation by transcriptional regulation of stemness genes including *HMGA1*, *c-MYC*, vascular endothelial growth factor A (*VEGF-A*), and Wnt family member 1 (*WNT1*) [18,34,55–59]. ARID3B knockout in immortalized first trimester human trophoblast cells results in reduced proliferation of these cells [18]. The *c-MYC* protooncogene has been identified as a proliferation factor in human cytotrophoblast cells and its level is reduced when cytotrophoblasts differentiate into syncytiotrophoblast [60].

To date, the role of LIN28-*let-7* miRNA axis in trophoblast cells has not been studied in vivo. The aim of this study was to demonstrate the role of LIN28-*let-7* axis in the regulation of proliferation-associated genes in trophoblast cells in vivo. We used sheep as an in vivo model to generate trophectoderm specific knockdown of LIN28A or LIN28B by infecting day 9-hatched blastocysts with shRNA-expressing lentiviral particles. This way, only the trophoblast cells will be infected by the lentiviral particles [61–63] and any phenotype will be due to knockdown of LIN28A or

LIN28B in trophoblast cells. We hypothesized that the LIN28-*let-7* miRNAs axis plays an important role in sheep trophoblast cell proliferation and conceptus elongation by regulating the expression of genes associated with cell proliferation including *IGF2BP1, IGF2BP2, IGF2BP3, HMGA1, ARID3B,* and *c-MYC*.

2. Results

2.1. LIN28 Knockdown in Trophectoderm Resulted in Reduced Proliferation of Trophoblast Cells and Lower Expression of Proliferation-Associated Genes

Day 9-hatched blastocysts were infected with lentiviral particles expressing shRNA to knockdown LIN28A (AKD) or LIN28B (BKD), or scramble control shRNA (SC). The blastocysts treated with lentiviral particles were surgically transferred to synchronized ewes at day 9 of estrus. The conceptuses were collected at day 16 of gestation and trophectoderm (TE) was separated from embryo. LIN28A and LIN28B mRNAs and proteins were quantified by real-time PCR and Western blot. LIN28A mRNA and protein was significantly reduced in AKD TE while LIN28B mRNA and protein was significantly reduced in BKD TE compared to SC (Figure 1A,B). As expected, due to reduced LIN28, *let-7* miRNAs (*let-7a, let-7b, let-7c, let-7d, let-7e, let-7f, let-7g, let-7i*) were significantly higher in AKD and BKD TE compared to SC, and there was no significant change in *let-7* miRNAs between AKD and BKD TE (Figure 1C). These results suggest that reduced LIN28A or LIN28B led to significant increase in *let-7* miRNAs.

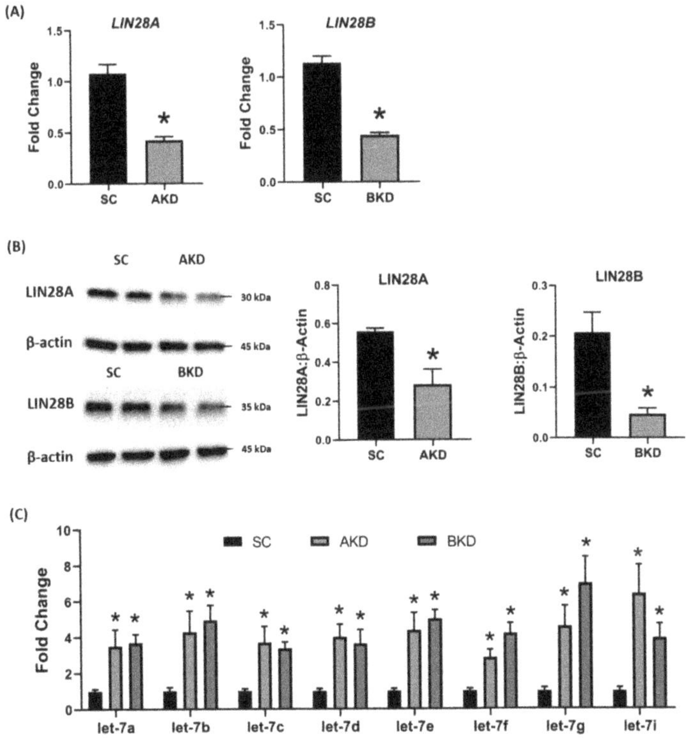

Figure 1. LIN28A or LIN28B knockdown and *let-7* miRNAs in day 16 sheep TE. (**A**) *LIN28A* and *LIN28B* mRNA in AKD ($n = 5$) and BKD ($n = 6$) day 16 TE compared to SC ($n = 6$). (**B**) Representative immunoblots for LIN28A, LIN28B, and β-actin in AKD, BKD, and SC day 16 TE, and densitometric analysis. (**C**) *Let-7* miRNAs in AKD and BKD day 16 TE and SC. * $p < 0.05$ vs. SC.

To determine the effect of LIN28 knockdown on conceptus elongation, we measured the length of TE. Knockdown of LIN28A or LIN28B resulted in a significant reduction in day 16 TE length compared to SC (Figure 2A,B). There was no significant difference in elongation of AKD vs. BKD TE. This data suggest knockdown of either LIN28A or LIN28B in vivo resulted in reduced proliferation of trophoblast cells.

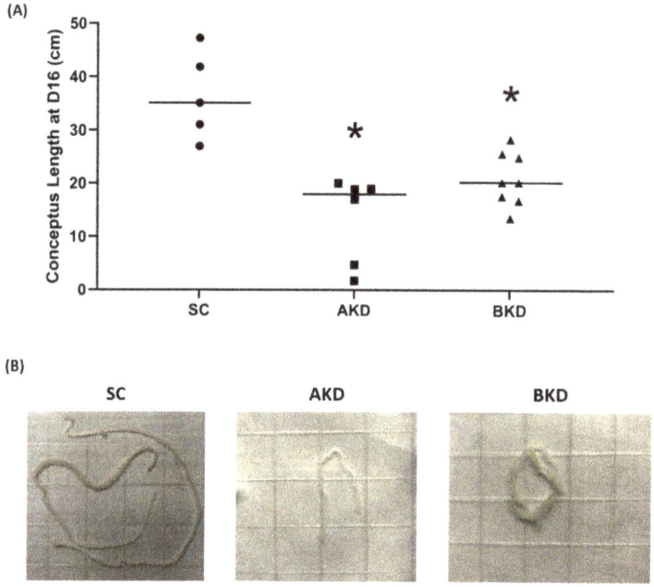

Figure 2. (**A**) Conceptus length at day 16 after LIN28A knockdown (AKD, $n = 6$) and LIN28B knockdown (BKD, $n = 8$) compared to scramble control (SC, $n = 5$). (**B**) Representative images of day 16 sheep conceptuses for AKD, BKD, and SC day 16 TE; * $p < 0.05$ vs. SC.

Due to the potential for reduced proliferation of trophoblast cells in the AKD and BKD conceptuses, we measured the mRNA and protein levels of *let-7*-regulated proliferation-associated genes. The mRNA and protein levels of IGF2BP1, IGF2BP2, IGF2BP3, HMGA1, ARID3B, and c-MYC were significantly reduced in AKD and BKD day 16 TE compared to SC (Figure 3, Figure 4). These results suggest that high *let-7* miRNAs in AKD and BKD TE led to a significant reduction in expression of IGF2BP1, IGF2BP2, IGF2BP3, HMGA1, ARID3B, and c-MYC, and the reduced proliferation of trophoblast cells in AKD and BKD TE is due to significantly reduced expression of these proliferation-associated genes.

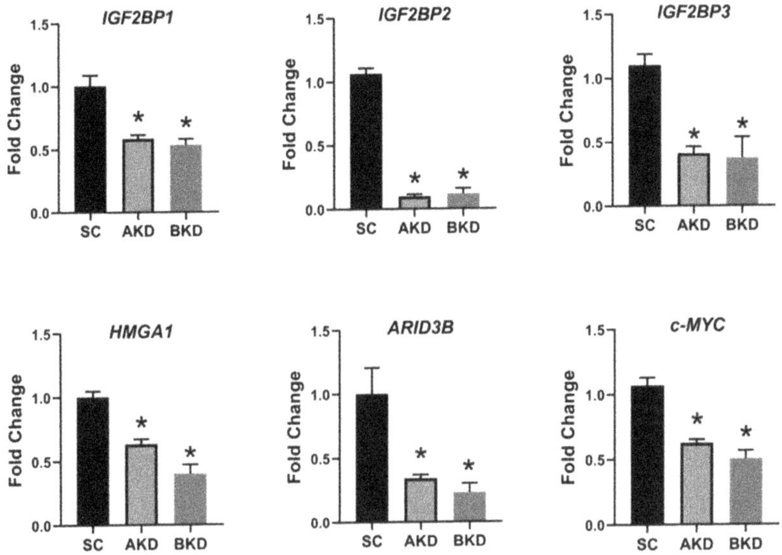

Figure 3. *IGF2BP1, IGF2BP2, IGF2BP3, HMGA1, ARID3B,* and *c-MYC* mRNA in AKD and BKD day 16 TE compared to SC ($n = 5$), * $p < 0.05$ vs. SC.

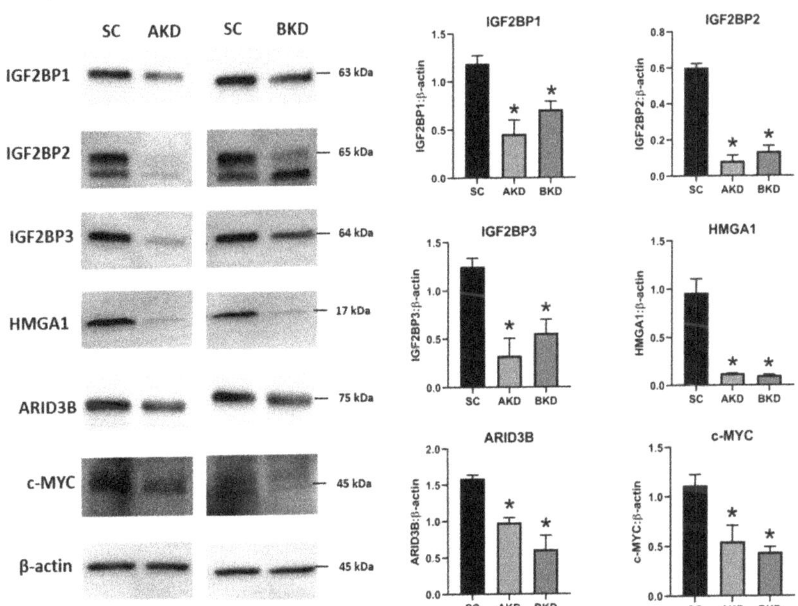

Figure 4. Representative immunoblots for IGF2BP1, IGF2BP2, IGF2BP3, HMGA1, ARID3B, c-MYC, and β-actin, and densitometric analysis of immunoblotting results in AKD and BKD day 16 sheep TE compared to SC ($n = 3$), * $p < 0.05$ vs. SC.

2.2. Ovine Trophoblast Cells Generated from Day 16 Trophectoderm Had a Significant Reduction in LIN28

To further investigate the regulation of ovine trophoblast cell proliferation by LIN28 in vitro, we used day 16 TE to generate ovine trophoblast cells. The day 16 TE was minced and plated in collagen-coated plates and was passaged to obtain a cell line. The cells used for further experiments were collected at passage 4–6, so we called these cells non-immortalized ovine trophoblast (OTR) cells. Interestingly, real-time PCR data showed that OTR cells had a significant reduction in *LIN28A* and *LIN28B* mRNAs compared to day 16 TE (Figure 5A). Densitometric analysis of Western blots showed that LIN28A and LIN28B proteins were also significantly reduced in OTR cells compared to day 16 TE (Figure 5B). Furthermore, real-time PCR data showed significant increase in *let-7* miRNAs (*let-7a, let-7b, let-7c, let-7d, let-7e, let-7f, let-7g, let-7i*) in OTR cells compared to day 16 TE (Figure 5C). The significantly reduced LIN28 and high *let-7* miRNAs in OTR cells after only 4–6 passages suggest that these cells differentiated to a different phenotype compared to trophoblast cells in day 16 TE.

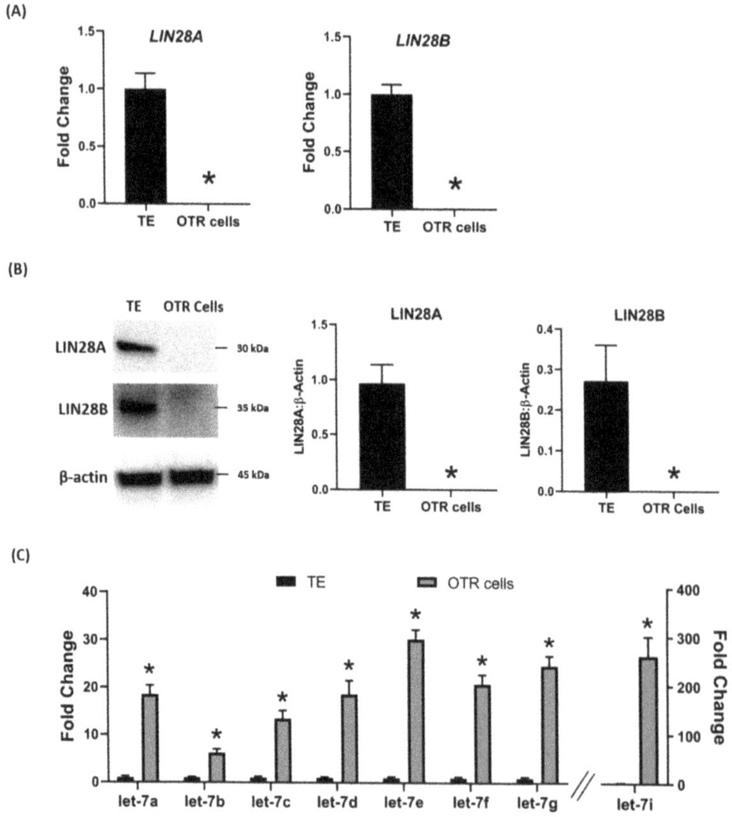

Figure 5. LIN28A, LIN28B, and *let-7* miRNAs in non-immortalized ovine trophoblast (OTR) cells. (**A**) *LIN28A* and *LIN28B* mRNA in OTR cells compared to day 16 sheep TE. (**B**) Representative immunoblots for LIN28A, LIN28B, and β-actin, and densitometric analysis of immunoblotting results in OTR cells compared to day 16 TE. (**C**) *Let-7* miRNAs in OTR cells compared to day 16 TE ($n = 3$), * $p < 0.05$ vs. TE.

The effect of low LIN28 and high *let-7* miRNAs on proliferation-associated genes in OTR cells was determined by measuring IGF2BP1, IGF2BP2, IGF2BP3, HMGA1, ARID3B, and c-MYC mRNAs and proteins. Real-time PCR showed that mRNA levels of *IGF2BP1, IGF2BP2, IGF2BP3, HMGA1, ARID3B,* and *c-MYC* were significantly reduced in OTR cells compared to day 16 TE (Figure 6). Densitometric analysis of Western blots revealed a significant reduction in protein levels of IGF2BP1, IGF2BP2, IGF2BP3, HMGA1, ARID3B, and c-MYC in OTR cells compared to day 16 TE (Figure 7). These results suggest that reduced LIN28 and high *let-7* miRNAs led to reduced expression of proliferation-associated genes in OTR cells.

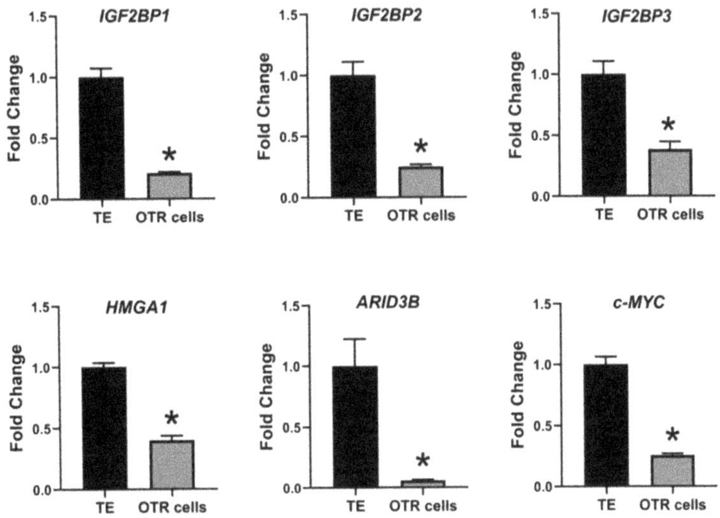

Figure 6. *IGF2BP1, IGF2BP2, IGF2BP3, HMGA1, ARID3B,* and *c-MYC* mRNA in OTR cells compared to day 16 TE ($n = 3$), * $p < 0.05$ vs. TE.

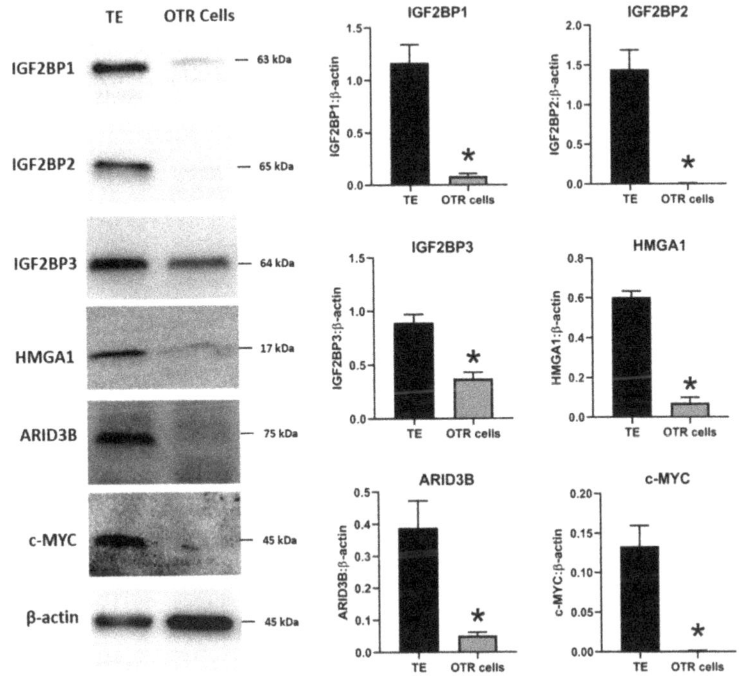

Figure 7. Representative immunoblots for IGF2BP1, IGF2BP2, IGF2BP3, HMGA1, ARID3B, c-MYC, and β-actin, and densitometric analysis of immunoblotting results in OTR cells compared to day 16 TE ($n = 3$), * $p < 0.05$ vs. TE.

The OTR cells originated from day 16 TE undergo senescence after only a few passages. Therefore, the OTR cells were immortalized by overexpressing human telomerase reverse transcriptase (hTERT) to keep them growing for further in vitro experiments. The newly generated immortalized cells were referred to as immortalized ovine trophoblast (iOTR) cells.

2.3. Overexpression of LIN28 in iOTR Cells Resulted in Increased Expression of Proliferation-Associated Genes

To determine if LIN28 overexpression will rescue the expression of proliferation-associated genes, the iOTR cells were infected with lentiviral particles to generate LIN28A knockin (AKI) or LIN28B knockin (BKI), or lentiviral particles with empty expression vector as expression vector control (EVC). Real-time PCR data showed that AKI iOTR cells had a significant increase in *LIN28A* mRNA while BKI iOTR cells had a significant increase in *LIN28B* mRNA compared to EVC (Figure 8A). The densitometric analysis of Western blots showed a significant increase in LIN28A protein in AKI and significant increase in LIN28B protein in BKI iOTR cells compared to EVC (Figure 8B). Moreover, the real-time PCR data showed that *let-7* miRNAs (*let-7a, let-7b, let-7c, let-7d, let-7e, let-7f, let-7g, let-7i*) were significantly reduced in both AKI and BKI iOTR cells compared to EVC (Figure 8C). These results suggest that increased expression of either LIN28A or LIN28B leads to reduction in *let-7* miRNAs.

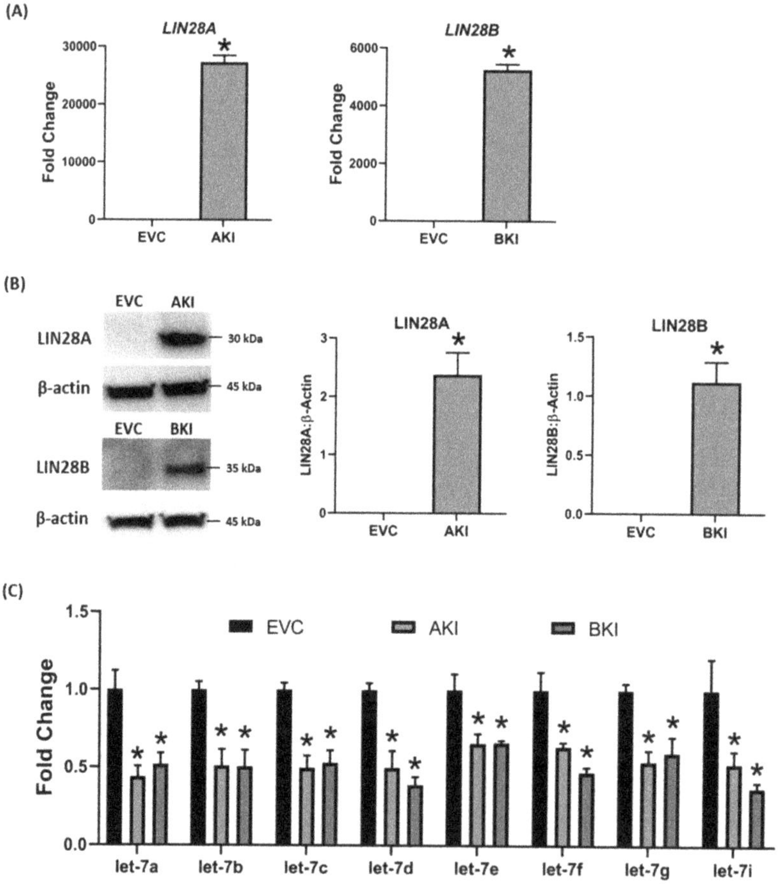

Figure 8. LIN28A, LIN28B, and *let-7* miRNAs in LIN28A knockin (AKI) and LIN28B knockin (BKI) immortalized ovine trophoblast (iOTR) cells. (**A**) *LIN28A* and *LIN28B* mRNA in AKI and BKI iOTR cells compared to expression vector control (EVC). (**B**) Representative immunoblots for LIN28A, LIN28B, and β-actin, and densitometric analysis of immunoblotting results in AKI and BKI iOTR cells compared to EVC. (**C**) *Let-7* miRNAs in AKI and BKI iOTR cells compared to EVC ($n = 3$), * $p < 0.05$ vs. EVC.

To determine the effect of LIN28 overexpression and reduction in *let-7* miRNAs on expression of proliferation-associated genes, real-time PCR and Western blot analysis were done. Real-time PCR showed that mRNA levels of *IGF2BP1, IGF2BP2, IGF2BP3, HMGA1, ARID3B,* and *c-MYC* were significantly increased in both AKI and BKI iOTR cells compared to EVC (Figure 9). Densitometric analysis of Western blots showed significant increase in IGF2BP1, IGF2BP2, IGF2BP3, HMGA1, ARID3B, and c-MYC proteins in both AKI and BKI iOTR cells compared to EVC (Figure 10). These results suggest that the expression of proliferation-associated genes in immortalized ovine trophoblast cells is regulated by the LIN28-*let-7* axis.

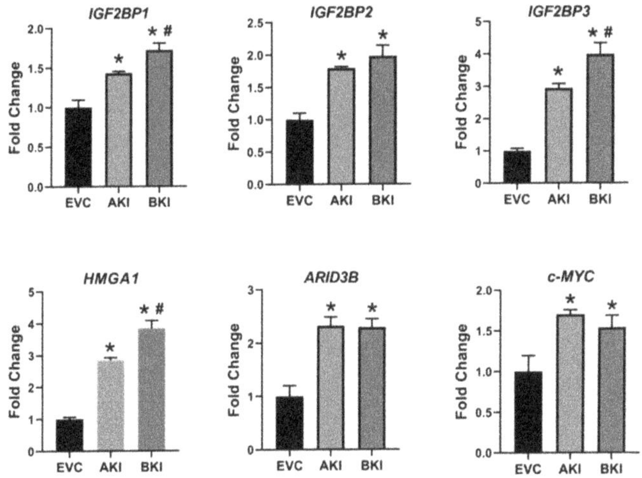

Figure 9. *IGF2BP1, IGF2BP2, IGF2BP3, HMGA1, ARID3B,* and *c-MYC* mRNA in AKI and BKI iOTR cells compared to EVC ($n = 3$), where * $p < 0.05$ vs. EVC and # $p < 0.05$ vs AKI.

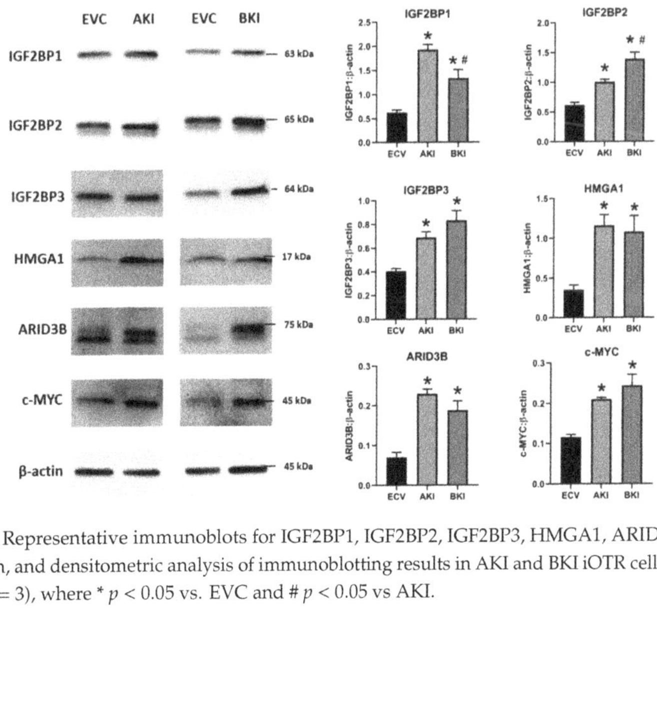

Figure 10. Representative immunoblots for IGF2BP1, IGF2BP2, IGF2BP3, HMGA1, ARID3B, c-MYC, and β-actin, and densitometric analysis of immunoblotting results in AKI and BKI iOTR cells compared to EVC ($n = 3$), where * $p < 0.05$ vs. EVC and # $p < 0.05$ vs AKI.

2.4. Overexpression of LIN28 Led to Significant Increase in Trophoblast Cell Proliferation

The role of LIN28-*let-7* miRNA axis on the functionality of iOTR cells was determined by measuring proliferation of AKI and BKI iOTR cells compared to EVC after 4 h, 24 h, 48 h, and 72 h. The results showed that proliferation of both AKI and BKI iOTR cells was significantly increased at 24 h, 48 h, and 72 h compared to EVC (Figure 11A). Furthermore, proliferation of BKI iOTR cells was not different at 24 h but was significantly higher at 48 h and 72 h compared to AKI iOTR cells (Figure 11A). The matrigel invasion assay showed that there was no significant change in the invasion index of AKI and BKI iOTR cells compared to EVC (Figure 11B). These results suggest that increased proliferation of AKI and BKI iOTR cells is due to increased expression of proliferation-associated genes in these cells compared to EVC.

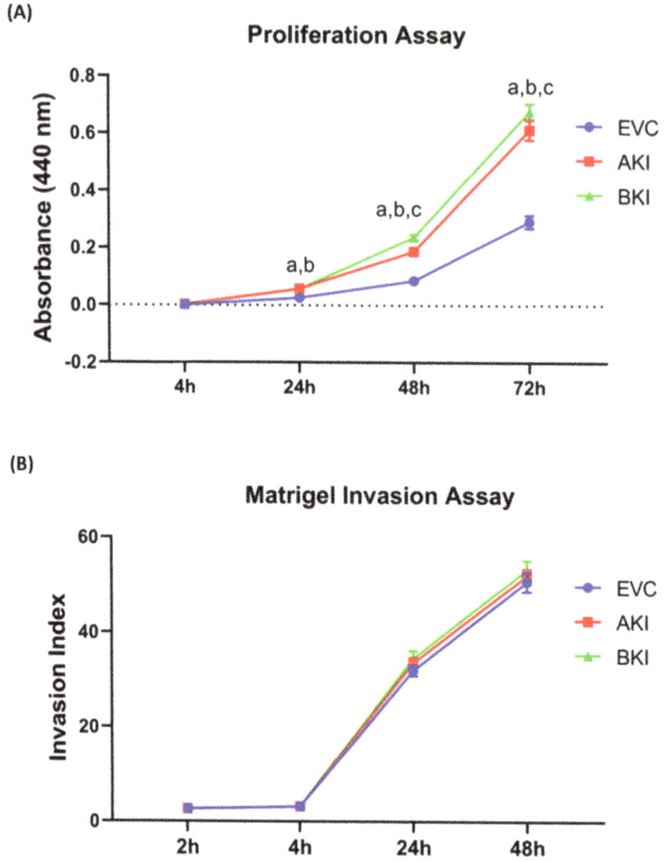

Figure 11. (**A**) Proliferation of AKI, BKI, and EVC iOTR cells (n = 4/treatment) was measured after 4 h, 24 h, 48 h, and 72 h using Quick Cell Proliferation Assay Kit. (**B**) Invasion of AKI, BKI, and EVC iOTR cells (n = 4) measured after 2 h, 4 h, 24 h, and 48 h using the Matrigel Invasion Assay Kit; where a, $p < 0.05$ for AKI vs. EVC; b, $p < 0.05$ for BKI vs. EVC; and c, $p < 0.05$ for BKI vs. AKI.

3. Discussion

The pluripotency factors LIN28A and LIN28B inhibit the maturation of *let-7* miRNAs [39,62]. Recently, we showed that both Lin28A and LIN28B were significantly decreased and levels of *let-7* miRNAs (*let-7a*, *let-7b*, *let-7c*, *let-7d*, *let-7e*, *let-7f*, *let-7g*, *let-7i*) were significantly increased in term human placentas from IUGR pregnancies compared to control pregnancies [18]. We further demonstrated that double knockout of LIN28A and LIN28B in immortalized first trimester human trophoblast (ACH-3P) cells resulted in more robust increase in *let-7* miRNAs compared to knockout of either LIN28A or LIN28B. Similarly, double knockin of LIN28A and LIN28B in Sw.71 cells led to more robust decrease in *let-7* miRNAs compared to knockin of either LIN28A or LIN28B [18]. In this study, we show that RNA interference (RNAi) of LIN28A or LIN28B in sheep TE in vivo resulted in significant increase in *let-7* miRNAs (*let-7a, let-7b, let-7c, let-7d, let-7e, let-7f, let-7g, let-7i*). Moreover, the conceptus elongation was significantly reduced after RNAi of LIN28A or LIN28B in TE.

Although animal models with global gene knockout or knockdown have been extensively and successfully used in many studies, it is difficult to exclude the effect of global gene manipulation while focusing on mechanisms involved in one tissue type or organ. Incubating the day 9-hatched sheep blastocyst with shRNA expressing lentiviral particles caused viral infection of only trophoblast cells while all other cells including the inner cell mass were spared of lentiviral infection [61]. Hence, significant reduction in LIN28A or LIN28B in day 16 TE conceptus was restricted to the trophoblast cells only. Conceptus elongation in sheep was due to rapid proliferation of trophoblast cells [7–9]; therefore, a significant reduction in conceptus length at day 16 after trophectoderm-specific LIN28A or LIN28B knockdown indicates reduced proliferation of trophoblast cells.

LIN28 is a part of a complex genetic pathway known to regulate multiple downstream targets [64]. Inhibition of biogenesis of mature *let-7* miRNAs is one of the main pathways through which LIN28 regulates the expression of its downstream targets [32,65]. *Let-7* miRNAs reduce expression of many genes by degrading their mRNAs or inhibiting translation [39]. Our results show that RNAi of LIN28A or LIN28B and resultant increase in *let-7* miRNAs in day 16 TE significantly reduced IGF2BP1, IGF2BP2, IGF2BP3, HMGA1, ARID3B, and c-MYC. Previous studies have shown the role of IGF2BP1, IGF2BP2, IGF2BP3, HMGA1, ARID3B, and c-MYC in cell proliferation [17,38–40,43,45,46,48,51], suggesting that reduced proliferation of trophoblast cells after LIN28A or LIN28B knockdown in TE is due to reduced expression of proliferation-associated genes (Figure 12).

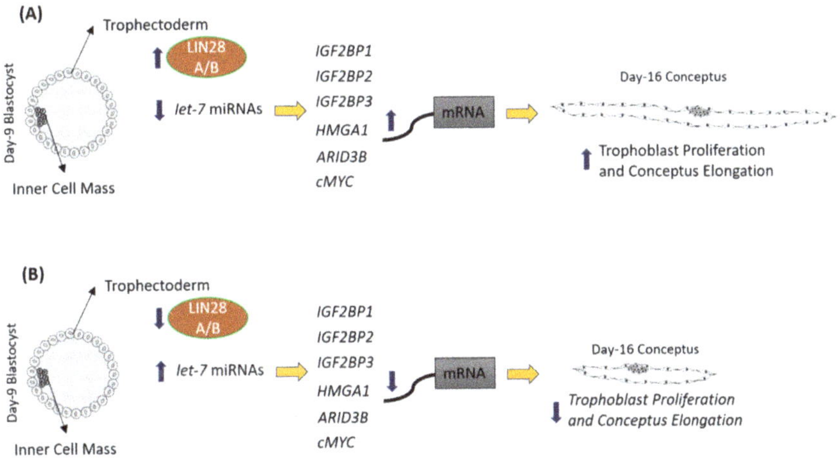

Figure 12. Graphical abstract. (**A**) In control day 9 TE, *let-7* miRNAs are low because of high LIN28A and LIN28B. Because of low levels of let-7 miRNAs, IGF2BP1, IGF2BP2, IGF2BP3, HMGA1, ARID3B, and c-MYC are higher leading to increased proliferation of trophoblast cells and hence conceptus elongation. (**B**) In LIN28A or LIN28B KD day 9 TE, *let-7* miRNAs are higher. Elevated let-7 miRNAs target *IGF2BP1, IGF2BP2, IGF2BP3, HMGA1, ARID3B,* and *c-MYC*, leading to reduced expression of these genes, leading to reduced proliferation of trophoblast cells and, hence, reduced conceptus elongation.

To further investigate the role of the LIN28-*let-7* miRNA axis in sheep trophoblast cells in vitro, OTR cells were generated from day 16 TE. Surprisingly, both LIN28A and LIN28B were depleted and the *let-7* miRNAs were significantly higher in OTR cells compared to day 16 TE. The senescence of OTR cells after only a few passages along with high *let-7* miRNAs suggests that these cells are a differentiated phenotype of trophoblast cells compared to day 16 TE. Furthermore, the expression of proliferation-associated genes including IGF2BP1, IGF2BP2, IGF2BP3, HMGA1, ARID3B, and c-MYC was also significantly reduced in OTR cells compared to day 16 TE. To overcome the OTR cell senescence, we generated immortalized ovine trophoblast (iOTR) cells by expressing hTERT in these cells. The OTR cells were differentiated and depleted LIN28 and immortalization with hTERT expression did not change LIN28 expression in iOTR cells. Immortalized human first trimester trophoblast cells (Sw.71 cells) were also generated by expressing hTERT in first trimester human trophoblast cells at passage 3 [66]. Sw.71 cells also had depleted LIN28A and LIN28B and high *let-7* miRNAs compared to other trophoblast-derived cell lines such as ACH-3P cells [18]. The iOTR cells generated in this study were similar to Sw.71 cells in terms of LIN28 and *let-7* miRNAs expression, which may be because both cell lines were generated by immortalizing the passaged trophoblast cells. The contrasting levels of LIN28, *let-7* miRNAs, and *let-7* miRNA target genes between day 16 TE and iOTR cells should be taken in consideration if using these cells for further studies.

To generate iOTR cells that were more similar to day 16 TE, we overexpressed LIN28A or LIN28B in iOTR cells. Overexpression of LIN28A or LIN28B in iOTR cells led to a significant decrease in *let-7* miRNAs and significant increase in expression of IGF2BP1, IGF2BP2, IGF2BP3, HMGA1, ARID3B, and c-MYC compared to control. The proliferation of both AKI and BKI iOTR cells was significantly increased compared to control; however, there was no change in cell invasion. We recently showed that knockout of LIN28A and LIN28B in ACH-3P cells reduced cell proliferation but did not affect cell invasion [19]. Both ACH-3P and iOTR cells have low invasion; therefore, it would be hard to see a further reduction in invasion. Moreover, cell proliferation and invasion are two distinct and critical processes during placental development. Reduced trophoblast cell proliferation without reduced invasion can be sufficient to cause impaired conceptus elongation and attachment. These results suggest that the LIN28-*let-7* miRNA axis plays a role in proliferation of immortalized trophoblast cells by regulating the expression of genes associated with cell proliferation. We suggest that the iOTR cells overexpressing LIN28A and LIN28B would be a better choice to study molecular mechanisms in ovine trophoblast cells compared to iOTR cells with depleted LIN28A and LIN28B.

To our knowledge, this is the first in vivo study defining the role of LIN28-*let*-7 miRNA axis in early placental development by trophoblast-specific RNAi of LIN28A or LIN28B. Due to a wide range of *let-7* miRNAs target genes, as well as the ability of LIN28 to directly bind the mRNA of different genes, LIN28 knockdown might be affecting many different genetic pathways in trophoblast cells, which are yet to be explored. Knockdown of LIN28A or LIN28B and high *let-7* miRNAs in TE led to reduced conceptus elongation, which can result in impaired placentation, fetal growth restriction, loss of pregnancy, and reduced fertility in domestic ruminants. We recently showed that term human placentas from IUGR pregnancies have low LIN28 and high *let-7* miRNAs [18]. MicroRNAs can be easily measured in tissue biopsies, blood, and other biological samples. Based on our studies, we suggest that low LIN28 or high *let-7* miRNAs in placenta could be detected in blood and be used as potential biomarkers for intrauterine growth restriction.

4. Materials and Methods

4.1. Lentivirus Vector Construction for shRNA Expression

Lentiviral infection was used to stably integrate and express shRNA targeting LIN28A or LIN28B mRNA in the host cell. Lentiviral vectors were constructed using the protocol previously described by Baker et al. [61]. Briefly, LIN28A targeting shRNA, LIN28B targeting shRNA, or scrambled control shRNA sequence (Table S1) were first cloned into the pLKO.1 vector (plasmid 10878,

Addgene, Cambridge, MA, USA), which contained the human U6 promoter upstream of cloning site for shRNA cassettes. The human U6 promoter and downstream LIN28A/LIN28B/SC shRNA sequence within pLKO.1 was PCR amplified using a forward primer with a 5′ XbaI restriction site(5′-TCTAGATTCACCGAGGGCCTATTTCCC-3′) and a reverse primer containing a 3′ XhoI restriction site (5′-GAATACTGCCATTTGTCTCGAGGTCG-3′). The resulting PCR amplicon was gel purified and cloned into the StrataClone PCR cloning vector using StrataClone PCR Cloning KIT (Agilent, Santa Clara, CA). The human U6 promoter and LIN28A/LIN28B/SC shRNA DNA fragment was digested from StrataClone PCR cloning vector using XbaI/XhoI restriction enzymes. Subsequently, the DNA fragment was ligated into the pLL3.7 vector also digested with XbaI/XhoI. Insertion of the human U6 promoter and LIN28A/LIN28B/SC shRNA sequence into pLL3.7 was verified by sanger sequencing.

4.2. Lentivirus Vector Construction for Overexpression of LIN28A and LIN28B

To overexpress LIN28A and LIN28B, pCDH lentiviral expression vector (System Biosciences, Palo Alto, CA, USA) was used. The mRNA was extracted from day 16 TE using RNeasy Mini Kit (Qiagen Inc. Germantown, MD, USA) following the manufacturer's protocol, and then reverse transcribed to cDNA using iScript cDNA synthesis kit (Bio-Rad Laboratories, Hercules, CA, USA). The cDNA was amplified using PCR primers for LIN28A or LIN28B (Supplementary Table S2). The PCR primers included restriction sites for NheI and SwaI restriction enzymes. The resulting PCR amplicons were gel purified and cloned into the StrataClone PCR cloning vector using StrataClone PCR Cloning KIT (Agilent, Santa Clara, CA). StrataClone vector with successful cloning of PCR product was double digested using NheI/SwaI. The double digested product was cloned in double digested pCDH vector and confirmed by sanger sequencing.

4.3. Lentiviral Vector for Immortalizing Passaged Ovine Trophoblast Cells

To immortalize the OTR cells, pLV-hTERT-IRES-hygro was used (Addgene, Watertown, MA, USA, Plasmid # 85140) [67]. The pLV-hTERT-IRES-hygro vector-based lentiviral particles expressed human telomerase reverse transcriptase (hTERT) in the infected cells.

4.4. Production of Lentiviral Particles

To generate lentiviral particles, three vectors were used including transfer vector (LL3.7 or pCDH or pLV-hTERT-IRES-hygro), packaging plasmid (psPAX2 from Addgene, Watertown, MA, USA, Plasmid # 12260), and envelope plasmid (pMD2.G from Addgene, Watertown, MA, USA, Plasmid # 12259). The 293FT cells (Invitrogen, Carlsbad, CA, USA) were cultured in dulbecco's modified eagle medium (DMEM) with high-glucose supplemented with 10% heat-inactivated fetal bovine serum (FBS) and 1x penicillin-streptomycin-amphotericin B (PSA) solution, at 37 °C and 5% CO_2. Then, 8.82 μg transfer vector DNA, 6.66 μg psPAX2 packaging plasmid DNA, and 2.70 μg pMD2.G envelope plasmid DNA was mixed with 180 μL of polyfect transfection reagent (Qiagen Inc., Germantown, MD, USA) and the final volume was brought up to 855 μL using DMEM high-glucose media without any supplements. The plasmids-polyfect mixture was incubated at room temperature for 10 min and then gently mixed in the media on 70-80 % confluent 293FT cells. Cells were incubated for 4–6 h at 37 °C and 5% CO2. After incubation time, the transfection media was replaced by fresh DMEM high-glucose media supplemented with 10% FBS and 1x PSA solution. After 72 h, the medium containing lentiviral particles was collected and ultra-centrifuged over a 20% sucrose cushion at 25,000 RPM for 2 h at 4 °C. LL3.7 vector-based lentiviral particles were resuspended in chemically defined medium-2 (CDM-2), whereas pCDH or pLV-hTERT-IRES-hygro vector-based lentiviral particles were resuspended in 1x PBS, aliquoted, and stored at −80 °C.

To infect the cells by pCDH or pLV-hTERT-IRES-hygro based-lentiviral particle, the median tissue culture infectious dose (TCID50) of lentiviral particles was calculated. The frozen viral aliquot was resuspended in 0.5-1 mL of appropriate media with 8 μg/mL polybrene. The target cells were incubated with lentiviral particles at multiplicity of infection (MOI) of10 for 24 h at 37 °C and 5% CO_2. After 72 h of culture, cells were selected with appropriate selection antibiotic. The LL3.7 vector-based lentiviral particles were tittered by infecting human embryonic kidney (HEK) cells and counting green fluorescent protein (GFP)-positive cells [61].

4.5. Blastocyst Collection and Transfer

All animal procedures were approved by Institutional Animal Care and Use Committee at Colorado State University, Fort Collins, Colorado, USA. Blastocysts collection and transfer was done following the procedure previously described by Baker at al [61]. A group of 12 ewes at day 6-12 of estrus were synchronized by two intramuscular injections of prostaglandin F-2α (PGF-2α) at 10 mg/dose given at interval of 4 h (Lutalyse, Pfizer, New York, NY). After 48 h of estrus synchronization, 4 ewes were separated to be used as recipients while 8 donor ewes were bred by intact rams. At day 9, donor ewes were euthanized using pentobarbital sodium (90 mg/kg IV, Pentasol, Vibrac, Fort Worth, TX, USA), and blastocysts were flushed from the uterus using DMEM-F-12 (1:1) medium supplemented with 0.25% BSA. The hatched blastocysts were infected with 100,000 shRNA expressing lentiviral particles in a 100 µL drop of CDM-2 media with 5 µg/mL polybrene (Sigma-Aldrich, St. Louis, MO, USA). Blastocysts with lentiviral particles were kept in incubator for 4–5 h at 5% CO2, 5% O2, and 38.5 °C. Overnight-fasted recipient ewes were sedated using ketamine (12.5 mg/kg IV, Ketacine, VetOne, Boise, ID, USA) and diazepam (0.125 mg/kg IV, Hospira, Lake Forest, IL, USA). Surgical procedure was performed under general anesthesia on 2 L/min O2 and 2–4% isoflurane (Fluriso, VetOne, Boise, IS, USA). A total of 23 blastocysts were transferred including 6 SC, 7 AKD, and 10 BKD. One blastocyst was transferred in each recipient.

4.6. Tissue Collection

For analysis of LIN28A or LIN28B knockdown and its effect on conceptus elongation, terminal surgeries were conducted on recipient ewes at 16 days gestational age (dGA), and tissues were collected. Conceptuses were flushed from the uterus using DMEM-F-12 (1:1) medium. After separating the embryo, trophectoderm length was measured, and both embryo and TE were snap frozen. TE samples were used to extract mRNA, miRNA, or proteins for further analysis.

4.7. Cell Lines

Day 16 TE from 3 non-infected pregnancies was minced in DMEM-F-12 (1:1) medium supplemented with 10% bovine serum albumin, 1x penicillin-streptomycin-amphotericin B solution, 10 µg/mL insulin, 0.1 mM non-essential amino acids, 2 mM glutamine, and 1 mM sodium pyruvate. The minced tissue was spun down at 1000 rpm for 5 min and the supernatant was incubated in a 100-mm collagen-treated tissue culture dish at 37 °C and 5% CO_2. After 24 h, the cells attached to the plate were washed and incubated with fresh complete medium. After 48–72 h, the cells were passaged and later collected at passage number 4–6 at 70–80% confluency to extract mRNA, miRNA, and proteins for further analysis. Western blot analysis for cytokeratin-7 (CK-7) was done to confirm the phenotype of OTR cells. To generate immortalized ovine trophoblast cells (iOTR cells), the OTR cells were infected with pLV-hTERT-IRES-hygro based lentiviral particles resuspended in complete DMEM-F12 (1:1) medium supplemented with 8 µg/mL polybrene transfection reagent. The media with viral particles was replaced with fresh media after 24 h. The cells were selected in complete DMEM-F12 (1:1) medium supplemented with 300–500 µg/mL hygromycin B (Sigma-Aldrich, St. Louis, MO, USA).

4.8. Overexpression of LIN28A and LIN28B

To overexpress *LIN28* genes in iOTR cells, pCDH-LIN28A or pCDH-LIN28B-based lentiviral particles were used to infect iOTR cells at 70–80% confluency in one well of a 12-well plate (Corning Inc., Corning, NY, USA). After 48–72 h, the infected cells were selected using 2–4 µg/mL puromycin. Successful gene knockin was confirmed using real-time PCR and Western blot analysis. We generated iOTR cells with knockin of LIN28A (AKI) or LIN28B (BKI) and iOTR cells infected with empty-expression vector-based lentiviral particles to use as expression vector control (EVC).

4.9. RNA Extraction and Real-Time PCR

For real-time PCR analysis, mRNA was isolated from day 16 sheep TE, OTR cells, and iOTR cells using RNeasy Mini Kit (Qiagen Inc. Germantown, MD, USA), following the manufacturer's protocol. The mRNA was reverse transcribed to cDNA using iScript cDNA synthesis kit (Bio-Rad Laboratories, Hercules, CA, USA). Real-time PCR reactions were run in triplicate in 384-well plates, using 10 μL reaction volume in each well. The reaction volume included 5 μL of 2x Light-Cycler 480 SYBR Green I Master (Roche Applied Science, Penzberg, Germany), 50 ng reverse-transcribed mRNA, and 1 μM of target-specific forward and reverse primers. Primer sequences used for real-time PCR are listed in Supplementary Table S2. PCR reactions were incubated in the Light-Cycler 480 PCR machine (Roche Applied Science, Penzberg, Germany) at the following cycling conditions: 95 °C for 10 min, 45 cycles of 95 °C for 30 s, 55 °C for 1 min, and 72 °C for 1 min. Relative mRNA levels were normalized using *RPS15*. For miRNA profiling, total RNA was extracted using a miRNeasy Mini Kit (Qiagen Inc. Germantown, MD, USA), following the manufacturer's protocol. Then, 300 ng total RNA was reverse-transcribed to cDNA using miScript RT II kit (Qiagen Inc. Germantown, MD, USA). Real-time PCR reactions were run in triplicate in 384-well plates, using 10 μL reaction volume in each well. The reaction volume included 5 μL of 2x QuantiTech SYBR Green Master Mix (Qiagen Inc. Germantown, MD, USA), 3 ng cDNA, 1x miScript universal primer (Qiagen Inc. Germantown, MD, USA), and 1x miScript assay for *let-7* miRNAs (*let-7a, let-7b, let-7c, let-7d, let-7e, let-7f, let-7g, let-7i*). These reactions were incubated in the Light-Cycler 480 PCR machine (Roche Applied Science, Penzberg, Germany) at the following cycling conditions: 95 °C for 15 min, 45 cycles of 94 °C for 15 s, 55 °C for 30 s, and 70 °C for 30 s. Relative miRNA levels were normalized using *SNORD-48*.

4.10. Protein Extraction and Western Blot

Western blot analysis was performed using whole cell lysate to quantify proteins in cells and tissue samples. For protein extraction, cell pellets were resuspended in 200–400 μL radioimmunoprecipitation assay (RIPA) buffer (20 mM Tris, 137 mM NaCl, 10% glycerol, 1% nonidet *p*-40, 3.5 mM sodium dodecyl sulfate (SDS), 1.2 mM sodium deoxycholate, 1.6 mM ethylenediaminetetraacetic acid (EDTA), pH 8) containing 1x protease/phosphate inhibitor cocktail (Sigma-Aldrich, St. Louis, MO, USA). Whole cell lysate was incubated on ice for 5 min and then centrifuged at 14,000 g for 5 min to remove cell debris. To extract protein from day 16 TE, the tissue was homogenized in RIPA buffer. Homogenized samples were sonicated using a Bioruptor Sonication System (Diagenode, Denville, NJ, USA) for 5 cycles of 30 s "ON" and 30 s "OFF". Sonicated samples were centrifuged at 14,000 g for 5 min to remove debris. Protein concentration was measured using the bicinchoninic acid (BCA) protein assay kit (ThermoFisher, Waltham, MA, USA). Protein was separated in 4–15% Bis-Tris gels (Bio-Rad Laboratories, Hercules, CA, USA) at 90 volts for 15 min and 125 volts for 60 min, and then transferred to 0.45 μm pore size nitrocellulose membrane (Bio-Rad Laboratories, Hercules, CA, USA) at 100 volts for 2 h at 4 °C. The membranes were then blocked in 5% non-fat dry milk solution in tris buffered saline with tween 20 (TBST) (50 mM Tris, 150 mM NaCl, 0.05% Tween 20, pH 7.6) for 1 h at room temperature. After blocking, the membranes were washed 3 times with 1x TBST for 5 min each, and then incubated at 4 °C overnight with specific primary antibody. After overnight incubation, the membranes were washed 3 times with 1x TBST for 5 min each. After washing, the membranes were incubated with appropriate secondary antibody conjugated to horseradish peroxidase for 1 h at room temperature. After removing the secondary antibody, the membranes were washed following the same procedure and developed using Super Signal WestDura Extended Duration Substrate (ThermoFisher, Waltham, MA, USA) and imaged using ChemiDoc XRS+ chemiluminescence system (Bio-Rad Laboratories, Hercules, CA, USA). The images were quantified using Image-Lab software (Bio-Rad Laboratories, Hercules, CA, USA). To normalize protein quantity, β-actin, α-tubulin, or glyceraldehyde 3-phosphate dehydrogenase (GAPDH), was used as loading control. Each experiment was repeated on three replicates. The antibodies used and their dilutions are listed in Table S3.

4.11. Cell Proliferation Assay

Cell proliferation was measured using Quick Cell Proliferation Assay Kit (Abcam, Cambridge, MA, USA) following manufacturer's protocol. This assay is based on cleavage of tetrazolium salt (WST-1) to formazan by mitochondrial dehydrogenases. EVC, AKI, and BKI iOTR cells were plated to a density of 1000 cells/100 μL in 96-well tissue culture plates, with four replicates of each cell type. After 4, 24, 48, or 96 h of plating the cells, 10 μL WST-1 reagent was added in each well followed by incubation for 2 h in standard culture conditions. Absorbance was measured using Cytation 3 Multi-Mode Reader (BioTek Instruments, Inc., VT, USA) at 440 nm with reference wavelength of 650 nm.

4.12. Matrigel Invasion Assay

Cell invasion was measured using Corning BioCoat Tumor Invasion System (Corning, New York, NY, USA) following manufacturer's protocol. EVC, AKI, and BKI iOTR cells were stained with CellTracker™ Green 5-chloromethylfluorescein diacetate (CMFDA) (Invitrogen, Carlsbad, CA, USA). Four replicates of each cell line were plated at a density of 10,000 cells/500 μL DMEM/F-12 (1:1) media without phenol red and fetal bovine serum, in a 24-multiwell insert plate with 8 μm pore size polyethylene terephthalate membrane coated with uniform layer of matrigel matrix. DMEM/F-12 (1:1) media without any cells was added in four wells to be used as blank. In the bottom wells, 750 μL of DMEM/F-12 (1:1) media with 10% fetal bovine serum was added. Plates were read at 2, 4, 24, and 48 h after plating the cells using Cytation 3 Multi-Mode Reader (BioTek Instruments, Inc., VT, USA) with top and bottom reading ability, at 492 nm excitation and 517 nm emission wavelengths. The invasion index was calculated based on relative fluorescent units (RFU) using the formula: (RFU of cells at the bottom/RFU of cells at top + RFU of cells at bottom) × 100.

4.13. Statistics

All data were analyzed using GraphPad Prism 7 Software. To determine significance of mRNAs, miRNAs, and proteins, t-test was used when comparing two groups and analysis of variance followed by Tukey's honestly significant difference (HSD) post hoc test was done when comparing three groups. The p values less than 0.05 were considered statistically significant. The error bars in the figures indicate standard error of the mean (SEM).

Supplementary Materials: Supplementary materials can be found at http://www.mdpi.com/1422-0067/21/7/2549/s1.

Author Contributions: A.A. and Q.A.W. designed research; A.A. performed research, analyzed the data, and wrote the manuscript; A.A. and R.V.A. performed surgical embryo transfer; Q.A.W., G.J.B., and R.V.A. contributed new reagents and analytic tools, and edited the manuscript. T.E.S. and M.D.S. helped in research design and edited the manuscript. All authors have read and agreed to the published version of the manuscript.

Funding: This project was supported by Agriculture and Food Research Initiative Competitive Grant No. 2017-67015-26460 from the United States Department of Agriculture (USDA) National Institute of Food and Agriculture. This work was supported by the USDA National Institute of Food and Agriculture, Hatch Project COL00293D, accession number 1021217.

Acknowledgments: The authors express their gratitude to Institute of International Education (IIE) and United States Education Foundation in Pakistan (USEFP) for funding doctoral studies of Asghar Ali at Colorado State University, through the Foreign Fulbright Scholarship Program. The authors are thankful to the maintenance crew at Animal Reproduction and Biotechnology lab, especially Richard Brandes, Greg Harding, and Joel Artzer. The authors also acknowledge Jennifer Barfield, Zella Brink, Dana Fuller, and Maria Marquez for collaborating with us.

Conflicts of Interest: The authors declare no conflicts of interest.

Abbreviations

OTR	non-immortalized ovine trophoblast cells
iOTR	immortalized ovine trophoblast cells
TE	trophectoderm
SC	scramble control shRNA
AKD	knockdown of LIN28A
BKD	knockdown of LIN28B
EVC	iOTR cells infected with empty lentiviral expression vectors
AKI	iOTR cells overexpressing LIN28A
BKI	iOTR cells overexpressing LIN28B

References

1. Knöfler, M.; Haider, S.; Saleh, L.; Pollheimer, J.; Gamage, T.K.J.B.; James, J. Human placenta and trophoblast development: key molecular mechanisms and model systems. *Cell. Mol. Life Sci.* **2019**, *76*, 3479–3496. [CrossRef] [PubMed]
2. Aplin, J.D. The cell biological basis of human implantation. *Baillieres Best Pract. Res. Clin. Obstet. Gynaecol.* **2000**, *14*, 757–764. [CrossRef] [PubMed]
3. Red-Horse, K.; Zhou, Y.; Genbacev, O.; Prakobphol, A.; Foulk, R.; McMaster, M.; Fisher, S.J. Trophoblast differentiation during embryo implantation and formation of the maternal-fetal interface. *J. Clin. Investig.* **2004**, *114*, 744–754. [CrossRef] [PubMed]
4. Brosens, I.; Pijnenborg, R.; Vercruysse, L.; Romero, R. The "Great Obstetrical Syndromes" are associated with disorders of deep placentation. *Am. J. Obstet. Gynecol.* **2011**, *204*, 193–201. [CrossRef] [PubMed]
5. Jauniaux, E.; Poston, L.; Burton, G.J. Placental-related diseases of pregnancy: involvement of oxidative stress and implications in human evolution. *Hum. Reprod. Update* **2006**, *12*, 747–755. [CrossRef]
6. Spencer, T.E.; Johnson, G.A.; Bazer, F.W.; Burghardt, R.C. Implantation mechanisms: insights from the sheep. *Reprod. Camb. Engl.* **2004**, *128*, 657–668. [CrossRef]
7. Rowson, L.E.; Moor, R.M. Development of the sheep conceptus during the first fourteen days. *J. Anat.* **1966**, *100*, 777–785.
8. Wintenberger-Torrés, S.; Fléchon, J.E. Ultrastructural evolution of the trophoblast cells of the pre-implantation sheep blastocyst from day 8 to day 18. *J. Anat.* **1974**, *118*, 143–153.
9. Wang, J.; Guillomot, M.; Hue, I. Cellular organization of the trophoblastic epithelium in elongating conceptuses of ruminants. *C. R. Biol.* **2009**, *332*, 986–997. [CrossRef]
10. Spencer, T.E.; Hansen, T.R. Implantation and Establishment of Pregnancy in Ruminants. *Adv. Anat. Embryol. Cell Biol.* **2015**, *216*, 105–135.
11. Shorten, P.R.; Ledgard, A.M.; Donnison, M.; Pfeffer, P.L.; McDonald, R.M.; Berg, D.K. A mathematical model of the interaction between bovine blastocyst developmental stage and progesterone-stimulated uterine factors on differential embryonic development observed on Day 15 of gestation. *J. Dairy Sci.* **2018**, *101*, 736–751. [CrossRef] [PubMed]
12. Farin, C.E.; Imakawa, K.; Hansen, T.R.; McDonnell, J.J.; Murphy, C.N.; Farin, P.W.; Roberts, R.M. Expression of trophoblastic interferon genes in sheep and cattle. *Biol. Reprod.* **1990**, *43*, 210–218. [CrossRef] [PubMed]
13. Thatcher, W.W.; Guzeloglu, A.; Mattos, R.; Binelli, M.; Hansen, T.R.; Pru, J.K. Uterine-conceptus interactions and reproductive failure in cattle. *Theriogenology* **2001**, *56*, 1435–1450. [CrossRef]
14. Roberts, R.M.; Chen, Y.; Ezashi, T.; Walker, A.M. Interferons and the maternal-conceptus dialog in mammals. *Semin. Cell Dev. Biol.* **2008**, *19*, 170–177. [CrossRef]
15. Aires, M.B.; Degaki, K.Y.; Dantzer, V.; Yamada, A.T. Bovine placentome development during early pregnancy. *Microscope* **2014**, *1*, 390–396.
16. Lopez-Tello, J.; Arias-Alvarez, M.; Gonzalez-Bulnes, A.; Sferuzzi-Perri, A.N. Models of Intrauterine growth restriction and fetal programming in rabbits. *Mol. Reprod. Dev.* **2019**, *86*, 1781–1809. [CrossRef]
17. Devor, E.J.; Reyes, H.D.; Santillan, D.A.; Santillan, M.K.; Onukwugha, C.; Goodheart, M.J.; Leslie, K.K. Placenta-Specific Protein 1: A Potential Key to Many Oncofetal-Placental OB/GYN Research Questions. *Obstet. Gynecol. Int.* **2014**, *2014*, 678984. [CrossRef]

18. Ali, A.; Anthony, R.V.; Bouma, G.J.; Winger, Q.A. LIN28-let-7 axis regulates genes in immortalized human trophoblast cells by targeting the ARID3B-complex. *FASEB J. Off. Publ. Fed. Am. Soc. Exp. Biol.* **2019**, *33*, 12348–12363. [CrossRef]
19. West, R.C.; McWhorter, E.S.; Ali, A.; Goetzman, L.N.; Russ, J.E.; Anthony, R.V.; Bouma, G.J.; Winger, Q.A. HMGA2 is regulated by LIN28 and BRCA1 in human placental cells. *Biol. Reprod.* **2019**, *100*, 227–238. [CrossRef]
20. Seabrook, J.L.; Cantlon, J.D.; Cooney, A.J.; McWhorter, E.E.; Fromme, B.A.; Bouma, G.J.; Anthony, R.V.; Winger, Q.A. Role of LIN28A in Mouse and Human Trophoblast Cell Differentiation. *Biol. Reprod.* **2013**, *89*, 95. [CrossRef]
21. Moss, E.G.; Tang, L. Conservation of the heterochronic regulator Lin-28, its developmental expression and microRNA complementary sites. *Dev. Biol.* **2003**, *258*, 432–442. [CrossRef]
22. Chan, H.W.; Lappas, M.; Yee, S.W.Y.; Vaswani, K.; Mitchell, M.D.; Rice, G.E. The expression of the let-7 miRNAs and Lin28 signalling pathway in human term gestational tissues. *Placenta* **2013**, *34*, 443–448. [CrossRef] [PubMed]
23. Feng, C.; Neumeister, V.; Ma, W.; Xu, J.; Lu, L.; Bordeaux, J.; Maihle, N.J.; Rimm, D.L.; Huang, Y. Lin28 regulates HER2 and promotes malignancy through multiple mechanisms. *Cell Cycle* **2012**, *11*, 2486–2494. [CrossRef] [PubMed]
24. Ma, W.; Ma, J.; Xu, J.; Qiao, C.; Branscum, A.; Cardenas, A.; Baron, A.T.; Schwartz, P.; Maihle, N.J.; Huang, Y. Lin28 regulates BMP4 and functions with Oct4 to affect ovarian tumor microenvironment. *Cell Cycle* **2013**, *12*, 88–97. [CrossRef]
25. Hagan, J.P.; Piskounova, E.; Gregory, R.I. Lin28 recruits the TUTase Zcchc11 to inhibit let-7 maturation in mouse embryonic stem cells. *Nat. Struct. Mol. Biol.* **2009**, *16*, 1021–1025. [CrossRef]
26. Heo, I.; Joo, C.; Cho, J.; Ha, M.; Han, J.; Kim, V.N. Lin28 mediates the terminal uridylation of let-7 precursor MicroRNA. *Mol. Cell* **2008**, *32*, 276–284. [CrossRef]
27. Heo, I.; Joo, C.; Kim, Y.-K.; Ha, M.; Yoon, M.-J.; Cho, J.; Yeom, K.-H.; Han, J.; Kim, V.N. TUT4 in concert with Lin28 suppresses microRNA biogenesis through pre-microRNA uridylation. *Cell* **2009**, *138*, 696–708. [CrossRef]
28. Piskounova, E.; Polytarchou, C.; Thornton, J.E.; LaPierre, R.J.; Pothoulakis, C.; Hagan, J.P.; Iliopoulos, D.; Gregory, R.I. Lin28A and Lin28B inhibit let-7 microRNA biogenesis by distinct mechanisms. *Cell* **2011**, *147*, 1066–1079. [CrossRef]
29. Rybak, A.; Fuchs, H.; Smirnova, L.; Brandt, C.; Pohl, E.E.; Nitsch, R.; Wulczyn, F.G. A feedback loop comprising lin-28 and let-7 controls pre-let-7 maturation during neural stem-cell commitment. *Nat. Cell Biol.* **2008**, *10*, 987–993. [CrossRef]
30. Newman, M.A.; Thomson, J.M.; Hammond, S.M. Lin-28 interaction with the Let-7 precursor loop mediates regulated microRNA processing. *RNA N. Y.* **2008**, *14*, 1539–1549. [CrossRef]
31. Ibarra, I.; Erlich, Y.; Muthuswamy, S.K.; Sachidanandam, R.; Hannon, G.J. A role for microRNAs in maintenance of mouse mammary epithelial progenitor cells. *Genes Dev.* **2007**, *21*, 3238–3243. [CrossRef]
32. Lee, H.; Han, S.; Kwon, C.S.; Lee, D. Biogenesis and regulation of the let-7 miRNAs and their functional implications. *Protein Cell* **2016**, *7*, 100–113. [CrossRef]
33. Viswanathan, S.R.; Daley, G.Q. Lin28: A MicroRNA Regulator with a Macro Role. *Cell* **2010**, *140*, 445–449. [CrossRef]
34. Liao, T.-T.; Hsu, W.-H.; Ho, C.-H.; Hwang, W.-L.; Lan, H.-Y.; Lo, T.; Chang, C.-C.; Tai, S.-K.; Yang, M.-H. let-7 Modulates Chromatin Configuration and Target Gene Repression through Regulation of the ARID3B Complex. *Cell Rep.* **2016**, *14*, 520–533. [CrossRef] [PubMed]
35. Rahman, M.M.; Qian, Z.R.; Wang, E.L.; Sultana, R.; Kudo, E.; Nakasono, M.; Hayashi, T.; Kakiuchi, S.; Sano, T. Frequent overexpression of HMGA1 and 2 in gastroenteropancreatic neuroendocrine tumours and its relationship to *let-7* downregulation. *Br. J. Cancer* **2009**, *100*, 501–510. [CrossRef] [PubMed]
36. Yang, X.; Cai, H.; Liang, Y.; Chen, L.; Wang, X.; Si, R.; Qu, K.; Jiang, Z.; Ma, B.; Miao, C.; et al. Inhibition of c-Myc by let-7b mimic reverses mutidrug resistance in gastric cancer cells. *Oncol. Rep.* **2015**, *33*, 1723–1730. [CrossRef] [PubMed]
37. Uchikura, Y.; Matsubara, K.; Matsubara, Y.; Mori, M. P34. Role of high-mobility group A1 protein in trophoblast invasion. *Pregnancy Hypertens. Int. J. Womens Cardiovasc. Health* **2015**, *5*, 243. [CrossRef]

38. Sampson, V.B.; Rong, N.H.; Han, J.; Yang, Q.; Aris, V.; Soteropoulos, P.; Petrelli, N.J.; Dunn, S.P.; Krueger, L.J. MicroRNA Let-7a Down-regulates MYC and Reverts MYC-Induced Growth in Burkitt Lymphoma Cells. *Cancer Res.* **2007**, *67*, 9762–9770. [CrossRef]
39. Boyerinas, B.; Park, S.-M.; Shomron, N.; Hedegaard, M.M.; Vinther, J.; Andersen, J.S.; Feig, C.; Xu, J.; Burge, C.B.; Peter, M.E. Identification of Let-7–Regulated Oncofetal Genes. *Cancer Res.* **2008**, *68*, 2587–2591. [CrossRef]
40. Bell, J.L.; Wächter, K.; Mühleck, B.; Pazaitis, N.; Köhn, M.; Lederer, M.; Hüttelmaier, S. Insulin-like growth factor 2 mRNA-binding proteins (IGF2BPs): Post-transcriptional drivers of cancer progression? *Cell. Mol. Life Sci.* **2013**, *70*, 2657–2675. [CrossRef]
41. JnBaptiste, C.K.; Gurtan, A.M.; Thai, K.K.; Lu, V.; Bhutkar, A.; Su, M.-J.; Rotem, A.; Jacks, T.; Sharp, P.A. Dicer loss and recovery induce an oncogenic switch driven by transcriptional activation of the oncofetal Imp1–3 family. *Genes Dev.* **2017**, *31*, 674–687. [CrossRef] [PubMed]
42. Degrauwe, N.; Suvà, M.-L.; Janiszewska, M.; Riggi, N.; Stamenkovic, I. IMPs: an RNA-binding protein family that provides a link between stem cell maintenance in normal development and cancer. *Genes Dev.* **2016**, *30*, 2459–2474. [CrossRef] [PubMed]
43. Zhao, W.; Lu, D.; Liu, L.; Cai, J.; Zhou, Y.; Yang, Y.; Zhang, Y.; Zhang, J. Insulin-like growth factor 2 mRNA binding protein 3 (IGF2BP3) promotes lung tumorigenesis via attenuating p53 stability. *Oncotarget* **2017**, *8*, 93672–93687. [CrossRef] [PubMed]
44. Li, Y.; Francia, G.; Zhang, J.-Y. p62/IMP2 stimulates cell migration and reduces cell adhesion in breast cancer. *Oncotarget* **2015**, *6*, 32656–32668. [CrossRef]
45. Mahapatra, L.; Andruska, N.; Mao, C.; Le, J.; Shapiro, D.J. A Novel IMP1 Inhibitor, BTYNB, Targets c-Myc and Inhibits Melanoma and Ovarian Cancer Cell Proliferation. *Transl. Oncol.* **2017**, *10*, 818–827. [CrossRef]
46. Zhou, Y.; Meng, X.; Chen, S.; Li, W.; Li, D.; Singer, R.; Gu, W. IMP1 regulates UCA1-mediated cell invasion through facilitating UCA1 decay and decreasing the sponge effect of UCA1 for miR-122-5p. *Breast Cancer Res. BCR* **2018**, *20*, 32. [CrossRef]
47. Schmiedel, D.; Tai, J.; Yamin, R.; Berhani, O.; Bauman, Y.; Mandelboim, O. The RNA binding protein IMP3 facilitates tumor immune escape by downregulating the stress-induced ligands ULPB2 and MICB. *eLife* **2016**, *5*, e13426. [CrossRef]
48. Lederer, M.; Bley, N.; Schleifer, C.; Hüttelmaier, S. The role of the oncofetal IGF2 mRNA-binding protein 3 (IGF2BP3) in cancer. *Semin. Cancer Biol.* **2014**, *29*, 3–12. [CrossRef]
49. Wu, C.; Ma, H.; Qi, G.; Chen, F.; Chu, J. Insulin-like growth factor II mRNA-binding protein 3 promotes cell proliferation, migration and invasion in human glioblastoma. *Onco. Targets Ther.* **2019**, *12*, 3661–3670. [CrossRef]
50. Wan, B.-S.; Cheng, M.; Zhang, L. Insulin-like growth factor 2 mRNA-binding protein 1 promotes cell proliferation via activation of AKT and is directly targeted by microRNA-494 in pancreatic cancer. *World J. Gastroenterol.* **2019**, *25*, 6063–6076. [CrossRef]
51. Cao, J.; Mu, Q.; Huang, H. The Roles of Insulin-Like Growth Factor 2 mRNA-Binding Protein 2 in Cancer and Cancer Stem Cells. Available online: https://www.hindawi.com/journals/sci/2018/4217259/ (accessed on 18 November 2019).
52. Brooks, K.; Burns, G.W.; Moraes, J.G.N.; Spencer, T.E. Analysis of the Uterine Epithelial and Conceptus Transcriptome and Luminal Fluid Proteome During the Peri-Implantation Period of Pregnancy in Sheep. *Biol. Reprod.* **2016**, *95*, 88. [CrossRef] [PubMed]
53. Uchikura, Y.; Matsubara, K.; Muto, Y.; Matsubara, Y.; Fujioka, T.; Matsumoto, T.; Sugiyama, T. Extranuclear Translocation of High-Mobility Group A1 Reduces the Invasion of Extravillous Trophoblasts Involved in the Pathogenesis of Preeclampsia: New Aspect of High-Mobility Group A1. *Reprod. Sci. Thousand Oaks Calif.* **2017**, *24*, 1630–1638. [CrossRef] [PubMed]
54. Kumar, P.; Luo, Y.; Tudela, C.; Alexander, J.M.; Mendelson, C.R. The c-Myc-regulated microRNA-17~92 (miR-17~92) and miR-106a~363 clusters target hCYP19A1 and hGCM1 to inhibit human trophoblast differentiation. *Mol. Cell. Biol.* **2013**, *33*, 1782–1796. [CrossRef] [PubMed]
55. Bobbs, A.; Gellerman, K.; Hallas, W.M.; Joseph, S.; Yang, C.; Kurkewich, J.; Cowden Dahl, K.D. ARID3B Directly Regulates Ovarian Cancer Promoting Genes. *PlLoS ONE* **2015**, *10*, e0131961. [CrossRef]

56. Roy, L.; Samyesudhas, S.J.; Carrasco, M.; Li, J.; Joseph, S.; Dahl, R.; Cowden Dahl, K.D. ARID3B increases ovarian tumor burden and is associated with a cancer stem cell gene signature. *Oncotarget* **2014**, *5*, 8355–8366. [CrossRef]
57. Ratliff, M.L.; Mishra, M.; Frank, M.B.; Guthridge, J.M.; Webb, C.F. The Transcription Factor ARID3a Is Important for In Vitro Differentiation of Human Hematopoietic Progenitors. *J. Immunol. Baltim. Md 1950* **2016**, *196*, 614–623. [CrossRef]
58. Habir, K.; Aeinehband, S.; Wermeling, F.; Malin, S. A Role for the Transcription Factor Arid3a in Mouse B2 Lymphocyte Expansion and Peritoneal B1a Generation. *Front. Immunol.* **2017**, *8*, 1387. [CrossRef]
59. Rhee, C.; Edwards, M.; Dang, C.; Harris, J.; Brown, M.; Kim, J.; Tucker, H.O. ARID3A is required for mammalian placenta development. *Dev. Biol.* **2017**, *422*, 83–91. [CrossRef]
60. Lala, N.; Girish, G.V.; Cloutier-Bosworth, A.; Lala, P.K. Mechanisms in decorin regulation of vascular endothelial growth factor-induced human trophoblast migration and acquisition of endothelial phenotype. *Biol. Reprod.* **2012**, *87*, 59. [CrossRef]
61. Baker, C.M.; Goetzmann, L.N.; Cantlon, J.D.; Jeckel, K.M.; Winger, Q.A.; Anthony, R.V. Development of ovine chorionic somatomammotropin hormone-deficient pregnancies. *Am. J. Physiol. Regul. Integr. Comp. Physiol.* **2016**, *310*, R837–R846. [CrossRef]
62. Jeckel, K.J.; Boyarko, A.C.; Bouma, G.J.; Winger, Q.A.; Anthony, R.V. Chorionic Somatomammotropin Impacts Early Fetal Growth and Placental Gene Expression. *J. Endocrinol.* **2018**, *237*, 301–310. [CrossRef] [PubMed]
63. Purcell, S.H.; Cantlon, J.D.; Wright, C.D.; Henkes, L.E.; Seidel, G.E.; Anthony, R.V. The Involvement of Proline-Rich 15 in Early Conceptus Development in Sheep. *Biol. Reprod.* **2009**, *81*, 1112–1121. [CrossRef] [PubMed]
64. Tsialikas, J.; Romer-Seibert, J. LIN28: roles and regulation in development and beyond. *Development* **2015**, *142*, 2397–2404. [CrossRef] [PubMed]
65. Su, J.-L.; Chen, P.-S.; Johansson, G.; Kuo, M.-L. Function and regulation of let-7 family microRNAs. *MicroRNA Shariqah United Arab Emir.* **2012**, *1*, 34–39. [CrossRef]
66. Straszewski-Chavez, S.L.; Abrahams, V.M.; Alvero, A.B.; Aldo, P.B.; Ma, Y.; Guller, S.; Romero, R.; Mor, G. The isolation and characterization of a novel telomerase immortalized first trimester trophoblast cell line, Swan 71. *Placenta* **2009**, *30*, 939–948. [CrossRef]
67. Hayer, A.; Shao, L.; Chung, M.; Joubert, L.-M.; Yang, H.W.; Tsai, F.-C.; Bisaria, A.; Betzig, E.; Meyer, T. Engulfed cadherin fingers are polarized junctional structures between collectively migrating endothelial cells. *Nat. Cell Biol.* **2016**, *18*, 1311–1323. [CrossRef]

International Journal of *Molecular Sciences*

MDPI

Article

Downregulation of Placental Amino Acid Transporter Expression and mTORC1 Signaling Activity Contributes to Fetal Growth Retardation in Diabetic Rats

Jie Xu [1,†], Jiao Wang [1,†], Yang Cao [1], Xiaotong Jia [1], Yujia Huang [1], Minghui Cai [1], Chunmei Lu [1,*] and Hui Zhu [1,2,*]

1 Department of Physiology, Harbin Medical University, Harbin 150081, China; xujie@ems.hrbmu.edu.cn (J.X.); wangj_84321@163.com (J.W.); caoyanghmu@163.com (Y.C.); jiaqiaoran@icloud.com (X.J.); woainimen924@sina.com (Y.H.); caiminghui@hrbmu.edu.cn (M.C.)

2 Laboratory of Medical Genetics, Harbin Medical University, and The Key Laboratory of Preservation of Human Genetic Resources and Disease Control in China, Chinese Ministry of Education, Harbin 150081, China

* Correspondence: luchunmei@hrbmu.edu.cn (C.L.); dzhuhui@aliyun.com (H.Z.); Tel./Fax: +86-451-8667-4538 (C.L. & H.Z.)

† These authors contributed equally to this work.

Received: 5 February 2020; Accepted: 5 March 2020; Published: 7 March 2020

Abstract: Alterations in placental transport may contribute to abnormal fetal intrauterine growth in pregnancies complicated by diabetes, but it is not clear whether the placental amino acid transport system is altered in diabetic pregnancies. We therefore studied the changes in the expressions of placental amino acid transporters in a rat model of diabetes induced by streptozotocin, and tested the effects of hyperglycemia on trophoblast amino acid transporter in vitro. Our results showed that the expressions for key isoforms of system L amino acid transporters were significantly reduced in the placentas of streptozotocin-induced diabetic pregnant rats, which was associated with the decreased birthweight in the rats. A decreased placental efficiency and decreased placental mammalian target of rapamycin (mTOR) complex 1 (mTORC1) activity were also found in the rat model. In addition, hyperglycemia in vitro could inhibit amino acid transporter expression and mTORC1 activity in human trophoblast. Inhibition of mTORC1 activity led to reduced amino acid transporter expression in placental trophoblast. We concluded that reduced placental mTORC1 activity during pregnancy resulted in decreased placental amino acid transporter expression and, subsequently, contributed to fetal intrauterine growth restriction in pregnancies complicated with diabetes.

Keywords: placenta; amino acid transporter; mammalian target of rapamycin; gestational diabetes

1. Introduction

Diabetes in pregnancy is considered a high-risk condition for maternal and neonatal morbidity and remains a significant medical challenge. Women with diabetes have a higher risk of preeclampsia, caesarean section, and fetal outcomes including congenital anomalies, stillbirth, and macrosomia. Moreover, women with diabetes at the time of conception have more adverse outcomes (i.e., birth defects, perinatal mortality, and morbidity) than women who develop gestational diabetes during pregnancy [1]. The adverse outcomes associated with diabetes in pregnancy are substantially associated with hyperglycemia. However, the molecular mechanism underlying how maternal hyperglycemia leads to abnormal fetal growth remains unclear.

Placental transfer of amino acids via amino acid transporters is essential for optimal fetal growth and development. Decrease in amino acid transport by the placenta is implicated as a potential cause of fetal growth restriction. There are a variety of such transporters, selective for various classes of amino acids. Na^+-dependent system A amino acid transporter is specific for neutral amino acids with short side chains such as glycine and alanine. System L amino acid transporter is Na^+-independent exchanger specific for large neutral amino acids. Studies have shown that placental system A activity is lower in intrauterine growth restriction (IUGR) compared with normal pregnancies [2,3]. However, alterations in the placental amino acid transporters in pregnancies complicated by diabetes are largely unknown.

The mammalian target of rapamycin (mTOR) signaling pathway has newly been suggested to be a nutrient sensor in the placenta. mTOR is a ubiquitously expressed serine/threonine protein kinase that exists in two complexes, mTOR complexes 1 (mTORC1) and 2 (mTORC2). Placental nutrient transporter expression is mainly regulated by mTORC1. We previously reported that mTORC1 might regulate placental glucose transport by altering the glucose transporter isoform-3 expression in placental trophoblast [4]. In addition, evidence has shown that mTORC1 could regulate placental amino acid transport by modulating the cell surface abundance of system A and system L transporter isoforms without affecting global protein expression in primary human trophoblast cells [5]. Alterations of placental mTORC1 activity has been found as being associated with adverse pregnancy outcomes. For example, studies have shown that placental mTORC1 activity was decreased in human IUGR [6] or animal model of IUGR [7], suggesting placental mTORC1 plays an important role in the development of abnormal fetal growth. Although dysregulation of placental mTOR has been well documented in IUGR pregnancies, it has rarely been studied in pregnancy complicated with diabetes.

It is speculated that mTORC1 may be activated in diabetes in association with increased placental amino acid transport [8]. Thus, we aimed to investigate the alteration of mTORC1 pathway and system L amino acid transporters in diabetic placentas and to study the impacts of high glucose on mTORC1 activity and system L amino acid transporter expression in placental trophoblast cells.

2. Results

2.1. STZ Induced Severe Diabetes in Rats

Maternal glucose concentration was markedly increased in the rats injected with streptozotocin (STZ). All the rats developed severe diabetes according to their blood glucose levels (greater than 300 mg/dl) in the STZ-induced diabetes (STZ-D) group (Figure 1A), and the maternal rats showed classical features of type 1 diabetes, such as polyphagia, polyuria, and polydipsia (data not shown). In addition, maternal diabetes resulted in adverse placental and fetal outcomes, as shown in Figure 1B.

2.2. Pregestational Diabetes Resulted in Fetal Growth Restriction and Decreased Placental Efficiency in Rats

Fetal weight was significantly decreased in STZ-D rats compared with normal rats, suggesting newborns from severe diabetic mothers presented intrauterine growth restriction (Figure 1C). Nonetheless, there was no significant difference in placental weight (Figure 1D). However, the fetal weight/placental weight ratio was significantly decreased in STZ-D rats compared with normal rats (Figure 1E). The data are summarized in Table 1.

Because the fetal weight/placental weight ratio was considered as an indicator of placental efficiency, our results suggested that severe pregestational diabetes led to decreased placental efficiency, which might contribute to fetal growth delay in diabetic pregnancies.

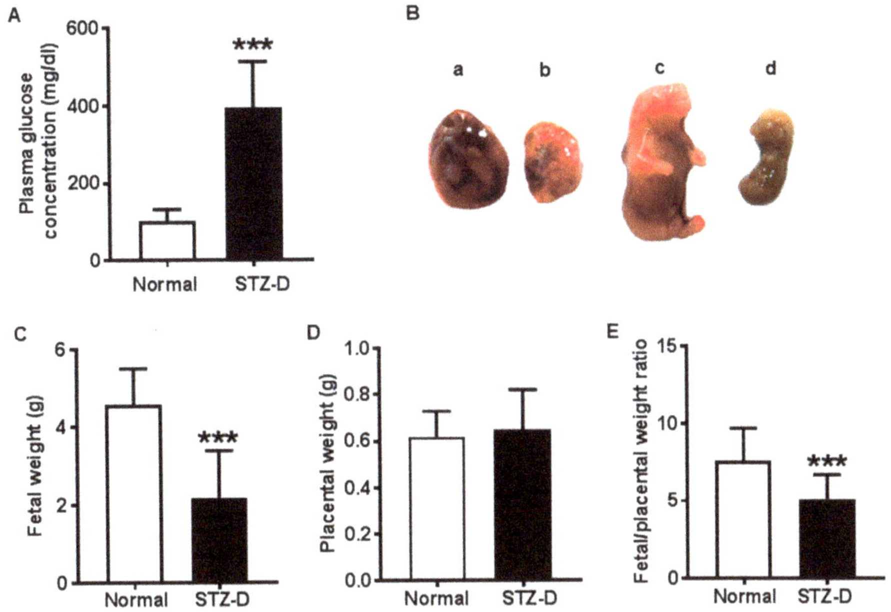

Figure 1. Pregestational diabetes led to fetal growth restriction and decreased placental efficiency in streptozotocin-induced diabetes (STZ-D) rats. (**A**) Maternal plasma glucose levels of normal and STZ-D rats (n = 13). (**B**) Representative pictures of abnormal growth of the fetus and placentas in STZ-D rats: a, normal placenta; b, STZ-D placenta; c, normal fetus; d, STZ-D fetus. (**C**) Birthweight from normal and STZ-D rats, showing that STZ-D pregnant rats had decreased birthweight compared with normal rats (n = 30). (**D**) Placental weight derived from normal and STZ-D rats (n = 30). (**E**) Fetal weight/placental weight ratio (n = 30) showing that STZ-D rats had decreased fetal to placental weight ratio. Data are expressed as mean ± SD. *** $p < 0.001$, normal versus STZ-D.

Table 1. Data of maternal glucose concentration, fetal and placental weight, and fetal/placental weight ratio in normal and STZ-D rats.

Group	Normal	STZ-D	n	p-Value
Maternal glucose concentration (mg/dL)	103 ± 29.78	395 ± 118.3	13	<0.001
Fetal weight (g)	4.57 ± 0.95	2.19 ± 1.21	30	<0.001
Placental weight (g)	0.62 ± 0.11	0.65 ± 0.17	30	0.42
Fetal/placental weight ratio	7.60 ± 2.01	5.11 ± 1.59	30	<0.001

The data are shown as mean ± standard deviation (SD).

2.3. *Pregestational Diabetes Resulted in Decreased Placental Amino Acid Transporter Expression in Rats*

Expression of placental system L amino acid transporter isoform LAT1 and LAT2 was assayed by quantitative-PCR and Western blot in normal and diabetic pregnant rats. Our results showed that relative mRNA expression of placental LAT1 and LAT2 was significantly reduced in STZ-D rats compared with normal pregnancies (Figure 2A). Protein level of placental LAT1 (0.95 ± 0.03 vs. 0.72 ± 0.07, normal vs. STZ-D, $p < 0.05$) and LAT2 (0.90 ± 0.03 vs. 0.79 ± 0.06, normal vs. STZ-D, $p < 0.05$) were also decreased in STZ-D pregnant rats (Figure 2B). LAT1 and LAT2 expression was also examined by immunohistochemistry staining in placental tissue sections. Consistent with Western blot data, pregestational diabetes caused reduction of LAT1 and LAT2 expression in the placentas (Figure 2C). These findings indicated that down-regulation of placental amino acid transporters was closely associated with fetal growth restriction and decreased placental efficiency in the rat model of severe gestational diabetes.

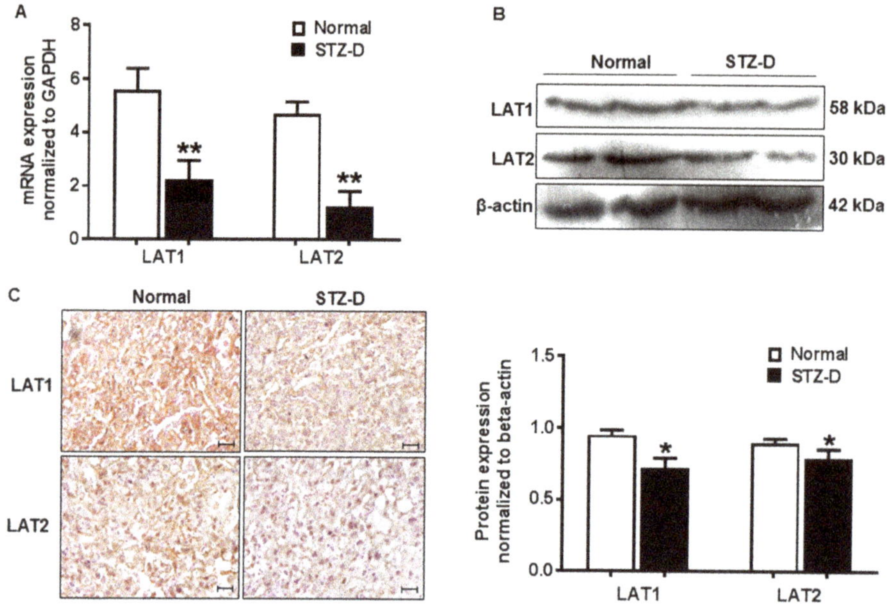

Figure 2. Pregestational diabetes resulted in decreased placental amino acid transporter expression in STZ-D rats. (**A**) Relative mRNA expression of system L amino acid transporter LAT1 and LAT2 detected by quantitative-PCR in placentas derived from normal and STZ-D pregnant rats (n = 3). (**B**) Expression of LAT1 and LAT2 detected by Western blots in placentas from normal and STZ-D rats. The bar graph shows the relative density of protein expression for LAT1 and LAT2 after normalization with β-actin expression in each sample. Data are mean ± SD from six normal and six STZ-D placentas. * $p < 0.05$, ** $p < 0.01$, normal versus STZ-D. (**C**) Representative immunostaining images of LAT1 and LAT2 expressions in tissue sections from normal and STZ-D placentas. Scale bar, 250 μm.

2.4. *Pregestational Diabetes Reduced Placental mTORC1 Activity in Rats*

To investigate the alterations of placental mTORC1 activity in diabetic pregnancies, expressions of phosphorylated S6 kinase1 (p-S6K1) and eukaryotic translation initiation factor 4E-bingding protein 1 (p-4EBP1), two down-stream regulators of mTORC1, were examined in normal and STZ-D placentas. As shown in Figure 3, placental p-4EBP1(Thr-37/46) (0.87 ± 0.06 vs. 0.69 ± 0.12, normal vs. STZ-D, $p < 0.05$) and p-S6k1(Thr-389) (0.72 ± 0.06 vs. 0.51 ± 0.09, normal vs. STZ-D, $p < 0.05$) expressions were significantly reduced in STZ-D rats compared with normal rats, which suggested that severe pregestational diabetes could decrease placental mTORC1 signaling activity.

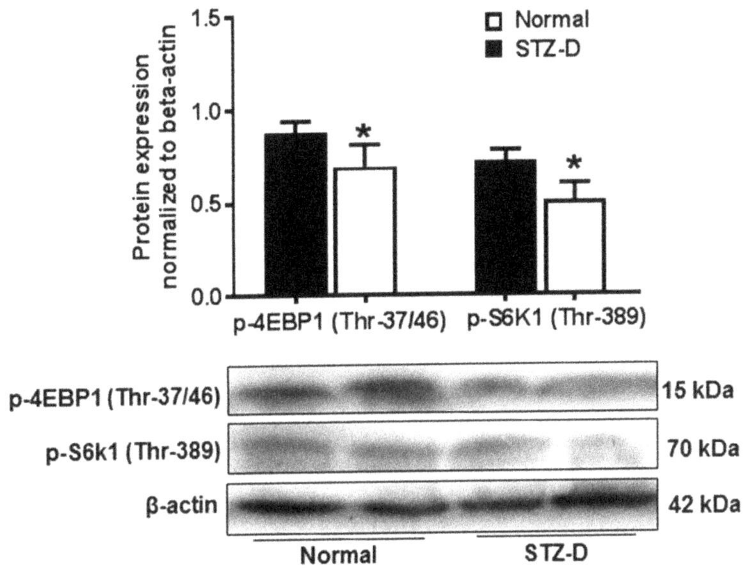

Figure 3. Placental mammalian target of rapamycin (mTOR) complex 1 (mTORC1) activity was decreased in STZ-D rats. Protein expression of p-4E-bingding protein 1 (4EBP1) and p-S6 kinase1 (S6K1) was detected by Western blot in placentas from normal and STZ-D pregnant rats. The bar graphs show relative expression after being normalized with β-actin expression in each sample. Gestational diabetes induced a decrease in p-4EBP1 and p-S6K1 expression. The lower panel shows representative blots for p-4EBP1 and p-S6K1. * $p < 0.05$, normal versus STZ-D. Data are from six independent experiments.

2.5. Hyperglycemia in Vitro Down-Regulated Amino Acid Transporter Expression and mTORC1 Activity in JEG-3 Trophoblast Cells

To study the regulation of hyperglycemia on placental amino acid transporter and mTORC1 activity in vitro, human trophoblast cell line JEG-3 was cultured with different concentrations of glucose, and expression of LAT1, LAT2, p-4EBP1(Thr-37/46), and p-S6k1(Thr-389) was examined by Western blot. Our results showed that LAT1 (1.02 ± 0.09 vs. 0.86 ± 0.06, 5 mM vs. 50 mM, $p < 0.05$) and LAT2 (0.89 ± 0.05 vs. 0.75 ± 0.06, 5 mM vs. 50 mM, $p < 0.05$) expression was significantly decreased in JEG-3 cells treated with 50 mM glucose compared to the cells cultured with 5 mM glucose (Figure 4A), which indicated the fact that high glucose would attenuate amino acid transporter expression in trophoblast cells. p-4EBP1(Thr-37/46) (0.97 ± 0.09 vs. 0.80 ± 0.05, 5 mM vs. 50 mM, $p < 0.05$) and p-S6k1 (Thr-389) (0.95 ± 0.14 vs. 0.78 ± 0.08, 5 mM vs. 50 mM, $p < 0.05$) expression was also reduced in cells treated with 50mM glucose compared with the cells cultured with 5mM glucose (Figure 4B), suggesting high glucose in vitro could inhibit mTORC1 activity in placental trophoblast cells. Together with the findings in the STZ-D rats, our data suggest that maternal diabetes could lead to decreased placental amino acid transporter expression and mTORC1 activity, which may contribute to the fetal intrauterine growth restriction in diabetic pregnancies.

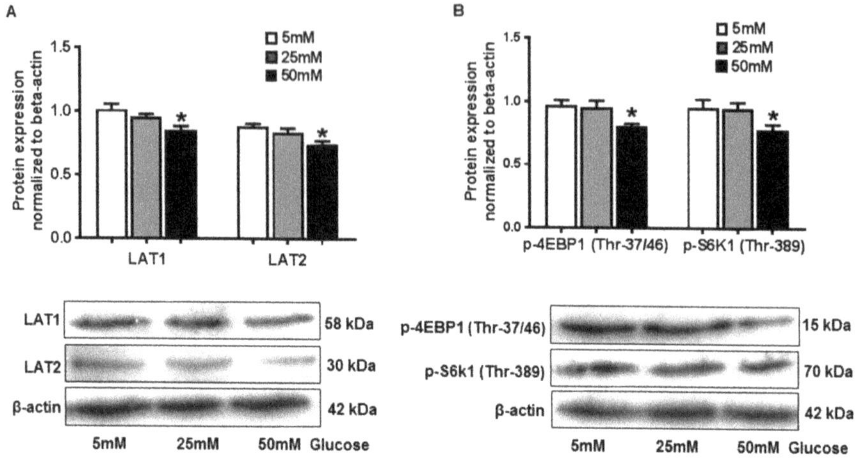

Figure 4. High glucose in vitro down-regulated amino acid transporter expression and mTORC1 activity in JEG-3 trophoblast cells. (**A**) Protein expression of LAT1 and LAT2 detected by Western blot in JEG-3 trophoblast cells cultured with different concentration of glucose. The bar graphs show relative expression after being normalized with β-actin expression in each sample. We found that 50mM glucose induced a significant decrease in LAT1 and LAT2 expression. The lower panel shows representative blots for LAT1 and LAT2. * $p < 0.05$, 5mM versus 50mM. Data are from five independent experiments. (**B**) Protein expression of p-4EBP1 and p-S6K1 detected by Western blot in JEG-3 trophoblast cells cultured with different concentration of glucose. The bar graphs show relative expression after being normalized with β-actin expression in each sample. We found that 50 mM glucose induced a significant decrease in p-4EBP1 and p-S6K1 expression. The lower panel shows representative blots for p-4EBP1 and p-S6K1. * $p < 0.05$, 5 mM versus 50 mM. Data are from five independent experiments.

2.6. Inhibition of mTORC1 Activity Resulted in Decreased Amino Acid Transporter Expression in JEG-3 Trophoblast Cells

To determine the effects of decreased mTORC1 activity on amino acid transporter in placental trophoblast cells, rapamycin, a specific mTORC1 inhibitor, was employed to inhibit mTORC1 activity in JEG-3 cells. Expression of LAT1 and LAT2 was measured by Western blot. As shown in Figure 5A, LAT1 (1.05 ± 0.14 vs. 0.55 ± 0.08, control vs. rapamycin, $p < 0.01$) and LAT2 (0.85 ± 0.12 vs. 0.60 ± 0.08, control vs. rapamycin, $p < 0.05$) expressions were both significantly reduced in rapamycin-treated cells compared with control cells.

mTORC1 inhibition was also conducted by gene silencing targeting raptor, a key component of mTORC1. Consistently, protein expressions of LAT1 and LAT2 were significantly decreased in cells treated with raptor small interfering RNAs (siRNAs) compared to scrambled and negative control cells (Figure 5B).

LAT1 and LAT2 expression was also examined by immunofluorescence staining in JEG-3 cells treated with raptor siRNAs. Consistent with Western blot data, LAT1 (Figure 6A) and LAT2 (Figure 6B) expressions were reduced in cells cultured with raptor siRNAs compared to negative controls. Our results suggested that amino acid transporter expression was regulated by placental mTORC1 signaling pathway, and that inhibition of mTORC1 activity could lead to decreased expression of amino acid transporter in human trophoblast cells.

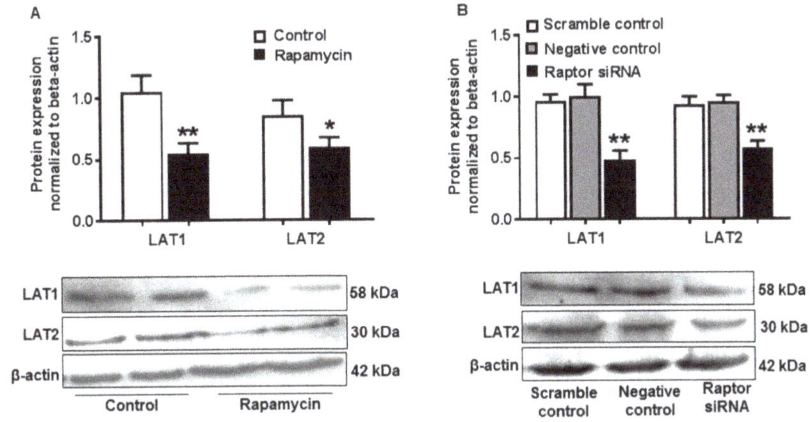

Figure 5. Inhibition of mTORC1 activity resulted in decreased amino acid transporter expression in JEG-3 trophoblast cells. (**A**) Protein expression of LAT1 and LAT2 assayed by Western blot in JEG-3 cells treated with or without rapamycin. The bar graphs show relative expression after being normalized with β-actin expression in each sample. Rapamycin induced a significant decrease in LAT1 and LAT2 expression. Data are from five independent experiments. * $p < 0.05$, ** $p < 0.01$, control versus rapamycin. (**B**) Expression of LAT1 and LAT2 detected by Western blot in JEG-3 cells treated with scrambled or negative or raptor siRNA. The bar graph shows data from five independent experiments. Raptor siRNA caused a significant reduction in LAT1 and LAT2 expression. ** $p < 0.01$, scrambled control or negative control versus raptor siRNA.

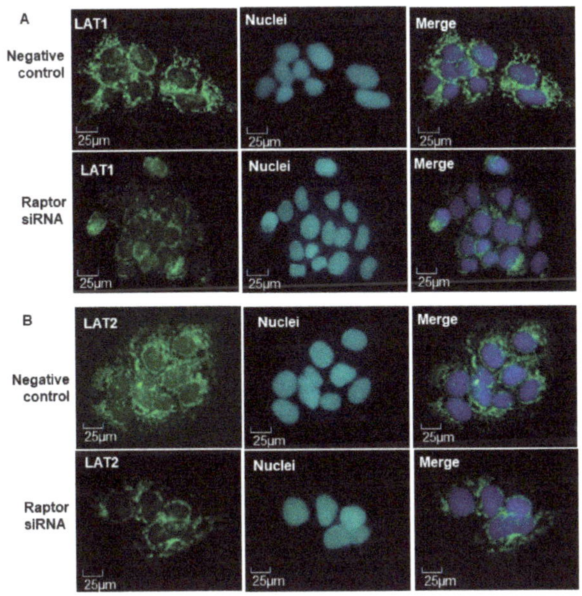

Figure 6. Representative imaging of immunofluorescent staining of LAT1 and LAT2 in JEG-3 trophoblast cells treated with negative or raptor siRNA. (**A**) LAT1 expression; (**B**) LAT2 expression. Consistent with Western blot results, raptor siRNA could reduce LAT1 and LAT2 expression in placental trophoblast.

3. Discussion

In the present study, we obtained interesting findings showing that fetal weight and fetal/placental weight ratio was significantly decreased in a rat model of STZ-induced diabetes compared to normal pregnancies, which confirmed that pregestational diabetes could lead to placental dysfunction and fetal intrauterine growth restriction.

The influence of maternal diabetes on fetal and placental growth might be related to the degree of gestational glucose intolerance, which could lead to opposite fetal outcomes. In some studies, an increased birthweight and placental weight was found in pregnancies complicated with mild diabetes [9], but limited fetal intrauterine growth, miscarriage stoics, and malformations were usually reported in pregnancies with severe diabetes [10], which was verified in our study. Weiss's study found that hyperglycemia in vitro could inhibit the proliferation of first-trimester trophoblast cells [11], which suggested that a reduced growth of the placenta might cause the fetal intrauterine growth delay in diabetes.

Placental efficiency, which is defined as fetal/placental weight ratio, is a useful predictor of those placentas that have adapted their nutrient transfer per gram placenta. A smaller ratio usually indicates that the placenta is less functionally efficient. Several studies have found that fetal/placental weight ratio was reduced in rodent models of IUGR [12–14], suggesting an inefficient placenta that failed to adapt its nutrient supply to meet the demands of the growing fetus. Additionally, it has been observed that fetal/placental weight ratios are associated with a variety of human pregnancy syndromes, including preeclampsia, preterm birth, and IUGR [15,16]. Therefore, our findings provide evidence that decreased placental efficiency plays important role in the fetal growth delay in pregestational diabetes mellitus.

Amino acids are crucial for the development of fetal tissue organs, and active transport of amino acid via placental amino acid transporters is the only way for fetuses to obtain amino acids from maternal circulation. Decreased expression or activity of the placental amino acid transporter system has been found in human IUGR or animal models [2,17], but few studies have been performed to investigate the alterations of placental amino acid transporters in pregnancies complicated with diabetes. Our results showed that the expression of placental amino acid transporters was reduced in the rat model of severe diabetes. In addition, using JEG-3 cell line as a human trophoblast cell model, we verified that the hyperglycemia in vitro could down regulate the expression of amino acid transporters in human trophoblast. These findings suggested that restricted fetal growth resulting from maternal diabetes was associated with the decreased placental amino acid transporter expression. Kuruvilla et al. found that the number of system A amino acid transporters on placental microvillus membrane (MVM) was decreased in diabetic pregnancies associated with macrosomia [18]. In contrast, Jansson et al. reported an increased activity of system A amino acid transporters on the placental MVM, regardless of the type of gestational diabetes [19]. The study methods and subjects may have contributed to the difference in the results. Recently, Ericsson et al. found that glucose injection during early gestation in rats had no significant effects on the expression of placental system A amino acid transporter, but that the transport activity of amino acid transporter was significantly decreased [20].

mTORC1 signaling pathway has been considered as a nutrient sensor in the placenta. It has been well studied that placental mTORC1 activity was decreased in IUGR model [6,21], but few studies were taken to investigate the placental mTROC1 changes in pregnancies complicated with diabetes. Our results found that the activity of placental mTORC1 was significantly reduced in a rat model with severe pregestational diabetes. Furthermore, the in vitro study performed in JEG-3 trophoblast cells confirmed that the high glucose could reduce the activity of mTORC1 signaling pathways. These results provided direct evidence revealing the alterations of placental mTORC1 pathway in diabetic pregnancies. Several studies have reported the mTORC1 changes in people with obesity, which is one of the high risks of diabetes. For example, some studies found that mTORC1 activity was promoted in obese people because of high insulin levels or increased inflammatory cytokines [22].

In contrast, Jansson et al. found that obesity could inhibit the activity of placental mTORC1 in a mouse model of obesity [23].

To determine the effects of decreased mTORC1 activity on placental amino acid transporters, we combined rapamycin treatment and gene silencing targeting raptor to inhibit mTORC1 signaling in JEG-3 cells. Our results showed that expression of system L amino acid transporters was reduced in trophoblast, of which mTORC1 signaling pathway was suppressed. It suggested that trophoblast mTORC1 could modulate amino acid transfer across the placenta by regulating the expression of key amino acid transporters. In contrast to Rosario's report that trophoblast mTORC1 regulation of amino acid transporters occurred mainly by modulation of the translocation of specific transporter isoforms between plasma membrane and cell interior [5], our results showed that mTORC1 could regulate the trophoblast amino acid transporter at the translational level.

In summary, our findings have significant impact on abnormal fetal growth in diabetic pregnancies. In the present study, we found that inefficient placenta was associated with limited fetal growth in a rat model of severe gestational diabetes. Furthermore, we found that placental amino acid transporter expression and mTORC1 activity was reduced in this model. We also demonstrated that decreased mTORC1 activity could lead to decreased amino acid transporter expression in placental trophoblast. These results provide evidence that decreased placental mTORC1 activity could lead to reduced placental amino acid transporter expression and, subsequently, contribute to limited fetal growth in severe diabetic pregnancies.

4. Materials and Methods

4.1. Animals

Adult female Sprague Dawley rats were given with a single intraperitoneal injection of streptozotocin (STZ) (65 mg/kg) to induce diabetes. At 3 days after STZ injection, plasma glucose concentrations were measured using a glucose analysis kit. The diabetic state was defined as a plasma glucose concentration exceeding 19.5 mmol/L (180 mg/dL). Both normal and diabetic females were mated overnight with normal male rats. Pregnancy was confirmed by the presence of sperm in vaginal smears. The same day that sperm appeared in vaginal smears was designated day 0 of gestation. All animals were housed under controlled temperature (23 °C) and a 12 h light/dark cycle, and they had free access to standard rat chow and tap water. The protocols for all animal experiments were approved by the Institutional Animal Care and Use Committee (IACUC) of Harbin Medical University (approval code: QC2017089; approval date: 1 October 2017).

4.2. Collection of Blood and Placenta Samples of Rat

Animals were killed by cervical dislocation at gestation days 21. Both offspring and placentas were carefully dissected out and immediately weighed. Thereafter, the placenta samples were immediately frozen in liquid nitrogen and kept at −80 °C until usage.

4.3. Human Trophoblast Cell Culture

Human trophoblast cell line JEG-3 was purchased from the cell bank of the Chinese Academy of Sciences. They were cultured in Dulbecco's modified Eagle's medium (Hyclone, China) containing 10% FBS and 1% penicillin/streptomycin, and incubated at standard culture conditions of 5% CO_2 in air at 37 °C.

4.4. mTORC1 Inhibition in Placental Trophoblast

Downregulation of mTORC1 activity was conducted as described in a previous study [4]. Briefly, human JEG-3 cells were cultured with 100 nM rapamycin (Gene Operation, USA), a specific mTORC1 inhibitor, or transfected of small interference RNAs (GenePharma, Shanghai, China) targeting raptor

(100 nM; sense: 5′-CGAGAUUGGACGACCAAAUTT-3′). At 48 h in culture, cells were collected for RNA and protein extraction.

4.5. RNA Isolation and Quantitative Real-Time PCR

Total RNA was extracted from placenta tissue and JEG-3 cells using the trizol reagent (Invitrogen, United Kingdom). cDNA was synthesized using the PrimeScript 1st Strand cDNA Synthesis Kit (Takara, Japan), following the manufacturer's instructions. Real-time PCRs were performed in 20 μL mixtures using the SYBR Premix Ex TaqTM II Kit (Takara, Japan). Gene expression was assayed in duplicate and normalized against β-actin. Relative expression values were calculated by the $^{\Delta\Delta}$CT method of relative quantification using Applied Biosystems 7500 Real-Time PCR System.

4.6. Western Blot Analysis

Protein expression of LAT1, LAT2, p-4EBP1 (Thr-37/46), and p-S6k1 (Thr-389) was examined by Western blotting. An aliquot of 20 μg of total protein was subject to electrophoresis then transferred to nitrocellulose membrane. After blocking, the membranes were probed with anti-human antibody against LAT1, LAT2, p-4EBP1 (Thr-37/46), or p-S6k1 (Thr-389) and followed by corresponding biotinylate conjugated secondary antibodies. Antibodies against LAT1 and LAT2 were obtained from Santa Cruz Biotechnology; p-4EBP1 (Thr-37/46) and p-S6k1 (Thr-389) antibodies were purchased from Cell Signaling Technology (Danvers, MA). The bound antibody was visualized with an enhanced chemiluminescent detection kit (Thermo Scientific). β-actin expression was used as loading control for each sample. The band densities were scanned and analyzed by Alpha Imager software (Alpha Innotech Corporation, CA, USA).

4.7. Immunohistochemistry

Expression of LAT1 and LAT2 was also examined by immunohistochemistry (IHC) staining in paraffin-embedded placental tissue sections. A standard IHC staining procedure was performed as previously described [24]. Briefly, a series of deparaffinization was carried out with xylene and ethanol alcohol. Antigen retrieval was performed by boiling tissue slides with 0.01 mol/L citric buffer. Hydrogen peroxide was used to quench the endogenous peroxidase activity. After blocking, the sections were incubated with primary monoclonal antibodies specific against human LAT1 or LAT2 overnight at 4 °C. Corresponding biotinylate conjugated secondary antibodies and ABC staining system were subsequently used, according to the manufacturer's instructions. Slides stained with the same antibody were all processed at the same time. Stained slides were reviewed under an Olympus microscope, and images were captured by a digital camera with PictureFrame computer software.

4.8. Immunofluorescence

LAT1 and LAT2 expression were examined by immunofluorescence in JEG-3 cells transfected with scramble or raptor siRNA. The cells were fixed with methanol at −20 °C for 5 min. After permeablilized by Triton X-100, cells were blocked in goat serum and then incubated with anti-LAT1 or anti-LAT2 antibody overnight at 4 °C. After several washes, cells were incubated with fluorescein isothiocyanate (FITC)-conjugated goat anti-rabbit IgG for 1 h at 37 °C, and then with dye 4,6-diamido-2-phenylindole (DAPI) for 15 min to counterstain nuclei. After three washes in PBS, coverslips were mounted in a drop of 30% glycerol. Images were acquired with a NikonE800 fluorescence microscope.

4.9. Data Presentation and Statistics

Data were presented as means ± SD. Statistical analysis was performed with unpaired *t*-test or AVOVA using GraphPad Prism 7 software. A probability level of less than 0.05 was considered statistically significant.

Author Contributions: Conceptualization, H.Z.; formal analysis, J.W.; funding acquisition, C.L.; investigation, J.X., Y.H., Y.C., and M.C.; methodology, J.W.; software, Y.H.; validation, X.J.; writing—original draft, J.X.; writing—review and editing, C.L. and H.Z. All authors have read and agreed to the published version of the manuscript.

Funding: This research received no external funding.

Acknowledgments: This work was funded by Heilongjiang Natural Science Foundation (QC2017089), University Nursing Program for Young Scholars with Creative Talents in Heilongjiang Province (UNPYSCT-2018054), China Postdoctoral Fund (2018M640304), Heilongjiang Postdoctoral Foundation (LBH-Z18108), Heilongjiang Postdoctoral Scientific Research Developmental Fund (LBH-Q18098), and Fund for Leading Talents Team of Heilongjiang Province.

Conflicts of Interest: The authors declare no conflict of interest.

References

1. Shand, A.W.; Bell, J.C.; McElduff, A.; Morris, J.; Roberts, C.L. Outcomes of pregnancies in women with pre-gestational diabetes mellitus and gestational diabetes mellitus; a population-based study in New South Wales, Australia, 1998–2002. *Diabet. Med.* **2008**, *25*, 708–715. [CrossRef] [PubMed]
2. Glazier, J.D.; Cetin, I.; Perugino, G.; Ronzoni, S.; Grey, A.M.; Mahendran, D.; Marconi, A.M.; Pardi, G.; Sibley, C.P. Association between the activity of the system a amino acid transporter in the microvillous plasma membrane of the human placenta and severity of fetal compromise in intrauterine growth restriction. *Pediatr. Res.* **1997**, *42*, 514–519. [CrossRef] [PubMed]
3. Jansson, T.; Ylven, K.; Wennergren, M.; Powell, T.L. Glucose transport and system A activity in syncytiotrophoblast microvillous and basal plasma membranes in intrauterine growth restriction. *Placenta* **2002**, *23*, 392–399. [CrossRef] [PubMed]
4. Xu, J.; Lu, C.; Wang, J.; Zhang, R.; Qian, X.; Zhu, H. Regulation of Human Trophoblast GLUT3 Glucose Transporter by Mammalian Target of Rapamycin Signaling. *Int. J. Mol. Sci.* **2015**, *16*, 13815–13828. [CrossRef] [PubMed]
5. Rosario, F.J.; Kanai, Y.; Powell, T.L.; Jansson, T. Mammalian target of rapamycin signalling modulates amino acid uptake by regulating transporter cell surface abundance in primary human trophoblast cells. *J. Physiol.* **2013**, *591*, 609–625. [CrossRef] [PubMed]
6. Roos, S.; Jansson, N.; Palmberg, I.; Saljo, K.; Powell, T.L.; Jansson, T. Mammalian target of rapamycin in the human placenta regulates leucine transport and is down-regulated in restricted fetal growth. *J. Physiol.* **2007**, *582*, 449–459. [CrossRef]
7. Arroyo, J.A.; Brown, L.D.; Galan, H.L. Placental mammalian target of rapamycin and related signaling pathways in an ovine model of intrauterine growth restriction. *Am. J. Obstet. Gynecol.* **2009**, *201*, 616.e1-7. [CrossRef]
8. Jansson, N.; Rosario, F.J.; Gaccioli, F.; Lager, S.; Jones, H.N.; Roos, S.; Jansson, T.; Powell, T.L. Activation of placental mTOR signaling and amino acid transporters in obese women giving birth to large babies. *J. Clin. Endocrinol. Metab.* **2013**, *98*, 105–113. [CrossRef]
9. Taricco, E.; Radaelli, T.; de Santis, M.S.N.; Cetin, I. Foetal and placental weights in relation to maternal characteristics in gestational diabetes. *Placenta* **2003**, *24*, 343–347. [CrossRef]
10. Vambergue, A.; Fajardy, I. Consequences of gestational and pregestational diabetes on placental function and birth weight. *World J. Diabetes* **2011**, *2*, 196–203. [CrossRef]
11. Weiss, U.; Cervar, M.; Puerstner, P.; Schmut, O.; Haas, J.; Mauschitz, R.; Arikan, G.; Desoye, G. Hyperglycaemia in vitro alters the proliferation and mitochondrial activity of the choriocarcinoma cell lines BeWo, JAR and JEG-3 as models for human first-trimester trophoblast. *Diabetologia* **2001**, *44*, 209–219. [CrossRef] [PubMed]
12. Jansson, N.; Pettersson, J.; Haafiz, A.; Ericsson, A.; Palmberg, I.; Tranberg, M.; Ganapathy, V.; Powell, T.L.; Jansson, T. Down-regulation of placental transport of amino acids precedes the development of intrauterine growth restriction in rats fed a low protein diet. *J. Physiol.* **2006**, *576*, 935–946.
13. Kusinski, L.C.; Stanley, J.L.; Dilworth, M.R.; Hirt, C.J.; Andersson, I.J.; Renshall, L.J.; Baker, B.C.; Baker, P.N.; Sibley, C.P.; Wareing, M.; et al. eNOS knockout mouse as a model of fetal growth restriction with an impaired uterine artery function and placental transport phenotype. *Am. J. Physiol. Regul. Integr. Comp. Physiol.* **2012**, *303*, R86–R93. [CrossRef] [PubMed]

14. Coan, P.M.; Vaughan, O.R.; Sekita, Y.; Finn, S.L.; Burton, G.J.; Constancia, M.; Fowden, A.L. Adaptations in placental phenotype support fetal growth during undernutrition of pregnant mice. *J. Physiol.* **2010**, *588*, 527–538. [CrossRef] [PubMed]
15. Wallace, J.M.; Horgan, G.W.; Bhattacharya, S. Placental weight and efficiency in relation to maternal body mass index and the risk of pregnancy complications in women delivering singleton babies. *Placenta* **2012**, *33*, 611–618. [CrossRef]
16. Matsuda, Y.; Ogawa, M.; Nakai, A.; Hayashi, M.; Satoh, S.; Matsubara, S. Fetal/Placental weight ratio in term Japanese pregnancy: Its difference among gender, parity, and infant growth. *Int. J. Med. Sci.* **2015**, *12*, 301–305. [CrossRef]
17. Rosario, F.J.; Jansson, N.; Kanai, Y.; Prasad, P.D.; Powell, T.L.; Jansson, T. Maternal protein restriction in the rat inhibits placental insulin, mTOR, and STAT3 signaling and down-regulates placental amino acid transporters. *Endocrinology* **2011**, *152*, 1119–1129. [CrossRef]
18. Kuruvilla, A.G.; D'Souza, S.W.; Glazier, J.D.; Mahendran, D.; Maresh, M.J.; Sibley, C.P. Altered activity of the system a amino acid transporter in microvillous membrane vesicles from placentas of macrosomic babies born to diabetic women. *J. Clin. Investig.* **1994**, *94*, 689–695. [CrossRef]
19. Jansson, T.; Ekstrand, Y.; Bjorn, C.; Wennergren, M.; Powell, T.L. Alterations in the activity of placental amino acid transporters in pregnancies complicated by diabetes. *Diabetes* **2002**, *51*, 2214–2219. [CrossRef]
20. Ericsson, A.; Saljo, K.; Sjostrand, E.; Jansson, N.; Prasad, P.D.; Powell, T.L.; Jansson, T. Brief hyperglycaemia in the early pregnant rat increases fetal weight at term by stimulating placental growth and affecting placental nutrient transport. *J. Physiol.* **2007**, *581*, 1323–1332. [CrossRef]
21. Mejia, C.; Lewis, J.; Jordan, C.; Mejia, J.; Ogden, C.; Monson, T.; Winden, D.; Watson, M.; Reynolds, P.R.; Arroyo, J.A. Decreased activation of placental mTOR family members is associated with the induction of intrauterine growth restriction by secondhand smoke in the mouse. *Cell Tissue Res.* **2017**, *367*, 387–395. [CrossRef] [PubMed]
22. Laplante, M.; Sabatini, D.M. mTOR signaling in growth control and disease. *Cell* **2012**, *149*, 274–293. [CrossRef] [PubMed]
23. Lager, S.; Samulesson, A.M.; Taylor, P.D.; Poston, L.; Powell, T.L.; Jansson, T. Diet-induced obesity in mice reduces placental efficiency and inhibits placental mTOR signaling. *Physiol. Rep.* **2014**, *2*, e00242. [CrossRef] [PubMed]
24. Xu, J.; Gu, Y.; Lewis, D.F.; Cooper, D.B.; McCathran, C.E.; Wang, Y. Downregulation of vitamin D receptor and miR-126-3p expression contributes to increased endothelial inflammatory response in preeclampsia. *Am. J. Reprod. Immunol.* **2019**, *82*, e13172. [CrossRef] [PubMed]

International Journal of
Molecular Sciences

MDPI

Article

Fetuin-A Inhibits Placental Cell Growth and Ciliogenesis in Gestational Diabetes Mellitus

Chia-Yih Wang [1,2], Mei-Tsz Su [3], Hui-ling Cheng [1,3], Pao-Lin Kuo [3] and Pei-Yin Tsai [3,*]

1 Department of Cell Biology and Anatomy, College of Medicine, National Cheng Kung University, Tainan 701, Taiwan; b89609046@gmail.com (C.-Y.W.); tomato4329@gmail.com (H.-l.C.)

2 Institute of Basic Medical Sciences, College of Medicine, National Cheng Kung University, Tainan 701, Taiwan

3 Department of Obstetrics and Gynecology, National Cheng Kung University Hospital, College of Medicine, National Cheng Kung University, Tainan 704, Taiwan; sumeitsz@mail.ncku.edu.tw (M.-T.S.); paolinkuo@gmail.com (P.-L.K.)

* Correspondence: tsaipy@mail.ncku.edu.tw; Tel.: +886-6-2353535 (ext. 5222); Fax: +886-6-2093007

Received: 9 September 2019; Accepted: 16 October 2019; Published: 21 October 2019

Abstract: Gestational diabetes mellitus (GDM) is a type of unbalanced glucose tolerance that occurs during pregnancy, which affects approximately 10% of pregnancies worldwide. Fetuin-A is associated with insulin resistance, and the concentration of circulating fetuin-A increases in women with GDM, however, the role of fetuin-A in the placenta remains unclear. In this study, we enrolled placental samples from twenty pregnant women with GDM and twenty non-GDM pregnant women and found that the abundance of fetuin-A was upregulated in terms of mRNA and protein levels. Fetuin-A inhibited placental cell growth by inducing apoptosis and inhibiting S phase entry. Irregular alignment of mitotic chromosomes and aberrant mitotic spindle poles were observed. In addition, centrosome amplification was induced by fetuin-A treatment, and these amplified centrosomes nucleated microtubules with disorganized microtubule arrays in placental cells. Furthermore, fetuin-A inhibited autophagy, and thus blocked the growth of the primary cilium, a cellular antenna that regulates placenta development and differentiation. Thus, our study uncovered the novel function of fetuin-A in regulating placental cell growth and ciliogenesis.

Keywords: fetuin-A; GDM; cell growth; centrosome; primary cilium; autophagy

1. Introduction

Gestational diabetes mellitus (GDM) is a type of unbalanced glucose tolerance that occurs during pregnancy, which affects approximately 10% of pregnancies worldwide [1]. The clinical and public health relevance of gestational diabetes mellitus has been widely debated due to its increasing incidence and resulting negative economic impact, and the potential for severe GDM-related pregnancy complications [2]. In addition, the American Diabetes Association (ADA) "Management of Diabetes in Pregnancy: Standards of Medical Care in Diabetes 2019" recommends diabetes care from preconception to postpartum [3]. Several risk factors, including obesity, history of previous GDM diagnosis, advanced maternal age, and gestational hypertension, have been implicated in the pathogenesis of GDM [4,5]. In general, specific risks of uncontrolled diabetes in pregnancy include spontaneous abortion, fetal anomalies, preeclampsia, fetal demise, macrosomia, neonatal hypoglycemia, and neonatal hyperbilirubinemia, among others. In addition, diabetes in pregnancy may increase the risk of obesity and type 2 diabetes in the offspring later in life [6,7]. Thus, it is important to understand the pathogenesis of GDM.

Fetuin-A belongs to the cystatin protease inhibitor superfamily [8]. It is the major human secretory protein derived from the liver and adipose tissue and performs several pathophysiological functions related to insulin sensitivity [9], glucose tolerance [10], and even soft tissue calcification [11]. In the liver and skeletal muscles, fetuin-A, per se, is an endogenous inhibitor of the insulin receptor [12] and is crucial for lipid-induced insulin resistance [13]. By binding to the β-subunit of the insulin receptor, fetuin-A inhibits the activity of the insulin receptor, followed by the blocking of insulin-stimulated GLUT4 translocation and Akt activation [14]. In addition, fetuin-A is also involved in inflammatory signaling [13]. Fetuin-A acts as an endogenous ligand for the innate immune Toll-like receptor (TLR)-4, thus promoting lipid-induced insulin resistance. Moreover, fetuin-A is associated with several metabolic disorders. High serum fetuin-A concentrations are observed in patients with several metabolic syndromes, including insulin resistance, fatty liver, and diabetes [10,15]. A recent study also showed that the circulating fetuin-A concentration increases in GDM women [16], however, the underlying molecular mechanism is still unclear.

The centrosome is the major microtubule organization center that orchestrates microtubule networks for proper cell migration and division [17]. It comprises mother and daughter centrioles and the surrounding pericentriolar materials (PCM). The duplication of the centrosome coordinates with DNA replication. During the S phase, each centriole functions as a platform for a new procentriole to grow. The duplicated centrosomes start to separate to the opposite site of the nucleus, followed by the establishment of mitotic spindle poles for proper chromosome segregation in the M phase. Thus, precise control of centrosome homeostasis is important to maintain cell growth and genomic instability [18,19].

The centrosome also contributes to the growth of the primary cilium [20]. The primary cilium is the cellular protrusion that receives chemical or mechanical signals for proper development and differentiation [21]. The primary cilium is composed of the central microtubule-built axoneme and the overlying ciliary membrane [22]. On the axoneme, intraflagellar transporters regulate cilia dynamics and functions via anterograde and retrograde transportations. Recent studies have also demonstrated that the primary cilia play important roles in placentation during early pregnancy [23].

In this study, we enrolled placental samples from twenty pregnant women with GDM and twenty non-GDM pregnant women and found that the abundance of fetuin-A was upregulated in terms of mRNA and protein levels. The upregulated fetuin-A impeded cell cycle progression and induced apoptosis. In addition, centrosome amplification with disorganized microtubule arrays was observed in fetuin-A-treated placental cells. Furthermore, fetuin-A inhibited autophagy, therefore, blocking the growth of the primary cilium. Thus, our study uncovered the effect of fetuin-A on the regulation of placental cell growth and ciliogenesis.

2. Results

2.1. The Expression of Fetuin-A Is Upregulated in the Placentas of Gestational Diabetes Mellitus (GDM) Patients

Pregnant women who suffer from gestational diabetes mellitus (GDM) show higher levels of fetuin-A in their circulation [24], however, little is known about the effect of fetuin-A in the placenta. Therefore, we enrolled placental samples from twenty pregnant women with GDM and twenty non-GDM pregnant women. In the first trimester, the mean value of the body mass index (BMI) in the control group was 23.40 and that in the GDM group was 27.06. In the third trimester, the mean BMI value in the control group was 26.80 and that in the GDM group was 30.38 (Table 1). The treatment strategy for women with GDM follows the guidelines of the ADA "Management of Diabetes in Pregnancy: Standards of Medical Care in Diabetes 2019", including lifestyle management, medical nutrition therapy, and pharmacological therapy [3]. The mRNA level of fetuin-A in the placenta, which was measured by real-time PCR, was increased in patients with GDM (Figure 1A, $p = 0.008$). To further confirm this, the protein level of fetuin-A was also analyzed by immunoblotting assay. The abundance of placental fetuin-A was higher in GDM patients than in non-GDM subjects (Figure 1B,C, $p = 0.008$). Thus, the expression of fetuin-A is upregulated in the placentas of patients with GDM. Next, we tested whether the upregulation of fetuin-A in the placenta was induced by glucose. The immortalized placental HTR8 cells were cultured with different concentrations of glucose for 72 h, and the expression of fetuin-A was examined. The abundance of fetuin-A was increased in a dose-dependent manner (Figure 1D,E, ** $p = 0.007$ and *** $p = 0.0002$). Thus, the expression of fetuin-A is induced by high glucose treatment in HTR8 cells.

Table 1. Characteristics of study population.

	Normal Control ($n = 20$)	GDM ($n = 20$)
Maternal age (years)	32.78 ± 4.2	33.56 ± 4.93
Nulliparity (%)	41.94	48
Gestational age at delivery (weeks)	38.43 ± 1.14	38 ± 1.2
Chinese Han ethnicity (%)	100	100
BMI 1st trimester (Kg/m^2)	23.4 ± 3.48	27.06 ± 5.85 **
BMI 3rd trimester (Kg/m^2)	26.8 ± 3.39	30.38 ± 6 **
Systolic blood pressure (mmHg)	123.04 ± 12.55	133.43 ± 19.73 *
Diastolic blood pressure (mmHg)	73.83 ± 10.01	81.71 ± 11.56 *
Glucose-AC (mg/dL)	78 ± 7.01	97.74 ± 28.22
Glucose-PC 1 hour (mg/dL)	134.83 ± 6.34	195.67 ± 55.1 *
Glucose-PC 2 hours (mg/dL)	126.67 ± 26.07	193.47 ± 71.5 *
	Neonatal outcome	
Birth weight (g)	3033.49 ± 643.81	3244.7 ± 562.75
Female sex (%)	44.12	37.04
1 min Apgar score	8.7 ± 0.61	8.75 ± 0.56
5 min Apgar score	9.81 ± 0.4	9.79 ± 0.42
Placenta weight	662 ± 128.69	695 ± 167.06

Data are presented as mean ± SD. * $p < 0.05$; ** $p < 0.01$, t test.

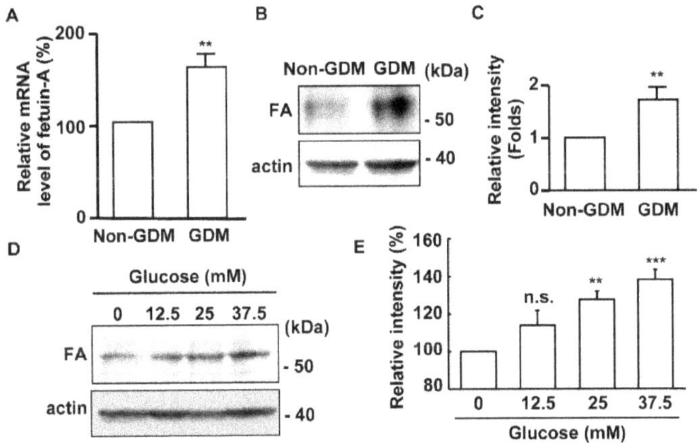

Figure 1. Fetuin-A is upregulated in the placentas of gestational diabetes mellitus (GDM) patients. (**A–C**) Fetuin-A is upregulated in the placentas of GDM patients: (**A**) quantification results of the fetuin-A mRNA level in the placentas of non-GDM and GDM women, (**B**) whole placenta extracts of non-GDM and GDM women were analyzed by immunoblot with antibodies against fetuin-A and actin, and (**C**) quantification results in (**B**). (**D,E**) Glucose induces the expression of fetuin-A in HTR8 cells: (**D**) whole cell extracts of fetuin-A-treated HTR8 cells were analyzed by immunoblot with antibodies against fetuin-A and actin and (**E**) quantitation of the relative intensity of fetuin-A in (**E**). n.s., no significance; ** $p < 0.01$ and *** $p < 0.001$.

2.2. Fetuin-A Inhibits Placental Cell Growth

The effect of fetuin-A on placental cell growth was examined. A previous study showed that treatment with 600 μg/mL of fetuin-A for 48 h inhibited primary extravillous trophoblast cell growth [25]. Therefore, we treated HTR8 cells with 600 μg/mL of fetuin-A for 24 or 48 h, and the cell numbers were counted. At 24 h after fetuin-A treatment, the cell numbers were significantly reduced, and treatment with fetuin-A for 48 h inhibited placental cell growth to the half maximal inhibitory concentration (IC50) (Figure 2A,B, Figure 2A: $p = 0.04$ and Figure 2B: $p = 0.0009$). Thus, the following experiments were performed by treating cells with 600 μg/mL of fetuin-A for 48 h. When checking the morphology of fetuin-A-treated cells, several apoptotic bodies were observed, suggesting that fetuin-A treatment might induce apoptosis. To further confirm this, the marker of apoptosis, cleaved-caspase-3, was checked. Upon fetuin-A treatment, the level of cleaved-caspase-3 increased significantly (Figure 2C,D). Thus, fetuin-A induces apoptosis in placental cells.

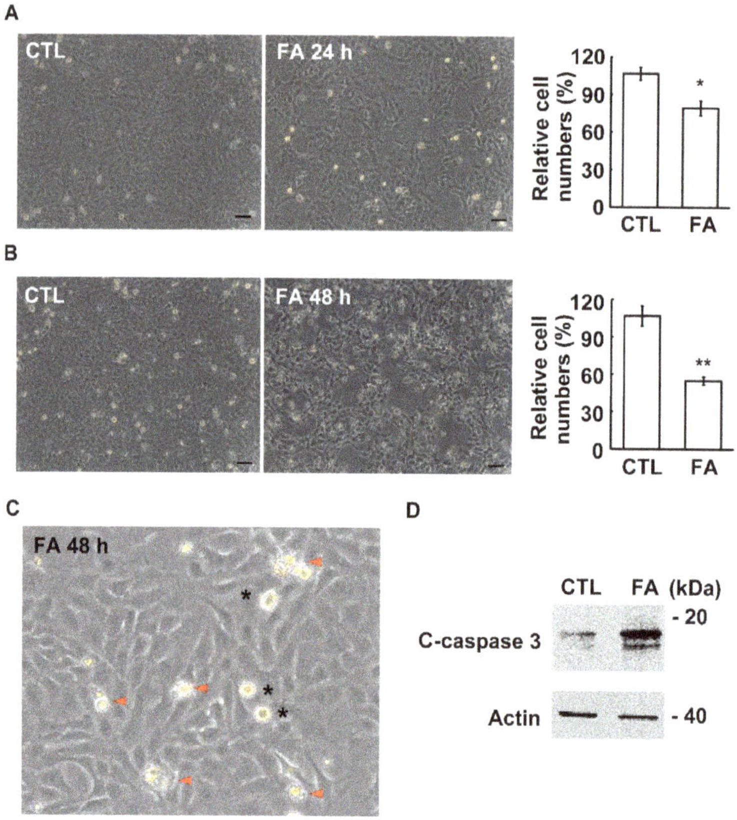

Figure 2. Fetuin-A inhibits HTR8 cell growth. (**A**,**B**) Fetuin-A inhibits HTR8 cell growth in a time-dependent manner. The cell numbers are shown as bright-field images (left panel) and quantification results (right panel) following treatment with 600 μg/mL of fetuin-A in HTR8 cells for 24 h (**A**) and 48 h (**B**). CTL: control and FA: fetuin-A. These results are the mean ± SD from three independent experiments. Scale bar 100 μM. (**C**,**D**) Fetuin-A induces apoptosis. (**C**) The apoptotic bodies (arrowhead in red) are observed upon treatment with 600 μg/mL of fetuin-A for 48 h in HTR8 cells. The mitotic cells are indicated by asterisks. The magnification is 400×. (**D**) Whole cell extracts of fetuin-A-treated HTR8 cell line were analyzed by immunoblot with antibodies against cleaved-caspase-3 (C-caspase-3) and actin. * $p < 0.05$ and ** $p < 0.01$.

To further study how fetuin-A affects cell growth, the ability of cells to enter into the S phase was examined by the EdU incorporation assay. Fetuin-A treatment reduced the population of cells with EdU positive signals (Figure 3A,B, $p = 6 \times 10^{-8}$). Then, the S phase related cyclins, including cyclin E and cyclin A, and the activation of CDK2 were examined. The abundance of cyclin A, but not cyclin E, and the level of phosphorylated CDK2 were reduced upon fetuin-A treatment (Figure 3C–E; Figure 3D, $p = 3 \times 10^{-6}$, and Figure 3E, $p = 0.002$). These data suggest that fetuin-A inhibits cyclin A-CDK2 activation, and thus leads to reduced S phase entry. Next, the ability of cells to enter the M phase was examined. Upon fetuin-A treatment, the mitotic index was reduced, indicating that the ability of cells to enter the M phase was reduced (Figure 4A, $p = 0.02$). In conclusion, fetuin-A inhibits placental cell growth.

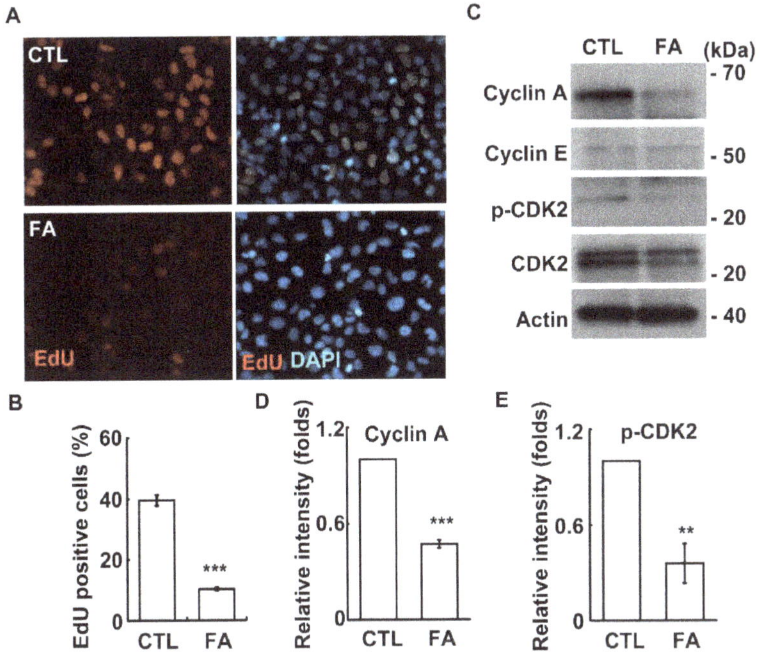

Figure 3. Fetuin-A inhibits S phase entry. (**A**,**B**) EdU incorporation was reduced in fetuin-A-treated HTR8 cells: (**A**) immunostaining of EdU (red) and DAPI (blue) in scramble control (CTL) or fetuin-A (FA) treated HTR8 cells and (**B**) quantification results of (**A**). The magnification is 200×. These results are the mean ± SD from three independent experiments and more than 1000 cells were counted in each individual group. (**C**–**E**) Fetuin-A inhibited cyclin A expression and CDK2 activation: (**C**) whole cell extracts of fetuin-A-treated HTR8 cells were analyzed by immunoblot with antibodies against cyclin A, cyclin E, CDK2, phosphorylated CDK2 at Thr160 (p-CDK2), and actin. (**D**,**E**) Quantitation of the relative intensity of cyclin A (**D**) and p-CDK2 in (**E**). ** $p < 0.01$ and *** $p < 0.001$.

2.3. *Fetuin-A Induces Centrosome Amplification*

The mitotic apparatus of the fetuin-A-treated cells was further examined. Normally, the mitotic spindle poles (γ-tubulin signals) orchestrate the mitotic spindle to align the duplicated chromosomes in the middle of the cells (Figure 4B, left panel), however, upon fetuin-A treatment, aberrant multiple mitotic spindle poles (cells with more than two γ-tubulin spots at M phase) were observed, accompanied by chromosome misalignment (Figure 4B, right panel and Figure 4C, $p = 0.003$). Thus, fetuin-A reduces M phase entry and disorganizes the mitotic apparatus in placental cells.

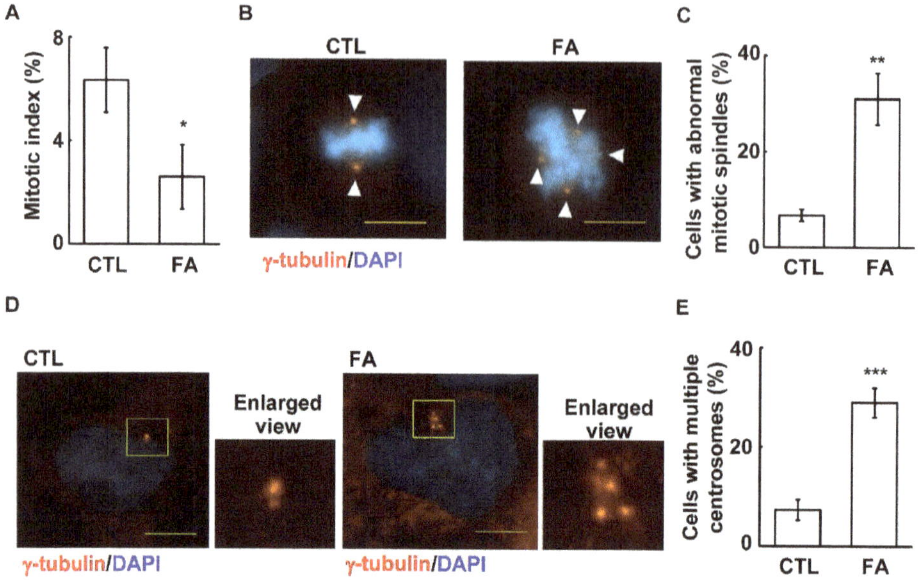

Figure 4. Fetuin-A induces centrosome amplification. (**A**) Fetuin-A inhibited cells entering the M phase. Quantification results of the mitotic index in the absence (CTL) or presence of fetuin-A (FA). These results are the mean ± SD from three independent experiments; more than 1000 cells were counted in each individual group. (**B**,**C**) Aberrant mitotic spindle poles were induced by fetuin-A treatment: (**B**) immunofluorescence staining showed increased mitotic spindle poles (γ-tubulin staining, as shown by the arrowhead) upon fetuin-A treatment and (**C**) quantification results of (**B**). (**D**,**E**) Fetuin-A induced centrosome amplification: (**D**) immunofluorescence staining showed increased γ-tubulin numbers upon fetuin-A treatment and (**E**) quantification results of (**D**). ** $p < 0.01$ and *** $p < 0.001$, results are the mean ± SD from three independent experiments, more than 100 cells were counted in each individual group. Scale bar 5 μM.

As centrosomes form the mitotic spindle poles for proper chromosome segregation, abnormal centrosomes might lead to the development of aberrant mitotic spindles. Thus, the centrosome numbers were counted by the staining of the marker of pericentriolar material, γ-tubulin, in fetuin-A-treated cells. When examining the centrosomal numbers, only one (before duplication) or two (after duplication) centrosomes were observed in control cells, however, the treatment of fetuin-A led to centrosome amplification, as shown by the presence of more than two γ-tubulin spots (Figure 4D,E, $p = 0.001$). Thus, fetuin-A induces centrosome amplification.

2.4. *Fetuin-A Leads to Disorganized Microtubule Nucleation*

Microtubule nucleation activity is mainly regulated by the centrosome, and fetuin-A leads to aberrant centrosome homeostasis. Thus, the microtubule nucleation ability was examined by the microtubule regrowth assay. Microtubules were disrupted by 1 h of nocodazole treatment. Then, the cells were washed by PBS and grown in the fresh medium for 10 min to allow the microtubule to regrow. Nocodazole treatment efficiently disrupted microtubule arrays in both the control and fetuin-A-treated cells (Figure 5A). After microtubule regrowth for 10 min, the microtubule concentrated around the centrosome and the array started to emanate from the centrosome to the periphery of the cells (Figure 5B, upper panel), however, in fetuin-A-treated cells, the emanated microtubule density was less than that of in the control cells, and these microtubule arrays did not extend to the cell periphery (Figure 5B, lower panel). Thus, fetuin-A treatment leads to the disorganization of microtubule nucleation.

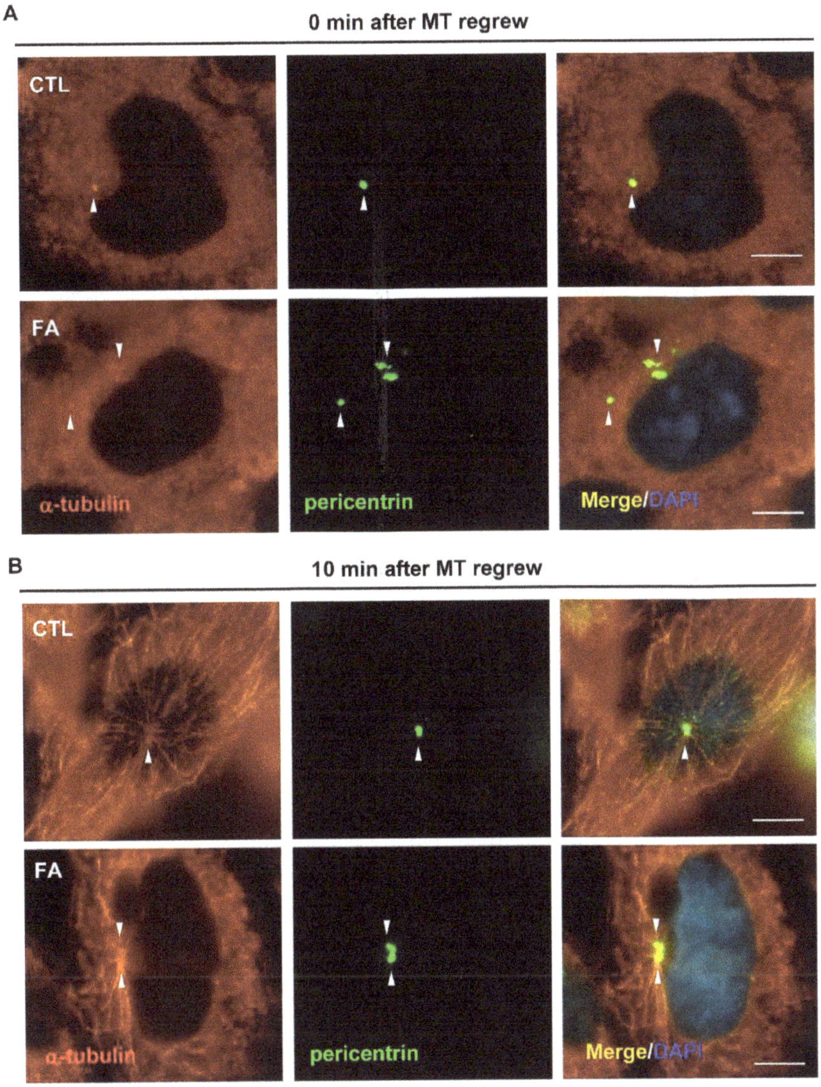

Figure 5. Fetuin-A treatment leads to disorganized microtubule arrays. Microtubules were depolymerized by nocodazole treatment (**A**), and, then, cells were cultured in the drug-free medium for 10 min (**B**), in the absence (CTL) or presence of fetuin-A (FA). Centrosomes and microtubules (MT) were immunostained with antibodies against pericentrin (centrosome marker) and γ-tubulin (microtubule marker). DNA was stained with DAPI. Scale bar 5 μM.

2.5. Fetuin-A Inhibits Primary Cilium Formation in Placental Cells

The primary cilium is important for maintaining placenta development. Therefore, we examined whether fetuin-A affects primary cilium formation. First, we examined whether placental HTR8 cells could grow primary cilia by forcing cells to enter the G0 phase under serum deprivation. At the G0 stage, the primary cilium, as shown by acetylated tubulin staining (axoneme marker), started to grow from the mother centriole (Cep164 staining) (Figure 6A, upper panel). Then, we examined other ciliary components. The ciliary membrane marker Arl13b and the intraflagella transporter marker IFT88 colocalized with the acetylated tubulin, suggesting that these cilia were intact. Then, we examined the effects of fetuin-A on ciliogenesis. Fetuin-A treatment inhibited primary cilia formation, as shown by the reduction of all ciliary markers (Figure 6B–E, $p = 0.005$). Thus, fetuin-A inhibits primary cilium formation in placental cells.

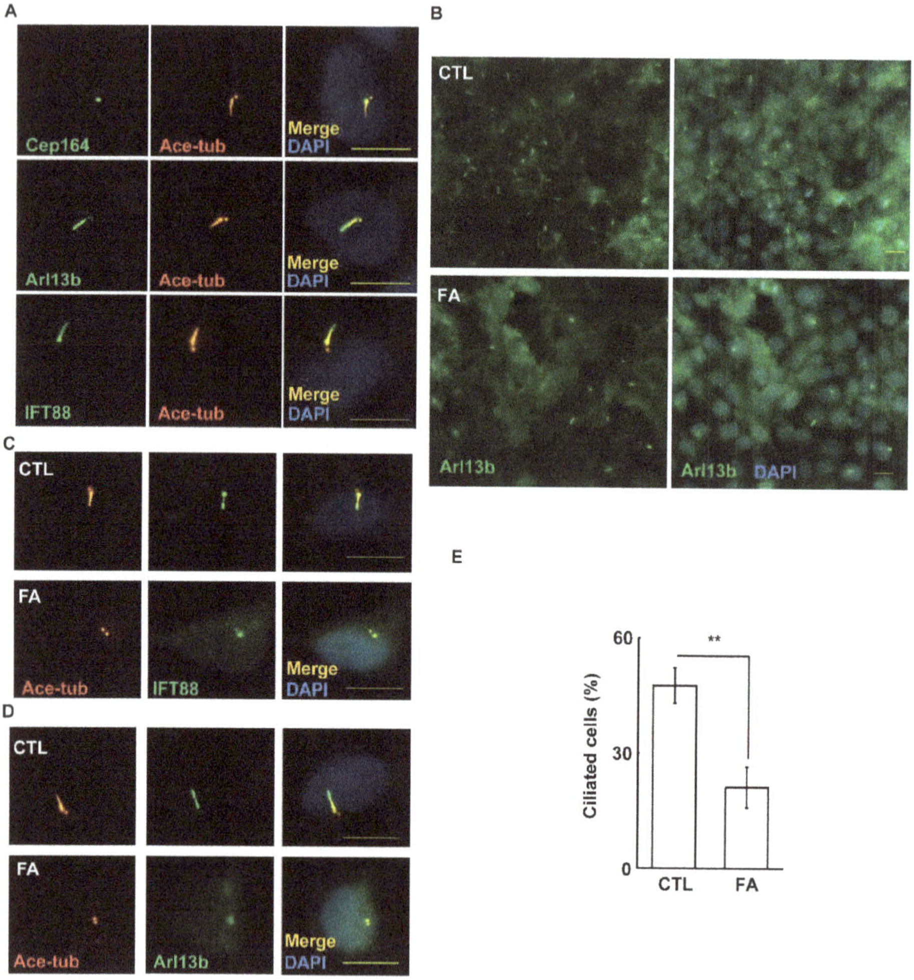

Figure 6. Fetuin-A inhibits ciliogenesis. (**A–D**) Primary cilia were examined in the absence (CTL) or presence of fetuin-A (FA) by immunostaining with antibodies against acetylated tubulin (Ace-tub), Cep164, IFT88, or Arl13b. (**E**) Quantification results of the frequency of ciliated HTR8 cells. ** $p < 0.01$, results are the mean ± SD from three independent experiments, more than 100 cells were counted in each individual group. DNA was stained with DAPI. Scale bar 10 µM.

2.6. Fetuin-A Inhibits Autophagic Flux

Autophagy promotes primary cilium formation under serum deprivation. Therefore, we examined whether fetuin-A affects autophagy. First, we examined the autophagy by immunostaining with an antibody against LC3. In the control cells, the LC3 signal was hardly detected, however, Fetuin-A treatment induced LC3 accumulation in the cytoplasm, suggesting that fetuin-A treatment affected autophagy (Figure 7A). The accumulation of LC3 signal in the cytoplasm might result from the acceleration of autophagy or defective autophagic flux, and thus the conversion of LC3 I to LC3 II was examined by immunoblotting assay. Upon fetuin-A treatment, the conversion of LC3 I to LC3 II was reduced, as shown by a lowered LC3 II to I ratio (Figure 7B, $p = 0.02$), suggesting that the autophagic flux was reduced. To further confirm this observation, the expression of Beclin1, the key enzyme in the initiation of autophagic flux, was examined. The expression of Beclin1 was reduced by fetuin-A treatment (Figure 7C). Thus, fetuin-A inhibits autophagic flux.

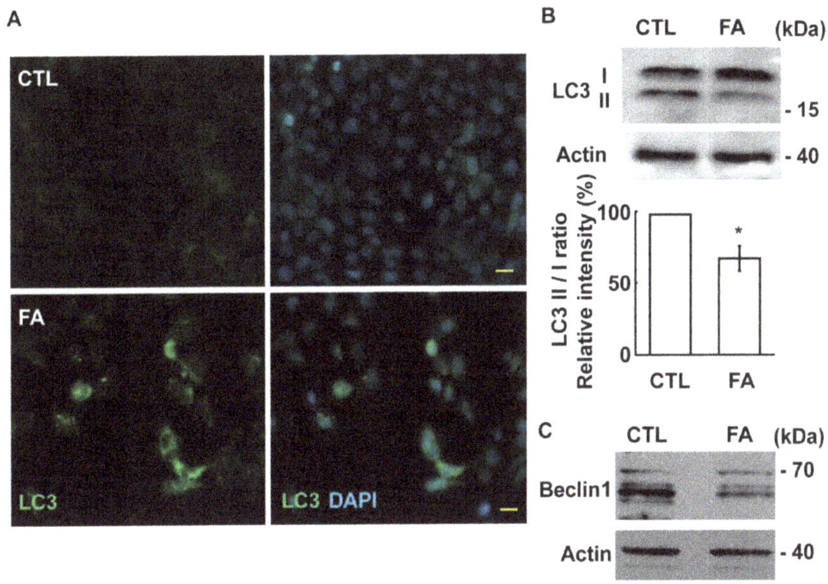

Figure 7. Fetuin-A inhibits autophagy. (**A**) Autophagy was examined by immunostaining with an antibody against LC3. DNA was stained with DAPI. Scale bar 10 μM. (**B**,**C**) Extracts of control (CTL) or fetuin-A (FA)-treated cells were analyzed by immunoblotting with antibodies against LC3, Beclin1, or actin. (**B**, lower panel) Quantification results of the relative intensity of the LC3 II to LC3 I ratio. * $p < 0.05$.

3. Discussion

In this study, we demonstrated that fetuin-A is relevant for cell growth and ciliogenesis in the placentas of GDM patients. Treatment with fetuin-A leads to abnormal centrosome amplification followed by the induction of aberrant mitotic spindle poles. In addition, fetuin-A also disorganizes the microtubule arrays. Furthermore, fetuin-A inhibits autophagy, followed by the reduction of primary cilium formation in placental cells. Taken together, fetuin-A affects the centrosome and autophagy, leading to defective placenta growth.

Circulating fetuin-A is upregulated in GDM patients [26], however, little is known about the effect of fetuin-A on the placenta. In this study, we showed that the placental fetuin-A concentration increases in GDM women. It is an important issue to clarify the source of placental fetuin-A. The elevated fetuin-A might be derived from maternal circulation. Additionally, it might also be synthesized locally at the placenta. In our study, we found that the mRNA level of fetuin-A increased in the placentas of GDM patients. In addition, a high glucose concentration induced the expression of fetuin-A in immortalized placental cells, suggesting that fetuin-A was also synthesized locally. Thus, fetuin-A might act on the placenta through paracrine or autocrine effects. Circulating maternal fetuin-A acts on the placenta via a paracrine effect. A high maternal glucose concentration also induces placental cells to produce more fetuin-A locally. Thus, both circulating and local placental fetuin-A coordinately affect placental development in GDM.

The primary cilium regulates development and differentiation and, additionally, this tiny organelle plays roles in female reproduction. Endocrine gland-derived vascular endothelial growth factor (EG-VEGF), the key endocrine factor for proper placentation, binds to its receptor on the primary cilium, triggering the downstream signaling cascade for trophoblast invasion [23,27]. The regulation of ciliogenesis is initiated by TTBK2 recruitment and CP110 removal from the mother centriole [28], and resorption of the primary cilium is triggered by HDAC6 activation [29]. Recent studies showed that the primary cilia is also regulated by the miR-200 family in the placenta [27]. In this sstudy, we showed that fetuin-A also inhibits ciliogenesis in placental cells. So far, it is still unclear how fetuin-A affects the placental cilia. The concentration of miR-200 family members is upregulated in women with GDM [30]. In addition, the concentrations of fetuin-A and miR-200 family members increase in nonalcoholic fatty liver disease patients [10,31]. Thus, we speculate that, in the placentas of GDM women, the fetuin-A might induce the expression of miR-200 family members, thus inhibiting the formation of the primary cilia, however, this hypothesis still needs to be tested.

In summary, we showed that fetuin-A inhibits placental cell growth and is required for the control of centrosome homeostasis and ciliogenesis, which are two important events for proper placentation.

4. Materials and Methods

4.1. Cell Culture

The immortalized HTR-8/SVneo (HTR8) cell line (ATCC CRL-3271), which was obtained from ATCC (Manassas, VA, USA). was derived by transfecting cells that grew out of chorionic villi explants of human first-trimester placentas with the SV40 large T antigen. HTR8 cells were grown in Roswell Park Memorial Institute (RPMI)-1640 medium supplemented with 10% fetal bovine serum. These cells were cultured in a humidified atmosphere at 5% CO_2 at 37 °C. Mycoplasma contamination was regularly examined by immunoblotting assay and immunofluorescence staining (DNA was stained by DAPI) according to the guidelines.

4.2. Study Population and Sample Collection

The experimental procedure was approved by the biosafety committee of the National Cheng Kung University. The human sample analysis was approved by the Institutional Review Board (IRB) of the National Cheng Kung University Hospital (NCKUH, no. A-ER-105-021; 19 February 2016). The expression level of fetuin-A in the placentas of women with GDM (n = 20) and healthy controls (n = 20) was measured in a case-control study. The diagnosis of gestational diabetes was made at 24 to 28 weeks of gestation via a two-hour 75 g oral glucose tolerance test according to the guidelines of the IADPSG (International Association of the Diabetes and Pregnancy Study Groups) and ADA (American Diabetes Association). The positive diagnosis criteria were one or more plasma glucose values that met or exceeded the following values: fasting glucose level > 92 mg/dL, 1 h glucose level > 180 mg/dL, and 2 h glucose level > 153 mg/dL. Only singleton pregnancies were included in the study.

4.3. Immunofluorescence Microscopy

Cells were grown on coverslips prior to the performance of experiments. Cells were fixed with cold methanol at −20 °C for 6 min. Then, the cells were washed with PBS three times followed by blocking with 5% BSA for 1 h. After blocking, cells were incubated with antibodies overnight at 4 °C. Then, cells were washed with PBS three times and incubated with FITC-conjugated and/or Cy3-conjugated secondary antibodies (Invitrogen, Carlsbad, CA, USA) for 1 h at room temperature in the dark. After extensive washing, the cover slips were mounted in 50% glycerol (in PBS) on glass slides. Cells were observed with an AxioImager M2 fluorescence microscope (Zeiss, Switzerland).

4.4. Microtubule Regrowth Assay

Cells were cultured in nocodazole (10 mg/mL) containing medium for 1 h to depolymerize the microtubules. Then, the nocodazole was removed by washing with PBS three times, and then cells were incubated in fresh medium at 37 °C for 10 min. After microtubule repolymerization, cells were fixed with cold methanol at −20 °C for 6 min, followed by immunofluorescence staining.

4.5. EdU Incorporation Assay

Cells were cocultured with commercially available EdU for 30 min according to the manufacturer's instructions (Invitrogen, Carlsbad, CA, USA). Then, these cells were observed with fluorescence EdU signals by an AxioImager M2 fluorescence microscope.

4.6. Cell Growth Assay

Cells were trypsinized and resuspended in PBS for cell number counting or they were centrifuged for further Western blot analysis after drug treatment. Centrifuged cells were further lysed with lysis buffer containing 0.5% NP-40, 300 mM NaCl, 1 mM EDTA, and the protease inhibitor cocktail (Roche, Mannheim, Germany) followed by centrifugation (15,000 rpm, 4 °C). The supernatant was collected and further analyzed by Western blot analysis.

4.7. Antibodies

The following antibodies were obtained commercially: anti-acetylated-tubulin (T7451), anti-γ-tubulin (T5326), and anti-α-tubulin (T9026) (Sigma, St. Louis, MO, USA); anti-LC3A/B (D3U4C) XP (#12741), anti-cyclin A2 (#4656), anti-cyclin E1 (HE12; #4129), anti-CDK2 (#2546), and anti-CDK2 phospho-Thr160 (#2561) were purchased from Cell Signaling (Beverly, MA, USA); anti-fetuin-A (ab137125), and anti-Beclin 1 antibody [EPR19662] (ab207612) were purchased from Abcam (Cambridge, UK); anti-actin (AC-15) (GTX26276) was purchased from Genetex (Irvine, CA, USA); anti-IFT88 (13967-1-AP) was purchased from Proteintech (Chicago, IL, USA); and anti-CEP164 (NBP1-81445) was purchased from Novus (Littleton, CO, USA).

4.8. Statistical Analysis

All data are presented as the mean ± S.D. from at least three independent experiments. In each individual experiment, more than 100 cells were counted in each experimental group. To show the statistical significance, unpaired two-tailed t-tests were used to show the differences between two groups. p-values of less than 0.05 were considered statistically significant.

Author Contributions: Conceptualization, C.-Y.W. and P.-Y.T.; methodology, P.-L.K. and M.-T.S.; validation, C.-Y.W. and P.-Y.T.; formal analysis, H.-l.C.; investigation, C.-Y.W. and H.-l.C.; resources, P.-Y.T.; data curation, M.-T.S. and H.-l.C.; writing—original draft preparation, C.-Y.W. and P.-Y.T.; visualization, C.-Y.W.; supervision, P.-Y.T.; project administration, P.-Y.T.; funding acquisition, C.-Y.W. and P.-Y.T.

Funding: This study was supported by grants from the Ministry of Science and Technology, MOST106-2320-B-006-056-MY3 to Chia-Yih Wang and MOST107-2314-B-006-042 to Pei-Yin Tsai. This study was supported by a grant from the National Cheng Kung University Hospital, NCKUH-10505024 to Pei-Yin Tsai.

Acknowledgments: We are grateful for the support from the Core Research Laboratory, College of Medicine, National Cheng Kung University.

Conflicts of Interest: The authors declare no conflict of interest. The funders had no role in the design of the study; in the collection, analyses, or interpretation of data; in the writing of the manuscript, or in the decision to publish the results.

References

1. Melchior, H.; Kurch-Bek, D.; Mund, M. The Prevalence of Gestational Diabetes. *Dtsch. Arztebl. Int.* **2017**, *114*, 412–418. [PubMed]

2. Chiefari, E.; Arcidiacono, B.; Foti, D.; Brunetti, A. Gestational diabetes mellitus: An updated overview. *J. Endocrinol. Investig.* **2017**, *40*, 899–909. [CrossRef] [PubMed]
3. American Diabetes Association. 14. Management of Diabetes in Pregnancy: Standards of Medical Care in Diabetes-2019. *Diabetes Care* **2019**, *42*, S165–S172. [CrossRef] [PubMed]
4. Bryson, C.L.; Ioannou, G.N.; Rulyak, S.J.; Critchlow, C. Association between gestational diabetes and pregnancy-induced hypertension. *Am. J. Epidemiol.* **2003**, *158*, 1148–1153. [CrossRef]
5. Zhang, C.; Rawal, S.; Chong, Y.S. Risk factors for gestational diabetes: Is prevention possible? *Diabetologia* **2016**, *59*, 1385–1390. [CrossRef]
6. Holmes, V.A.; Young, I.S.; Patterson, C.C.; Pearson, D.W.; Walker, J.D.; Maresh, M.J.; McCance, D.R. Diabetes and Pre-eclampsia Intervention Trial Study Group. Optimal glycemic control, pre-eclampsia, and gestational hypertension in women with type 1 diabetes in the diabetes and pre-eclampsia intervention trial. *Diabetes Care* **2011**, *34*, 1683–1688. [CrossRef]
7. Dabelea, D.; Hanson, R.L.; Lindsay, R.S.; Pettitt, D.J.; Imperatore, G.; Gabir, M.M.; Roumain, J.; Bennett, P.H.; Knowler, W.C. Intrauterine exposure to diabetes conveys risks for type 2 diabetes and obesity: A study of discordant sibships. *Diabetes* **2000**, *49*, 2208–2211. [CrossRef]
8. Mori, K.; Emoto, M.; Inaba, M. Fetuin-A: A multifunctional protein. *Recent Pat. Endocr. Metab. Immune Drug Discov.* **2011**, *5*, 124–146. [CrossRef]
9. Stefan, N.; Hennige, A.M.; Staiger, H.; Machann, J.; Schick, F.; Krober, S.M.; Machicao, F.; Fritsche, A.; Haring, H.U. Alpha2-Heremans-Schmid glycoprotein/fetuin-A is associated with insulin resistance and fat accumulation in the liver in humans. *Diabetes Care* **2006**, *29*, 853–857. [CrossRef]
10. Ou, H.Y.; Yang, Y.C.; Wu, H.T.; Wu, J.S.; Lu, F.H.; Chang, C.J. Increased fetuin-A concentrations in impaired glucose tolerance with or without nonalcoholic fatty liver disease, but not impaired fasting glucose. *J. Clin. Endocrinol. Metab.* **2012**, *97*, 4717–4723. [CrossRef]
11. Jahnen-Dechent, W.; Heiss, A.; Schafer, C.; Ketteler, M. Fetuin-A regulation of calcified matrix metabolism. *Circ. Res.* **2011**, *108*, 1494–1509. [CrossRef] [PubMed]
12. Srinivas, P.R.; Wagner, A.S.; Reddy, L.V.; Deutsch, D.D.; Leon, M.A.; Goustin, A.S.; Grunberger, G. Serum alpha 2-HS-glycoprotein is an inhibitor of the human insulin receptor at the tyrosine kinase level. *Mol. Endocrinol.* **1993**, *7*, 1445–1455. [PubMed]
13. Pal, D.; Dasgupta, S.; Kundu, R.; Maitra, S.; Das, G.; Mukhopadhyay, S.; Ray, S.; Majumdar, S.S.; Bhattacharya, S. Fetuin-A acts as an endogenous ligand of TLR4 to promote lipid-induced insulin resistance. *Nat. Med.* **2012**, *18*, 1279–1285. [CrossRef] [PubMed]
14. Goustin, A.S.; Derar, N.; Abou-Samra, A.B. Ahsg-fetuin blocks the metabolic arm of insulin action through its interaction with the 95-kD beta-subunit of the insulin receptor. *Cell. Signal.* **2013**, *25*, 981–988. [CrossRef] [PubMed]
15. Ou, H.Y.; Yang, Y.C.; Wu, H.T.; Wu, J.S.; Lu, F.H.; Chang, C.J. Serum fetuin-A concentrations are elevated in subjects with impaired glucose tolerance and newly diagnosed type 2 diabetes. *Clin. Endocrinol.* **2011**, *75*, 450–455. [CrossRef]
16. Iyidir, O.T.; Degertekin, C.K.; Yilmaz, B.A.; Altinova, A.E.; Toruner, F.B.; Bozkurt, N.; Ayvaz, G.; Akturk, M. Serum levels of fetuin A are increased in women with gestational diabetes mellitus. *Arch. Gynecol. Obstet.* **2015**, *291*, 933–937. [CrossRef]
17. Nigg, E.A.; Holland, A.J. Once and only once: Mechanisms of centriole duplication and their deregulation in disease. *Nat. Rev. Mol. Cell Biol.* **2018**, *19*, 297–312. [CrossRef]
18. Chen, T.Y.; Syu, J.S.; Han, T.Y.; Cheng, H.L.; Lu, F.I.; Wang, C.Y. Cell Cycle-Dependent Localization of Dynactin Subunit p150 glued at Centrosome. *J. Cell. Biochem.* **2015**, *116*, 2049–2060. [CrossRef]
19. Chen, T.Y.; Syu, J.S.; Lin, T.C.; Cheng, H.L.; Lu, F.L.; Wang, C.Y. Chloroquine alleviates etoposide-induced centrosome amplification by inhibiting CDK2 in adrenocortical tumor cells. *Oncogenesis* **2015**, *4*, e180. [CrossRef]
20. Lattao, R.; Kovacs, L.; Glover, D.M. The Centrioles, Centrosomes, Basal Bodies, and Cilia of Drosophila melanogaster. *Genetics* **2017**, *206*, 33–53. [CrossRef]
21. Chen, T.Y.; Lien, W.C.; Cheng, H.L.; Kuan, T.S.; Sheu, S.Y.; Wang, C.Y. Chloroquine inhibits human retina pigmented epithelial cell growth and microtubule nucleation by downregulating p150(glued). *J. Cell. Physiol.* **2019**, *234*, 10445–10457. [CrossRef] [PubMed]
22. Johnson, C.A.; Malicki, J.J. The Nuclear Arsenal of Cilia. *Dev. Cell* **2019**, *49*, 161–170. [CrossRef] [PubMed]

23. Wang, C.Y.; Tsai, H.L.; Syu, J.S.; Chen, T.Y.; Su, M.T. Primary Cilium-Regulated EG-VEGF Signaling Facilitates Trophoblast Invasion. *J. Cell. Physiol.* **2017**, *232*, 1467–1477. [CrossRef] [PubMed]
24. Kalabay, L.; Cseh, K.; Pajor, A.; Baranyi, E.; Csakany, G.M.; Melczer, Z.; Speer, G.; Kovacs, M.; Siller, G.; Karadi, I.; et al. Correlation of maternal serum fetuin/alpha2-HS-glycoprotein concentration with maternal insulin resistance and anthropometric parameters of neonates in normal pregnancy and gestational diabetes. *Eur. J. Endocrinol.* **2002**, *147*, 243–248. [CrossRef] [PubMed]
25. Gomez, L.M.; Anton, L.; Srinivas, S.K.; Elovitz, M.A.; Parry, S. Effects of increased fetuin-A in human trophoblast cells and associated pregnancy outcomes. *Am. J. Obstet. Gynecol.* **2012**, *207*, 484.e1–484.e8. [CrossRef] [PubMed]
26. Kralisch, S.; Hoffmann, A.; Lossner, U.; Kratzsch, J.; Bluher, M.; Stumvoll, M.; Fasshauer, M.; Ebert, T. Regulation of the novel adipokines/hepatokines fetuin A and fetuin B in gestational diabetes mellitus. *Metab. Clin. Exp.* **2017**, *68*, 88–94. [CrossRef]
27. Wang, C.Y.; Tsai, P.Y.; Chen, T.Y.; Tsai, H.L.; Kuo, P.L.; Su, M.T. Elevated miR-200a and miR-141 inhibit endocrine gland-derived vascular endothelial growth factor expression and ciliogenesis in preeclampsia. *J. Physiol.* **2019**, *597*, 3069–3083. [CrossRef]
28. Goetz, S.C.; Liem, K.F., Jr.; Anderson, K.V. The spinocerebellar ataxia-associated gene Tau tubulin kinase 2 controls the initiation of ciliogenesis. *Cell* **2012**, *151*, 847–858. [CrossRef]
29. Pugacheva, E.N.; Jablonski, S.A.; Hartman, T.R.; Henske, E.P.; Golemis, E.A. HEF1-dependent Aurora A activation induces disassembly of the primary cilium. *Cell* **2007**, *129*, 1351–1363. [CrossRef]
30. Guarino, E.; Delli Poggi, C.; Grieco, G.E.; Cenci, V.; Ceccarelli, E.; Crisci, I.; Sebastiani, G.; Dotta, F. Circulating MicroRNAs as Biomarkers of Gestational Diabetes Mellitus: Updates and Perspectives. *Int. J. Endocrinol.* **2018**, *2018*, 6380463. [CrossRef]
31. Kong, L.; Zhu, J.; Han, W.; Jiang, X.; Xu, M.; Zhao, Y.; Dong, Q.; Pang, Z.; Guan, Q.; Gao, L.; et al. Significance of serum microRNAs in pre-diabetes and newly diagnosed type 2 diabetes: A clinical study. *Acta Diabetol.* **2011**, *48*, 61–69. [CrossRef] [PubMed]

MDPI
St. Alban-Anlage 66
4052 Basel
Switzerland
Tel. +41 61 683 77 34
Fax +41 61 302 89 18
www.mdpi.com

International Journal of Molecular Sciences Editorial Office
E-mail: ijms@mdpi.com
www.mdpi.com/journal/ijms